Lecture Notes in Computer Science 6829

Commenced Publication in 1973
Founding and Former Series Editors:
Gerhard Goos, Juris Hartmanis, and Jan van Leeuwen

Kyung-Hyune Rhee   DaeHun Nyang (Eds.)

Information Security and Cryptology - ICISC 2010

13th International Conference
Seoul, Korea, December 1-3, 2010
Revised Selected Papers

 Springer

Volume Editors

Kyung-Hyune Rhee
Pukyong National University
Department of IT Convergence Application Engineering
599-1 Daeyeon 3-Dong Namgu, Busan 608-737, Republic of Korea
E-mail: khrhee@pknu.ac.kr

DaeHun Nyang
INHA University
Department of Computer Science and Information Technology
253 Yonghyun-dong, Nam-gu, Incheon 402-751, Republic of Korea
E-mail: nyang@inha.ac.kr

ISSN 0302-9743   e-ISSN 1611-3349
ISBN 978-3-642-24208-3   e-ISBN 978-3-642-24209-0
DOI 10.1007/978-3-642-24209-0
Springer Heidelberg Dordrecht London New York

Library of Congress Control Number: 2011936884

CR Subject Classification (1998): E.3, K.6.5, C.2, D.4.6, G.2.1, E.4

LNCS Sublibrary: SL 4 – Security and Cryptology

Typesetting: Camera-ready by author, data conversion by Scientific Publishing Services, Chennai, India

Printed on acid-free paper

Springer is part of Springer Science+Business Media (www.springer.com)

Preface

ICISC 2010, the 13th International Conference on Information Security and Cryptology, was held in Seoul, Korea, during December 1–3, 2010. It was organized by the Korea Institute of Information Security and Cryptology (KIISC). The aim of this conference was to provide a forum for the presentation of new results in research, development, and applications in the field of information security and cryptology. It also intended to be a place where research information can be exchanged.

The conference received 99 submissions from 27 countries, covering all areas of information security and cryptology. The review and selection processes were carried out in two stages by the Program Committee (PC) of 64 prominent experts via online meetings through the iChair Web server. First, each paper was blind reviewed by at least three PC members, and papers co-authored by the PC members were reviewed by at least five PC members. Second, individual review reports were revealed to PC members, and detailed interactive discussion on each paper followed. Through this process, the PC finally selected 28 papers from 16 countries. The acceptance rate was 28.2%. The authors of selected papers had a few weeks to prepare for their final versions based on the comments received from the reviewers. These revised papers were not subject to editorial review and the authors bear full responsibility for their contents.

The conference featured one tutorial and two invited talks. The tutorial was delivered by Tatsuaki Okamoto from NTT Information Sharing Platform Laboratories. The invited speakers were Sakir Sezer from ECIT SoC Research Division and Giuseppe Ateniese from The Johns Hopkins University.

There are many people who contributed to the success of ICISC 2010. We would like to thank all the authors who submitted papers to this conference. We are deeply grateful to all 64 members of the PC, especially to those who shepherded conditionally accepted papers. It was a truly nice experience to work with such talented and hard-working researchers. We wish to thank all the external reviewers for assisting the PC in their particular areas of expertise. We would also like to thank the iChair developers for allowing us to use their software.

Finally, we would like to thank all the participants of the conference who made this event an intellectually stimulating one through their active contribution and all organizing members who nicely managed the conference.

December 2010   Kyung-Hyune Rhee
                                                                                      DaeHun Nyang

Organization

ICISC 2010, The 13th Annual International Conference on Information Security and Cryptology, was held during December 1–3, 2010, at Chung-Ang University, Seoul, Korea, and organized by Korea Institute of Information Security and Cryptology (KIISC) (http://www.kiisc.or.kr) in cooperation with the Ministry of Public Administration and Security (MOPAS) (http://www.mopas.go.kr)

General Chair

Jong In Lim   KIISC, Korea

Program Co-chairs

Kyung-Hyune Rhee	Pukyong National University, Korea
DaeHun Nyang	INHA university, Korea

Program Committee

Joonsang Baek	Institute for Infocomm Research, Singapore
Alex Biryukov	University of Luxembourg, Luxembourg
Seongtaek Chee	Attached Institute of ETRI, Korea
Jung Hee Cheon	Seoul National University, Korea
Yongwha Chung	Korea University, Korea
Paolo Milani Comparetti	Vienna University of Technology, Austria
Frëdëric Cuppens	Telecom Bretagne, France
Paolo D'Arco	University of Salerno, Italy
Bart De Decker	Katholieke Universiteit Leuven, Belgium
David Galindo	University of Luxembourg, Luxembourg
Philippe Golle	Palo Alto Research Center, USA
Vipul Goyal	UCLA, USA
Louis Granboulan	EADS Innovation Works, France
Matthew Green	Independent Security Evaluators, USA
JaeCheol Ha	Hoseo University, Korea
Dong-Guk Han	Kookmin University, Korea
Martin Hell	Lunds Universitet, Sweden
Deukjo Hong	Attached Institute of ETRI, Korea
Jin Hong	Seoul National University, Korea
Seokhie Hong	Korea University, Korea
Jung Yeon Hwang	ETRI, Korea
David Jao	University of Waterloo, Canada

Organizing Committee Chair

Sehyun Park   Chung-Ang University, Seoul, Korea

Organizing Committee Vice-Chair

Eul Gyu Im   Hanyang University, Seoul, Korea

Organizing Committee

Jong-Soo Jang   ETRI, Korea
Dong-Kyue Kim   Hanyang University, Korea
Daesung Kwon   Attached Institute of ETRI, Korea
Im-Yeong Lee   Soonchunhyang University, Korea
Mun-Kyu Lee   Inha University, Korea
Yongsu Park   Hanyang University, Korea
Jungtaek Seo   Attached Institute of ETRI, Korea
Kyung-Ah Shim   NIMS, Korea
Yoo-jae Won   KISA, Korea
Jeong Hyun Yi   Soongsil University, Korea

External Reviewers

Anna Lisa Ferrara   Hyung Tae Lee
Antonio Muñoz   Hyun-Min Kim
Benedikt Gierlichs   Isaac Agudo
Bin Zhang   Ivica Nikolic
Bo Zhu   Jae Ahn Hyun
Bonwook Koo   Jae Hong Seo
Cheol-Min Park   Jae Woo Seo
Chunhua Su   Jangseung Kim
Claudio Soriente   Jeonil Kang
Daeseon Choi   Jheng-Hong Tu
Daesung Kwon   Ji Young Chun
Eun Sung Lee   Jihye Kim
Gunil Ma   Joaquin Garcia-Alfaro
Hakan Seyalioglu   Johann Großschädl
Hans Loehr   Joseph K. Liu
Hee-Seok Kim   Jun Pang
HongTae Kim   Jung Youl Park
Hsi-Chung Lin   Kai Yuen Cheong
Hugo Jonker   Kazuhiko Minematsu
Hyeong-Chan Lee   Kazumasa Omote
Hyun-Dong So   Kenneth Matheis

Lingling Xu
Luca Davi
Maki Shigeri
Martin Ågren
Ming Duan
Mingwu Zhang
Minkyu Kim
Moonsung Lee
MyungKeun Yoon
Myungsun Kim
Nicky Mouha
Özgür Dagdelen
Paul Stankovski
Pedro Peris-Lopez
Pieter Verhaeghe
Ping Wang
Qiping Lin
Ralf-Philipp Weinmann
Rishiraj Bhattacharyya
Ruei-Hau Hsu
Ryoji Ohta
Sangrae Cho
Sasa Radomirovic
Sebastian Faust
Seog Chung Seo
Somindu C. Ramanna
Sanjay Bhattacherjee
Subhabrata Samajder
Sung-Kyung Kim

Sungwook Eom
Sung-Wook Lee
Taechan Kim
Takashi Nishide
Takeshi Kawabata
Taku Hayashi
Tamer AbuHmed
Teruo Saito
Thomas Baigneres
Thomas Schneider
Ton van Deursen
Vincent Naessens
Wei-Chih Lien
Xingwen Zhao
Xinyi Huang
Yanjiang Yang
Yasufumi Hashimoto
Yeonkyu Kim
Ying Qiu
Young-In Cho
YoungJae Maeng
Young-Ran Lee
YoungSeob Cho
Younho Lee
Youn-Taek Young
Yu Sasaki
Kenji Ohkuma
Zayabaatar
Zheng Gong

Sponsoring Institutions

National Security Research Institute (NSRI)
Electronics and Telecommunications Research Institute (ETRI)
Korea Internet & Security Agency (KISA)
Korean Federation of Science and Technology Societies (KOFST)
Chung-Ang University Home Network Research Center (CAU HNRC)
Chungnam National University Internet Intrusion Response Technology
    Research Center (CNU IIRTRC)
Korea University Center for Information Security Technologies (KU CIST)

Table of Contents

Cryptanalysis

Cryptographic Algorithms

Implementation

Network and Mobile Security

Symmetric Key Cryptography

Cryptographic Protocols

Side Channel Attack

Analysis of Nonparametric Estimation Methods for Mutual Information Analysis

Alexandre Venelli[1,2]

[1] IML - ERISCS Université de la Méditerranée,
Case 907, 163 Avenue de Luminy
13288 Marseille Cedex 09, France
[2] Vault-IC France, an INSIDE Contactless Company
Avenue de la Victoire, Z.I. Rousset,
13790 Rousset, France
avenelli@insidefr.com

Abstract. Mutual Information Analysis (MIA) is a side-channel attack introduced recently. It uses mutual information, a known information theory notion, as a side-channel distinguisher. Most previous attacks use parametric statistical tests and the attacker assumes that the distribution family of the targeted side-channel leakage information is known. On the contrary, MIA is a generic attack that assumes the least possible about the underlying hardware specifications. For example, an attacker should not have to guess a linear power model and combine it with a parametric test, like the Pearson correlation factor. Mutual information is considered to be very powerful however it is difficult to estimate. Results of MIA can therefore be unreliable and even bias. Several efficient parametric estimators of mutual information are proposed in the literature. They are obviously very efficient when the distribution is correctly guessed. However, we loose the original goal of MIA which is to assume the least possible about the attacked devices. Hence, nonparametric estimators of mutual information should be considered in more details and, in particular, their efficiency in the side-channel context. We review some of the most powerful nonparametric methods and compare their performance with state-of-the-art side-channel distinguishers.

Keywords: Side-channel analysis, mutual information analysis, entropy estimation, nonparametric statistics.

1 Introduction

Side-channel analysis is a technique that uses information leaked by a physical implementation of cryptographic algorithms. The concept of using side-channel information to break a cryptosystem was introduced by Kocher [12]. In his paper, Kocher analyses differences in the computation time of certain cryptographic operations that depend on a secret. On embedded devices, monitoring the power consumption or recording the electromagnetic radiations is easy to realize and is very revealing of the computations executed by the system. Statistical tests

K.-H. Rhee and D. Nyang (Eds.): ICISC 2010, LNCS 6829, pp. 1–15, 2011.

are used in side-channel cryptanalysis so that an attacker does not need to know precise implementation details in order to extract secret keys. Generally, once the side-channel information is recorded from a device, the attacker post processes it and evaluates it by some statistical analysis. In 1999, Kocher et al. [13] introduced the concept of Differential Power Analysis (DPA), a side-channel attack that uses the difference of means as statistical test. The attacker makes a key hypothesis and partitions side-channel measurements into two sets depending on the value of a key-dependent computation in the cryptographic algorithm. Then, the adversary computes the difference of means of the two sets for each key hypothesis. If the difference shows distinct peaks, the corresponding key hypothesis is assumed to be correct.

A lot of research in the side-channel domain consist in proposing relevant statistical tests to enhance the results of these attacks. In 2004, Brier et al. [4] introduce the use of the Pearson correlation factor as statistical test. The corresponding side-channel attack is called Correlation Power Analysis (CPA). This correlation factor seems to give the best results on the vast majority of embedded devices. This is mostly due to the technology Complementary Metal Oxyde Semiconductor (CMOS) that is used in the industry to build smart cards. It is commonly assumed that the number of bits of a bus or an internal register, that flip at a given time, is linearly proportional to the current absorption of the device [17]. This supposition seems correct on most CMOS systems. As the CPA finds the linear dependencies between power consumption curves and a leakage function based on a key guess and a plaintext value, it is very powerful.

Making an assumption on the power consumption characteristic details of a device can be considered a strong hypothesis for an attacker. In 2008, Gierlichs et al. [8] propose a side-channel attack effective without any knowledge or restrictive assumption about the power model of the device, i.e. the relationships between the power consumption of the device and its processed data. The attack is called Mutual Information Analysis (MIA) and uses mutual information as a side-channel distinguisher. When the CPA only records linear relations, the estimation of mutual information does not need to have assumptions about the dependencies of the variables. Even if the MIA is more generic, in practice the CPA often performs better on CMOS logic. However for devices using special types of logic, as the dual-rail logic [10,5], the assumption that the relation between processed data and power consumption is linear should not hold.

The poor performance of the MIA compared to the CPA [8,20,24,29] may not be inherent to its properties but due to its inefficient estimation in most cases. The MIA, as presented in the original paper [8], uses histograms, the less effective method to estimate mutual information. Different authors [24,16] propose more efficient techniques to estimate mutual information, however the techniques make assumptions on the power model of the device, i.e. parametric techniques. In this paper, we focus our analysis on nonparametric methods for mutual information estimation. We then review and evaluate the efficiency of each method when applied with MIA. Finally, we compare state-of-the-art side-channel distinguishers with the most performant nonparametric estimators of mutual information.

Section 2 summarizes the fundamentals of information theory as well as introduces generalized mutual information. In Section 3, we study some of the most used statistical tests of the the side-channel literature. Section 4 reviews classical methods of estimation of mutual information, and more particularly, nonparametric methods. We evaluate the different techniques of estimation in the context of side-channel analysis on various setups in Section 5. Section 6 concludes the article.

2   Information Theory Framework

Shannon in [26] laid down foundations of information theory in communication systems. The entropy in a signal corresponds to the quantity of information it contains. In the context of cryptanalysis and more particularly side-channel attacks, one is interested in how much information is generated from a cryptographic device. If the device leaks information when it processes a secret, an attacker could recover the leakage through side-channel analysis and hence obtain information, e.g. bits of the secret. Mutual information is a measure closely related to entropy. It is a special case of the notion of relative entropy which records something close to a distance between two distribution functions.

2.1   Basics on Probability Theory

Let X be a random variable which takes on a finite set of values $\{x_1, x_2, \ldots, x_n\}$. Let $\mathbb{P}(X = x_i)$ be the probability distribution of X. Hence, the function $f : x \mapsto \mathbb{P}(X = x)$ is often called the probability density function (pdf) of X. Similarly, we define the function $F : x \mapsto \mathbb{P}(X \leq x)$ as the cumulative distribution function (cdf) of X.

The entropy of X is defined as

$$H(X) = -\sum_{x} f(x) \log(f(x)).$$

Let $H(X)$ and $H(Y)$ be the entropy of X and Y respectively. The joint entropy of X and Y is defined as

$$H(X, Y) = -\sum_{x,y} \mathbb{P}(X = x, Y = y) \log(\mathbb{P}(X = x, Y = y)).$$

The conditional entropy of X given Y, noted $H(X|Y)$, is defined as

$$H(X \mid Y) = \sum_{y} \mathbb{P}(Y = y) H(X \mid Y = y), \text{ with}$$

$$H(X \mid Y = y) = -\sum_{x} \mathbb{P}(X = x, Y = y) \log(\mathbb{P}(X = x, Y = y)).$$

The mutual information $I(X;Y)$ quantifies the amount of information between two variables X and Y. It is defined as

$$I(X;Y) = H(X) - H(X \mid Y).$$

Mutual information is in fact a special case of the Kullback-Leibler (KL) divergence [15]. This divergence measures the dissimilarity between two distributions. Let f and g be two pdf of a random variable X. The KL divergence, also called relative entropy, is then defined as

$$D_{\mathrm{KL}}(f \parallel g) = \sum_x f(x) \log \frac{f(x)}{g(x)}.$$

The mutual information can then be described as

$$I(X;Y) = D_{\mathrm{KL}}(f(x,y) \parallel f(x)f(y)).$$

2.2   Generalized Mutual Information

Let X be a discrete random variable as previously defined. The Rényi entropy [25] of order α is defined as

$$H_\alpha(X) = \begin{cases} \frac{1}{1-\alpha} \log \sum_x f(x)^\alpha & \text{for } \alpha \geq 0, \alpha \neq 1 \\ - \sum_x f(x) \log f(x) & \text{for } \alpha = 1. \end{cases}$$

The entropy of Shannon corresponds to $H_1(X)$. With the previous definition of Rényi entropy, we can introduce the quantity

$$I_\alpha(X;Y) = H_\alpha(X) + H_\alpha(Y) - H_\alpha(X,Y).$$

The quantity I_α has the following property:

$$I_\alpha \geq 0 \quad \text{if and only if } \alpha = 0 \text{ or } 1.$$

The value I_α only corresponds to the classical definition of mutual information in these two cases. However in [23, Basic Theorem, Ch. 3], the authors consider the case $\alpha = 2$. Using the collision entropy H_2, they call the quantity $I_2(X;Y)$ Generalized Mutual Information (GMI) where either the random variable X or Y is uniformly distributed. In this case, the GMI and the classical mutual information are both strictly positive and measure both the independence between two variables. The GMI is particularly interesting as there is a more efficient method of estimation based on kernel estimators (Sec. 4.3) [22].

3   Classical Side-Channel Distinguishers

3.1   Differential Side-Channel Model

Let K be a random variable representing a part of the secret. Let X be a random variable representing a part of the input, or output, of the cryptographic algorithm. Suppose an attacker wants to target an intermediate value computed with the function F that takes as parameters X and K. Let L be a random variable representing the side-channel leakage generated by the computation of $F(X, K)$. In practice, the attacker is only able to obtain N realizations of the random variable L, noted $V_L = (l_1, \ldots, l_N)$, as he inputs N different values of X, noted $V_X = (x_1, \ldots, x_N)$. Using a distinguisher function D, he combines these two vectors plus an hypothesis on the value of the secret k'. If the distinguisher D is relevant and if the leakage vector V_L brings enough information on $F(X, K)$, then the correct value k taken by K can be recovered. In the literature, some worked on creating a model for $F(X, K)$. For example, taking the Hamming weight of the output of F [18], the Hamming distance [4] or simply its value [8] was considered. Other researches were conducted on the distinguisher function D that plays a fundamental role in the attack. Depending on the choice, the function is able to extract more or less information from the side-channel leakages. We briefly review in the following the statistical tests used as function D proposed in the literature.

3.2   Difference of Means

Kocher et al. [13] proposed the concept of differential side-channel attack in 1999. In their original paper, the authors use a Difference of Means (DoM) as distinguisher function. It is in fact a simplified student T-test, a well-known statistical test. For simplicity reasons, we suppose the function $F(X, K)$ only outputs the least significant bit of the result. Let k' be an hypothesis on the secret. The attacker can form two sets:

$$G_0 = \{L \mid F(x_j, k') = 0\} \quad \text{and} \quad G_1 = \{L \mid F(x_j, k') = 1\}.$$

Finally, he computes the difference of means between the two partitions as:

$$\Delta_{k'} = \frac{\sum_{l \in G_0} l}{|G_0|} - \frac{\sum_{l \in G_1} l}{|G_1|}.$$

If the attacker detects a significant difference between the two sets, he can suppose that the hypothesis k' is correct.

3.3   Pearson Correlation Factor

Introduced by Brier et al. [4] in the context of side-channel analysis, the Pearson correlation factor, also called Pearson rho or product-moment correlation, measures linear dependencies between two variables X and L. The authors called the attack Correlation Power Analysis (CPA). In practice, if the attacker is only able to obtain N realizations of the leakage function, then the formula is:

$$\rho_{k'}(X, L) = \frac{N \sum_i l_i F(x_i, k') - (\sum_i l_i \sum_i F(x_i, k'))}{\sqrt{N \sum_i l_i^2 - (\sum_i l_i)^2} \sqrt{N \sum_i F(x_i, k')^2 - (\sum_i F(x_i, k'))^2}}.$$

Pearson correlation calculations are based on the assumption that both X and L values are sampled from a normal distribution. Hence, Pearson rho is part of parametric tests. On the contrary, methods that do not assume a particular distribution family for the data are said to be nonparametric.

3.4   Cluster Analysis

Differential Cluster Analysis (DCA) was introduced in [3]. It uses classical cluster analysis statistics in the side-channel analysis context. The principle of cluster analysis is to group similar objects into respective categories, i.e. clusters, and then use a statistical method in order to discover structures in the observed data. In side-channel analysis, the clusters often correspond to the outputs of the attacked intermediate value, which is similar to the mutual information technique presented Sec. 4. Amongst the statistical function used to characterize clusters proposed in [3], the use of variance seems particularly suited.

3.5   Nonparametric Correlation Statistics

Nonparametric tests make no assumptions about the distribution parameters of the variables. They do not rely on the estimation of parameters such as the mean or the standard deviation. Therefore, they are often called parameter-free or distribution-free methods. The most commonly used nonparametric equivalents to Pearson correlation factor are Spearman R, Kendall tau and coefficient Gamma. The coefficient Gamma [9] is similar to Kendall tau and is not very relevant in our analysis.

The use of the Spearman R has been proposed in [2]. Spearman R assumes that the variables are on a rank ordered scale. If several values of the variables are equal, which is the case in the context of side-channel analysis, the formula for Spearman R is the same as for Pearson's rho. The rank of identical values is the mean of their respective ranks.

Kendall tau [11] is similar in terms of results to Spearman R. However, its computation and its statistical meaning is different. Kendall measures the degree of relationships between variables whereas Pearson and Spearman test the null hypothesis that there is no relationships between variables. There is different versions of Kendall statistic. In our context, one should use the coefficient that makes adjustments for tied values:

$$\tau_b = \frac{N_c - N_d}{\sqrt{(N(N-1)/2 - t)(N(N-1)/2 - u)}},$$

where N_c is the number of pairs ranked in the same order on both variables, N_d is the number of pairs ranked differently on the variables, t is the number of tied values in the first variable, u is the number of tied values in the second and N is the number of observations.

In [29], the authors propose to use other nonparametric statistics: the Kolmogorov-Smirnov (K-S) test and the Cramér-von Mises (CVM) test. These tests are very similar to the DCA and the mutual information analysis. Indeed, the data is placed in different clusters, each typically covering a range of values of the attacked intermediate value. Let $F_X(x)$ and $F_L(x)$ be the empirical cumulative distribution functions of the sample populations X and L. The K-S test between the variables X and L is:

$$D_{\mathrm{KS}}(X \parallel L) = \sup_x |F_X(x) - F_L(x)|.$$

The CVM test is defined similarly as:

$$D_{\mathrm{CVM}}(X \parallel L) = \sum_x (F_X(x) - F_L(x))^2.$$

4   Estimators of Mutual Information

Gierlichs et al. in [8] propose the use of mutual information as side-channel distinguisher in an attack called Mutual Information Analysis (MIA). The authors present this method as an interesting alternative to the powerful CPA as the attacker does not have to assume a particular power consumption model for the targeted device (Sec. 3.1). Indeed, mutual information records both linear and non-linear relationships between variables while CPA only measures linear ones. In theory, MIA should be considered more generic as the attacker makes less assumptions about the device. However in practice, the results of MIA are not good compared to CPA [29,24,20]. In fact, the efficiency of MIA is closely related to its chosen estimator of mutual information. Some authors studied parametric estimation methods and their efficiency combined with MIA [24,7,16]. On the contrary, nonparametric estimators are not thoroughly researched [28], although they fit the original purpose of MIA more suitably.

4.1   Parametric vs Nonparametric Estimation

There are two basic approaches to estimation: parametric and nonparametric. In this paper, we restrict ourselves to the nonparametric field. Parametric estimation makes assumptions about the regression function that describes the relationship between dependent variables. Therefore, the density function will assume that the data are from a known family of distributions, such as normal, and the parameters of the function are then optimized by fitting the model to the data set. Nonparametric estimation, by contrast, is a statistical method that has no meaningful associated parameters. There is often no reliable measure used for the choice of the parameters. However, this type of estimation seems more suitable to the original purpose of the MIA: a generic side-channel attack that makes the less assumptions possible. Hence, this paper seeks to introduce efficient nonparametric pdf estimation methods in the context of side-channel analysis.

4.2   Histogram-Based Estimator

The most simple and time efficient method to estimate pdf is using histograms. An histogram consists in a partition of the range of values of each variables into

b discrete bins of equal length. The pdf of each bin is estimated by the relative frequency of occurrence of samples in the bin. Let X be a random variable with N realizations. The b partitions are defined as: $a_i = [o + ih, o + (i+1)h]$ where o the value of the origin, h is the width of the bins and $i = 0, \ldots, b-1$. Let k_i be the number the measurements of X that lie in the interval a_i. The pdf f_i of X can be approximated as

$$\hat{f}_i = \frac{k_i}{N}.$$

As this method is nonparametric, its parameters are not easily determined. The choice of the number of bins b or their width can be non-trivial. In any case, the partitioning must be the same for both variables. Even if Histogram-based Estimation (HE) is computationally efficient, its results contain more statistical errors than other methods.

4.3   Kernel Estimator

Kernel Density Estimation (KDE) constructs a smooth estimate of the density by centering kernel functions at data samples [19]. The kernels weight the distance of each points in the sample to the reference point depending on the form of the kernel function and according to a given bandwidth h. In KDE, h plays a similar role as b in HE. In fact, the uniform kernel function forms an histogram. Gaussian kernels are most commonly used and we use them as well in this study. Let $\{x_1, \ldots, n_N\}$ be N realizations of the random variable X. The pdf estimate using a Gaussian kernel is given by:

$$\hat{f}(x) = \frac{1}{N} \frac{1}{h\sqrt{2\pi}} \sum_{i=1}^{N} \exp\left(-\frac{(x-x_i)^2}{2h^2}\right).$$

This estimation method is quite costly in computational time. Kernel estimators are considered to be very good for density estimation of one-dimensional data however it is not always the case for mutual information estimation.

4.4 k-Nearest Neighbor Estimator

Kraskov et al. [14] present a new estimator based on distances of k-Nearest Neighbors (KNN) to estimate densities. The authors consider a bivariate sample and, for each reference point, a distance length is computed so that k neighbors are within this distance length noted $\epsilon(i)$ for a reference point i. The number of points with distance $\epsilon(i)/2$ gives the estimate of the joint density at the point i. The distance is then projected into each variable subspace to estimate the marginal density of each variable. The estimation of MI using KNN depends on the choice of k. In [14] the authors explain that statistical errors increase when k decreases. In practice, we should use $k > 1$, however if k is too large, systematic errors can outweigh the decrease of statistical errors. KNN gives good results with less statistical errors than previous methods but with a computationally heavy algorithm [21].

4.5   B-Spline Estimator

In [6] Daub et al. introduce the use of B-spline functions as entropy estimators. A B-spline curve is a generalized Bézier curve. It is specified by the parameters:

- the degree d, or order $k = d + 1$, so that each segment of the piecewise polynomial curve has degree d or less,
- a sequence of $m + 1$ numbers, $t_0, \ldots, t_m$, called knot vector, such that $t_i \leq t_{i+1}, \forall i \in \{1, \ldots, m - 1\}$,
- control points, $b_0, \ldots, b_n$.

A B-spline curve is defined in terms of B-spline basis functions. The i-th basis function of degree d, noted $B_{i,d}$, defined by the knot vector $t_0, \ldots, t_m$ is defined by the Cox-de Boor recursion formula as follows:

$$B_{i,0}(z) = \begin{cases} 1 & \text{if } t_i \leq z < t_{i+1} \\ 0 & \text{otherwise.} \end{cases}$$

$$B_{i,d}(z) = \frac{z - t_i}{t_{i+d} - t_i} B_{i,d-1}(z) + \frac{t_{i+d+1} - z}{t_{i+d+1} - t_{i+1}} B_{i+1,d-1}(z),$$

for $i = 0, \ldots, n$ and $d \geq 1$. Finally, the property:

$$\sum_{i=0}^{n} B_{i,d}(z) = 1,$$

for any value of z, makes B-spline basis functions suitable as a pdf estimator. This estimator is noted BSE. More details on the use and advantages of BSE in the side-channel context are available in [28].

5   Experimental Analysis

We analyze in this section the practical efficiency of nonparametric estimators of mutual information in the context of side-channel attacks. We compare their performances with state-of-the-art proposed side-channel distinguishers:

- classical parametric test, CPA (Sec. 3.3),
- nonparametric tests, SPE (Sec. 3.5), CVM (Sec. 3.5),
- cluster analysis, DCA with variance as criterion function (Sec. 3.4),
- mutual information with parametric estimation, Cumulant-based Estimator (CE) [16] which is the most powerful parametric estimator,
- mutual information with nonparametric estimation, GMIA (Sec. 2.2), HE (Sec. 4.2), KDE (Sec. 4.3), KNN (Sec. 4.4), BSE (Sec. 4.5).

In order to compare the efficiency of side-channel attacks, we use common metrics proposed in the literature [27]. Guessed Entropy (GE) is the average position of the correct key hypothesis in the sorted vector of hypothesis at the end of the attack. Results using another metric are presented in Appendix A.

Attacks are performed on two different setups: the publicly available power curves of DPA Contest 2008/2009 [30] of a DES implementation and curves acquired on an Atmel STK600 board with an Atmel AVR ATmega2561 [1] of a multi-precision multiplication algorithm.

On DPA Contest 2008/2009 curves of a DES, the intermediate value targeted is the output the SBox in the last round. For the attacks using mutual information with nonparametric estimation, we consider no power model, i.e. the value of the data. The Hamming weight model is used for the other attacks. Each attack is performed on 135 sets of 600 power curves in order to average the results. We evaluate each distinguishers and present the results in Fig. 1. Attacks can be assigned to different groups depending on their efficiency. The best seems to be CE, CPA and SPE which are all parametric tests. The following attacks are KDE, DCA and CVM amongst which are the first mutual information nonparametric ones. The BSE is next, followed by KNN, GMIA and HE.

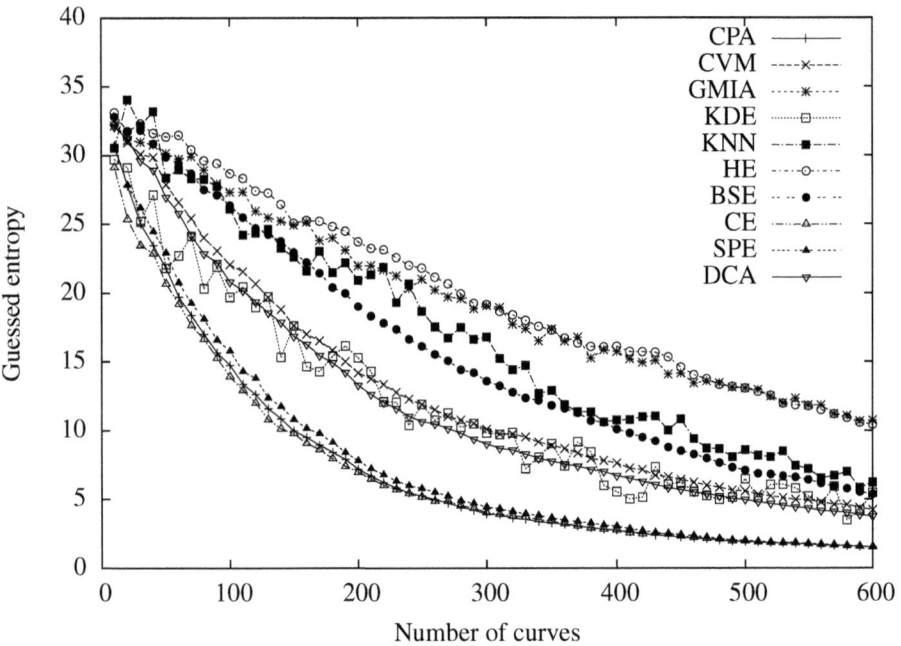

Fig. 1. Guessed entropy results on DPA Contest 2008/2009 curves of a DES

The same attacks are also performed on curves acquired on a STK600 development board with a 8-bit Atmel AVR ATmega2561. This setup is not particularly well suited to perform side-channel attacks. Therefore the power traces contains significantly more noise than the DPA Contest ones. We attack a column-wise multi-precision multiplication algorithm implemented in software. The targeted values are the intermediate 8-bit multiplications $x_i \times y_j$ considering one of the multiplicand known by the attacker. As with the previous setup, we do not consider a power model for the attacks using mutual information with nonparametric estimation. The other attacks assume the Hamming weight model. Each attack is performed on 20 sets of 2000 power curves. We obtain a slightly different performance from several attacks (Fig. 2). As previously, the most powerful attacks are still CE, CPA and SPE. However BSE seems to perform much better and is at the same level as CVM, DCA and KDE. The estimators BSE and KDE are the most efficient nonparametric methods but BSE is more computationally efficient than KDE, hence an more interesting choice.

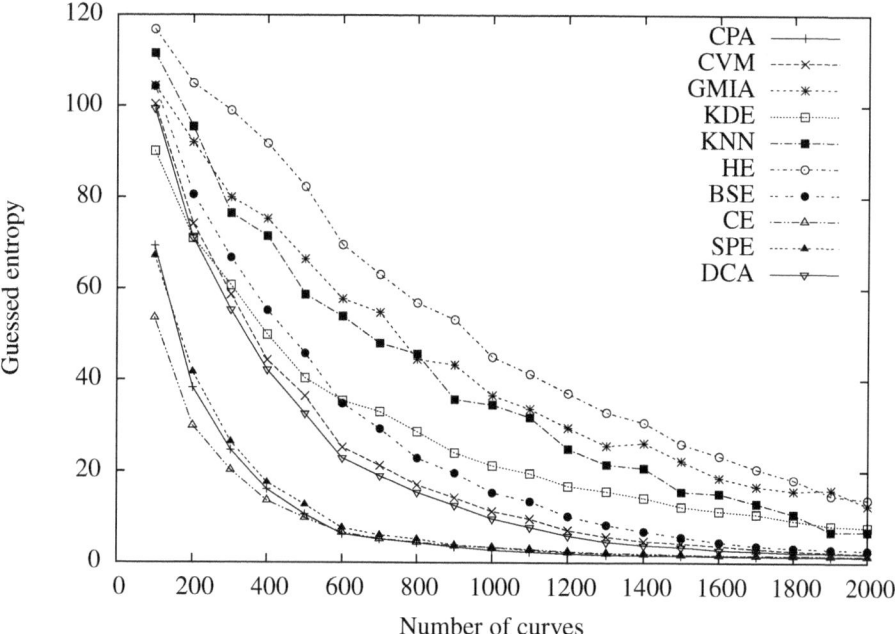

Fig. 2. Guessed entropy results on STK600 curves of a multi-precision multiplication

With this overall comparison of state-of-the-art side-channel distinguishers, we can note differences between classical statistical tests performance and their efficiency in the side-channel context. For example, the KNN estimator should be less subject to statistical errors than BSE or KDE. However it performs worse in this scenario. Classical parametric tests are still amongst the most powerful in most cases. In particular the recently presented Cumulant-based Estimator [16], a parametric estimator of mutual information, is very interesting. These experimental analysis also show the gain obtained when using efficient nonparametric estimators of mutual information. Even if the MIA attack is not the most powerful, its performance is greatly improved compared to the classical histogram estimator that has been used in the literature as a reference.

6   Conclusion

In this paper, we review some of the most statistically powerful nonparametric estimators of mutual information in the context of side-channel analysis. The distinction between parametric and nonparametric methods is important and should be clearly made when comparing side-channel distinguishers efficiency. Depending on the supposed knowledge of the adversary, one of these two classes of attacks needs to be considered. We also note that, in terms of performance, nonparametric estimation in MIA is not as bad as previously thought. The KDE and BSE estimators perform quite well for an acceptable computational overhead in the case of BSE. Even if the study is done on CMOS devices, we can expect a similar improvement of performance on different types of logic when using efficient nonparametric methods.

References

1. ATMEL: ATmega 2561 Data Sheet,
   http://www.atmel.com/dyn/resources/prod_documents/doc2549.pdf
2. Batina, L., Gierlichs, B., Lemke-Rust, K.: Comparative Evaluation of Rank Correlation Based DPA on an AES Prototype Chip. In: Wu, T.-C., Lei, C.-L., Rijmen, V., Lee, D.-T. (eds.) ISC 2008. LNCS, vol. 5222, pp. 341–354. Springer, Heidelberg (2008)
3. Batina, L., Gierlichs, B., Lemke-Rust, K.: Differential Cluster Analysis. In: Clavier, C., Gaj, K. (eds.) CHES 2009. LNCS, vol. 5747, pp. 112–127. Springer, Heidelberg (2009)
4. Brier, E., Clavier, C., Olivier, F.: Correlation Power Analysis with a Leakage Model. In: Joye, M., Quisquater, J.-J. (eds.) CHES 2004. LNCS, vol. 3156, pp. 16–29. Springer, Heidelberg (2004)
5. Chen, Z., Zhou, Y.: Dual-Rail Random Switching Logic: A Countermeasure to Reduce Side Channel Leakage. In: Goubin, L., Matsui, M. (eds.) CHES 2006. LNCS, vol. 4249, pp. 242–254. Springer, Heidelberg (2006)
6. Daub, C., Steuer, R., Selbig, J., Kloska, S.: Estimating Mutual Information Using B-spline Functions - an Improved Similarity Measure for Analysing Gene Expression Data. BMC Bioinformatics 5, 118 (2004)
7. Flament, F., Guilley, S., Danger, J., Elaabid, M., Maghrebi, H., Sauvage, L.: About Probability Density Function Estimation for Side Channel Analysis. In: COSADE 2010 (2010)
8. Gierlichs, B., Batina, L., Tuyls, P., Preneel, B.: Mutual Information Analysis. In: Oswald, E., Rohatgi, P. (eds.) CHES 2008. LNCS, vol. 5154, pp. 426–442. Springer, Heidelberg (2008)
9. Goodman, L., Kruskal, W.: Measures of Association for Cross Classifications. II: Further Discussion and References. Journal of the American Statistical Association 49, 732–764 (1954)

10. Guilley, S., Hoogvorst, P., Mathieu, Y., Pacalet, R.: The "Backend Duplication" Method. In: Rao, J.R., Sunar, B. (eds.) CHES 2005. LNCS, vol. 3659, pp. 383–397. Springer, Heidelberg (2005)
11. Kendall, M.: A New Measure of Rank Correlation. Biometrika 30, 1–2 (1938)
12. Kocher, P.: Timing Attacks on Implementations of Diffie-Hellman, RSA, DSS, and Other Systems. In: Koblitz, N. (ed.) CRYPTO 1996. LNCS, vol. 1109, pp. 104–113. Springer, Heidelberg (1996)
13. Kocher, P., Jaffe, J., Jun, B.: Differential Power Analysis. In: Wiener, M. (ed.) CRYPTO 1999. LNCS, vol. 1666, pp. 388–397. Springer, Heidelberg (1999)
14. Kraskov, A., Stogbauer, H., Grassberger, P.: Estimating Mutual Information. Physical Review E 69, 66138 (2004)
15. Kullback, S., Leibler, R.: On Information and Sufficiency. The Annals of Matematical Statistics 22, 79–86 (1951)
16. Lee, T.H., Berthier, M.: Mutual Information Analysis under the View of Higher-Order Statistics. In: Echizen, I., Kunihiro, N., Sasaki, R. (eds.) IWSEC 2010. LNCS, vol. 6434, pp. 285–300. Springer, Heidelberg (2010)
17. Messerges, T.S., Dabbish, E.A., Sloan, R.H.: Investigations of Power Analysis Attacks on Smartcards. In: USENIX Workshop on Smartcard Technology. pp. 151–162 (1999)
18. Messerges, T.S., Dabbish, E.A., Sloan, R.H.: Power Analysis Attacks of Modular Exponentiation in Smartcards. In: Koç, Ç.K., Paar, C. (eds.) CHES 1999. LNCS, vol. 1717, pp. 144–157. Springer, Heidelberg (1999)
19. Moon, Y.I., Rajagopalan, B., Lall, U.: Estimation of Mutual Information using Kernel Density Estimators. Physical Review E 52(3), 2318–2321 (1995)
20. Moradi, A., Mousavi, N., Paar, C., Salmasizadeh, M.: A Comparative Study of Mutual Information Analysis under a Gaussian Assumption. In: Youm, H.Y., Yung, M. (eds.) WISA 2009. LNCS, vol. 5932, pp. 193–205. Springer, Heidelberg (2009)
21. Papana, A., Kugiumtzis, D.: Evaluation of Mutual Information Estimators on Nonlinear Dynamic Systems. Nonlinear Phenomena in Complex Systems 11, 225–232 (2008)
22. Pompe, B., Heilfort, M.: On the Concept of the Generalized Mutual Information Function and Efficient Algorithms for Calculating it (1995)
23. Pompe, B., Physik, F.: Measuring Statistical Dependences in a Time Series. Journal of Statistical Physics 73, 587–610 (1993)
24. Prouff, E., Rivain, M.: Theoretical and Practical Aspects of Mutual Information Based Side Channel Analysis. In: Abdalla, M., Pointcheval, D., Fouque, P.-A., Vergnaud, D. (eds.) ACNS 2009. LNCS, vol. 5536, pp. 499–518. Springer, Heidelberg (2009)
25. Rényi, A.: On Measures of Information and Entropy. In: Proceedings of the 4th Berkeley Symposium on Mathematics, Statistics and Probability, vol. 1, pp. 547–561 (1961)
26. Shannon, C.: A Mathematical Theory of Communication. The Bell System Technical Journal 27, 379–423 (1948)
27. Standaert, F.X., Gierlichs, B., Verbauwhede, I.: Partition vs. Comparison Side-Channel Distinguishers: An Empirical Evaluation of Statistical Tests for Univariate Side-Channel Attacks against Two Unprotected CMOS Devices. In: Lee, P.J., Cheon, J.H. (eds.) ICISC 2008. LNCS, vol. 5461, pp. 253–267. Springer, Heidelberg (2009)

28. Venelli, A.: Efficient Entropy Estimation for Mutual Information Analysis Using B-Splines. In: Samarati, P., Tunstall, M., Posegga, J., Markantonakis, K., Sauveron, D. (eds.) WISTP 2010. LNCS, vol. 6033, pp. 17–30. Springer, Heidelberg (2010)
29. Veyrat-Charvillon, N., Standaert, F.: Mutual Information Analysis: How, When and Why? In: Clavier, C., Gaj, K. (eds.) CHES 2009. LNCS, vol. 5747, pp. 429–443. Springer, Heidelberg (2009)
30. VLSI research group and TELECOM ParisTech: The DPA Contest (2008/2009), http://www.dpacontest.org

A   First-order Success Rate Results

We present here the results of the success rate metric on the two platforms detailed in Sec. 5. First-order success rate is, for a given number of curves, the probability that the correct key hypothesis is ranked first in the sorted vector of hypothesis. These results are consistent with the guessed entropy metric presented in Sec. 5.

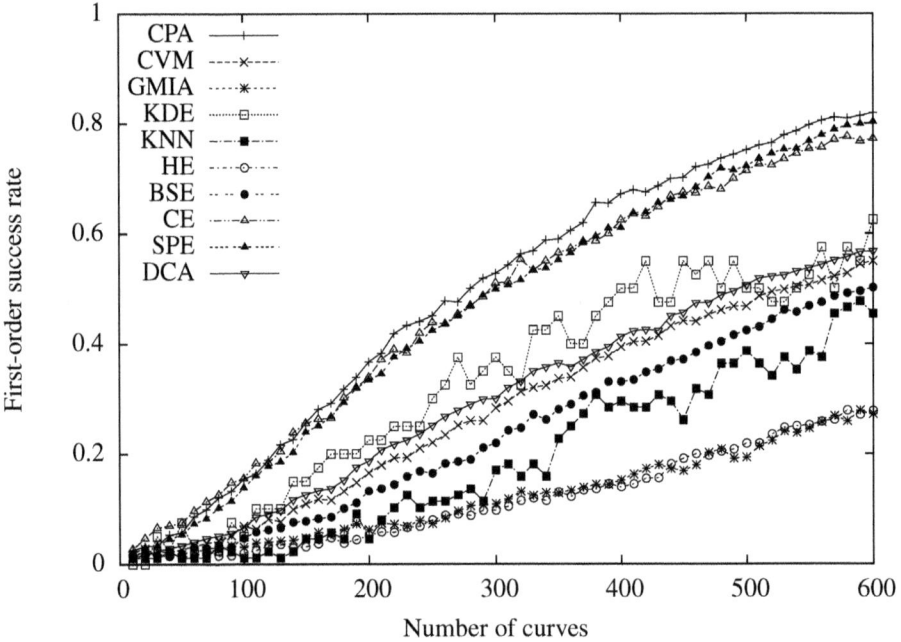

Fig. 3. First-order success rate on DPA Contest 2008/2009 curves of a DES

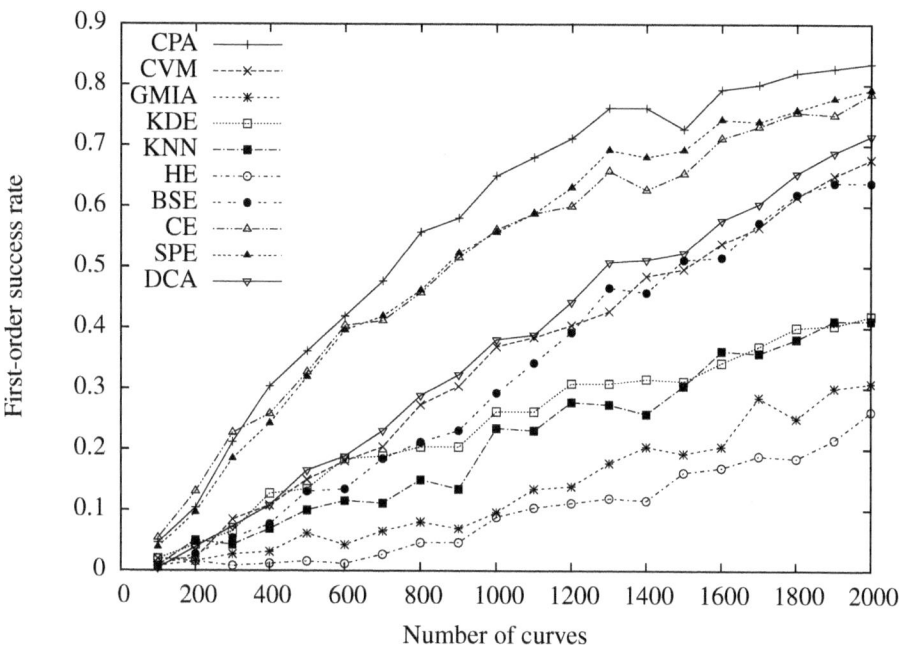

Fig. 4. First-order success rate results on STK600 curves of a multi-precision multiplication

Bias Analysis of a Certain Problem
with Applications to E0 and Shannon Cipher

Yi Lu* and Yvo Desmedt**

Department of Computer Science
University College London, UK

Abstract. Bias analysis is an important problem in cryptanalysis. When
the critical bias can be expressed by the XOR of many terms, it is well-
known that we can compute the bias of their sum by the famous Piling-up
lemma assuming all the terms are independent. In this paper, we consider
the terms of the sum are dependent and we study above bias problem.
More precisely, let each term be a Boolean function of a variable over
$GF(2)^n$. We assume the distribution D of the XOR of k variables is
known, each variable is uniformly distributed individually, and moreover, the XOR of k variables and $(k-1)$ variables all are independent.
We give a simple expression for the bias of the sum of k Boolean functions. It takes time $O(kn \cdot 2^n)$ to compute the bias, while under the
independence assumption, it takes time $O(k \cdot 2^n)$ to compute by Piling-
up lemma. We further compare the general bias in our problem with the
bias in the independent case. It is remarkable to note that the former can
differ significantly from the latter. As application, we apply our results
to cryptanalysis of two real examples, Bluetooth encryption standard
E0 and Shannon cipher, which show a strongly biased and weakly biased D respectively. For E0, our analysis allows to make the best known
key-recovery attack with precomputation, time and data complexities
$O(2^{37})$. For Shannon cipher, our analysis verifies the validity of the estimated complexity $O(2^{107})$ of the previous distinguishing attack [5]. As
comparison, we also studied a variant of Shannon cipher, which shows
much stronger dependency within the internal states. We gave a distinguishing attack on the Shannon variant with reduced complexity $O(2^{93})$.

Keywords: linear cryptanalysis, bias, Piling-up lemma, E0, Shannon
cipher.

1 Introduction

In linear cryptanalysis [8], bias analysis is an important problem. A large bias will
make distinguishing attacks or even key-recovery attacks possible. The cryptanalyst is thus often faced with the problem of trying to find a possibly large bias

* Funded by British Telecommunications under Grant Number ML858284/CT506918.
** Funded by EPSRC EP/C538285/1, by BT, (as BT Chair of Information Security),
and by RCIS (AIST).

K.-H. Rhee and D. Nyang (Eds.): ICISC 2010, LNCS 6829, pp. 16–28, 2011.

for a cipher. When the critical bias can be expressed by the XOR of many terms and when there shows no clear dependency between them (though we are also sure that they are not really independent), it is convenient and common to assume these terms are independent and use the famous Piling-up lemma [8] to combine the individual bias of each term to get the total bias estimate. This independence assumption, however, is not always appropriate to approximate the total bias. In [6], it was shown that in the context of block ciphers, the approximation by Piling-up lemma sometimes can differ considerably from the real value of the bias.

In this paper, we take the dependency[1] between the terms of the sum into account and study the above bias problem. More precisely, let each term be a Boolean function[2] of a variable over $GF(2)^n$. We assume the distribution of the XOR of k variables is known, each variable is uniformly distributed individually, and moreover, the XOR of k variables and $(k-1)$ variables all are independent. We give a simple expression for the bias of the sum of k Boolean functions. It takes time $O(kn \cdot 2^n)$ to compute the bias. It grows linearly in k and is practical for modest n. In contrast, under the independence assumption, it needs $O(k \cdot 2^n)$ time to compute the bias by Piling-up lemma. Note that a special case of our result was given in [9] previously; [9] considered the bias of our problem when all the functions are the same and the XOR of all variables is the constant of the all zero vector.

Furthermore, we compare the general bias in our problem with the bias in the independent case. We note that the former can differ significantly from the latter: 1) if one function is balanced, then the real bias is always no smaller than the bias in the independence case; 2) if no functions are balanced, then the real bias can be smaller than the bias in the independence case, which implies that the convenient independence assumption sometimes *over-estimates* the real bias; 3) if all functions are the same and k is even, then the real bias can be negative, while the bias in the independence case is never negative.

As one application, our result is applied to analyze precisely the bias for the famous encryption standard E0, which is used in the short-range wireless radio standard Bluetooth [2]. We observe that the dependency between the involved variables is strong. We show that if the multiple polynomial of the related LFSR feedback polynomial has even weight, we have four largest biases; if the weight is odd, we have two largest biases. In comparison, according to the traditional bias estimate approach based on independence assumption, it was believed that there are two largest biases [7], regardless of the multiple polynomial weight. We demonstrate that the two biases used in the key-recovery attack [7] are both *under-estimated* and should be doubled instead. This allows to make the best known key-recovery attack on E0, with precomputation, time and data complexities $O(2^{37})$.

Another application is a recently proposed stream cipher Shannon [11] designed by Qualcomm group [10]. We revisit the keystream bias of Shannon

---

[1] The bias problem which considered dependency of another form was studied in [3].
[2] Not necessarily the same Boolean function.

studied in [5]. We show that unlike E0, the dependency within Shannon internal states is weak. The estimated complexity $O(2^{107})$ of the distinguishing attack [5], which assumed that the internal states were independent, is still valid. Meanwhile, we study a variant of Shannon cipher, which shows much stronger dependency within the internal states. A distinguishing attack on the Shannon variant with reduced complexity $O(2^{93})$ is proposed. Note that if the internal states were assumed to be independent, the keystream bias of this Shannon variant remains the same as that of the Shannon cipher and so does the attack complexity $O(2^{107})$.

The rest of the paper is organized as follows. Section 2 introduces our main bias problem. In Section 3, we apply our results to Bluetooth E0, which allows to make the best key-recovery attack known so far. We demonstrate another application of our theory to Shannon cipher and a Shannon variant in Section 4. Finally we conclude in Section 5.

2   Our Bias Problem

In linear cryptanalysis [8], the following problem is frequently encountered to the cryptanalyst: given Boolean functions $f_1, f_2 : GF(2)^n \rightarrow GF(2)$, what is the bias δ of $f_1(a) \oplus f_2(b)$? Here, the bias of a Boolean variable X refers to $\Pr(X = 0) - \Pr(X = 1)$. Due to the famous Piling-up Lemma [8], this problem has a very simple solution when the inputs a and b are independent and uniformly distributed: $\delta = \delta_1 \cdot \delta_2$, where δ_1, δ_2 is the bias of of $f_1(a)$, $f_2(b)$ respectively.

In this paper, we are interested in the above problem for the case that a and b are dependent. We will study the case that a, b is uniformly distributed individually and the only dependency relation between them is that $a \oplus b$ is not uniformly distributed (but a and $a \oplus b$ are independent). Formally speaking, given $f_1, f_2 : GF(2)^n \rightarrow GF(2)$ and a distribution D over $GF(2)^n$, we study the bias δ of $f_1(a) \oplus f_2(b)$ assuming that the uniformly distributed n-bit a, b satisfy that a and $a \oplus b$ are independent and that $a \oplus b$ complies with the given distribution D. Note that if D is a uniform distribution, from our assumption we can deduce that a and $a \oplus b$ are independent and uniformly distributed. Hence, a and b are independent and uniformly distributed. Our problem then degrades to the old well-known problem and we have $\delta = \delta_1 \cdot \delta_2$.

Algorithm 1. Computing the bias δ of $f_1(a) \oplus f_2(b)$

$u \leftarrow 0$
for all n-bit a **do**
  for all n-bit x **do**
    $b \leftarrow a \oplus x$
    $u \leftarrow u + D(x) \cdot (-1)^{f_1(a) \oplus f_2(b)}$
  end for
end for
output $\delta \leftarrow u/2^n$

Clearly, δ can be computed by a naive computation as shown in Algorithm 1. It needs time $O(2^{2n})$. Note that the bias δ as computed by Algorithm 1 can be expressed by

$$\delta = \frac{1}{2^n} \sum_a (-1)^{f_1(a)} \sum_b (-1)^{f_2(b)} D(a \oplus b). \tag{1}$$

Define functions $g_1, g_2 : GF(2)^n \to \{1, -1\}$ as follows

$$g_1(x) = (-1)^{f_1(x)} \tag{2}$$
$$g_2(x) = (-1)^{f_2(x)} \tag{3}$$

We can rewrite Eq.(1) by

$$\delta = \frac{1}{2^n} \sum_a g_1(a) \cdot \sum_b g_2(b) \cdot D(a \oplus b) \tag{4}$$

$$= \frac{1}{2^n} \sum_a g_1(a) \cdot (g_2 \otimes D)(a) \tag{5}$$

$$= \frac{1}{2^n} (g_1 \otimes g_2 \otimes D)(0) \tag{6}$$

where $\otimes$ denotes the convolution. As convolution can be computed efficiently by Fast Walsh Transform (FWT), we finally have proved the following result.

Theorem 1. *Given $f_1, f_2 : GF(2)^n \to GF(2)$ and a distribution D over $GF(2)^n$, assuming that the uniformly distributed n-bit a, b satisfy that $a, a \oplus b$ are independent and that $a \oplus b$ complies with the given distribution D, the bias δ of $f_1(a) \oplus f_2(b)$ can be expressed by*

$$\delta = \frac{1}{2^{2n}} \sum_x \widehat{g_1}(x) \cdot \widehat{g_2}(x) \cdot \widehat{D}(x),$$

where the Walsh Transform $\widehat{F}$ of F is defined by $\widehat{F}(x) = \sum_y (-1)^{x \cdot y} F(y)$ and g_1, g_2 are defined in Eq.(2), Eq.(3) respectively.

Note that it needs $O(n \times 2^n)$ time to compute δ by above theorem, while it needs $O(2^n)$ time to compute δ under the independence assumption by Piling-up lemma. Moreover, Theorem 1 can be easily generalized to the following result.

Corollary 1. *Given $f_1, f_2, \ldots, f_k : GF(2)^n \to GF(2)$ and a distribution D over $GF(2)^n$, assuming that the uniformly distributed n-bit $a_1, a_2, \ldots, a_k$ satisfy that $a_1, \ldots, a_{k-1}, a_1 \oplus \cdots \oplus a_k$ are independent and that $a_1 \oplus \cdots \oplus a_k$ complies with the given distribution D, the bias δ of $f_1(a_1) \oplus \cdots \oplus f_k(a_k)$ can be expressed by*

$$\delta = \frac{1}{2^{kn}} \sum_x \widehat{g_1}(x) \cdot \widehat{g_2}(x) \cdots \widehat{g_k}(x) \cdot \widehat{D}(x),$$

where $g_i(x) = (-1)^{f_i(x)}$ for $i = 1, \ldots, k$.

It needs $O(kn \times 2^n)$ time to compute δ. It grows linearly in k. It is practical for modest n, for example $n \leq 32$. Without our result, the naive computation needs $O(2^{kn})$ time. In contrast, note that under the independence assumption, it takes time $O(k \times 2^n)$ to compute the bias by Piling-up lemma.

Next, let us see how the distribution D affects the bias δ. First, if D is a uniform distribution, we can deduce that a_i's are independent and uniformly distributed. On one hand, Piling-up lemma directly tells us that $\delta = \delta_1 \cdot \delta_2 \cdots \delta_k$, where δ_i denotes the bias of f_i. On the other hand, from the assumption D is uniform distribution, we know that $\widehat{D}(x) = 1$ for $x = 0$ and $\widehat{D}(x) = 0$ for $x \neq 0$. By Corollary 1, we have

$$\delta = \frac{1}{2^{kn}} \widehat{g_1}(0) \cdot \widehat{g_2}(0) \cdots \widehat{g_k}(0)$$

As $\delta_i = \frac{1}{2^n} \widehat{g_i}(0)$, we also have $\delta = \delta_1 \cdots \delta_k$. Thus we have seen that our result incorporates the Piling-up lemma as a special case.

Second, if $a_1 \oplus a_2 \oplus \cdots \oplus a_k = a_0$ for a fixed n-bit a_0, that is, $D(a_0) = 1$, then $\widehat{D}(x) = 1$ if $x \cdot a_0 = 0$ and $\widehat{D}(x) = -1$ if $x \cdot a_0 = 1$. By Corollary 1, now we have

$$\delta = \frac{1}{2^{kn}} \left(\sum_{x:a_0 \cdot x = 0} \widehat{g_1}(x) \cdots \widehat{g_k}(x) - \sum_{x:a_0 \cdot x = 1} \widehat{g_1}(x) \cdots \widehat{g_k}(x) \right). \tag{7}$$

Note that when $a_0 = 0$, we have

$$\delta = \frac{1}{2^{kn}} \sum_{x} \widehat{g_1}(x) \cdots \widehat{g_k}(x). \tag{8}$$

The above result (8) with $f_1 = f_2 = f_3 = f_4$ and $k = 4$ was shown in [9].

Thirdly, let $\beta = \max_{x \neq 0} \widehat{D}(x)$ be the largest bias of D. And we consider a general D with $\beta \neq \pm 1$. Note that regardless of D, we always have $\widehat{D}(0) = 1$. From Corollary 1,

$$\delta = \frac{1}{2^{kn}} \left(\widehat{g_1}(0) \cdots \widehat{g_k}(0) + \sum_{x:0 < |\widehat{D}(x)| < 1} \widehat{g_1}(x) \cdots \widehat{g_k}(x) \cdot \widehat{D}(x) \right). \tag{9}$$

In case that D is a weakly biased distribution, the second addend on the right-hand side of Eq.(9) is negligible compared with the first addend (because $\beta \ll 1$ and the number of biases which are roughly on the same order of magnitude is also very small). Thus, we can have the approximation below

$$\delta \approx \frac{1}{2^{kn}} \widehat{g_1}(0) \cdots \widehat{g_k}(0) = \delta_1 \cdots \delta_k.$$

If D is not weakly biased, the bias δ cannot be approximated by the bias in the independence case. In Section 3, Section 4, we will demonstrate with an example of strongly biased and weakly biased D respectively.

When comparing the value of the bias δ in our problem with the bias in the independent case, we point out three important remarks below.

Remark 1. If $\delta_i = 0$ for $i \in \{1, \ldots, k\}$ (or equivalently f_i is balanced), then it is easy to see that δ in our problem is always no smaller than the bias in the independent case.

Examples: see Section 3.

Remark 2. If $\delta_i \neq 0$ for all $i = 1, \ldots, k$, it is possible to have $\delta < \delta_1 \cdots \delta_k$.

Examples: see Section 3.

   This implies that the independence assumption, which is so often used for convenience, sometimes would *over-estimate* the real bias.

Remark 3. If $f_1 = \cdots = f_k$ and k is even, it is possible to have $\delta < 0$. In comparison, note that the bias in the independent case can never be negative.

As an illustrative example to Remark 3, consider $k = 2, f_1 = f_2$ with $f_1(0) = f_1(2) = f_1(3) = 1$ and $f_1(1) = 0$, $D(0) = D(1) = 1/8, D(2) = D(3) = 3/8$. We can check that $\delta = -3/16$ and the bias in the independent case is $\delta_1^2 = 1/4$.

3   Application One: E0

In this section, we will see that our proposed new tool in Section 2 is applicable to a precise bias analysis of the famous encryption standard E0, which is used in the short-range wireless radio standard Bluetooth [2]. E0 uses a 128-bit secret key. It uses four (regularly clocked) LFSRs of 128 bits in total and a 4-bit Finite State Memory (FSM). The FSM updates its 4-bit state at each clock by the outputs of the LFSRs, and the FSM outputs one bit c_t^0 out of its 4-bit state. The keystream bit z_t at each clock is computed as

$$z_t = x_t^1 \oplus x_t^2 \oplus x_t^3 \oplus x_t^4 \oplus c_t^0, \tag{10}$$

where x_t^i denotes the output of LFSR R_i. For a complete and detailed description, see [2].

   In [7], the bias within a consecutive sequence of $\{c_t^0\}$ up to 26 bits was systematically analyzed. Let M be a 26 bit vector. It was shown [7] that when $M = (11111)_2$ or $M = (100001)_2$ (in binary form), the absolute value of the bias of $M \cdot c_t^0 c_{t-1}^0 \cdots c_{t-25}^0$ is the largest $\frac{25}{256}$. Let $\beta_i(x)$ be the feedback polynomial of LFSR R_i and $\beta_i(x)$ is a primitive polynomial. It is well-known that the the equivalent LFSR, which can generate the same sequence of $x_t^1 \oplus x_t^2 \oplus x_t^3 \oplus x_t^4$, has the feedback polynomial $\beta(x) = \prod_{i=1}^4 \beta_i(x)$. Let $p(x) = x^{p_1} \oplus x^{p_2} \oplus \cdots \oplus x^{p_w}$ be the multiple polynomial of $\beta(x)$ with degree d and weight w, where $d = p_w > \ldots > p_2 > p_1 = 0$. We have the following equality always holds for all t_0,

$$\bigoplus_{i=1}^{w} (x_{t_0+p_i}^1 \oplus x_{t_0+p_i}^2 \oplus x_{t_0+p_i}^3 \oplus x_{t_0+p_i}^4) = 0. \tag{11}$$

Therefore, we have the following equality always holds for all t_0,

$$\bigoplus_{i=1}^{w} M \cdot (z_{t_0+p_i}, \ldots, z_{t_0+p_i+25}) = \bigoplus_{i=1}^{w} M \cdot (c_{t_0+p_i}^0, \ldots, c_{t_0+p_i+25}^0) \tag{12}$$

by the keystream generation function. According to [7], the exact bias δ_0 of $M \cdot (c_{t_0}^0, \ldots, c_{t_0+25}^0)$ can be calculated. From this, the total bias δ of the sum of w such terms, which is

$$\bigoplus_{i=1}^{w} M \cdot (c_{t_0+p_i}^0, \ldots, c_{t_0+p_i+25}^0), \tag{13}$$

was deduced in [7] to be $\delta = \delta_0^w$ by the Piling-up lemma. Therefore, by Eq.(12), [7] concluded that E0 keystream output has a bias of δ_0^w, which was then used to mount the best known attacks on one-level E0 [7].

3.1   Our Analysis

Note that the above application of the Piling-up lemma to deduce $\delta = \delta_0^w$ is based on the assumption that with $i = 1, 2, 3, 4$,

$$x_{t_0+p_1+1}^i, \ldots, x_{t_0+p_1+24}^i,$$
$$x_{t_0+p_2+1}^i, \ldots, x_{t_0+p_2+24}^i,$$
$$\vdots$$
$$x_{t_0+p_w+1}^i, \ldots, x_{t_0+p_w+24}^i,$$

together with the FSM states at $t = t_0+p_1+1, t_0+p_2+1, \ldots, t_0+p_w+1$ all are independent. Furthermore, it was formally proved in [7] that for $w = 1$ this is true assuming the initial states are random and uniformly distributed. In fact, we can check the following equality always holds:

$$\bigoplus_{j=1}^{w} x_{t_0+p_j+1}^i, \bigoplus_{j=1}^{w} x_{t_0+p_j+2}^i, \ldots, \bigoplus_{j=1}^{w} x_{t_0+p_j+24}^i = \mathbf{0}, \tag{14}$$

for $i = 1, 2, 3, 4$, where $\mathbf{0}$ denotes the all zero vector. This implies that the above independence assumption is wrong and the Piling-up lemma is not appropriate to use to deduce the bias δ of (13). On the other hand, in order to apply Corollary 1 in Section 2 and calculate the real bias δ of (13), we can use (14) to deduce the relevant distribution D as follows.

We let D represents the distribution of the $4 * 24 + 4 = 100$ bit vector, in which, the least significant 4 bits consists of the XOR of the FSM state at $t = t_0 + p_i + 1$ with $i = 1, \ldots, w$, and the most significant 96 bits consists of the XOR of $x_{t_0+p_i+1}^1, \ldots, x_{t_0+p_i+24}^1, x_{t_0+p_i+1}^2, \ldots, x_{t_0+p_i+24}^2, x_{t_0+p_i+1}^3, \ldots, x_{t_0+p_i+24}^3,$ $x_{t_0+p_i+1}^4, \ldots, x_{t_0+p_i+24}^4$ at $i = 1, \ldots, w$ respectively. Assuming that the FSM states at $t = t_0 + p_1 + 1, t_0 + p_2 + 1, \ldots, t_0 + p_w + 1$ are random and uniformly distributed, from (14) we deduce that $D(0) = D(1) = D(2) = \cdots = D(15) = \frac{1}{16}$. It is easy to know the Walsh coefficients of D: $\widehat{D}(x) = 1$ for all x whose least significant 4 bits are zeros (denoted by $LSB_4(x) = 0$) and $\widehat{D}(x) = 0$ otherwise. Define $f : GF(2)^{100} \to GF(2)$, which maps the FSM state at $t = t_0 + 1$ and 4

LFSR outputs at $t = t_0 + 1, \ldots, t_0 + 24$ to $M \cdot (c_{t_0}^0, \ldots, c_{t_0+25}^0)$. Consequently, following Corollary 1, we compute the bias δ of (13) as

$$\delta = \frac{1}{2^{100w}} \sum_{x \in GF(2)^{100} : LSB_4(x) = 0} (\widehat{g}(x))^w,$$

where $g(x) = (-1)^{f(x)}$. From Section 2, we know that $\delta \approx \delta_0^w$ if D is a very weakly biased distribution. However, the Walsh spectrum of D indicates that this approximation may not be appropriate.

3.2   Our Results

Due to limited computation power, we only compute for all 8-bit M, where $f : GF(2)^\ell \to GF(2)$ and $\ell \leq 28$. Table 1 – Table 4 compare the real bias δ with δ_0^w for $M = (11111)_2, M = (100001)_2, M = (10111)_2, M = (110001)_2, M = (1011)_2,$ $w = 2, \ldots, 6$ ('X' denotes no bias). Our computation shows that the bias can be roughly approximated by using these largest $\widehat{g}(x)^w$, where $\widehat{D}(x) = 1$. Thus, the number and the sign of these largest $\widehat{g}(x)$, where $\widehat{D}(x) = 1$ determine the value of the bias. Of all 8-bit M's, we found out that when $\widehat{D}(x) = 1$, the largest $|\widehat{g}(x)|$ is achieved with $M = (11111)_2, (100001)_2, (10111)_2, (110001)_2$. Furthermore, each of above M has 6 positive and 2 negative, 2 positive and 6 negative, 4 positive and 4 negative, 4 positive and 4 negative largest respectively. Consequently, as shown in Table 1 – Table 4, it is easy to see that when w is even, the largest bias (up to 8 bits) is achieved with $M = (11111)_2, (100001)_2, (10111)_2, (110001)_2$; if w is odd, the largest bias (up to 8 bits) is achieved with $M = (11111)_2, (100001)_2$ only. This is in clear contrast to the result [7] of the traditional bias estimate approach based on independence assumption, which concluded that $M = (11111)_2, (100001)_2$ are the only largest biases (up to 26 bits) regardless of parity of w.

Meanwhile, we give examples here to illustrate our important remarks in Section 2. Table 4 shows a good example to Remark 1, where the bias approximation by Piling-up lemma shows no bias but our result proves wrong. With regards

Table 1. Comparison of δ with δ_0^w for $w = 2, \ldots, 6$, where $M = (11111)_2$

w	2	3	4	5	6
$log_2\|\delta\|$	-3	-8	-10.5	-14.7	-17
$log_2\|\delta_0^w\|$	-6.7	-10	-13.4	-16.7	-20

Table 2. Comparison of δ with δ_0^w for $w = 2, \ldots, 6$, where $M = (100001)_2$

w	2	3	4	5	6
$log_2\|\delta\|$	-2.6	-7	-10.4	-14.7	-17
$log_2\|\delta_0^w\|$	-6.7	-10	-13.4	-16.7	-20

Table 3. Comparison of δ with δ_0^w for $w = 2, \ldots, 6$, where $M = (10111)_2$

w	2	3	4	5	6
$log_2|\delta|$	-3	X	-10.2	X	-17
$log_2|\delta_0^w|$	-12	-18	-24	-30	-36

Table 4. Comparison of δ with δ_0^w for $w = 2, \ldots, 6$, where $M = (110001)_2$

w	2	3	4	5	6
$log_2|\delta|$	-2.6	-12.1	-10.2	-22.7	-17
$log_2|\delta_0^w|$	X	X	X	X	X

Table 5. Comparison of our improved key-recovery attack on one-level E0 with previous attacks

attack	precomputation	time	data	memory
[1]	X	$2^{67.6}$	$2^{23.1}$	$2^{46.1}$
[4]	2^{28}	2^{49}	$2^{23.4}$	2^{37}
[7]	2^{37}	2^{39}	2^{39}	2^{27}
this paper	2^{37}	2^{37}	2^{37}	2^{27}

to Remark 2, the bias with $w = 3$ or 5 in Table 3 shows that the independence assumption *over-estimates* the real bias here.

Based on our above bias analysis, we can now improve the best known key-recovery attack [7] on one-level E0 as follows. Note that to recover the shortest LFSR R_1 in [7], the multiple polynomial of $\prod_{i=2}^{4} \beta_i(x)$ is used rather than the multiple polynomial of $\beta(x)$. This affects the relevant distribution D as well as the bias. Assuming that the involved state of R_1 and the involved state of FSM are random and uniformly distributed, D is uniformly distributed over 25+4=29 bits rather than over 4 bits as mentioned in Section 3.1. Similar analysis shows that the bias is $2^{-15.7}$ for $w = 5$ with $M = (11111)_2, M = (100001)_2$. Finally, Table 5 compares our improved attack with the previous attacks [1, 4, 7]. This is the best key-recovery attack on E0 known so far with precomputation, time and data complexities $O(2^{37})$.

4   Application Two: Shannon Cipher

Shannon [11] is a recently proposed synchronous stream cipher designed by G. Rose et al. from Qualcomm [10]. It has been designed according to Profile 1A of the ECRYPT call for stream cipher primitives, and it uses a secret key of up to 256 bits. The internal state uses a single nonlinear feedback shift register. This shift register state at time $t \geq 0$ consists of 16 elements s_{t+i} $(i = 0, \ldots, 15)$ of

32-bits. Let z_t denotes the 32-bit output of Shannon at time t. The following important relation was identified and formally proved in [5]:

$$(z_t \lll 1) \oplus z_{t+16}$$
$$= f_1(s_{t+21} \oplus s_{t+22} \oplus K) \oplus f_2((s_{t+11} \oplus s_{t+24}) \lll 1) \oplus$$
$$f_1(s_{t+25} \oplus s_{t+26} \oplus K) \oplus f_2((s_{t+15} \oplus s_{t+28}) \lll 1) \oplus$$
$$f_2(s_{t+19} \oplus s_{t+32}) \oplus f_2((s_{t+3} \oplus s_{t+16}) \lll 1), \tag{15}$$

where $f_1, f_2 : GF(2^{32}) \rightarrow GF(2^{32})$ are nonlinear and defined in [11], and K is a 32-bit secret constant derived from the initialization process. Treating each term in the right-hand sum (15) as independent ones, [5] uses Piling-up lemma to compute the largest bias of (15) from the individual bias of each term. The largest bias (15) was shown in [5] to be 2^{-56}. In total, 32 such equally largest biases makes the complexity $O(2^{107})$ for the distinguishing attack (see [5]).

4.1   Our Analysis on Shannon Cipher and a Shannon Variant

Here, we want to analyze the influence of the dependency between the terms in the right-hand sum (15) to the total bias. Our starting point is that (15) can be viewed as the sum of three items, $f_1(s_{t+21} \oplus s_{t+22} \oplus K) \oplus f_1(s_{t+25} \oplus s_{t+26} \oplus K)$, $f_2((s_{t+11} \oplus s_{t+24}) \lll 1) \oplus f_2((s_{t+15} \oplus s_{t+28}) \lll 1)$ and $f_2(s_{t+19} \oplus s_{t+32}) \oplus f_2((s_{t+3} \oplus s_{t+16}) \lll 1)$. For each item, the inputs are dependent as explained below. The input difference to the two terms of the first item is

$$(s_{t+21} \oplus s_{t+22} \oplus K) \oplus (s_{t+25} \oplus s_{t+26} \oplus K) = (s_{t+21} \oplus s_{t+25}) \oplus (s_{t+22} \oplus s_{t+26}). \tag{16}$$

As the Shannon keystream output is defined as $z_t = s_{t+9} \oplus s_{t+13} \oplus f_2(s_{t+3} \oplus s_{t+16})$, we have $s_{t+9} \oplus s_{t+13} = z_t \oplus f_2(s_{t+3} \oplus s_{t+16})$. We rewrite (16) by

$$z_{t+12} \oplus z_{t+13} \oplus f_2(s_{t+15} \oplus s_{t+28}) \oplus f_2(s_{t+16} \oplus s_{t+29}) \tag{17}$$

Given the keystream, we consider $z_{t+12} \oplus z_{t+13}$ as a known value (denoted by c_{t+12}). Let D_{f_2} be the distribution of f_2 assuming a uniform distribution of the input. From (17), the distribution $D(x)$ of the input difference to the item $f_1(s_{t+21} \oplus s_{t+22} \oplus K) \oplus f_1(s_{t+25} \oplus s_{t+26} \oplus K)$ can be expressed by $D(x) = D_{f_2} \otimes D_{f_2}(x \oplus c_{t+12})$, where $\otimes$ denotes convolution. Similarly, we express the distribution $D'(x)$ of the input difference to the item $f_2((s_{t+11} \oplus s_{t+24}) \lll 1) \oplus f_2((s_{t+15} \oplus s_{t+28}) \lll 1)$ by $D'(x) = D_{f_2} \otimes D_{f_2}((x \ggg 1) \oplus \alpha_{t+2})$, where $\alpha_{t+2} = z_{t+2} \oplus z_{t+15}$ is known from the keystream.

For the bias pattern (also called output mask) $0x410a4a1$ used in [5], we use Corollary 1 to compute the bias for $f_1(s_{t+21} \oplus s_{t+22} \oplus K) \oplus f_1(s_{t+25} \oplus s_{t+26} \oplus K)$ and $f_2((s_{t+11} \oplus s_{t+24}) \lll 1) \oplus f_2((s_{t+15} \oplus s_{t+28}) \lll 1)$ respectively. We found that D, D' are very weakly biased, and they show no significant fluctuations over the value of c_{t+12}, α_{t+2} respectively. We conclude that the dependency of the internal states of Shannon is weak and the keystream bias (15) can be approximated[3] with Piling-up lemma, which was estimated as $O(2^{-56})$ in [5]. Thus, the estimated complexity $O(2^{107})$ of the distinguishing attack [5], which uses 32 such largest biases, is still valid.

Furthermore, we consider a variant of Shannon when both f_1, f_2 only uses one element from the shift register as the input instead of two in the Shannon cipher. For convenience of our discussion, we consider the variant of removing the second s term in the inputs of both f_1, f_2 (K is still present). The keystream bias relation (15) for our Shannon variant then becomes

$$(z_t \lll 1) \oplus z_{t+16} = f_1(s_{t+21} \oplus K) \oplus f_1(s_{t+25} \oplus K) \oplus$$
$$f_2(s_{t+11} \lll 1) \oplus f_2(s_{t+15} \lll 1) \oplus f_2(s_{t+19}) \oplus f_2(s_{t+3} \lll 1). \qquad (18)$$

Under the independence assumption, this will not change the keystream bias. However, note that the involved distributions of the input difference change now. We can see that now $D(x) = D_{f_2}(x \oplus z_{t+12})$, $D'(x) = D_{f_2}((x \ggg 1) \oplus z_{t+2})$, $D'' = f_1 \otimes f_2$. Given a bias pattern, we can use Corollary 1 to compute the individual bias (denote by $\delta_1, \delta_2, \delta_3$ respectively) of $f_1(s_{t+21} \oplus K) \oplus f_1(s_{t+25} \oplus K)$, $f_2(s_{t+11} \lll 1) \oplus f_2(s_{t+15} \lll 1)$ and $f_2(s_{t+19}) \oplus f_2(s_{t+3} \lll 1)$. With the bias pattern $M = 0x9292949$ and $z_{t+12} = 0x80$, $\delta_1 \approx 2^{-18}$; in contrast, the independence assumption makes $\delta_1 = 0$. Meanwhile, we have[4] $\delta_2 \approx \delta_3 \approx 2^{-16}$, which are approximately the same as using Piling-up lemma to compute. Thus, given $z_{t+12} = 0x80$, the bias of $M \cdot ((z_t \lll 1) \oplus z_{t+16})$ is $2^{-18-16-16} = 2^{-50}$. This translates to a distinguishing attack with complexity $O(2^{100})$. Moreover, we found 15 other bias patterns[5] with corresponding z_{t+12}'s, which all have $\delta_1 \approx 2^{-19}, \delta_2 \approx \delta_3 \approx 2^{-16}$. Considering all these biases, we have a distinguishing attack with complexity $2^{(19+16+16)*2}/16 = 2^{98}$.

Additionally, we can prove that given the values of M and z_{t+12}, δ_1 remains the same for the bias patterns $M \lll a$ and $z_{t+12} \lll a$, where $a = 0, \ldots, 31$. We give a brief sketch of proof here. In order to prove above statement, it suffices to show that $\delta_1 = \delta_1'$, where δ_1' denotes the corresponding bias for $M' = M \lll 1$. Let $g_1(x) = (-1)^{M \cdot f_1(x)}$ and $g_1'(x) = (-1)^{M' \cdot f_1(x)}$. First, we have $g_1(x) = g_1'(x \lll 1)$ for all x. From this we deduce $\widehat{g_1}(x) = \widehat{g_1'}(x \lll 1)$ for all x. On the other hand, we use the property that $f_2(x \lll a) = f_2(x) \lll a$ for all x (see [5]) to deduce $D_{f_2}(y) = D_{f_2}(y \lll 1)$ for all y. So, $D_{f_2}(x \oplus z_{t+12}) = D_{f_2}((x \oplus z_{t+12}) \lll 1)$. Finally we apply Corollary 1 to conclude $\delta_1 = \delta_1'$. Hence,

---

[3] Note that the input difference to another item $f_2(s_{t+19} \oplus s_{t+32}) \oplus f_2((s_{t+3} \oplus s_{t+16}) \lll 1)$ is $((s_{t+3} \lll 1) \oplus s_{t+19}) \oplus ((s_{t+16} \lll 1) \oplus s_{t+32})$. From [5], the following holds $(s_t \lll 1) \oplus s_{t+16} = f_1(s_{t+12} \oplus s_{t+13} \oplus K) \oplus f_2((s_{t+2} \oplus s_{t+15}) \lll 1)$. Thus the distribution D'' of the input difference to this item is $D'' = D_{f_1} \otimes D_{f_1} \otimes D_{f_2} \otimes D_{f_2}$. From D_{f_1}, D_{f_2} we deduce that D'' is too flat and we can approximate the bias of this item by Piling-up lemma.

[4] Here δ_2 fluctuates insignificantly over the value of z_{t+2}.

[5] They are 0x424a425, 0x420a525, 0x420a4a5, 0x420a425, 0x414a4a5, 0x414a4a1, 0x410a4a5, 0x9294949, 0x9292929, 0x9290949, 0x1292929, 0x1012929, 0x292949, 0x292941, 0x24a525.

each bias pattern can be expanded to 32 ones of equal bias, and we can further decrease the complexity of our attack on Shannon variant to $2^{98}/32 = 2^{93}$.

5   Conclusion

In this paper, we study the bias problem of the XOR of many Boolean function outputs, whose inputs are dependent. When all inputs are independent, our bias problem degrades to the old well-known problem and can be solved by Piling-up lemma. We give a simple expression to compute the bias efficiently. It takes time $O(kn \cdot 2^n)$. It turns out that our result generalizes the previous work of [9]. Furthermore, we note that the general bias in our problem can differ significantly from the bias in the independent case. As a general guideline, we note that when the distribution D of the XOR of the involved variables is weakly biased, the bias can be approximated by using the Piling-up lemma; otherwise, it is not appropriate to use the Piling-up lemma to compute the bias. As application, we demonstrate with two real examples, E0 and Shannon cipher, which have a strongly biased and weakly biased D respectively. For E0, our analysis allows to make the best known key-recovery attack with precomputation, time and data complexities $O(2^{37})$. For Shannon cipher, our analysis verifies the validity of the estimated complexity $O(2^{107})$ of the previous distinguishing attack [5]. As comparison, we also studied a variant of Shannon cipher, which shows much stronger dependency within the internal states. We gave a distinguishing attack on the Shannon variant with reduced complexity $O(2^{93})$.

References

1. Armknecht, F., Krause, M.: Algebraic attacks on combiners with memory. In: Boneh, D. (ed.) CRYPTO 2003. LNCS, vol. 2729, pp. 162–175. Springer, Heidelberg (2003)
2. Bluetooth specification, http://www.bluetooth.org
3. Canteaut, A., Naya-Plasencia, M.: Computing the biases of parity-check relations (2009), http://arxiv.org/abs/0904.4412
4. Courtois, N.T.: Fast algebraic attacks on stream ciphers with linear feedback. In: Boneh, D. (ed.) CRYPTO 2003. LNCS, vol. 2729, pp. 176–194. Springer, Heidelberg (2003)
5. Hakala, R.M., Nyberg, K.: Linear distinguishing attack on shannon. In: Mu, Y., Susilo, W., Seberry, J. (eds.) ACISP 2008. LNCS, vol. 5107, pp. 297–305. Springer, Heidelberg (2008)
6. Kukorelly, Z.: The piling-up lemma and dependent random variables. In: Walker, M. (ed.) Cryptography and Coding 1999. LNCS, vol. 1746, pp. 186–190. Springer, Heidelberg (1999)
7. Lu, Y., Vaudenay, S.: Faster correlation attack on bluetooth keystream generator E0. In: Franklin, M. (ed.) CRYPTO 2004. LNCS, vol. 3152, pp. 407–425. Springer, Heidelberg (2004)

8. Matsui, M.: Linear cryptanalysis method for DES cipher. In: Helleseth, T. (ed.) EUROCRYPT 1993. LNCS, vol. 765, pp. 386–397. Springer, Heidelberg (1994)
9. Molland, H., Helleseth, T.: An improved correlation attack against irregular clocked and filtered keystream generators. In: Franklin, M. (ed.) CRYPTO 2004. LNCS, vol. 3152, pp. 373–389. Springer, Heidelberg (2004)
10. Qualcomm, http://www.qualcomm.com/
11. Rose, G., Hawkes, P., Paddon, M., McDonald, C., Vries, M.: Design and Primitive Specification for Shannon, Symmetric Cryptography (2007)

Known and Chosen Key Differential Distinguishers for Block Ciphers

Ivica Nikolić[1,*], Josef Pieprzyk[2], Przemysław Sokołowski[2,3], and Ron Steinfeld[2]

[1] University of Luxembourg, Luxembourg
[2] Macquarie University, Australia
[3] Adam Mickiewicz University, Poland
ivica.nikolic@uni.lu,
{josef.pieprzyk,przemyslaw.sokolowski,ron.steinfeld}@mq.edu.au

Abstract. In this paper we investigate the differential properties of block ciphers in hash function modes of operation. First we show the impact of differential trails for block ciphers on collision attacks for various hash function constructions based on block ciphers. Further, we prove the lower bound for finding a pair that follows some truncated differential in case of a random permutation. Then we present open-key differential distinguishers for some well known round-reduced block ciphers.

Keywords: Block cipher, differential attack, open-key distinguisher, Crypton, Hierocrypt, SAFER++, Square.

1 Introduction

Block ciphers play an important role in symmetric cryptography providing the basic tool for encryption. They are the oldest and most scrutinized cryptographic tool. Consequently, they are the most trusted cryptographic algorithms that are often used as the underlying tool to construct other cryptographic algorithms. One such application of block ciphers is for building compression functions for the hash functions.

There are many constructions (also called hash function *modes*) for turning a block cipher into a compression function. Probably the most popular is the well-known Davies-Meyer mode. Preneel et al. in [27] have considered all possible modes that can be defined for a single application of n-bit block cipher in order to produce an n-bit compression function. They have found that there are 12 modes that are resistant against generic attacks. Later these findings have been formally proven in [7]. To make hash functions resistant against the birthday-paradox attack, it is better to use double-block modes. Basic double-block modes have been proposed in [8,14,20]. Note that the Tandem-DM mode has been proven to be collision resistant in [12], while a weakness in MDC-2 was found in [17].

Proofs of security of the above modes are performed under the assumption that the underlying block cipher is ideal. This assumption is not satisfied if the

---

* The work was done while this author was visiting Macquarie University.

K.-H. Rhee and D. Nyang (Eds.): ICISC 2010, LNCS 6829, pp. 29–48, 2011.
© Springer-Verlag Berlin Heidelberg 2011

cipher is used to build hash functions. Note that the ideal cipher is related to the concept of pseudo-random permutation, where the adversary does not know the cryptographic key. Clearly, for hash function constructions based on block ciphers, the adversary fully controls the key.

Biham and Shamir introduced differential analysis in [3] and successfully analyzed DES. The idea is to follow the propagation of a difference in the state of the cipher throughout consecutive rounds. When the input-output differences can be predicted with a sufficiently high probability, then the cipher can be distinguished from a pseudo-random permutation. This concept can trivially be adjusted for the case, where the adversary knows/controls the key of the cipher (open-key differential distinguishers). The goal of adversary in this case would be to find an input-output pair of differences for the cipher that can be predicted with a probability higher than in a random permutation.

Unlike in the secret-key model, where the complexity of an attack is usually bounded by the size of the key space (i.e. 2^k for a k-bit key), the attacks in the open-key model are bounded by the size of the state space (i.e. 2^n for an n-bit state). Therefore, some of the published attacks in the secret-key model (precisely, the attacks with a complexity higher than 2^n) become worse than simple generic attacks, when applied in the open-key model.

Our Contributions. We investigate the impact of block cipher open-key differential distinguishers on hash function modes of operation. Our main contributions can be summarized as follows:

1. For a variety of hash function modes based on block-ciphers, we determine which collision finding attack variants (collisions, pseudo collisions, semi-free start collisions, or free start collisions) are feasible, assuming that the adversary is given a specific differential trail for the underlying block cipher in the open-key model. We target all Preneel-Govaerts-Vandewalle (PGV) single-block-length compression modes, as well as four double-block-length modes.

2. We examine several well known block ciphers (Crypton, Hierocrypt-3, SAFER++, Square, and generic Feistel ciphers) and for each of them, we present new known-key and chosen-key differential distinguishers - see Table 1. Our distinguishers use the rebound attack [25] as a starting point, but we obtain substantial improvements in the number of attacked rounds by exploiting some cipher-specific properties that allow us to manipulate bits of the subkeys (a similar technique was used in the context of analysing the Whirlpool function [21]). In the chosen-key model, for substitution-permutation (SP) ciphers, we obtain an explicit formula for the number of additional rounds that can be attacked for free, when the cipher has an invertible key schedule.

3. To show the efficiency of our distinguishers, we give a proof of a lower bound on the complexity of differential distinguishers in the case of a black-box random permutation. Although this bound has been used for a while (mainly as an upper bound, e.g. in [13] it is called a limited-birthday distinguisher), as far as we know, it has never been formally proved.

Table 1. Summary of attacks on the ciphers examined in the paper. The "Encryptions" column gives the expected number of encriptions in the case of a SP cipher, while the "Lower bound" column – the expected number of encryptions required in the case of a random permutation. In case of n-bit Feistel cipher r is a number of covered rounds, and 2^c is the complexity of some differential attack.

Cipher	Distinguisher	Rounds	Encryptions	Lower bound	Reference
Crypton	Known-key	7	2^{48}	2^{61}	Section 5.1
	Chosen-key	9	2^{48}	2^{61}	Section 5.1
Hierocrypt-3	Known-key	3.5	2^{48}	2^{61}	Section 5.1
	Chosen-key	4.5	2^{48}	2^{61}	Section 5.1
SAFER++	Known-key	6.5	2^{120}	2^{128}	Section 5.2
	Chosen-key	6.5	2^{112}	2^{128}	Section 5.2
Square	Known-key	7	2^{48}	2^{61}	Section 5.1
	Chosen-key	8	2^{48}	2^{61}	Section 5.1
n-bit Feistel	Differential attack	r	2^c		
with k-bit key	Known-key	$r+2$	2^c		Section 5.3
	Chosen-key	$r+\lfloor\frac{2k}{n}\rfloor$	2^c		Section 5.3

Organization. The paper is organized as follows. In Section 2 we define the open-key distinguishers and review techniques for constructing differential trails. In Section 3, we present our findings about the impact of block cipher differential trails on the security of hash function modes. Section 4 contains our lower bound on the complexity of differential distinguishers for black-box random permutations. In Section 5, we present our cipher specific known-key and chosen-key differential distinguishers for various block ciphers. Section 6 concludes the paper.

2 Preliminaries

2.1 Open-Key Distinguishers for Block Ciphers

A distinguisher is one of the weakest cryptographic attacks that can be launched against a secret-key cipher. In this attack, there are two oracles: one that simulates the cipher for which the cryptographic key has been chosen at random and the other simulates a truly random permutation. The adversary can query both oracles and their task is to decide which oracle is the cipher (or random permutation). The attack is considered to be successful if the number of queries required to make a correct decision is below a well defined level.

The idea of open-key distinguishers was introduced by Knudsen and Rijmen in [18] for analysis of AES and a class of Feistel ciphers. They examined the security of these block ciphers in *a model where the adversary knows the key*. Later, the same approach was used in the attack on 8-round reduced AES-128 [13] and for analysis of Rijndael with large blocks [26], where the authors defined a new security notion for a known-key cipher. The idea of chosen-key distinguishers was introduced in the attack on the full-round AES-256 [5]. This time *the adversary* is assumed to have *a full control over the key.* A chosen-key attack was launched on 8-round reduced AES-128 in [6].

Both the known-key and chosen-key distinguishers are collectively known *open-key distinguishers.* The adversary has the knowledge of the key or even can choose a value of the key. To succeed, the adversary has to discover some property of the attacked cipher that holds with a probability higher than for a random permutation.

Differential distinguishers in the open-key model are defined in similar way as in the secret-key model. The adversary builds a differential trail $(\Delta_P, \Delta_K) \to \Delta_2$ for the block cipher $E_K(P)$. In other words, he finds a pair[1] of plaintexts (P_1, P_2) and a pair of keys (K_1, K_2), together known as a differential pair, such that $P_1 \oplus P_2 = \Delta_P$, $K_1 \oplus K_2 = \Delta_K$ and $E_{K_1}(P_1) \oplus E_{K_2}(P_2) = \Delta_2$. The pair (Δ_P, Δ_K) is the input difference, while Δ_2 is the output difference. At least one of Δ_P and Δ_K has to be non-zero. For example, the trails given in [6,13,26] have differences only in the plaintext, while the trail from [5] has differences in both the key and the plaintext.

2.2   Design of Differential Distinguishers for Block Ciphers

We will focus our analysis on substitution-permutation (SP) block ciphers. Each round of such ciphers consists of two types of transformations: 1) a non-linear layer of S-boxes, and 2) a linear-diffusion layer (LD). The non-linear layer operates on bytes, i.e. the inputs to the S-boxes are bytes of the state. The linear-diffusion layer may apply different transformations such as multiplications of the columns/rows of the state matrix by a fixed diffusion matrix, transpositions of rows/columns, rotations of elements of the state matrix, subkey additions, and others.

Differential trails for ciphers are given as a sequence of input-output word differences of each transformation of the state. Since SP ciphers are usually byte-oriented, these trails can be given as a sequence of active bytes, i.e. bytes that have differences. Depending on the properties of the S-box layer and the linear-diffusion layer, the adversary can built two types of trails.

The first type is a *standard* differential trail, where the exact values of the input-output differences for each layer and for each round of the trail are fixed. The probability of these trails depends on the differential properties of the S-boxes, i.e. the probability that a given input difference to the S-box will produce a given output difference. Note that when these differences are fixed, then the trail in the linear-diffusion layer holds with probability 1.

The second type is a *truncated* differential trail [16]. In this trail only the position of the active bytes is important, while the actual difference values are ignored. Since, the S-box operates on a single byte, it means it cannot change an active byte to a non-active and vice-versa. Hence the adversary concentrates only on the linear-diffusion layer and finds the probability of a particular configuration of input-output active bytes.

---

[1] Actually the adversary can build many pairs of plaintexts and keys.

Although for SP ciphers the truncated differential approach is common, further in our analysis we will use both types of differential trails, together with trails with a difference in the plaintext only.

2.3 Techniques for Differential Trail Constructions

A major improvement in the analysis of SP cryptographic algorithms was the introduction of the rebound attack [25]. The idea is as follows. If we assume that the adversary controls the input to the S-boxes, then any input-output difference [2] to this layer can be obtained for free (simple table lookups). In other words, when Δ_1, Δ_2 are fixed, then it is easy to find x such that $S(x + \Delta_1) \oplus S(x) = \Delta_2$. In two consecutive middle rounds the adversary first fixes both the input differences of the LD layer in the first round, and the output differences of the LD layer of the second round. Then he goes forward through the first LD layer and backwards through the second LD layer. He ends up with fully determined differences, since the layers are linear. In between there is only one S-box layer (composed of a number of S-boxes), which can be passed for free when the adversary fixes the values, i.e. when he finds the proper solutions x of the above equation. Therefore, at the beginning of the first, and at the end of the second middle round, not only the differences, but now also the values have been fixed. The rounds that precede and follow the two middle rounds are passed probabilistically.

The technique of the rebound attack was improved with the Super-Sbox cryptanalysis [11,13,21]. When the round diffusion is incomplete then two layers of S-boxes can be passed for free using a precomputed lookup tables. The idea is similar to the one of the original rebound attack, but bigger lookup tables are used.

The key can be used to gain an additional degree of freedom, which in return can lead to more S-box layers passed for free. When the adversary controls the key, then the rebound attack can be extended to one or two additional rounds, depending on the size of the key. The subkey (roundkey) is xored in each round of the cipher. The first S-box layer can be passed for free using the previous rebound technique (by fixing not only the difference, but the exact values as well). The second S-box layer can be passed for free as well if the adversary controls the input values to this layer by solving the appropriate equations. These values can be manipulated with the subkey, i.e. the adversary can choose a proper subkey such that the inputs to the S-box layer can be of arbitrary value (yet, their difference is fixed). Hence, the adversary can pass the second S-box layer for free if he controls the subkey of this round. Let us explain the idea on an example (See Fig.1). Let $\Delta_1 \to \Delta_2 \to \Delta_3$

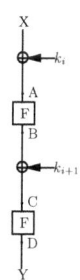

Fig. 1. Chosen-key distinguisher for SP ciphers

---

[2] Only half of the input-output differences are possible, but for each of them there are two different input values, hence on average it is true.

be an arbitrary two-round differential trail. First the adversary finds (with the rebound attack) a pair of states that satisfies the differential trail of the first round, i.e. he finds a pair $(A, A \oplus \Delta_1)$ that produces $(B, B \oplus \Delta_2)$ on the output. Then independently, he finds a pair of states for the second round, i.e. he finds $(C, C \oplus \Delta_2)$ that produces the output $(D, D \oplus \Delta_3)$. In the last step he has to fix a proper subkey k_{i+1} for the second round, which will connect the output of the first round and the input of the second round. To do so, the adversary fixes $k_{i+1} = B \oplus C$, and as the result he obtains a pair of states $(A \oplus k_i, A \oplus k_i \oplus \Delta_1)$ that satisfy the two round differential trail.

Similarly, the adversary can pass more S-box layers when he controls the subkeys of these layers. An obvious requirement for the subkeys of these additional rounds is that they need to be independent. Otherwise, a change in a subkey in one round will change the value of a subkey in another round, which might lead to incorrect input values for the S-box layer of this second round. A second requirement is an invertible key schedule. Since the adversary controls the values of the subkeys of some middle rounds, he has to be able to produce the values of the subkeys of the rounds that precede and follow these rounds, hence he has to find the master key from the fixed subkeys. It is important to notice that this technique requires a negligible memory.

2.4 Building the Differential Trails

For each of the techniques discussed above, the adversary first builds a trail that may have a plenty of active S-boxes in some middle rounds and a few at the ends of the trail. Then, a pair of values that follows the differential trail only in these middle rounds is found with complexity 1. The rest of the rounds, before and after the middle rounds, are covered probabilistically since the adversary has no degree of freedom left.

Finding the optimal differential trails with no difference in the key can be done automatically since the ciphers considered in this paper are byte-oriented with a block size of 16 bytes. This leads to a search space of 2^{16} possible starting values.

Some of the ciphers are based on the so-called wide trail strategy [10], and provide an efficient method for estimating the probability of the best round-reduced standard differential trails. These estimation are based on the differential properties of the S-boxes and the diffusion properties of the LD layers, which are often maximum distance separable mappings.

3 Impact of Block Cipher Known Key Differential Trails on Hash Modes

The most popular design of cryptographic hash is based on iterative use of a compression function. This construction is also known as the Merkle-Damgård (MD) structure. Early compression functions were using block ciphers as the main building block. Assume that we have a single instance of a block cipher

$E_K(P)$ and wish to design a compression function that takes a $2n$-bit input (H, M) and outputs a n-bit string $F(H, M)$. This problem has been investigated in [7,27] and it has been shown that there are 12 structures (modes) that are secure. An example of one such structure is the well-known Davies-Meyer (DM) mode that is defined as $F(H, M) = E_M(H) \oplus H$ (see mode 5 in Table 2), where H and M are the chaining value and the message, respectively.

In this work, we consider four types of collision attacks against the compression functions:

1. Collisions - for a fixed chaining value H_0, the adversary tries to find two distinct messages M_1, M_2 such that $F(H_0, M_1) = F(H_0, M_2)$.
2. Pseudo collisions - for a message M, the adversary wishes to find two distinct chaining values H_1, H_2 such that $F(H_1, M) = F(H_2, M)$.
3. Semi-free start collisions - the adversary attempts to find two distinct messages M_1, M_2 and a chaining value H such that $F(H, M_1) = F(H, M_2)$.
4. Free start collisions - the adversary tries to find two distinct chaining values H_1, H_2, and two distinct messages M_1, M_2 such that $F(H_1, M_1) = F(H_2, M_2)$.

We investigate the resistance of compression functions based on block ciphers against the attacks described above. We assume that the adversary can build a differential trail for the cipher with differences not only in the plaintext, or in the key, but also in both the plaintext and the key. For example, for the DM compression function, this means the adversary can find a pair of chaining values (H_1, H_2) and a pair of messages (M_1, M_2) (possibly in one of the pairs the two values are equal) such that $H_1 \oplus H_2 = \Delta_H, M_1 \oplus M_2 = \Delta_M$ and $F(H_1, M_1) \oplus F(H_2, M_2) = \Delta_H \oplus \Delta_2$. Hence, when the adversary can build *some* trail, i.e. when he cannot control the exact values of the differences Δ_H, Δ_2, then he can find a differential distinguisher for the DM compression function. On the other hand, when the adversary can build a *specific* trail for the cipher with a difference in the plaintext (H is the plaintext input to the cipher), such that $\Delta_H \oplus \Delta_2 = 0$, then he can find: 1) free-start collisions, if $\Delta_M, \Delta_H \neq 0$, 2) pseudo-collisions, if $\Delta_M = 0, \Delta_H \neq 0$, 3) collisions or semi-free start collisions, if $\Delta_M \neq 0, \Delta_H = \Delta_2 = 0$ (note that this implies that there are key collisions in the cipher since in DM, the message is the key).

The same approach can be applied to the other 11 modes. We try to find the all possible collision attacks under the assumption that the adversary can control the relation between the input and the output differences of a trail in the cipher. Our findings are presented in Table 2.

Often the block size of a cipher is too small to be secure in the compression mode. Hence, there is a class of compression functions, also called double-block-length ones, whose output size is two times bigger than the block size of the underlying cipher. We investigate the security of such functions proposed by Lai-Massey in [20], Hirose in [14] and Bracht et al. in [8]. Our results are presented in Table 3.

Table 2. The first column consists of numbers from [7]. The entries in the columns plaintext, key, plaintext and key show the best collision attacks for the modes when there is difference only in the plaintext, only in the key or both in the plaintext and key, respectively. The abbreviations C, PC, SFSC, FSC stand for collision, pseudo-collision, semi-free start collision, free start collision, respectively.

mode (i)	h'	plaintext	key	plaintext and key
1	$E_h(m) \oplus m$	C, SFSC	PC[a]	FSC
2	$E_h(h \oplus m) \oplus h \oplus m$	C, SFSC	PC	PC, FSC
3	$E_h(m) \oplus h \oplus m$	C, SFSC	PC	FSC
4	$E_h(h \oplus m) \oplus m$	C, SFSC	PC	PC, FSC
5	$E_m(h) \oplus h$	PC	C[a], SFSC[a]	FSC
6	$E_m(h \oplus m) \oplus h \oplus m$	PC	FSC	C, SFSC, FSC
7	$E_m(h) \oplus h \oplus m$	PC	C, SFSC	FSC
8	$E_m(h \oplus m) \oplus h$	PC	FSC	C, SFSC, FSC
9	$E_{h \oplus m}(m) \oplus m$	FSC	PC[a]	C, SFSC, FSC
10	$E_{h \oplus m}(h) \oplus h$	FSC	C[a], SFSC[a]	PC, FSC
11	$E_{h \oplus m}(m) \oplus h$	FSC	PC	C, SFSC, FSC
12	$E_{h \oplus m}(h) \oplus m$	FSC	C, SFSC	C, PC, FSC

[a] When key collisions exist in the cipher.

Table 3. In the first column A-DM, T-DM, DBL and MDC-2 are abbreviations of Abrest DM, Tandem DM, Double-Block-Length and Modification Detection Code 2 respectively (see [20] for the first two, [14] for the third and [8] for the last). The abbreviations C, PC, SFSC, FSC stand for collision, pseudo-collision, semi-free start collision, free start collision, respectively.

mode	(h', g')	plaintext	key	plaintext and key
A-DM	$h' = E_{g,m}(h) \oplus h$ $g' = E_{m,h}(\bar{g}) \oplus g$	FSC	C, SFSC	PC, FSC
T-DM	$h' = E_{g,m}(h) \oplus h$ $g' = E_{m,E_{g,m}(h)}(g) \oplus g$	FSC	C, SFSC	PC, FSC
DBL	$h' = E_{h\|m}(g \oplus c) \oplus g \oplus c$ $g' = E_{h\|m}(g) \oplus g$	PC	C, PC, SFSC, FSC	PC, FSC
MDC-2	$h' = (E_h(m) \oplus m)^L \| (E_g(m) \oplus m)^R$ $g' = (E_g(m) \oplus m)^L \| (E_h(m) \oplus m)^R$	C, SFSC	PC[a]	FSC

[a] When key collisions exist in the cipher.

Although we have analyzed the collision resistance of the above modes, the differential trails for the underlying ciphers in the open-key model can be used as a standalone cryptanalytical result for the compression functions.

4   Lower Bound on Complexity of Differential Distinguisher for Random Permutations

In this section we present a lower bound on the complexity of differential distinguishers for a black-box random permutation. This allows us to fairly compare our cipher-specific distinguisher complexities in Section 2.2 to the best possible black-box distinguisher. Although a similar *upper* bound has been used before (see, e.g. [13]), our result proves that it is indeed close to the best possible. To our knowledge, such a lower bound has not been published before, and may be of independent interest.

When the key is fixed, a block cipher becomes a permutation. An open-key differential distinguisher with no difference in the key is valid if the complexity of finding a differential pair is less than the complexity of finding such pair in a random permutation. When the input and output differences are fully fixed, in n-bit random permutation the complexity of finding a differential pair is 2^n, hence any differential distinguisher with a probability higher than 2^{-n} is valid. When the input difference is fixed, and the output difference can take values from a set of the cardinality 2^c, then for a random permutation, a differential pair can be found after performing 2^{n-c} encryptions. The general case when both the input and the output differences are taken from sets of fixed cardinalities, is discussed in the following lemma.

Lemma 1. *Let D_I, D_O denote subsets of $\{0,1\}^n$, which are closed under $\oplus$, i.e. $x \oplus y \in D_I$ (respectively D_O) for $x, y \in D_I$ (resp. D_O). For any attacker making queries to a random n-bit permutation π and its inverse π^{-1}, the complexity (measured in expected number of oracle queries) of finding a pair of inputs (x, y), where $x \oplus y \in D_I, |D_I| = 2^{c_I}$, such that $\pi(x) \oplus \pi(y) \in D_O, |D_O| = 2^{c_O}$, is lower bounded as $Q \geq \min(2^{\frac{n}{2}-2}, 2^{n-(c_I+c_O)-3})$.*

Proof. Since D_I and D_O are closed under $\oplus$, we may partition $\{0,1\}^n$ into input sets $A_1, \ldots, A_N$, where each $|A_i| = |D_I| = 2^{c_I}$, $N = \frac{2^n}{|D_I|} = 2^{n-c_I}$, such that $x \oplus y \in D_I$ for $x, y \in A_i$ for $i = 1, \ldots, N$. Similarly, we have a partition into output sets $B_1, \ldots, B_M$ where $|B_j| = |D_O| = 2^{c_O}$, $M = \frac{2^n}{|D_O|} = 2^{n-c_O}$ for all $j = 1, \ldots, M$.

Let us define the following game G_0: attacker $\mathcal{A}$ has an access to a random permutation oracle $\pi \colon \{0,1\}^n \to \{0,1\}^n$ and its inverse π^{-1}, making a total of q queries to these oracles.

In the following games G_k ($k = 0, 1, 2$), let E_k be the following event: $\mathcal{A}$ finds $x \neq y$ with $x, y \in A_i$ and $\pi(x), \pi(y) \in B_j$ for some i, j while interacting with game G_k.

We show below the following upper bound:

$$\Pr(E_0) \leq \frac{q^2}{2^n} + \frac{q}{2^{n-(c_O+c_I)}}. \tag{1}$$

Before we explain the formal proof, we remark that the intuition for this result is as follows. The first term $\frac{q^2}{2^n}$ is the upper bound on the collision probability error due to the fact that we simplify the problem by replacing the random permutation π with a random function. The last term arise because at each query to π (resp. π^{-1}) which is in some input set A_i (resp. output set B_j) there are at most 2^{c_I} points in A_i whose image under π is already defined (resp. at most 2^{c_O} points in B_j whose image under π^{-1} is already defined), thus occupying at most 2^{c_I} out of the 2^{n-c_O} output sets (resp. at most 2^{c_O} out of the 2^{n-c_I} input sets).

We first show that (1) implies the claimed expected complexity bound. In game G_0, let T denote the random variable defined as the number of oracle queries until the event E_0 occurs. We lower bound the expected value $Q = \mathrm{E}(T)$ as follows. Let $p(q)$ denote the right hand side of (1), and let q^* be such that $p(q^*) = \frac{1}{2}$. Since $\Pr(T \leq q) \leq p(q)$, we have

$$Q \geq \sum_{q>q^*} \Pr(T = q) \cdot q \geq q^* \cdot \Pr(T > q^*) \geq \frac{q^*}{2}. \tag{2}$$

Now, for $i \in \{1, 2\}$, let q_i denote the value of q such that the ith term on the right hand side of (1) is equal to $\frac{1}{4}$. Since there are 2 terms in (1), we may take $\min(q_1, q_2)$ as lower bound for q^*. Since $q_1 = 2^{\frac{n}{2}-1}$ and $q_2 = 2^{n-(c_I+c_O)-2}$, the claimed lower bound on Q follows.

It remains to prove (1). We will do this by building a chain of games, starting with G_0, which are similar until bad is set (for further details of this methodology see for example [2]).

First define a game G_1 to be similar to G_0 except that the permutation π is replaced by a relation $P \subset \{0,1\}^n \times \{0,1\}^n$ that is injective and functional, but not necessary defined in whole domain. According to naming convention in [2] relation P is called partial permutation, whereas injectivity and functional conditions together are named "permutation constraint". Initially P is empty and through execution of G_1 its values are being sampled randomly with respect to "permutation constraint". Whenever $P(x)$ (resp. $P^{-1}(y)$) is needed first it is checked if P (resp. P^{-1}) is defined on x (resp. y). If this is the case then appropriate value is returned, otherwise $P(x)$ (resp. $P^{-1}(y)$) is sampled uniformly at random from $\overline{img(P)}$ (resp. $\overline{img(P^{-1})}$), where $\overline{img(P)}$ is complement of image of P. Because the sampling is the same as in Game G_0, we have

$$\Pr(E_0) = \Pr(E_1). \tag{3}$$

Next we define game G_2 which is the same as G_1 except "permutation constraint" for P does not need to be fulfilled. That means the values $P(x)$ (resp. $P^{-1}(y)$) are sampled at random from $\{0,1\}^n$, but the game stops immediately when the "permutation constraint" is not satisfied. Unless the "permutation constraint" is violated by the occurrence of a collision between a new output value returned by P and a previous output value of P or input value queried to P^{-1} (resp. a collision between a new output value returned by P^{-1} and an previous output value of P^{-1} or input value queried to P), the games G_1 and

G_2 proceed identically. Since at each query there are at most q previous P (resp. P^{-1}) output values already defined, we have

$$|\Pr(E_2) - \Pr(E_1)| \leq \frac{q^2}{2^n}. \tag{4}$$

At this stage we stop building chain of games and we upper bound the probability $\Pr(E_2)$ directly. We claim that

$$\Pr(E_2) \leq \frac{q}{2^{n-(c_O+c_I)}}. \tag{5}$$

Let x denote the qth query of the attacker, define the following variables for $i = 1, \ldots, N$ and $j = 1, \ldots, M$:

- a_i^F = number of P oracle queries made so far which are in A_i,
- a_i^R = number of P^{-1} oracle answers given so far which fell in A_i,
- b_j^F = number of P^{-1} oracle queries made so far which are in B_j,
- b_j^R = number of P oracle answers given so far which fell in B_j.

Suppose that x is a query to P and that $x \in A_i$ for some i. We have so far $a_i^F + a_i^R$ points in A_i whose B_j sets are already defined. Hence the event E_2 will occur only if the uniformly random (in $\{0,1\}^n$) answer of P falls in one of those output sets, so it will happen in this query with probability $\leq \frac{a_i^F + a_i^R}{M} \leq \frac{|D_I|}{M} = \frac{1}{2^{n-(c_I+c_O)}}$, using $a_i^F + a_i^R \leq |D_I|$ (since the game did not stop so far). Similarly, if x is a query to P^{-1} and $x \in B_j$ for some j, then E_2 will occur in this query with probability $\leq \frac{b_j^F + b_j^R}{N} \leq \frac{|D_O|}{N} = \frac{1}{2^{n-(c_I+c_O)}}$. It follows that E_2 occurs among the first q queries with probability bounded by (5), as claimed. This completes the proof of the Lemma.   ∎

5   Differential Trails for Specific Block Ciphers

We have searched for differential trails in the following ciphers: Crypton, Hierocrypt-3, SAFER++, and Square. Specifically, we have tried to build standard and/or truncated trails, which can be used in a rebound-type attack. For some of the ciphers, the probabilities for the both standard and truncated differential trails were higher than in a random permutation. In this case, only the trails (which are usually truncated) with higher probability are presented.

The trails for the chosen-key distinguishers were built upon the trails for the known-key distinguishers by increasing the number of the full active middle rounds which can be covered for free when a proper subkey is fixed. When n-bit key is used, with an invertible key schedule that produces s-bit subkeys, then the chosen-key distinguisher has $\lfloor \frac{n}{s} \rfloor$ more rounds then the known-key distinguisher.

Due to space limitation, we will not give a full description of the attacked ciphers, but rather, introduce them briefly using the original notions and definitions from the source papers.

5.1   Crypton, Hierocrypt-3 and Square

Crypton [22], Hierocrypt-3 [28], and Square [9] are 128-bit SP block ciphers and have a various number of internal rounds depending on the length of the key. The best published attacks in the secret-key model are on 8 rounds of Crypton [15], 3-3.5 rounds of Hierocrypt-3 [1], and 8 rounds of Square [19].

The internal state of each cipher can be seen as 4×4 matrix of bytes, while a round consists of three types of transformations of the state: 1) byte-wise application of a non-linear S-box, 2) matrix-wise linear-diffusion (LD) layer that applies different linear transformations of various bytes of the matrix to introduce a sufficient diffusion among the bytes of the state, 3) subkey addition – a simple xor of the round key to the matrix. A round of Crypton consists of an S-box layer γ, LD layer composed of two transforms π and τ, and subkey addition σ. Hierocrypt-3 has six round transforms: two S-box layers $[S]$, two LD layers $[MDS_L]$ and $[MDS_H]$, and two subkey additions $[AK]$. A round of Square consists of four transforms: S-box layer γ, LD layer with two transforms θ and π, and a subkey addition σ. It is important to notice that all three ciphers have a non-linear, but invertible, key schedule. The 256-bit key versions of Crypton and Hierocrypt-3, have a key schedule such that each two consecutive 128-bit subkeys are independent.

For each cipher, we can build 7-round truncated differential trails (7 S-box layer trail in case of Hierocript), that have a full active state in the middle round, but only a few active S-boxes in the rest of the 3+3 rounds (S-box layers of Hierocript). These trails can be used to construct known-key distinguishers on 7 rounds of the ciphers, based on the rebound technique. Since the ciphers have invertible key schedules, we can increase the number of attacked rounds by switching from the known-key to the chosen-key attacks and using the degrees of freedom of the subkeys. Hence, we can construct a chosen-key differential distinguisher on 8 rounds of Crypton with 128-bit keys, and 9 rounds of Crypton with 256-bit keys (the additional round comes from extra 128-bit freedom of the key; the chosen-key has $\lfloor \frac{256}{128} \rfloor = 2$ more rounds than the known-key, see Section 2.3). For Hierocrypt-3, the result is a chosen-key distinguisher on 8 S-box layers = 4 rounds for 128-bit keys, and on 9 S-box layers=4.5 rounds for 256-bit keys. Square only supports 128-bit keys, hence the chosen-key distinguisher works on 8 rounds, which is indeed the total number of rounds of this cipher.

The trails used in the chosen-key distinguishers for 9, 4.5 and 8 rounds of Crypton, Hierocrypt-3, and Square, respectively are given in the Appendix A. Since the middle full-active state round(s) are covered by the rebound attack and by fixing the subkeys used in these rounds, we can assume that the probability of the trails in these rounds is 1. Hence, we count only the probability of the rest of the rounds. In each of the three trails, we have twice 2^{-24} – that is the probability that the linear-diffusion transformation will turn four active bytes into one active byte. The probability of the trail in the rest of the layers is 1. Therefore, to find pairs of plaintexts and ciphertexts that will follow the truncated differential trails, one has to start with 2^{48} pairs of states that pass the middle rounds (each pair can be build with negligible complexity). Out of 2^{48}

pairs, 2^{24} will produce four-to-one active byte in the first half of the trail, leading to a plaintext difference as the one in the trail. Out of these 2^{24}, one will produce four-to-one active in the second half of the trail and a ciphertext difference as the one in the trail. Now, let us try to compare our complexity of 2^{48} encryptions to the complexity in a case of a random permutation. By Lemma 1, to find this complexity we have to find the cardinalities of the plaintext and the ciphertext differences in the truncated trails. Although some of the plaintext/ciphertext differences in the trails have full active states, they are obtained by a linear transformation of some state with a four active bytes. Hence the cardinalities in all cases are $2^{4\cdot8} = 2^{32}$, and the complexity of producing a pair for a random permutation, that follows the trails, is at least $\min(2^{\frac{128}{2}-2}, 2^{128-(32+32)-3}) = 2^{61}$ encryptions.

To test the correctness of our results, we have constructed a chosen-key distinguisher on mCrypton [23], which has the same design as Crypton, but instead of bytes (8-bit words), it works with nibbles (4-bit words), and uses a non-invertible key schedule. The above distinguishers for Crypton can easily be applied to a modified version of mCrypton with a (invertible) key schedule identical to the one of Crypton. The chosen-key distinguisher for 9 rounds of this modified mCrypton was implemented on a PC, and a differential pair was found. The results are given in Appendix B.

5.2 SAFER++

SAFER++ [24] is a 128-bit SP block cipher. The version with 128-bit key has 7 rounds and the best published attack works for 5.5 rounds [4]. A round of SAFER++ consists of: 1) a byte-wise subkey addition, 2) a byte-wise S-box layer, 3) a byte-wise subkey addition, and 4) a state-wise linear-diffusion layer in the form of four 4-PHT. The subkey additions are modular and xor, and two different S-boxes are used. After the last round, there is an extra subkey addition. The key schedule is linear.

When the subkeys are fixed, then the S-box layer can be merged with the subkey additions to form another S-box layer, with the same input and output size. In other words, the subkey addition together with S-box and the subkey addition can be seen simply as some S-box (since the bytes of the subkeys are different, the S-boxes are also different). Hence, we can assume that a round of the cipher is composed of an S-box layer and a linear-diffusion layer, and all the additions in the cipher are modular.

Our automatic search for the best round-reduced standard differentials has found that there exist only two three-round trails with 10 active S-boxes (the rest of the trails have more than 10 active S-boxes). The first trail has 4,2,4 while the second has 2,3, and 5 active S-boxes in the first, the second, and the third round, respectively. We have used two 4-2-4 trails in our standard differential attack (see Fig.2). We attack 6.5 rounds of SAFER++, which is the full cipher, except for the first round, where the three transforms: subkey addition, S-box and subkey addition, are missing. As far as we know this is the first rebound attack with standard differentials. Therefore, we will describe it in more details.

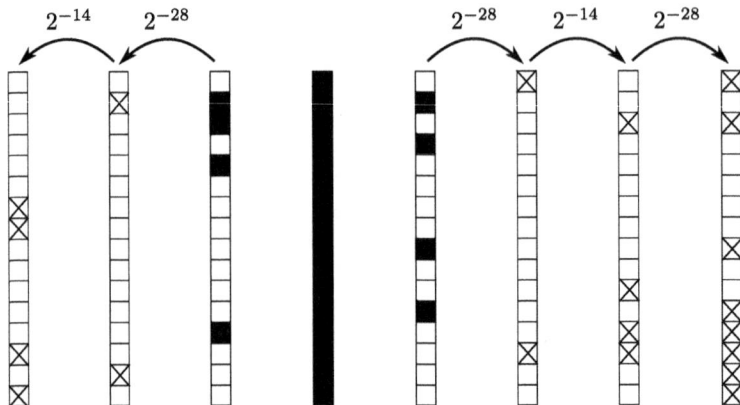

Fig. 2. Standard differential trail for 6.5 rounds of SAFER++ for the chosen-key distinguisher and 128-bit key. The first round is without the S-box layers, crossed square represents fixed 8-bit difference. The detailed trail is given in Fig.7 of the Appendix A.

First, to cancel the effects of the last extra subkey addition, we fix the MSB of the bytes 1, 3, 9, 12, 13, 14, 15, 16 of the last subkey to zero, while the values for the other bits of the subkey are randomly chosen. Then, from the mentioned subkey, we find the value of the master key, and the values for all remaining subkeys. Now we are ready to start the rebound attack.

We assign differences to the bytes 2, 3, 5, 13 (and no difference to other bytes) of the state before the linear layer in round 3. The differences should be such that after the linear layer all bytes are active (this holds for almost any assigned values). Similarly, we assign differences to the bytes 2, 4, 9, 12 of the state after the linear layer in round 4, go backwards through the linear layer and obtain a full active state. In between the top and the bottom active states, there is only the S-box layer, hence we match the differences through this layer, i.e. we fix the values of the bytes such that all the input differences produce all the output differences. Since the values of the full state have been fix, the rest of the rounds are passed probabilistically. There are 2, 4, 4, 2, 4 active S-boxes (16 in total) in the rounds 2, 3, 5, 6, 7, respectively.

If we assume that the differential propagation through all of the S-boxes occurs with the probability 2^{-7} then the complexity of the whole attack is $2^{7\cdot16} = 2^{112}$ encryptions. Note that for a fixed key, we have 2^{64} starting values for the rebound attack. We can choose different keys (such that the last subkey has the MSB of the mentioned above bytes fixed to zero) to get the necessary number of starting pairs for the differential attack. Since the input and output differences of the differential pair are fully fixed, such a pair in a random permutation can be found with 2^{128} encryptions.

5.3   Feistel Ciphers

Feistel ciphers with a SP round function can have a number of rounds covered for free in the known and chosen-key differential attacks. When the key is known, the S-box layers of two consecutive rounds can be attacked independently since the round function uses only half of the input. For a given two-round differential, first a pair of input states that satisfy the differential of the first round function is fixed, and then a pair of states of the second round function. Therefore, in an known-key attack, any differential trail can be extended by two additional rounds (this should not be confused with the distinguishers on 7-round Feistel ciphers proposed in [18]).

Assume that the adversary can control the key in a Feistel cipher. As the size of the input to the round function and the size of the round key are (usually) half as big as in the SP ciphers, the number of rounds that can be attacked for free is twice as big as for the SP ciphers. Let us examine the possibility of obtaining a pair of states for a three-round differential. Let n-bit Feistel cipher has an invertible key schedule that generates $\frac{n}{2}$-bit subkeys. To find a pair of states that follows some three-round differential:

$$(\Delta_1^L, \Delta_1^R) \rightarrow (\Delta_2^L, \Delta_2^R) \rightarrow (\Delta_3^L, \Delta_3^R) \rightarrow (\Delta_4^L, \Delta_4^R)$$

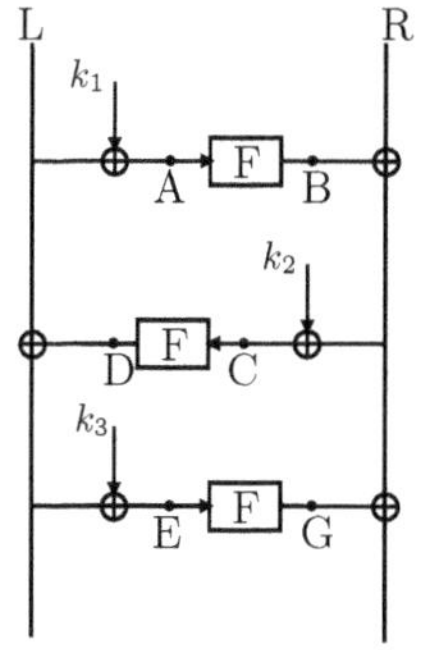

Fig. 3. Chosen-key distinguisher for 3-round Feistel ciphers

(the pair of states is $(L, R), (L \oplus \Delta_1^L, R \oplus \Delta_1^R)$), the adversary builds, as in the rebound attack, three pairs of states, separately for each round, that satisfy the one-round differentials, i.e. he finds the values A, C, E, such that

$$F(A) \oplus F(A \oplus \Delta_1^L) = \Delta_1^R \oplus \Delta_2^L,$$
$$F(C) \oplus F(C \oplus \Delta_2^L) = \Delta_2^R \oplus \Delta_3^L,$$
$$F(E) \oplus F(E \oplus \Delta_3^L) = \Delta_3^R \oplus \Delta_4^L.$$

Let $F(A) = B, F(C) = D, F(E) = G$. Then, in order to connect these three one-round differentials, the following conditions for the subkeys k_1, k_2, k_3 apply:

$$L \oplus k_1 = A,$$
$$R \oplus B \oplus k_2 = C,$$
$$L \oplus D \oplus k_3 = E.$$

From the first and the third equation, we get the relation $k_1 \oplus k_3 = A \oplus D \oplus E$ (note that the adversary does not control the values of A, D, E because they are fixed by the rebound attack). To satisfy this relation, the keys k_1, k_3 have to be independent (or be linearly dependent – but this is not common for ciphers). Once this is satisfied, the solution (L, R, k_1, k_2, k_3) for the system can be found in linear time. Hence in general, the master key has to be at least $\frac{3n}{2}$-bit long.

A similar analysis applies to cases when a higher number of rounds has to be covered for free. The only difference is that the resulting system has more equations. When r rounds are fixed, the system has r equation and $r + 2$ unknowns: $L, R, k_1, \ldots, k_r$. In order to find the solution in linear time, for any invertible key schedule, the subkeys have to be independent. Hence, to attack an additional r rounds of a n-bit Feistel cipher the key has to be at least $\frac{rn}{2}$-bit long.

6   Conclusions

We have examined the application of the differential trails in analysis of ciphers that are used for compression function constructions. We have considered both the known-key and chosen-key models. Specifically, we have analyzed the collision resistance of all compression functions based on single block ciphers as well as the four known double-block compression functions, when specific differential trails for the underlying ciphers can be built. Furthermore, we have presented differential distinguishers for Crypton, Hierocrypt-3, SAFER++, and Square. For these ciphers, we have shown that when the attack model is switched from secret-key to open-key, the number of rounds that can be attacked increases. We have given as well a formal proof of lower bound of constructing pair that follow a truncated trail in the case of a random permutation. Our results are summarized in Table 1.

The area of open-key distinguishers is largely unexplored. Finding similar distinguishers based on related-key differentials remains an open problem.

Acknowledgement. The authors would like to thank anonymous reviewers for their helpful comments.

Ivica Nikolić is supported by the Fonds National de la Recherche Luxembourg grant TR-PHD-BFR07-031. Josef Pieprzyk and Ron Steinfeld are supported by Australian Research Council grant DP0987734. Przemysław Sokołowski is supported by cotutelle Macquarie University Research Excellence Scholarship (cotutelle MQRES) and partially supported by Ministry of Science and Higher Education grant N N206 2701 33, 2007-2010.

References

1. Barreto, P.S.L.M., Rijmen, V., Nakahara Jr., J., Preneel, B., Vandewalle, J., Kim, H.Y.: Improved SQUARE Attacks Against Reduced-Round HIEROCRYPT. In: Matsui, M. (ed.) FSE 2001. LNCS, vol. 2355, pp. 165–173. Springer, Heidelberg (2002)
2. Bellare, M., Rogaway, P.: The Security of Triple Encryption and a Framework for Code-Based Game-Playing Proofs. In: Vaudenay, S. (ed.) EUROCRYPT 2006. LNCS, vol. 4004, pp. 409–426. Springer, Heidelberg (2006)
3. Biham, E., Shamir, A.: Differential cryptanalysis of DES-like cryptosystems. In: Menezes, A., Vanstone, S.A. (eds.) CRYPTO 1990. LNCS, vol. 537, pp. 2–21. Springer, Heidelberg (1991)
4. Biryukov, A., Cannière, C.D., Dellkrantz, G.: Cryptanalysis of SAFER++. In: Boneh, D. (ed.) CRYPTO 2003. LNCS, vol. 2729, pp. 195–211. Springer, Heidelberg (2003)
5. Biryukov, A., Khovratovich, D., Nikolić, I.: Distinguisher and Related-Key Attack on the Full AES-256. In: Halevi, S. (ed.) CRYPTO 2009. LNCS, vol. 5677, pp. 231–249. Springer, Heidelberg (2009)

6. Biryukov, A., Nikolić, I.: A New Security Analysis of AES-128. In: CRYPTO 2009 rump session (2009)
7. Black, J., Rogaway, P., Shrimpton, T.: Black-box analysis of the block-cipher-based hash-function constructions from PGV. In: Yung, M. (ed.) CRYPTO 2002. LNCS, vol. 2442, pp. 320–335. Springer, Heidelberg (2002)
8. Brachtl, B.O., Coppersmith, D., Hyden, M.M., Matyas Jr., S.M., Meyer, C.H.W., Oseas, J., Pilpel, S., Schilling, M.: Data authentication using modification detection codes based on a public one way encryption function. US Patent no. 4,908,861. Assigned to IBM. Filed (August 28, 1987) (March 13, 1990)
9. Daemen, J., Knudsen, L.R., Rijmen, V.: The Block Cipher Square. In: Biham, E. (ed.) FSE 1997. LNCS, vol. 1267, pp. 149–165. Springer, Heidelberg (1997)
10. Daemen, J., Rijmen, V.: The Wide Trail Design Strategy. In: Honary, B. (ed.) Cryptography and Coding 2001. LNCS, vol. 2260, pp. 222–238. Springer, Heidelberg (2001)
11. Daemen, J., Rijmen, V.: Understanding Two-Round Differentials in AES. In: Prisco, R.D., Yung, M. (eds.) SCN 2006. LNCS, vol. 4116, pp. 78–94. Springer, Heidelberg (2006)
12. Fleischmann, E., Gorski, M., Lucks, S.: On the Security of Tandem-DM. In: Dunkelman, O. (ed.) FSE 2009. LNCS, vol. 5665, pp. 84–103. Springer, Heidelberg (2009)
13. Gilbert, H., Peyrin, T.: Super-Sbox Cryptanalysis: Improved Attacks for AES-like permutations. In: Hong, S., Iwata, T. (eds.) FSE 2010. LNCS, vol. 6147, pp. 365–383. Springer, Heidelberg (2010); (to appear 2009)
14. Hirose, S.: Some Plausible Constructions of Double-Block-Length Hash Functions. In: Robshaw, M.J.B. (ed.) FSE 2006. LNCS, vol. 4047, pp. 210–225. Springer, Heidelberg (2006)
15. Kim, J., Hong, S., Lee, S., Song, J.H., Yang, H.: Truncated Differential Attacks on 8-Round CRYPTON. In: Lim, J.-I., Lee, D.-H. (eds.) ICISC 2003. LNCS, vol. 2971, pp. 446–456. Springer, Heidelberg (2004)
16. Knudsen, L.R.: Truncated and Higher Order Differentials. In: Preneel, B. (ed.) FSE 1994. LNCS, vol. 1008, pp. 196–211. Springer, Heidelberg (1995)
17. Knudsen, L.R., Mendel, F., Rechberger, C., Thomsen, S.S.: Cryptanalysis of MDC-2. In: Joux, A. (ed.) EUROCRYPT 2009. LNCS, vol. 5479, pp. 106–120. Springer, Heidelberg (2009)
18. Knudsen, L.R., Rijmen, V.: Known-Key Distinguishers for Some Block Ciphers. In: Kurosawa, K. (ed.) ASIACRYPT 2007. LNCS, vol. 4833, pp. 315–324. Springer, Heidelberg (2007)
19. Koo, B., Yeom, Y., Song, J.: Related-Key Boomerang Attack on Block Cipher SQUARE. Cryptology ePrint Archive, Report 2010/073 (2010), http://eprint.iacr.org/2010/073.pdf
20. Lai, X., Massey, J.L.: Hash Function Based on Block Ciphers. In: Rueppel, R.A. (ed.) EUROCRYPT 1992. LNCS, vol. 658, pp. 55–70. Springer, Heidelberg (1993)
21. Lamberger, M., Mendel, F., Rechberger, C., Rijmen, V., Schläffer, M.: Rebound Distinguishers: Results on the Full Whirlpool Compression Function. In: Matsui, M. (ed.) ASIACRYPT 2009. LNCS, vol. 5912, pp. 126–143. Springer, Heidelberg (2009)
22. Lim, C.H.: A Revised Version of Crypton - Crypton V1.0. In: Knudsen, L.R. (ed.) FSE 1999. LNCS, vol. 1636, pp. 31–45. Springer, Heidelberg (1999)
23. Lim, C.H., Korkishko, T.: mCrypton - A Lightweight Block Cipher for Security of Low-Cost RFID Tags and Sensors. In: Song, J., Kwon, T., Yung, M. (eds.) WISA 2005. LNCS, vol. 3786, pp. 243–258. Springer, Heidelberg (2006)

24. Massey, J., Khachatrian, G., Kuregian, M.: Nomination of SAFER++ as Candidate Algorithm for the New European Schemes for Signatures, Integrity, and Encryption (NESSIE). In: First Open NESSIE Workshop (November 2000)
25. Mendel, F., Rechberger, C., Schläffer, M., Thomsen, S.S.: The Rebound Attack: Cryptanalysis of Reduced Whirlpool and Grøstl. In: Dunkelman, O. (ed.) FSE 2009. LNCS, vol. 5665, pp. 260–276. Springer, Heidelberg (2009)
26. Minier, M., Phan, R.C.-W., Pousse, B.: Distinguishers for Ciphers and Known Key Attack against Rijndael with Large Blocks. In: Preneel, B. (ed.) AFRICACRYPT 2009. LNCS, vol. 5580, pp. 60–76. Springer, Heidelberg (2009)
27. Preneel, B., Govaerts, R., Vandewalle, J.: Hash Functions Based on Block Ciphers: A Synthetic Approach. In: Stinson, D.R. (ed.) CRYPTO 1993. LNCS, vol. 773, pp. 368–378. Springer, Heidelberg (1994)
28. Toshiba Corporation. Specification of Hierocrypt-3. submitted to the First Open NESSIE Workshop, Leuven, Belgium (November 13-14, 2000)

A   Differential Trails for Crypton, Hierocrypt-3, SAFER++ and Square

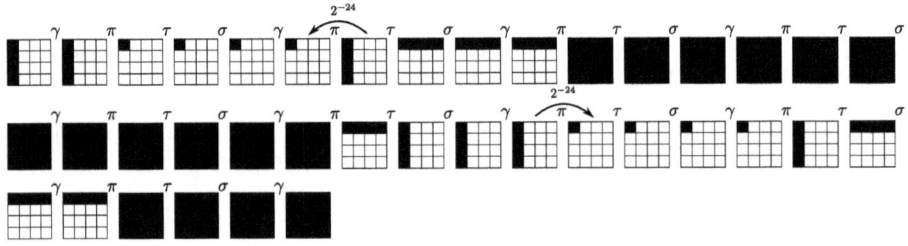

Fig. 4. Truncated differential trail for 9 rounds of Crypton for chosen-key distinguisher and 256-bit key

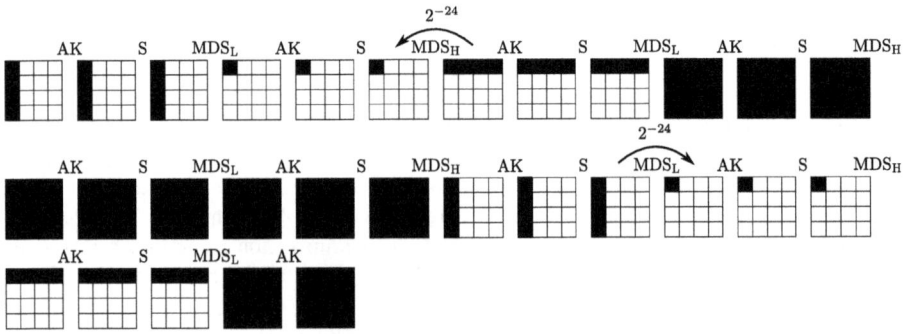

Fig. 5. Truncated differential trail for 4.5 rounds of Hierocrypt for chosen-key distinguisher and 256-bit key

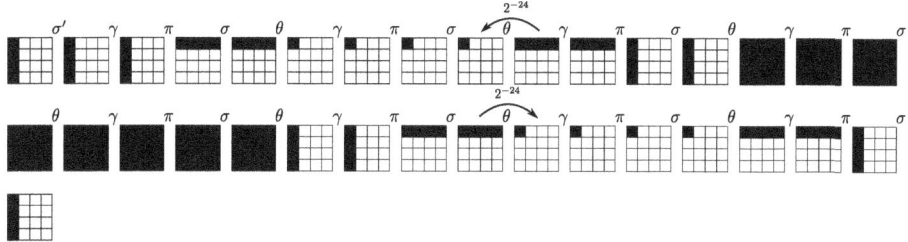

Fig. 6. Truncated differential trail for 8 rounds of Square for chosen-key distinguisher $(\sigma' = \sigma(\theta(k_0)))$

B   Truncated Differential Trail for Modified mCrypton

The key scheduling in the test implementation of mCrypton has been adopted from Crypton and has following way:

Let K be a 128-bit encryption key and $K = k_0 \ldots k_{31}$ where each k_i is four-bit nibble for $i = 0, \ldots, 31$. At first two temporal values U and V are derived from K so that $U[i] = k_{8i}k_{8i+2}k_{8i+4}k_{8i+6}$ and $V[i] = k_{8i+1}k_{8i+3}k_{8i+5}k_{8i+7}$ for $i = 0, 1, 2, 3$. Next for $U' = \gamma(U)$ and $V' = \gamma(V)$ the eight expanded keys are evaluated as:

$$E[i] = \bigoplus_{j \neq i} U'[j] \qquad\qquad E[i+4] = \bigoplus_{j \neq i} V'[j]$$

for $i = 0, 1, 2, 3$ with use of which the 13 subkeys for each encryption round are generated according to the following procedure:

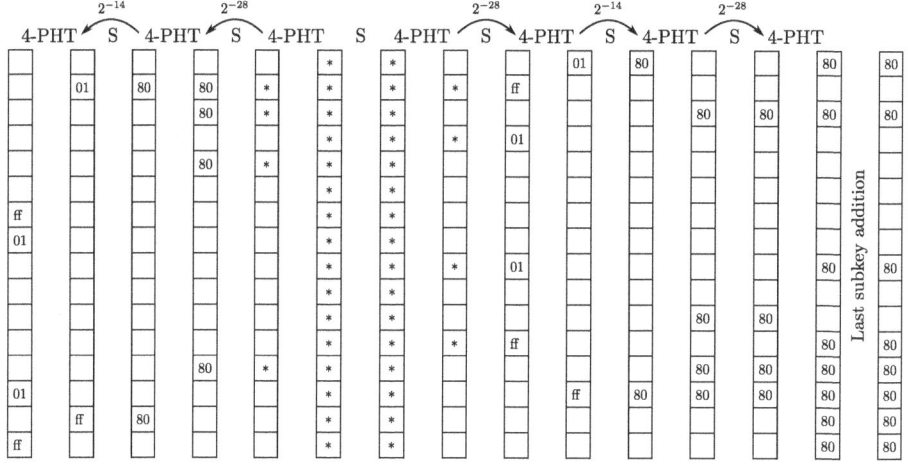

Fig. 7. Standard differential trail for 6.5 rounds of SAFER++ for chosen-key distinguisher and 128-bit key

1. for the first and the second round:

$$K_1[i] = E[i] \oplus C[0] \oplus MC_i \qquad K_2 = E[i+4] \oplus C[1] \oplus MC_i$$

for $i = 0,1,2,3$,

2. for the remaining eleven rounds ($r = 2,\ldots,12$) two steps are executed alternatively:

(a) for r even:

$$\{E[0], E[1], E[2], E[3]\} \leftarrow \{E[1]^{\lll 12}, E[2]^{\lll 8}, E[3]^{\lll_b 3}, E[0]^{\lll_b 3}\},$$
$$K_r[i] = E[i] \oplus C[r] \oplus MC_i,$$

(b) for r odd:

$$\{E[4], E[5], E[6], E[7]\} \leftarrow \{E[7]^{\lll_b 1}, E[4]^{\lll_b 1}, E[3]^{\lll 4}, E[0]^{\lll 8}\},$$
$$K_r[i] = E[i+4] \oplus C[r] \oplus MC_i,$$

for $i = 0,1,2,3$,

where $C[0] = \mathtt{f53a}$, $C[k] = C[k-1] + \mathtt{f372} \bmod 2^{16}$ for $k = 1,\ldots,12$, $MC_0 = \mathtt{acac}$, $MC_k = MC_{k-1}^{\lll_b 1}$ for $i = 0,1,2,3$ and $\lll_b a$ represents bit-left-rotation by a bits within each four-bit nibble.

Example of a differential trail for mCrypton:

i	$A[i]$	$D[i]$	$D_\gamma[i]$	$D_{\pi\circ\gamma}[i]$	$D_{\tau\circ\pi\circ\gamma}[i]$	$K[i]$
1	01d3	61b9	c000	c000	c000	9822
	7701	e71f	c000	0000	0000	e615
	1b65	4bb3	8000	0000	0000	a7a2
	ea3b	cf45	4000	0000	0000	dd34
2	11ac	c000	9000	8000	8991	e08d
	9137	0000	0000	0000	0000	ca42
	59d2	0000	0000	9000	0000	a541
	e714	0000	0000	1000	0000	2bc7
3	7952	8991	9afd	88b5	8991	3534
	ca28	0000	0000	9a7c	8a2a	b8b0
	6a3d	0000	0000	92ed	b7ed	c819
	68a0	0000	0000	1ad9	5cd9	e29a
4	6cb1	8991	d54b	d451	d119	5957
	ad6f	8a2a	5e93	1ead	4e1a	ced2
	f9b8	b7ed	1142	11fc	5afb	f360
	1682	5cd9	db24	9abe	1dce	db24
5	86f0	d119	16a7	1fda	1ec5	3d50
	e0d0	4e1a	bc48	ee6b	fe11	38ea
	0070	5afb	e9b4	c14d	d643	5440
	0d07	1dce	2295	5132	abd2	bf17

i	$A[i]$	$D[i]$	$D_\gamma[i]$	$D_{\pi\circ\gamma}[i]$	$D_{\tau\circ\pi\circ\gamma}[i]$	$K[i]$
6	0005	1ec5	e827	ea6f	e000	9e0c
	0066	fe11	ca6e	0000	a000	8aea
	0000	d643	a26d	0000	6000	b046
	7230	abd2	6a4b	0000	f000	bfb1
7	de7f	e000	e000	e000	e000	3e40
	b737	a000	c000	0000	0000	178d
	0d4d	6000	a000	0000	0000	1f6c
	3302	f000	6000	0000	0000	5e47
8	1b50	e000	6000	6000	6426	98c2
	f0ec	0000	0000	4000	0000	dab9
	1712	0000	0000	2000	0000	40a9
	9247	0000	0000	6000	0000	50a7
9	773d	6426	f9cc	e984	edb7	3334
	eeed	0000	0000	d94c	9918	e680
	002a	0000	0000	b1cc	84cc	6b92
	bbc0	0000	0000	78c8	4cc8	5b99
10	85d0	edb7	7299	0000	0000	13c3
	9026	9918	a8dd	0000	0000	8209
	9e52	84cc	1cfe	0000	0000	883a
	5ea9	4cc8	6271	0000	0000	8c1e

Fig. 8. The columns in the table represent: i - round number, $A[i]$ - value of the state in round i, $D[i]$ - difference between two states in round i, $D_\gamma[i]$ - difference between two states after γ in round i, $D_{\pi\circ\gamma}[i]$ - difference between two states after $\pi\circ\gamma$ in round i, $D_{\tau\circ\pi\circ\gamma}[i]$ - difference between two states after $\tau\circ\pi\circ\gamma$ in round i, $K[i]$ - subkey in round i. The trail was obtained for $K = \mathtt{679ff202d5834e529d9cf7013a4d8218}$.

Related-Key Attack on the Full HIGHT

Bonwook Koo, Deukjo Hong, and Daesung Kwon

The Attached Institute of ETRI
P.O. Box 1, Yuseong-Gu, Daejeon, Korea
{bwkoo,hongdj,ds_kwon}@ensec.re.kr

Abstract. HIGHT is a lightweight block cipher, proposed in CHES 2006 , and on the process of ISO/IEC 18033-3 standardization. It is a 32-round Feistel-like block cipher with 64-bit block and 128-bit key. In this paper, we present the first attack on the full HIGHT using related-key rectangle attack with $2^{123.169}$ encryptions, $2^{57.84}$ data, and 4 related keys. Our related-key rectangle attack is valid for 2^{126} weak keys and this attack can be easily extended to an attack for the full key space faster than an exhaustive key searching using 4 related keys.

We observe that an "add-difference" of master keys is propagated to an add-difference of subkeys with probability 1, so we can find 3-round local collisions of HIGHT by considering an add-difference as a relation of keys. Exploiting these local collisions and "over-simplified" structure of key-schedule, we construct a new 15.5-round related-key differential trail with relatively high probability. We construct a 24-round related-key rectangle distinguisher with probability $2^{-117.68}$ from an 8.5-round and a 15.5-round related-key truncated differential trail with local collisions by applying the ladder switch technique, and then suggest an attack on full rounds of HIGHT with this distinguisher. Our result implies that HIGHT cannot be regarded as an instantiation of the ideal cipher used in some provably secure schemes.

Keywords: HIGHT, Block cipher, Cryptanalysis, Related-key rectangle attack.

1   Introduction

In designing a block cipher, a strong key schedule has not been a main consideration. However, for recent years, the related-key attacks exploiting a weakness of a key schedule have provided interesting results [3,4,5,7]. Most of them indicate that simple structure of a key schedule causes weakness useful for certain attacks. We have KASUMI and AES as such examples. KASUMI, known as the A5/3 algorithm for GSM security, has a linear key schedule. It is fully broken by the related-key rectangle attack [2] and practically broken by the related-key sandwich attack [7]. The key schedules of AES have a lot of symmetry in their structures and use at most four S-boxes to generate a 128-bit subkey for a round. So the full rounds of AES-192 and AES-256 are attacked by the related-key amplified boomerang and the related-key boomerang attacks, respectively [4]. A

K.-H. Rhee and D. Nyang (Eds.): ICISC 2010, LNCS 6829, pp. 49–67, 2011.
© Springer-Verlag Berlin Heidelberg 2011

practical key recovery attack on 13 out of 14 rounds of AES-256, which has been recently proposed, also uses related keys [5].

HIGHT [8] is a block cipher which has a linear(in a modular addition point of view) key schedule with few propagations. It was proposed at CHES 2006 for lightweight computing environments such as radio frequency identifications (RFID). Also, HIGHT is a block cipher standard approved by Telecommunications Technology Association (TTA) of Korea and international standardization activities are in progress to include the HIGHT into ISO/IEC 18033-3 [9]. HIGHT is a 32-round block cipher in 8-branch type II generalized Feistel structure with 64-bit block and 128-bit key. The round functions of HIGHT is designed with bit-wise exclusive OR, addition modulo 2^8, and rotations. Such design aspects make HIGHT more efficient than most existing block ciphers including AES-128 on hardware implementation. The designers of HIGHT analyzed its security against various attacks including related-key attacks and they concluded that at least 20 rounds of HIGHT is secure against these attacks. But at ICISC 2007, Lu et al. presented some cryptanalytic results on the HIGHT reduced to 25, 26, and 28 rounds with or without initial and final whitening key additions, using impossible differential, related-key rectangle, related-key impossible differential attacks [10]. Moreover, at ACISP 2009, Özen et al. improved the attack results of ICISC 2007 into an impossible differential attack on 26 rounds of HIGHT and a related-key impossible differential attack on 31 rounds of HIGHT [13]. At CANS 2009, Zhang et al. pointed out an error in the 12-round saturation distinguisher introduced by designers of HIGHT and gave a saturation attack on 22 rounds of HIGHT with initial and final whitening keys using 17-round saturation distinguisher [15].

In this paper, we present a related-key attack on the full HIGHT slightly faster than the exhaustive key search. The attack consists of a related-key rectangle attack for a quarter of key space and an exhaustive key searching for the rest three-quarter of key space in the related-key attack model. Our related-key rectangle attack uses a 24-round related-key rectangle distinguisher with probability $2^{-117.68}$. This distinguisher is constructed from an 8.5-round($E0$) and a 15.5-round($E1$) related-key truncated differential trail by combining them with the ladder switch technique and $E1$ is a combination of three local collisions. The local collision is a related-key differential trail whose input and output differences are zero and in our attack, and we find two types of 4-round local collision and combine them alternately by using the byte-wise rotational property of subkey positioning. Every subkey byte is defined by a modular addition of a byte of encryption key and a predefined constant, so we give an *add-difference* for relation of keys to avoid paying probability for generating subkey differences by key schedule. For $E0$, we modify a known related-key differential trail [10,11] to avoid a flaw shown in Appendix A and transform it into related-key truncated differential trail to reduce data complexity. So we construct a related-key rectangle attack for a quarter of key space with $2^{123.17}$ time and $2^{57.84}$ data and an attack for whole key space with $2^{125.833}$ time and $2^{57.84}$ data. The time and data complexities for attacking HIGHT is given in the Table 1.

Table 1. Summary of the attacks on HIGHT(Imp.:Impossible, Diff.:Differential, Rel.:Related, Rec.:Rectangle, Wek.:Weak Key)

Rounds	Attack	Complexities		References
		Data	Time	
18	Imp. Diff.	$2^{46.8}$	$2^{109.2}$	[8]
22	Saturation	$2^{62.04}$	$2^{118.71}$	[15]
25	Imp. Diff.	2^{60}	$2^{126.78}$	[10]
26	Imp. Diff.	2^{61}	$2^{119.53}$	[13]
26	Rel.-Key Rec.	$2^{51.2}$	$2^{120.41}$	[10]
28	Rel.-Key Imp.	2^{60}	$2^{125.54}$	[10]
31	Rel.-Key Imp.	2^{63}	$2^{127.28}$	[13]
32	Rel.-Key Rec. for Wek.	$2^{57.84}$	$2^{123.17}$	This paper
32	Rel.-Key attack	$2^{57.84}$	$2^{125.833}$	This paper

To our knowledge, this is the first cryptanalytic result on the full rounds of HIGHT. Our attack has very high complexity but clearly, it shows that HIGHT does not reach the security goal in the related-key attack model, required for a block cipher which has a 64-bit block and a 128-bit key. It is also an evidence that HIGHT cannot be regarded as an instantiation of an ideal cipher. Namely, HIGHT would not be used as a substitute for an ideal cipher in applications which are provably secure based on ideal cipher, e.g., some block-cipher-based hash function schemes.

This paper is organized as follows. Section 2 gives the specifications of HIGHT. In Section 3, two types of local collisions of HIGHT and its probability are introduced and a weak key space is classified. In Section 4, a 24-round related-key rectangle distinguisher of HIGHT is presented with its separation into $E0$ and $E1$ and estimation of its probability. Attack procedure and complexity analysis are shown in Section 5. Finally, Section 6 concludes this paper. In Appendix A, some flaws in calculating a probability of differential trail include key addition is pointed out. A complexity of exhaustive key searching in related-key model is presented in Appendix B. An overall view of our attack is depicted in Appendix C.

2   Description of HIGHT

Throughout this paper, we use the following notations.

- $\oplus$: bitwise exclusive OR(XOR)
- $\boxplus$: addition modulo 2^8
- $\Delta(\nabla)$: a notation of xor-difference, an xor-difference Δx indicates that a pair (x, x') is defined by $x' = x \oplus \Delta x$
- $\Delta^+(\nabla^+)$: a notation of add-difference, an add-difference $\Delta^+ x$ indicates that a pair (x, x') is defined by $x' = x \boxplus \Delta^+ x$

- $X[i_1, i_2, ..., i_n]$: concatenation of $X[i_1]$, $X[i_2]$, ..., and $X[i_n]$
- $MSB_i(X)$: the most significant i bits of a string X
- $LSB_i(X)$: the least significant i bits of a string X
- $(\Delta X, \Delta Y) \overset{\boxplus}{\to} \Delta Z$: an event that $(x \boxplus y) \oplus (x' \boxplus y') = \Delta Z$, where $x \oplus x' = \Delta X$ and $y \oplus y' = \Delta Y$

HIGHT takes a 64-bit plaintext P and a 128-bit key K, and its 32-round encryption procedure produces a 64-bit ciphertext C. From now on, we present any 64-bit variable A and any 128-bit variable B as a tuple of eight bytes $(A[7], ..., A[1], A[0])$ and a tuple of sixteen bytes $(B[15], ..., B[1], B[0])$.

The key schedule produces 128 8-bit subkeys $SK[0], ..., SK[127]$ from a 128-bit key $K = (K[15], ..., K[0])$: for $0 \le i \le 7$ and $0 \le j \le 7$,

$$\begin{cases} SK[16i + j] \leftarrow K[j - i \bmod 8] \boxplus \delta[16i + j], \\ SK[16i + j + 8] \leftarrow K[(j - i \bmod 8) + 8] \boxplus \delta[16i + j + 8], \end{cases}$$

where $\delta[0], ..., \delta[127]$ are public constants.

Let $X_{i-1} = (X_{i-1}[7], ..., X_{i-1}[0])$ and $X_i = (X_i[7], ..., X_i[0])$ be the input and output of the round $i - 1$ for $1 \le i \le 32$, respectively, where 'round i' denotes the $(i + 1)$-th round(i.e. round 0 implies the first round).

The encryption procedure of HIGHT is as follows.

1. Initial Transformation:

$$X_0[0] \leftarrow P[0] \boxplus K[12]; X_0[2] \leftarrow P[2] \oplus K[13];$$
$$X_0[4] \leftarrow P[4] \boxplus K[14]; X_0[6] \leftarrow P[6] \oplus K[15];$$
$$X_0[1] \leftarrow P[1]; X_0[3] \leftarrow P[3]; X_0[5] \leftarrow P[5]; X_0[7] \leftarrow P[7].$$

2. Round Iteration for $1 \le i \le 32$:

$$X_i[0] \leftarrow X_{i-1}[7] \oplus (F_0(X_{i-1}[6]) \boxplus SK[4i - 1]);$$
$$X_i[2] \leftarrow X_{i-1}[1] \boxplus (F_1(X_{i-1}[0]) \oplus SK[4i - 2]);$$
$$X_i[4] \leftarrow X_{i-1}[3] \oplus (F_0(X_{i-1}[2]) \boxplus SK[4i - 3]);$$
$$X_i[6] \leftarrow X_{i-1}[5] \boxplus (F_1(X_{i-1}[4]) \oplus SK[4i - 4]);$$
$$X_i[1] \leftarrow X_{i-1}[0]; X_i[3] \leftarrow X_{i-1}[2]; X_i[5] \leftarrow X_{i-1}[4]; X_i[7] \leftarrow X_{i-1}[6],$$

where bijective linear functions F_0 and F_1 are defined by

$$\begin{cases} F_0(x) = x^{\lll 1} \oplus x^{\lll 2} \oplus x^{\lll 7}, \\ F_1(x) = x^{\lll 3} \oplus x^{\lll 4} \oplus x^{\lll 6}. \end{cases}$$

3. Final Transformation:

$$C[0] \leftarrow X_{32}[1] \boxplus K[0]; C[2] \leftarrow X_{32}[3] \oplus K[1];$$
$$C[4] \leftarrow X_{32}[5] \boxplus K[2]; C[6] \leftarrow X_{32}[7] \oplus K[3];$$
$$C[1] \leftarrow X_{32}[2]; C[3] \leftarrow X_{32}[4]; C[5] \leftarrow X_{32}[6]; C[7] \leftarrow X_{32}[0].$$

3   Local Collisions in HIGHT

Local collision is firstly introduced by Chabaud et al. in [6] for finding collisions in SHA-0 hash function using differential cryptanalysis. In block cipher cryptanalysis, if a difference caused only by a subkey difference is eliminated by other subkey differences with some probability a few rounds later, we call this property a local collision in block cipher. In HIGHT, we observe that there are two types of local collision which are depicted in Fig. 1.

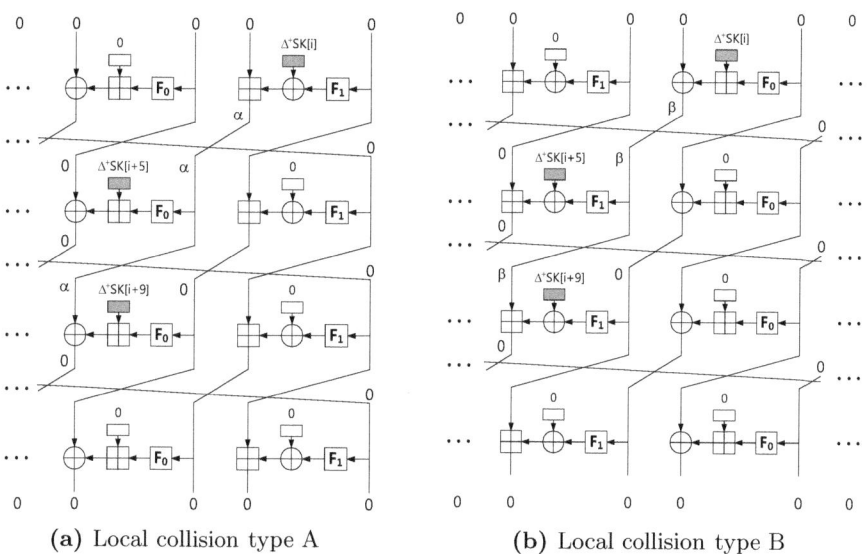

(a) Local collision type A   (b) Local collision type B

Fig. 1. Local collisions in HIGHT

3.1   Probabilities of Local Collisions

Fig. 1-(a) shows how the only nonzero differences $\Delta^+ SK[i]$, $\Delta^+ SK[i+5]$, and $\Delta^+ SK[i+9]$ in the form of the local collision type A lead to zero output differences of the round. Its probability is computed as $r = \sum_\alpha r_1(\alpha) r_2(\alpha) r_3(\alpha)$ where

$$r_1(\alpha) = \Pr(((X \oplus Y) + Z) \oplus ((X \oplus (Y + \Delta^+ SK[i])) + Z) = \alpha),$$
$$r_2(\alpha) = \Pr((X + Y) \oplus ((X \oplus F_0(\alpha)) + (Y + \Delta^+ SK[i+5])) = 0),$$
$$r_3(\alpha) = \Pr((X + Y) \oplus (X + (Y + \Delta^+ SK[i+9])) = \alpha).$$

Similarly, Fig. 1-(b) shows how the only nonzero differences $\Delta^+ SK[i]$, $\Delta^+ SK[i+5]$, and $\Delta^+ SK[i+9]$ in the form of the local collision type B lead to zero output differences of the round. Its probability is computed as $s = \sum_\beta s_1(\beta) s_2(\beta) s_3(\beta)$ where

$$s_1(\beta) = \Pr((X + Y) \oplus (X + (Y + \Delta^+ SK[i])) = \beta),$$
$$s_2(\beta) = \Pr(((X \oplus Y) + Z) \oplus (((X \oplus F_1(\beta)) \oplus (Y + \Delta^+ SK[i + 5])) + Z) = 0),$$
$$s_3(\beta) = \Pr(((X \oplus Y) + Z) \oplus ((X \oplus (Y + \Delta^+ SK[i + 9])) + (Z \oplus \beta)) = 0).$$

By an exhaustive computation, we can see the expected values of r and s when $(\Delta^+ SK[i], \Delta^+ SK[i + 5], \Delta^+ SK[i + 9]) = (\texttt{0x10}, \texttt{0x68}, \texttt{0x10})$ are $2^{-5.1420}$ and 2^{-8}, respectively, and we the following 8 possibilities of $(\Delta^+ SK[i], \Delta^+ SK[i + 5], \Delta^+ SK[i + 9])$ yielding the same probabilities:

$(\texttt{0x10}, \texttt{0x68}, \texttt{0x10}), (\texttt{0x10}, \texttt{0x68}, \texttt{0xf0}), (\texttt{0xf0}, \texttt{0x68}, \texttt{0x10}), (\texttt{0xf0}, \texttt{0x68}, \texttt{0xf0}),$

$(\texttt{0x10}, \texttt{0x98}, \texttt{0x10}), (\texttt{0x10}, \texttt{0x98}, \texttt{0xf0}), (\texttt{0xf0}, \texttt{0x98}, \texttt{0x10}), (\texttt{0xf0}, \texttt{0x98}, \texttt{0xf0}).$

We observed that the probability r that the local collision type A with $(\Delta^+ SK[i], \Delta^+ SK[i + 5], \Delta^+ SK[i + 9]) = (\texttt{0x10}, \texttt{0x68}, \texttt{0x10})$ occurs is nonzero only when $\alpha = \texttt{0x10}$ or $\texttt{0x30}$. Under the observation, the probability r is actually $2^{-4.67807}$, $2^{-5.41504}$, or $2^{-6.41504}$. So, we regard $2^{-6.41504}$ as a lower bound of r.

Similarly, we observed that the probability s that the local collision type B with $(\Delta^+ SK[i], \Delta^+ SK[i + 5], \Delta^+ SK[i + 9]) = (\texttt{0x10}, \texttt{0x68}, \texttt{0x10})$ occurs is nonzero only when $\beta = \texttt{0x70}$. Especially, for $\beta = \texttt{0x70}$, $s_2(\beta)$ is nonzero only when

$$SK[i + 5] \in T = \{x \vee \texttt{0x18} | x \in \mathrm{GF}(2^8)\}.$$

When $SK[i + 5] \in T$, the local collision of type B occurs with the probability $s = 2^{-6}$. Otherwise, it does not occur. Note that the fraction of T in $\mathrm{GF}(2^8)$ is $1/4$.

3.2   Local Collisions to a Long Differential Trail

We can use a sequence of local collisions, 'type A – type B – type A' to construct a 12-round related-key differential trail. Let i be a multiple of 4(i.e. $\Delta^+ SK[i]$ be the right most subkey difference of round $i/4$). If $\Delta^+ SK[i]$, $\Delta^+ SK[i + 5]$, and $\Delta^+ SK[i+9]$ are induced by the only nonzero add-differences $\Delta^+ K[j_1]$, $\Delta^+ K[j_2]$, and $\Delta^+ K[j_3]$ of master-key bytes, then by rotational property of key schedule,

$$\Delta^+ K[j_1] = \Delta^+ SK[i] = \Delta^+ SK[i + 17] = \Delta^+ SK[i + 34],$$
$$\Delta^+ K[j_2] = \Delta^+ SK[i + 5] = \Delta^+ SK[i + 22] = \Delta^+ SK[i + 39],$$
$$\Delta^+ K[j_3] = \Delta^+ SK[i + 9] = \Delta^+ SK[i + 26] = \Delta^+ SK[i + 43],$$

and differences of other subkeys are all zero if differences of other master key bytes are zero.

Therefore, if there exist nonzero add-differences $\Delta^+ K[j_1]$, $\Delta^+ K[j_2]$, and $\Delta^+ K[j_3]$ such that the probabilities p_1, p_2, and p_3 of local collisions from $i/4$ to $(i/4 + 3)$-th round, from $(i/4 + 4)$ to $(i/4 + 7)$-th round, and from $(i/4 + 8)$ to $(i/4 + 11)$-th round are all nonzero, then we can find a 12-round related-key differential trail of HIGHT with probability $p_1 \times p_2 \times p_3$ by combining them sequentially.

From the arguments in Section 3.1, when we take $(\Delta^+ K[j_1], \Delta^+ K[j_2], \Delta^+ K[j_3])$ $= (\texttt{0x10}, \texttt{0x68}, \texttt{0x10})$, the 12-round related-key differential trail is valid only for a quarter of the whole key space and its probability is lower bounded by $2^{-18.83008}$.

4   Related-Key Rectangle Distinguisher for 24 Rounds of HIGHT

4.1   Related-Key Rectangle Distinguisher

A rectangle distinguisher assumes that a block cipher $E_K : \{0,1\}^n \to \{0,1\}^n$ with an arbitrary key K can be represented by a composition of two sub-ciphers $E0_K$ and $E1_K$, i.e. $E_K = E1_K \circ E0_K$, where n is the bit-length of block. Our approach to construct a related-key rectangle distinguisher is somewhat different from previous works in the point that we use xor-difference for plaintexts or ciphertexts and add-difference for keys.

Assume that we have two related-key differentials for $E0$ and $E1$ with the following probabilities

$$p = \Pr[E0_K(P) \oplus E0_{K \boxplus \Delta^+ K}(P \oplus \Delta P) = \Delta Y], \tag{1}$$
$$q = \Pr[E1_K(Y) \oplus E1_{K \boxplus \nabla^+ K}(Y \oplus \nabla Y) = \nabla C]. \tag{2}$$

We consider four encryption oracles with 4 related keys denoted by E_{K1}, E_{K2}, E_{K3}, and E_{K4} and the relations between keys are as follows,

$$K2 = K1 \boxplus \Delta^+ K, \qquad K4 = K3 \boxplus \Delta^+ K,$$
$$K3 = K1 \boxplus \nabla^+ K, \qquad K4 = K2 \boxplus \nabla^+ K.$$

For a plaintext quartet (P_1, P_2, P_3, P_4) such that $P_1 \oplus P_2 = P_3 \oplus P_4 = \Delta P$, let $Y_i = E0_{Ki}(P_i)$ and $C_i = E_{Ki}(P_i) = E1_{Ki}(Y_i)$ for $1 \le i \le 4$. If the event $Y_1 \oplus Y_2 = Y_3 \oplus Y_4 = \Delta Y$ and the event $Y_1 \oplus Y_3 = \nabla Y$ occur, we obtain $Y_2 \oplus Y_4 = \nabla Y$ because

$$Y_2 \oplus Y_4 = (Y_2 \oplus Y_1) \oplus (Y_1 \oplus Y_3) \oplus (Y_3 \oplus Y_4)$$
$$= \Delta Y \oplus \nabla Y \oplus \Delta Y = \nabla Y.$$

Therefore, for a randomly chosen plaintext quartet (P_1, P_2, P_3, P_4) such that $P_1 \oplus P_2 = P_3 \oplus P_4 = \Delta P$, we have $C_1 \oplus C_3 = C_2 \oplus C_4 = \nabla C$ with the probability $p^2 \cdot 2^{-n} \cdot q^2$, from (1) and (2). If there exist more than two values for ΔY and ∇Y, the probability is amplified to

$$\hat{p}^2 \cdot 2^{-n} \cdot \hat{q}^2, \quad \text{where} \quad \hat{p}^2 = \sum_{\Delta Y} p^2 \quad \text{and} \quad \hat{q}^2 = \sum_{\nabla Y} q^2. \tag{3}$$

Our attack assumes more than two values for ΔP so our probability calculation in the next section would be slightly differ from (3).

4.2   Related-Key Rectangle Distinguisher of HIGHT

Related-key differential trail for $E0$ is based on a trail introduced in [10,11] but significantly modified to avoid the flaw explained in Appendix A and changed into truncated differential to reduce the data complexity. Related-key differential trail for $E1$ includes three local collisions as described in Section 3.1.

We define $E0$ and $E1$ by partial rounds from round 3 to round 10.5 and round 10.5 to round 26, respectively(0.5 round implies computation of 2 round functions out of 4 round functions in a round). The input and output bytes to $E0$ and $E1$ and corresponding differences are described in Table 2, where the $\mathcal{A}$, $\mathcal{B}$, and $\mathcal{C}$ are defined by sets of hexadecimal values as follows,

$$\mathcal{A} = \{14, 1c, 24, 2c, 34, 3c, 54, 5c, 64, 6c, 74, 7c, d4, dc, e4, ec\},$$
$$\mathcal{B} = \{14, 1c, 24, 2c, 34, 3c, 54, 5c, 64, 6c, 74, 7c, d4, dc, e4, ec, f4, fc\},$$
$$\mathcal{C} = \{10, 30, 70, f0\}.$$

Table 2. Byte positions and differences of both inputs and outputs for distinguishers of $E0$ and $E1$

$E0$	Pos- itions	Input	$(X_3[7], X_3[6], X_3[5], X_3[4], X_3[3], X_3[2], X_3[1], X_3[0])$
		Output	$(X_{12}[7], X_{11}[5], X_{10}[3], X_{10}[2], X_{11}[2], X_{12}[2], X_{13}[2], X_{13}[1])$
	Differ- ences	Input	(0x0, 0x0, $\mathcal{A}$, 0x80, 0x0, 0x0, 0x0, 0x0)
		Output	(0x0, 0x0, 0x0, $\mathcal{B}$, 0x80, 0x0, 0x0, 0x0)
$\Delta^+ K$ (0x0, 0x0, 0x0, 0x0, 0x0, 0x0, 0x0, 0x0, 0x0, 0x0, 0x0, 0x0, 0x0, 0x0, 0x80, 0x0)			
$E1$	Pos- itions	Input	$(X_{12}[7], X_{11}[5], X_{10}[3], X_{10}[2], X_{11}[2], X_{12}[2], X_{13}[2], X_{13}[1])$
		Output	$(X_{27}[7], X_{27}[6], X_{27}[5], X_{27}[4], X_{27}[3], X_{27}[2], X_{27}[1], X_{27}[0])$
	Differ- ences	Input	(0x0, 0x0, 0x0, 0x0, 0x0, 0x0, 0x0, $\mathcal{C}$)
		Output	(0x0, 0x0, 0x0, 0x0, 0x0, 0x0, 0x0, 0x0)
$\nabla^+ K$ (0x0, 0x0, 0x0, 0x0, 0x0, 0x0, 0x10, 0x0, 0x0, 0x0, 0x68, 0x0, 0x0, 0x0, 0x10, 0x0)			

The related-key differential trails for $E0$ and $E1$ are depicted in the the following fig. 5 in Appendix C.

Probability of Related-Key Differential Trail for $E0$. Our attack begins with gathering plaintext pairs which satisfy $\Delta X_3[5] \in \mathcal{A}$, $\Delta X_3[4] = 0x80$, and $\Delta X_3[0, 1, 2, 3, 6, 7] = 0x0$. Let a_i denote each element in $\mathcal{A}$, where $i = 0, 1, ..., 15$. The number of pairs such that $\Delta X_3[5] = a_i$ is same to the number of pairs such that $\Delta X_3[5] = a_j$ for all $0 \leq i, j \leq 15$. Let u_i denote the probabilities that $\Delta X_3[5] = a_i$ and $\Delta X_4[6] = 0$ for $i = 0, 1, ..., 15$, then

$$u_i = \Pr[\Delta X_4[6] = 0 | \Delta X_3[5] = a_i] \times \Pr[\Delta X_3[5] = a_i],$$

and since $\Pr[\Delta X_3[5] = a_i] = 2^{-4}$ for all $i = 0, 1, ..., 15$, the probability $\bar{u}$ such that $\Delta X_4[6] = 0$, among prepared pairs is calculated by

$$\bar{u} = \sum_{i=0}^{15} u_i = \frac{1}{16} \sum_{i=0}^{15} \Pr[\Delta X_4[6] = 0 | \Delta X_3[5] = a_i] > 2^{-4.09312}.$$

Let b_i denote each element in $\mathcal{B}$ and v_i denote the probabilities that $\Delta X_{10}[2] = b_i$ for $i = 0, 1, ..., 17$, then the probability $\hat{p}^2$ that both related-key differential trails for $E0$ are satisfied with the same output difference is calculated by

$$\hat{p}^2 = \bar{u}^2 \cdot \sum_{i=0}^{17} v_i^2 > 2^{-8.18624} \times 2^{-3.83007} > 2^{-12.017}.$$

Probability of Related-Key Differential Trail for $E1$ For each element c_i in $\mathcal{C}$, let the probabilities w_i be defined by

$$w_i = \Pr[\Delta X_{14}[2] = 0 | \Delta X_{13}[1] = c_i],$$

for $i = 0, 1, 2, 3$, then $\Delta X_{14}[0, 1, ..., 8] = 0$ with probabilities w_i. Both w_3 and w_4 are 2^{-3} for all $SK[52]$, whereas w_1 and w_2 are among 2^{-1}, 2^{-2}, and 2^{-3} according to $SK[52]$. So the lower-bound of $w_i (i = (0, 1, 2, 3)$ is 2^{-3}.

Since we assume that $\nabla^+ K[1] = $ 0x10, $\nabla^+ K[5] = $ 0x68, and $\nabla^+ K[9] = $ 0x10, we can calculate a nonzero probability q such that three local collisions occur sequentially as described in Section 3.1. As we know that both the first and the third local collisions during round 15~18 and round 23~26 are of type A and their probabilities are bounded below by $2^{-6.41504}$ and the second local collision during round 19~22 is of type B and its probability is 2^{-6}, the probability q such that related-key differential trail from round 15 to 26 is calculated by

$$2^{-6.41504-6-6.41504} = 2^{-18.83008} < q.$$

Hence, the probability $\hat{q}^2$ that both related-key differential trails for $E1$ are satisfied is calculated by

$$\hat{q}^2 = \sum_{i=0}^{3} w_i^2 \cdot q^2 \geq 2^{-4} \times 2^{-37.66016} = 2^{-41.66016} > 2^{-41.661}.$$

Therefore, we have a 24-round related-key rectangle distinguisher with the probability

$$\hat{p}^2 \cdot 2^{-64} \cdot \hat{q}^2 \geq 2^{-12.017-64-41.661} = 2^{-117.678} > 2^{-117.68}.$$

The probabilities occurring by additions between differences are computed by exhaustive counting with PC. By experiments on PC, we make sure that suggested probabilities of related-key differential trail for $E0$ and $E1$ are lower bounds of the actual ratio of right pairs for $E0$ and $E1$ respectively, under the assumption that plaintexts and related keys are randomly chosen.

5   Related-Key Rectangle Attack for the Full Rounds of HIGHT

In this section, we describe the attack for the full rounds of HIGHT by using the 24-round related-key rectangle distinguisher explained in Section 4.1 for round 3 to round 26. However, note that the distinguisher is valid only for a quarter of the key space. So, we apply a related-key rectangle attack for a quarter of the key space and an exhaustive key search for the other part of the key space. The outline of our attack is as follows.

1. **Related-key rectangle attack**: We denote the set of the key quartets by $\mathcal{K}_1$ such that the 24-round related-key rectangle distinguisher in Section 4.1 is valid. Assuming that we are given a key quartet from $\mathcal{K}_1$, we perform a related-key rectangle attack which consists of the following phases.

   (a) **Constructing the plaintext set**: We construct the plaintext set $\mathcal{S}$ for extracting the plaintext quartets required for the related-key rectangle distinguisher.

   (b) **Guessing and filtering**: Let Z_1 be required key bits to check whether a plaintext quartet from the plaintext set $\mathcal{S}$ satisfies the input differences of the distinguisher. We guess a value z_1 for Z_1 and select the plaintext quartets from $\mathcal{S}$ satisfying the input differences of the distinguisher with z_1. Then, we discard the quartets whose ciphertext differences do not match with the output differences of the distinguisher.

   (c) **Counting and sorting**: Let Z_2 be required key bits to check whether a surviving quartet satisfies the output differences of the distinguisher. For each candidate (z_1, z_2) for (Z_1, Z_2), we count the number of quartets satisfying the output differences of the distinguisher and restore it to the counter $t_{(z_1, z_2)}$. We sort the list of (z_1, z_2) according to $t_{(z_1, z_2)}$.

   (d) **Searching with the list**: We exhaustively search for the remained key bits for candidates with remarkably high $t_{(z_1, z_2)}$ until a right key quartet is found. If no right key quartet is found, go to (b) **Guessing and filtering phase**.

2. **Exhaustive key searching**: We denote the key space of HIGHT by $\mathcal{K}$ and let $\mathcal{K}_2 = \mathcal{K} \backslash \mathcal{K}_1$. Unless we find a right key quartet in $\mathcal{K}_1$ in the way of the related-key rectangle attack phase, we try to search it exhaustively for $\mathcal{K}_2$ in the way described in Appendix B.

5.1   Attack Procedure

Let sets $\mathcal{D}, \mathcal{E}, \mathcal{F}$, and $\mathcal{G}$ be defined by

$$\mathcal{D} = \{x \vee \mathtt{0x18} | \forall x \in \mathrm{GF}(2^8)\},$$
$$\mathcal{E} = \{\mathtt{0x00}, \mathtt{0x20}, \mathtt{0x40}, \mathtt{0x60}, \mathtt{0x80}, \mathtt{0xa0}, \mathtt{0xc0}, \mathtt{0xe0}\},$$
$$\mathcal{F} = \{\mathtt{0x18}, \mathtt{0x28}, \mathtt{0x38}, ..., \mathtt{0xf8}\},$$
$$\mathcal{G} = \{\mathtt{0x10}, \mathtt{0x30}, \mathtt{0x70}, \mathtt{0xf0}\}.$$

Constructing the plaintext set

1. Choose $58657 \approx 2^{15.84}$ structures S_i of 2^{40} plaintexts ${}_l^i P$ each, $i = 1, 2, ...,$ 58657, $l = 1, 2, ..., 2^{40}$, where in each structure, the 0, 6, 7-th bytes of ${}_l^i P$ are fixed, and the remaining 5 bytes take all the possible values. Obtain the ciphertexts ${}_l^i C$, ${}_l^i C^*$, ${}_l^i C'$, and ${}_l^i C'^*$ of ${}_l^i P$ encrypted with four related keys $K1$, $K2$, $K3$, and $K4$ respectively, where keys have a relation described in Table 2 of Section 4.1.

Guessing and filtering

2. Guess the 9 bytes $K1[0, 1, 2, 5, 6, 10, 12, 13, 14]$ such that $SK1[82] \in \mathcal{D}$ and do as follows, where SKi is subkey bytes produced by a secret key Ki.

   (a) Compute the subkeys $SK1[0, 1, 2, 5, 6, 10]$, and their related subkeys. Partially encrypt plaintext bytes ${}_l^i P[1, 2, 3, 4, 5]$ for each ${}_l^i P$ through partial rounds 0, 1, and 2 with 5 guessed subkey bytes and its related subkey bytes to get the following sets of intermediate values,

$$\{{}_l^i X_2[3], {}_l^i X_3[5], {}_l^i X_3[6], {}_l^i X_2[6], {}_l^i X_2[7]\},$$
$$\{{}_l^i X_2^*[3], {}_l^i X_3^*[5], {}_l^i X_3^*[6], {}_l^i X_2^*[6], {}_l^i X_2^*[7]\},$$
$$\{{}_l^i X_2'[3], {}_l^i X_3'[5], {}_l^i X_3'[6], {}_l^i X_2'[6], {}_l^i X_2'[7]\},$$
$$\{{}_l^i X_2'^*[3], {}_l^i X_3'^*[5], {}_l^i X_3'^*[6], {}_l^i X_2'^*[6], {}_l^i X_2'^*[7]\},$$

   for all $1 \leq i \leq 2^{15.84}$ and $1 \leq l \leq 2^{40}$.

   (b) Find all pairs $({}_l^i P, {}_u^i P)$ such that ${}_l^i X_2[3] \oplus {}_u^i X_2^*[3] = \texttt{0x80}$, ${}_l^i X_3[6] \oplus {}_u^i X_3^*[6] = {}_l^i X_2[6] \oplus {}_u^i X_2^*[6] = {}_l^i X_2[7] \oplus {}_u^i X_2^*[7] = \texttt{0x0}$, and ${}_l^i X_3[5] \oplus {}_u^i X_3^*[5] \in \mathcal{A}$ and store the corresponding ciphertext pairs $({}_l^i C, {}_u^i C^*)$ encrypted with each $K1$ and $K2$ in a hash table $\mathcal{H}$, for all $1 \leq i \leq 2^{15.84}$.

   (c) Find all pairs $({}_v^j P, {}_w^j P)$ such that ${}_v^j X_2'[3] \oplus {}_w^j X_2'^*[3] = \texttt{0x80}$, ${}_w^j X_3'^*[6] = {}_v^j X_2'[6] \oplus {}_w^j X_2'^*[6] = {}_v^j X_2'[7] \oplus {}_w^j X_2'^*[7] = \texttt{0x0}$, and ${}_v^j X_3'[5] \oplus {}_w^j X_3'^*[5] \in \mathcal{A}$ and store the corresponding ciphertext pairs $({}_v^j C', {}_w^j C'^*)$ encrypted with each $K3$ and $K4$ in a hash table $\mathcal{I}$, for all $1 \leq j \leq 2^{15.84}$.

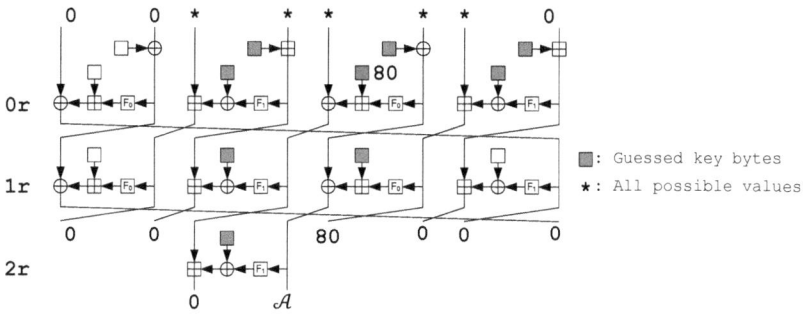

Fig. 2. Constructing plaintext sets and choosing pairs

(d) Store all quartets $(^i_lC, ^i_uC^*, ^j_vC', ^j_wC'^*)$ defined by all pairs $(^i_lC, ^i_uC^*) \in \mathcal{H}$ and $(^j_vC', ^j_wC'^*) \in \mathcal{I}$, for all $1 \leq i, j \leq 2^{15.84}$ and $1 \leq l, u, v, w \leq 2^{40}$ in a hash table $\mathcal{J}$.

(e) For all quartets $(^i_lC, ^i_uC^*, ^j_vC', ^j_wC'^*)$ in $\mathcal{J}$, do filtering by the following steps. In each steps, discard the quartets which do not satisfy the conditions and if less than 3 quartets are remained, then go to Step 2 with another key guessing.

  i. Check if $^i_lC[0,1] \oplus ^j_vC'[0,1] = 0$, $^i_uC^*[0,1] \oplus ^j_wC'^*[0,1] = 0$, $^i_lC[2] \oplus ^j_uC'[2] \in \mathcal{E}$, and $^i_vC^*[2] \oplus ^j_wC'^*[2] \in \mathcal{E}$ (2^{-42} filtering).

  ii. Compute and check if $^i_lX_{31}[3] \oplus ^j_uX'_{31}[3] \in \mathcal{F}$, and $^i_vX^*_{31}[3] \oplus ^j_wX'^*_{31}[3] \in \mathcal{F}$ (2^{-8} filtering).

  iii. Check if $\Pr[(F_0(^i_lC[6] \oplus ^j_uC'[6]), 0) \overset{\boxplus}{\rightarrow} {}^i_lC[7] \oplus ^j_uC'[7]] > 0$ and $\Pr[(F_0(^i_vC^*[6] \oplus ^j_wC'^*[6]), 0) \overset{\boxplus}{\rightarrow} {}^i_vC^*[7] \oplus ^j_wC'^*[7]] > 0$($2^{-5.65514}$ filtering).

  iv. Compute $\Delta T = {}^i_lX_{30}[4] \oplus ^j_uX'_{30}[4]$ and $\Delta T' = {}^i_vX^*_{30}[4] \oplus ^j_wX'^*_{30}[4]$, and check if $\Pr[(F_1(\Delta T), 0) \overset{\boxplus}{\rightarrow} {}^i_lC[6] \oplus ^j_uC'[6]] > 0$ and $\Pr[(F_1(\Delta T'), 0) \overset{\boxplus}{\rightarrow} {}^i_vC^*[6] \oplus ^j_wC'^*[6]] > 0$ ($2^{-5.65514}$ filtering).

  v. Compute and check if $\Pr[(F_1(^i_lX_{31}[4] \oplus ^j_uX'_{31}[4]), \Delta T) \overset{\boxplus}{\rightarrow} {}^i_lC[5] \oplus ^j_uC'[5]] > 0$ and $\Pr[(F_1(^i_vX^*_{31}[4] \oplus ^j_wX'^*_{31}[4]), \Delta T') \overset{\boxplus}{\rightarrow} {}^i_vC^*[5] \oplus ^j_wC'^*[5]] > 0$ ($2^{-4.69704}$ filtering).

  vi. If 3 or more quartets $(^i_lC, ^i_uC^*, ^j_vC', ^j_wC'^*)$ remained, record them and go to Step 3; otherwise, go to Step 2 with another guess.

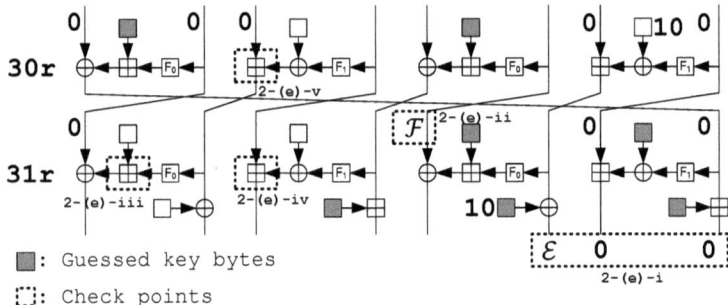

Fig. 3. Filtering of wrong quartets. The numbers nearby the check points indicate the corresponding steps from 2-(e)-i to 2-(e)-v.

Counting and sorting

3. In the following step 4 to step 12, discard the quartets with each key bytes guessed in each step which do not satisfy conditions. In each steps, if less than 3 quartets are remained, then go to Step 2 with another key guessing.

4. Guess the $LSB_4(K1[15])$ to compute $LSB_4(SK1[126])$ and its related keys. For each key and remained quartet, check (a) $LSB_4(^i_lX_{29}[3] \oplus ^j_uX'_{29}[3]) = 0$ and (b) $LSB_4(^i_vX^*_{29}[3] \oplus ^j_wX'^*_{29}[3]) = 0$.

5. Guess $MSB_4(K1[15])$ to compute $MSB_4(SK1[126])$ and its related keys. For each key and remained quartet, check (a) $MSB_4(^i_l X_{29}[3] \oplus ^j_u X'_{29}[3]) = 0$ and (b) $MSB_4(^i_v X^*_{29}[3] \oplus ^j_w X'^*_{29}[3]) = 0$.
6. Guess $LSB_4(K1[9])$ to compute $LSB_4(SK1[120])$ and its related keys. For each key and remained quartet, check (a) $LSB_4(^i_l X_{30}[1] \oplus ^j_u X'_{30}[1]) = 0$ and (b) $LSB_4(^i_v X^*_{30}[1] \oplus ^j_w X'^*_{30}[1]) = 0$.
7. Guess $MSB_4(K1[9])$ to compute $MSB_4(SK1[120])$ and its related keys. For each key and remained quartet, check (a) $MSB_4(^i_l X_{30}[1] \oplus ^j_u X'_{30}[1]) = 0$ and (b) $MSB_4(^i_v X^*_{30}[1] \oplus ^j_w X'^*_{30}[1]) = 0$.
8. Without key guessing, check if $^i_l X_{28}[0] \oplus ^j_u X'_{28}[0] \in \mathcal{G}$ and $^i_v X^*_{28}[0] \oplus ^j_w X'^*_{28}[0] \in \mathcal{G}$.
9. Without key guessing, check if $^i_l X_{28}[1] \oplus ^j_u X'_{28}[1] = 0$ and $^i_v X^*_{28}[1] \oplus ^j_w X'^*_{28}[1] = 0$.
10. Guess $LSB_4(K1[3])$ and $LSB_4(K1[11])$ to compute $LSB_4(SK1[122])$ and its related keys. For each key and remained quartet, check (a) $LSB_4(^i_l X_{30}[5] \oplus ^j_u X'_{30}[5]) = 0$ and (b) $LSB_4(^i_v X^*_{30}[5] \oplus ^j_w X'^*_{30}[5]) = 0$.
11. Guess $MSB_4(K1[3])$ and $MSB_4(K1[11])$ to compute $MSB_4(SK1[122])$ and its related keys. For each key and remained quartet, check (a) $MSB_4(^i_l X_{30}[5] \oplus ^j_u X'_{30}[5]) = 0$ and (b) $MSB_4(^i_v X^*_{30}[5] \oplus ^j_w X'^*_{30}[5]) = 0$.
12. Guess $K1[8]$ to compute $SK1[127]$ and its related keys. For each key and remained quartet, check (a) $^i_l X_{31}[7] \oplus ^j_u X'_{31}[7] = 0$ and (b) $^i_v X^*_{31}[7] \oplus ^j_w X'^*_{31}[7] = 0$.

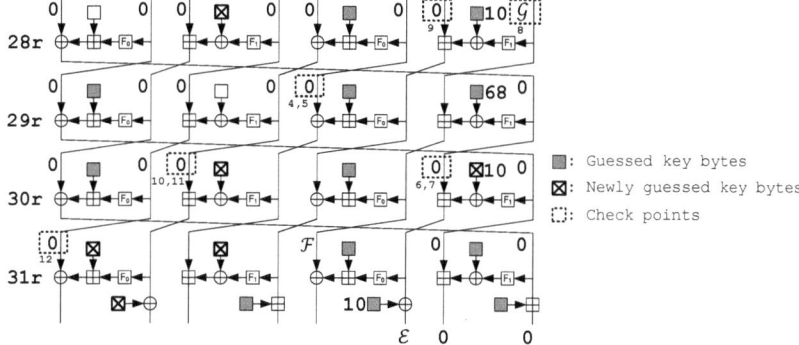

Fig. 4. Key counting procedure. The numbers nearby the check points indicate the corresponding steps from 4 to 12.

Searching with the list

13. If there exist a recorded $K1[0, 1, 2, 3, 5, 6, 8, 9, 10, 11, 12, 13, 14, 15]$ who have 3 or more remaining quartets $(^i_l C, ^i_v C^*, ^j_u C', ^j_w C'^*)$, then exhaustively search the remaining two key bytes for $K1[4]$ and $K1[7]$ with more than two known plaintexts and its corresponding ciphertexts. If a 128-bit key is suggested, output it and its related keys as the keys of encryption oracles of the full rounds of HIGHT, otherwise go to Step 2 with another guess.

An overall view of our related-key rectangle attack is shown in Fig. 5 in Appendix C.

5.2   Complexity Analysis

The probability of 24-round related-key rectangle distinguisher used in our attack is $2^{117.68}$. So, we form the plaintext structures such that we are given $2^{119.68}$ quartets satisfying the input differences of the distinguisher and expect approximately 4 right quartets after guessing $K1[0, 1, 2, 5, 6, 10, 12, 13, 14]$.

Since we use structures S_i includes 2^{40} plaintexts and 16 kinds of input differences of distinguisher(set $\mathcal{A}$ which is defined in Section 4.2) is assumed, we have 2^{44} pairs per structure and we can consider 2^{2m+88} quartets from 2^m structures thus the number of required structures is $2^m = 2^{15.84}$. Hence, our attack requires $2^{15.84+40} = 2^{55.84}$ plaintexts and encrypts them with four encryption oracles defined by four related keys to get $2^{57.84}$ ciphertexts and that is the data complexity of our attack.

In the first step of the attack procedure in the above section, the number of queries to four encryption oracles with related keys is $2^{57.84}$ but this is negligible in total complexity.

We guess 9 bytes of $K1[0, 1, 2, 5, 6, 10, 12, 13, 14]$ in step 2 with a restriction on $K1[6]$ that $SK1[82] \in \mathcal{D}$, so total number of guessed key is 2^{70}. Moreover, an attack procedure and tested quartets when 9 bytes of K are guessed as $K1$ is identical to an attack procedure and tested quartets when 9 bytes of $K \boxplus \Delta^+ K$ are guessed as $K1$, so the total number of key guessing for $K1[0, 1, 2, 5, 6, 10, 12, 13, 14]$ is reduced to 2^{69}.

From step 2-(a) to 2-(d), we explain how we make $2^{119.68}$ quartets from $2^{15.84}$ structures and step (e), we explain how we filter out the wrong quartets. For steps from 2-(a) to 2-(d), we partially encrypt all plaintexts in each structure for 6/128 HIGHT with 4 related keys and choose the pairs satisfying the input differences of distinguisher, so these steps require

$$4 \times 6 \times 2^{-7} \times 2^{55.840+69} \approx 2^{122.425}$$

encryptions and yields two hash tables $\mathcal{H}$ and $\mathcal{I}$ of $2^{59.84}$ ciphertext pairs, respectively. In step 2-(e), we partially decrypt for 1/128 HIGHT and 2/128 HIGHT to compute $({}_l^i X_{31}[3] \oplus {}_u^j X'_{31}[3], \, {}_v^i X^*_{31}[3] \oplus {}_w^j X'^*_{31}[3])$ and $(\Delta T, \Delta T')$, respectively, so these step requires

$$3 \times 2^{-7} \times 2^{57.840+69} \approx 2^{121.425}$$

decryptions.

In step 2-(e)-i, we check that two bytes of ciphertext differences are 0, and another one byte of ciphertext difference is equal to one of 8 elements in $\mathcal{E}$, so the filtering ratio of this step is $2^{-2 \times 2 \times 8 - 2 \times 5} = 2^{-42}$.

In step 2-(e)-ii, ${}_l^i X_{31}[3] \oplus {}_u^j X'_{31}[3]$ and ${}_v^i X^*_{31}[3] \oplus {}_w^j X'^*_{31}[3]$ must be one of 16 elements in $\mathcal{F}$, because $SK1[120] \oplus SK3[120] = SK2[120] \oplus SK4[120] = \texttt{0x10}$, and $SK1[116] \oplus SK3[116] = SK2[116] \oplus SK4[116] = \texttt{0x68}$, so the filtering ratio of this step is $2^{-2 \times 4} = 2^{-8}$.

Since the average ratio of Δx and Δz which satisfy that $\Pr[(\Delta x, 0) \xrightarrow{\boxplus} \Delta z] > 0$ is $2^{-2.82757}$, the filtering ratios of step 2-(e)-iii and iv are both $2^{-5.65514}$, and since the average ratio of Δx, Δy, and Δz which satisfy that $\Pr[(\Delta x, \Delta y) \xrightarrow{\boxplus} \Delta z] > 0$ is $2^{-2.34852}$, the filtering ratio of step 2-(e)-v is $2^{-4.69704}$. Therefore, after filtering steps,

$$2^{119.68-42-8-2\times 5.65514-4.69704} = 2^{53.67268} < 2^{53.68}$$

quartets are left in average.

From step 3 to step 12 are key searching steps with $2^{53.68}$ quartets. The time complexities and number of remained quartets for each step are calculated in Table 3.

Table 3. Complexities of key searching steps

Step	Key guess (bit)	# of Quartets to test	Time Complexity	Elimination Ratio	# of remaining quartets	Key guess (sum,key bit)	# of quartet per a
4-(a)	4	$2^{53.68}$	$2^{1+53.68+4-7}=2^{50.68}$	2^{-4}	$2^{53.68+4-4}=2^{53.68}$	4	$2^{53.68-4}=2^{49.68}$
4-(b)	0	$2^{53.68}$	$2^{1+53.68-7}=2^{46.68}$	2^{-4}	$2^{53.68-4}=2^{49.68}$	4	$2^{49.68-4}=2^{45.68}$
5-(a)	4	$2^{49.68}$	$2^{1+49.68+4-8}=2^{46.68}$	2^{-4}	$2^{49.68+4-4}=2^{49.68}$	8	$2^{49.68-8}=2^{41.68}$
5-(b)	0	$2^{49.68}$	$2^{1+49.68-8}=2^{42.68}$	2^{-4}	$2^{49.68-4}=2^{45.68}$	8	$2^{45.68-8}=2^{37.68}$
6-(a)	4	$2^{45.68}$	$2^{1+45.68+4-7}=2^{43.68}$	$2^{-1.5}$	$2^{45.68+4-1.5}=2^{47.18}$	12	$2^{47.18-12}=2^{35.18}$
6-(b)	0	$2^{47.18}$	$2^{1+47.18-7}=2^{42.18}$	$2^{-1.5}$	$2^{47.18-1.5}=2^{46.68}$	12	$2^{46.68-12}=2^{34.68}$
7-(a)	4	$2^{46.68}$	$2^{1+46.68+4-7}=2^{44.68}$	$2^{-1.5}$	$2^{46.68+4-1.5}=2^{48.18}$	16	$2^{48.18-16}=2^{32.18}$
7-(b)	0	$2^{48.18}$	$2^{1+48.18-7}=2^{43.18}$	$2^{-1.5}$	$2^{48.18-1.5}=2^{47.68}$	16	$2^{46.68-16}=2^{30.68}$
8	0	$2^{47.68}$	$2^{2+47.68-7}=2^{42.68}$	2^{-4}	$2^{47.68-4}=2^{43.68}$	16	$2^{43.68-16}=2^{27.68}$
9	0	$2^{43.68}$	$2^{2+43.68-7}=2^{38.68}$	2^{-16}	$2^{43.68-16}=2^{27.68}$	16	$2^{27.68-16}=2^{11.68}$
10-(a)	8	$2^{27.68}$	$2^{1+27.68+8-7}=2^{29.68}$	$2^{-2.58}$	$2^{27.68+8-2.58}=2^{33.1}$	24	$2^{33.1-24}=2^{9.1}$
10-(b)	0	$2^{33.1}$	$2^{1+33.1-7}=2^{27.1}$	$2^{-2.58}$	$2^{33.1-2.58}=2^{30.52}$	24	$2^{30.52-24}=2^{6.52}$
11-(a)	8	$2^{30.52}$	$2^{1+30.52+8-7}=2^{32.52}$	$2^{-2.58}$	$2^{30.52+8-2.58}=2^{35.94}$	32	$2^{35.94-32}=2^{3.94}$
11-(b)	0	$2^{35.94}$	$2^{1+35.94-7}=2^{29.94}$	$2^{-2.58}$	$2^{35.94-2.58}=2^{33.36}$	32	$2^{33.36-32}=2^{1.36}$
12-(a)	8	$2^{33.36}$	$2^{1+33.36+8-7}=2^{35.36}$	$2^{-5.17}$	$2^{33.36+8-5.17}=2^{36.19}$	40	$2^{35.19-40}=2^{-4.81}$
12-(b)	0	$2^{36.19}$	$2^{1+35.19-7}=2^{30.19}$	$2^{-5.17}$	$2^{36.19-5.17}=2^{31.02}$	40	$2^{31.02-40}=2^{-8.98}$
Total	40		$2^{50.9001}$		$2^{30.74}$	40	$2^{31.02-40}=2^{-8.98}$

Time complexities in Table 3 except step 8 and step 9 are calculated by the early abort technique [11] so these steps are divided into two sub-steps which check that each pair of ciphertexts or intermediate values is valid for a right quartet. The time complexities for each steps are calculated by the multiplications of the number of partially decrypted ciphertexts, the number of remaining quartets from the previous step, the number of guessed keys, and the ratio of partial rounds to HIGHT encryption.

In steps 6, 7, 10, 11, and 12, although we check 8-bit difference. Since a part of quartets are already discarded in the filtering steps by some conditions for the 8-bit difference, we can eliminate with the remaining ratios.

The number of remaining quartets is calculated by multiplication of the number of remaining quartets from the previous step, the number of guessed keys in current step, and the elimination ratio of current step. Using this, we check that how many quartets are counted for each guessed key in average, and if this ratio is significantly less than 1, we can conclude that the right quartets and right key are distinguished from wrong quartets and wrong keys. After step 12, total number of guessed keys is 2^{40} and the number of remaining quartets is $2^{31.02}$, so we expect

$$2^{31.02-40} = 2^{-8.98}$$

quartets are remained for a key in average while more than 3 quartets are remained if guessed key is right key. Thus we have to test quartets until step 12, and the computational complexity from step 4 to step 12 is $2^{50.9001}$.

Therefore, the computational complexity of our related-key rectangle attack for a quarter of key space is

$$2^{122.425} + 2^{121.425} + 2^{69+50.9001} < 2^{123.169},$$

and since the computational complexity of exhaustive key searching for the remaining part of key space is $3 \times 2^{124} = 2^{125.585}$, the total computational complexity of our related-key attack is

$$2^{123.169} + 2^{125.585} = 2^{125.833}.$$

6   Conclusions

In this paper, we find a 24-round related-key rectangle distinguisher using a local-collision property of 4 rounds of HIGHT and extremely deep ladder switch technique when add-differences are used for a relations of keys. This distinguisher can be regards as a 25-round distinguisher because the distinguisher is followed by one round truncated differential trail with probability 1. Based on this distinguisher, we present a related-key rectangle attack on the full rounds of HIGHT for a large weak key space and we consider a related-key attack which is valid for whole key space faster than 2^{126} encryptions required for the exhaustive key searching with 4 related keys. Time complexity of our attack is very marginal and seems to be hard to realize to extract the secret key bits. However, our result gives an evidence for the fact that HIGHT cannot be regarded as an instantiation of the ideal cipher.

References

1. Biham, E.: How to Forge DES-Enhanced Messages in 2^{28} Steps. CS 884 (August 1996)
2. Biham, E., Dunkelman, O., Keller, N.: A Related-Key Rectangle Attack on the Full KASUMI. In: Roy, B. (ed.) ASIACRYPT 2005. LNCS, vol. 3788, pp. 443–461. Springer, Heidelberg (2005)
3. Biryukov, A., Dunkelman, O., Keller, N., Khovratovich, D., Shamir, A.: Key Recovery Attacks of Practical Complexity on AES Variants With Up To 10 Rounds. To appear in EUROCRYPT 2010, Available at Cryptology ePrint Archive, Report 2009/374 (2010), http://eprint.iacr.org/2009/374
4. Biryukov, A., Khovratovich, D.: Related-key cryptanalysis of the full AES-192 and AES-256. In: Matsui, M. (ed.) ASIACRYPT 2009. LNCS, vol. 5912, pp. 1–18. Springer, Heidelberg (2009)

5. Biryukov, A., Khovratovich, D.: Feasible Attack on the 13-round AES-256. Cryptology ePrint Archive, Report 2010/257
6. Chabaud, F., Joux, A.: Differential Collisions in SHA-0. In: Krawczyk, H. (ed.) CRYPTO 1998. LNCS, vol. 1462, pp. 56–71. Springer, Heidelberg (1998)
7. Dunkelman, O., Keller, N., Shamir, A.: A Practical-Time Related-Key Attack on the KASUMI Cryptosystem Used in GSM and 3G Telephony. In: Rabin, T. (ed.) CRYPTO 2010. LNCS, vol. 6223, pp. 393–410. Springer, Heidelberg (2010)
8. Hong, D., Sung, J., Hong, S., Kim, J., Lee, S., Koo, B.-S., Lee, C., Chang, D., Lee, J., Jeong, K., Kim, H., Kim, J., Chee, S.: HIGHT: A New Block Cipher Suitable for Low-Resource Device. In: Goubin, L., Matsui, M. (eds.) CHES 2006. LNCS, vol. 4249, pp. 46–59. Springer, Heidelberg (2006)
9. International Organization for Standardization. ISO/IEC 18033-3:2005. Information technology – Security techniques – Encryption algorithms – Part 3: Block ciphers (2005)
10. Lu, J.: Cryptanalysis of reduced versions of the HIGHT block cipher from CHES 2006. In: Nam, K., Rhee, K. (eds.) ICISC 2007. LNCS, vol. 4817, pp. 11–26. Springer, Heidelberg (2007)
11. Lu, J.: Cryptanalysis of Block Ciphers. PhD thesis, Royal Holloway, University of London, England (July 2008)
12. Lipmaa, H., Moriai, S.: Efficient Algorithms for Computing Differential Properties of Addition. In: Matsui, M. (ed.) FSE 2001. LNCS, vol. 2355, pp. 336–350. Springer, Heidelberg (2002)
13. Özen, O., Varıcı, K., Tezcan, C., Kocair, Ç.: Lightweight block ciphers revisited: Cryptanalysis of reduced round PRESENT and HIGHT. In: Boyd, C., González Nieto, J. (eds.) ACISP 2009. LNCS, vol. 5594, pp. 90–107. Springer, Heidelberg (2009)
14. Vaudenay, S.: When is an Algorithm Legally Broken? Early Symmetric Crypto (ESC) Seminar (January 14, 2010)
15. Zhang, P., Sun, B., Li, C.: Saturation Attack on the Block Cipher HIGHT. In: Garay, J.A., Miyaji, A., Otsuka, A. (eds.) CANS 2009. LNCS, vol. 5888, pp. 76–86. Springer, Heidelberg (2009)

A   Some Flaws in Previous Attack on Reduced Rounds of HIGHT

As mentioned in Section 1, Lu et al. present a related-key rectangle attack on 26 rounds of HIGHT which uses two related-key differential trails for 10 rounds(for $E0$) and 8 rounds(for $E1$) of HIGHT, respectively. Their 10-round related-key differential trail for $E0$ covers rounds from 3 to 12, with the following input and output differences,

$$(0x2a, 0x43, 0x80, 0x0, 0x0, 0x0, 0x0, 0x0) \longrightarrow (0x0, ?, ?, 0x80, 0x0, 0x0, 0x0, 0x0),$$

where the relation of the key is $\Delta K[2] = 0x80$ and $\Delta K[i] = 0x0$ for $i = 0, 1, 3, 4, 5, ..., 15$. They compute the amplified probability $2^{-19.98}$ of $E0$ for some possible values for positions marked by '?', and this probability is computed based on the fact that the probability of the first 1-round differential trail,

$$(0\text{x}2\text{a}, 0\text{x}43, 0\text{x}80, 0\text{x}0, 0\text{x}0, ..., 0\text{x}0) \rightarrow (0\text{x}43, 0\text{x}80, 0\text{x}0, 0\text{x}0, ..., 0\text{x}0, 0\text{x}0)$$

is 2^{-3}. This probability arise from the following equation (4).

$$\Pr_{x,k}[(x \boxplus k) \oplus ((x \oplus 0\text{x}2\text{a}) \boxplus k) = 0\text{x}2\text{a}] = 2^{-3}. \qquad (4)$$

Here, we can observe that for a fixed k, the probability in equation (4) can be different from 2^{-3}, moreover, the probability in equation (4) is 0 for 148 out of 256 keys. Also, the probability of the related-key differential trail for $E1$ has the same flaws. The number of k such that the probability of the first round of $E1$ is 0 is 158.

In block cipher cryptanalysis, a target secret key is assumed to be fixed, so in some cases the related-key differential trails for $E0$ and $E1$ in [10] are not satisfied with suggested probabilities. Thus, the related-key rectangle attack in [10] is regarded as an attack valid only for weak keys.

B   Exhaustive Key Searching in the Related-Key Model

The validity of an attack on a cipher is usually proved through comparison of the complexities with those of an exhaustive key searching in the same attack model. Let $f_1, ..., f_{t-1}$ be simple bijective relations. We assume that we are given t distinct encryption oracles $E_{f_0(K)}, E_{f_1(K)}, ..., E_{f_{t-1}(K)}$ for a related-key tuple $(f_0(K), f_1(K), ..., f_{t-1}(K))$, where f_0 is the identity function. We also assume that the encryption oracle has the block size n and the key space $\mathcal{K}$ has 2^k elements. Especially, we focus on the case of $k/2 \leq n < k$ because our target is HIGHT. In this setting, the exhaustive key searching consists of the following phase.

1. Choose and fix two plaintexts P and P^*, and get the ciphertexts C_i and C_i^* for each encryption oracle $E_{f_i(K)}$.
2. Repeat the following phases.
   (a) Randomly pick one K' of key candidates, compute $C = E_{K'}(P)$, and check whether there exists a C_i such that $C = C_i$.
   (b) If such C_i is found, compute $C^* = E_{K'}(P^*)$, and check whether $C_i^* = C^*$.
   (c) If a match $(C, C^*) = (C_i, C_i^*)$ is found, halt and output K' as the right value of $f_i(K)$. Otherwise, discard $K', f_1^{-1}(K'), ..., f_{t-1}^{-1}(K')$ from search space, and go to (a) and try again.

This is not new and similar approaches were mentioned in [1,14]. We can expect a match is found with $2^{k-\log_2 t}$ trials. The match yields one of $f_0(K), f_1(K), ..., f_{t-1}(K)$ with a high probability. So, the time complexity of this attack is dominated by $2^{k-\log_2 t}$ encryptions. Therefore our attack is forced to have the time complexity less than 2^{126}, since we use $t = 4$ related keys of $k = 128$ bits.

C   An Overall View of Our Related-Key Rectangle Attack

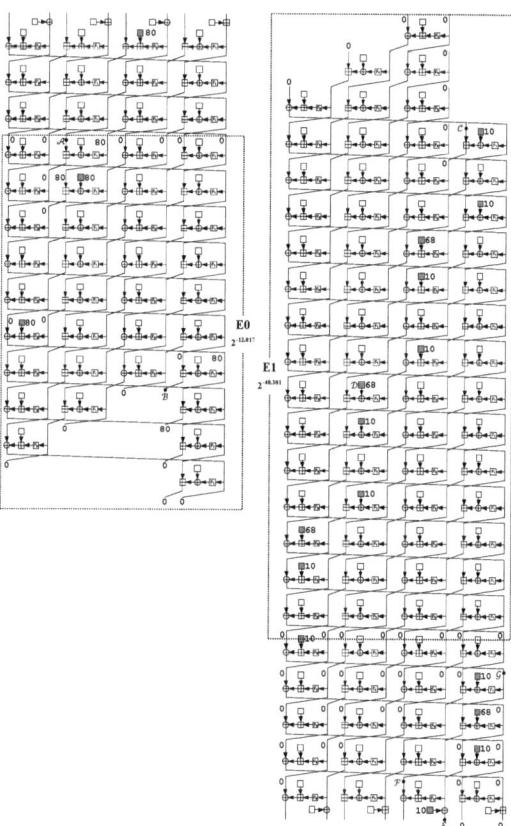

Fig. 5. Related-key differential trails of HIGHT

Preimage Attacks against PKC98-Hash and HAS-V

Yu Sasaki[1], Florian Mendel[2,*], and Kazumaro Aoki[1,2]

[1] NTT, 3-9-11 Midoricho, Musashino-shi, Tokyo 180-8585 Japan
[2] Institute for Applied Information Processing and Communications (IAIK),
Graz University of Technology, Inffeldgasse 16a, A-8010 Graz, Austria

Abstract. We propose preimage attacks against PKC98-Hash and HAS-V. PKC98-Hash is a 160-bit hash function proposed at PKC 1998, and HAS-V, a hash function proposed at SAC 2000, can produce hash values of $128 + 32k$ ($k = 0, 1, \ldots, 6$) bits. These hash functions adopt the Merkle-Damgård and Davies-Meyer constructions. One unique characteristic of these hash functions is that their step functions are not injective with a fixed message. We utilize this property to mount preimage attacks against these hash functions. Note that these attacks can work for an arbitrary number of steps. The best proposed attacks generate preimages of PKC98-Hash and HAS-V-320 in 2^{96} and 2^{256} compression function computations with negligible memory, respectively. This is the first preimage attack against the full PKC98-Hash function.

Keywords: PKC98-Hash, HAS-V, preimage, Davies-Meyer, non-injective step function.

1 Introduction

Cryptographic hash functions (hereafter, simply referred to as hash functions) are indispensable in achieving secure systems such as digital signatures and cryptographic protocols. However, since collisions were found [1] for the widely used hash function MD5 [2], analytic methods against hash functions have been greatly improved and the security of existing hash functions has become doubtful. The security of hash functions is usually evaluated by demonstrating cryptanalysis. Therefore, it is important to analyze hash functions using various structures to understand the security of the existing hash functions and to design secure hash functions.

There are many security requirements for hash functions depending on the usage. Specifically, the following three properties are the most important: collision resistance, second preimage resistance, and preimage resistance.

A hash function was proposed at PKC 1998 [3] without a specific name, and we call it *PKC98-Hash* as it was referred to in [4]. The hash function adopts special features such as a message-dependent rotation. However, the most unique

---

* The work in this paper has been supported by the Austrian Science Fund (FWF), project P21936.

K.-H. Rhee and D. Nyang (Eds.): ICISC 2010, LNCS 6829, pp. 68–91, 2011.

characteristic is that a step function is not injective. This design is very different from many other hash functions adopting a Davies-Meyer mode [5, Algorithm 9.42], which is a block-cipher based construction. Let us compare the behaviors of these two designs when a message input is fixed. When two different chaining variables are input to the standard Davies-Meyer mode, they never collide except for the last feed-forward computation. On the other hand, PKC98-Hash compresses the data in every step. From this observation, the following are interesting objectives on PKC98-Hash from the view of hash function design.

1. Obtaining collisions seem easier than in the standard Davies-Meyer construction.
2. Is it possible to use this property to mount a preimage attack?
3. Is there any advantage to this structure compared to the standard Davies-Meyer mode?

History of the cryptanalysis against PKC98-Hash shows that the first perception is correct. Han *et al.* [6] showed that Boolean functions used in PKC98-Hash do not satisfy the Strict Avalanche Criterion (SAC) which is intended to be satisfied by the designers, and a collision can be found in 2^{30} compression function computations *(hereafter, if the unit of computational complexity is expressed in terms of compression function computations, we omit it for the sake of simplicity. Otherwise, we explicitly write the unit)* if these functions are replaced with any function satisfying the SAC. Chang *et al.* [7] extend this attack and show that a collision can be found in $2^{37.13}$ by making a good differential path for the non-SAC functions. They also showed an example of a pair of colliding messages. Moreover, Mendel *et al.* [8] reduces the complexity to $2^{20.5}$ using the collision finding techniques described in [1].

While several collision attacks have been proposed, only one preimage attack has been proposed so far [4]. It applies a recently developed framework of the meet-in-the-middle preimage attack [9,10,11] to PKC98-Hash, and finds an attack against 80 steps out of 96 steps. However, note that this attack includes a technically vague point[1].

To obtain deep knowledge of a hash construction such as PKC98-Hash, more hash functions with a similar structure should be analyzed. HAS-V [12] is a hash function for this case. HAS-V was proposed as a candidate to meet the requirement of KCDSA [13], which is one of the ISO standard digital signature algorithms. It can output $128 + 32k$ ($k = 0, 1, \ldots, 6$) bits of hash values, and its compression function consists of two copies of the Davies-Meyer compression function. For HAS-V-320, Mendel *et al.* [14] showed that pseudo-near-collisions can be found with a complexity of 1, and preimages can be found with a complexity of 2^{162}. Mouha *et al.* [15] showed some results on finding collisions for simplified variants of HAS-V.

---

[1] The paper simply indicates "step^{-1}" in Section 5.3, and how to compute the inverse of the step function is not mentioned. Because the step function is not injective, we doubt that the step function can easily be inverted consistently with the message schedule.

Table 1. Summary of Current Results Compared to Previous Best Results

Reference	Target	No. steps	Attack Type	Complexity Time	Memory (words)	Preimage length (blocks)
[4]	PKC98-Hash	80	PPI	2^{144}	Negligible	—
			PI	2^{152}	7×2^{16}	2
Sect. 3.1	PKC98-Hash	**Full**	**PPI**	$\mathbf{2^{128}}$	**Negligible**	—
Sect. 3.2	PKC98-Hash	**Full**	**PI**	$\mathbf{2^{96}}$†	**Negligible**	1
Sect. B.1	PKC98-Hash	**Full**	**PI**	$\mathbf{2^{128}}$	**Negligible**	34
[14]	HAS-V-$(128+32k)$ $(k \geq 2)$	Full	PPI	2^{160}	Negligible	—
			PI	2^{162}	20×2^{160}	161
Sect. 4.1	HAS-V-320	Full	**PI**	2^{256}	**Negligible**	1
Sect. B.3	HAS-V-256	Full	**PI**	$\mathbf{2^{249.2}}$	**Negligible**	1
	HAS-V-288	Full	**PI**	$\mathbf{2^{253.5}}$	**Negligible**	1
Sect. 4.2	HAS-V-320	Full	**PPI**	$\mathbf{2^{128}}$	**Negligible**	—
Sect. 4.3	HAS-V-**128 + 32k** $(k < 6)$	Full	**PPI**	$\mathbf{2^{96}}$	**Negligible**	—

PPI: Pseudo-preimage attack, PI: Preimage attack
Bold-face fonts indicate the best results.
†: Success probability of this attack is $1 - e^{-1}$.

In this paper, we show how to exploit the non-injective property of the step function of PKC98-Hash, and propose the first preimage attack against the full specification. Moreover, this attack can work for any number of steps. The best proposed attack generates 1-block preimages with a complexity of 2^{96}. However, for a randomly given target hash value, the attack can generate preimages with a probability of $1 - e^{-1} \approx 0.63$ at most due to the effect of the message-dependent rotation. Note that, as a second-preimage attack, this attack succeeds with high probability if the given first-preimage is two blocks or more. We also show that by combining the non-injective property with a multi-collision technique [16], we can efficiently convert from pseudo-preimages to a preimage. This guarantees that for any given target hash value, preimages can be computed with a complexity of 2^{128}. Second, we apply the techniques to HAS-V. As a result, we can find preimages of one block long with only a negligible sized memory. Moreover, we can find pseudo-preimages of HAS-V-160 and HAS-V-128 while the previous work could not obtain any results on these output sizes. Again, this attack can work for any number of steps.

The analytic approach is totally different from the previous meet-in-the-middle preimage attack [4]. Table 1 summarizes the current results compared to the previous best results. These results show the weakness of non-injective step functions. Note that the attack in Section B.2 shows that the choice of the feed-forward operation sometimes makes the difference in security. It may be useful for hash function designers to take this result into account.

2   Preliminaries

2.1   Specification of PKC98-Hash

PKC98-Hash was proposed by Shin *et al.* [3]. Note that there are unclear parts in [3, Section 3] and we basically follow the interpretation by [4, Section 2.2].

PKC98-Hash is an iterated hash function based on the Merkle-Damgård design principle [5, Algorithm 9.25]. It processes 512-bit message blocks and produces a 160-bit hash value. To ensure that the message length is a multiple of 512 bits, an unambiguous padding method is applied. We refer to [3] for the description of the padding method. Let $M = M_0 \| M_1 \| \cdots \| M_{t-1}$ be a t-block message (after padding). Hash value $H (= H_t)$ is computed by the iteration of compression function CF : $\{0,1\}^{160} \times \{0,1\}^{512} \rightarrow \{0,1\}^{160}$ using recurrence $H_{i+1} \leftarrow \text{CF}(H_i, M_i)$ $(i = 0, \ldots, t-1)$, where H_0 is a predefined initial value. The compression function of PKC98-Hash adopts the Davies-Meyer structure [5, Algorithm 9.42]. It basically consists of two parts: the state update transformation and the message schedule.

Message Schedule. The 512-bit input M_i is divided into 16 words $X_0 \| X_1 \| \cdots \| X_{15}$ of 32 bits. X_j $(j = 16, 17, \ldots, 23)$ is computed with $X_j \leftarrow (X_{j-16} \oplus X_{j-14} \oplus X_{j-9} \oplus X_{j-4})^{\lll 1}$, where $\lll b$ denotes b-bit left rotation. The message input of each step is determined by $w_j \leftarrow X_{\rho^r(j \bmod 24)}$ $(j = 0, 1, \ldots, 95)$, where ρ is a permutation defined in Table 2 and $r \leftarrow \lfloor j/24 \rfloor$.

Table 2. Permutation ρ in PKC98-Hash

j	0	1	2	3	4	5	6	7	8	9	10	11	12	13	14	15	16	17	18	19	20	21	22	23
$\rho(j)$	4	21	17	1	23	18	12	10	5	16	8	0	20	3	22	6	11	19	15	2	7	14	9	13

State Update Transformation. The state update transformation starts from a (fixed) initial value of five 32-bit words and updates them in 96 steps. In each step one message word w_j is used to update the five state variables A_j, B_j, C_j, D_j, and E_j as shown below:

$$\begin{cases} A_{j+1} \leftarrow E_j \\ B_{j+1} \leftarrow (f_r(A_j, B_j, C_j, D_j, E_j) + w_j + K_r)^{\lll s_j} \\ C_{j+1} \leftarrow B_j^{\lll 10} \\ D_{j+1} \leftarrow C_j \\ E_{j+1} \leftarrow D_j \end{cases}, \qquad (1)$$

where r is $\lfloor j/24 \rfloor$, K_r is a constant defined by the specification, and s_j is $X_{\rho^{3-r}(j \bmod 24)} \bmod 2^5$. $f_r(A, B, C, D, E)$ is defined as

$$\begin{cases} f_0 = (A \wedge B) \oplus (C \wedge D) \oplus (B \wedge C \wedge D) \oplus E \\ f_1 = B \oplus ((D \wedge E) \vee (A \wedge C)) \\ f_2 = A \oplus (B \wedge (A \oplus D)) \oplus (((A \wedge D) \oplus C) \vee E) \end{cases}, \qquad (2)$$

where $f_3 = f_1$ and "$\wedge$," "$\oplus$," and "$\vee$" represent bitwise logical AND, XOR, and OR, respectively. We sometimes omit the AND symbol.

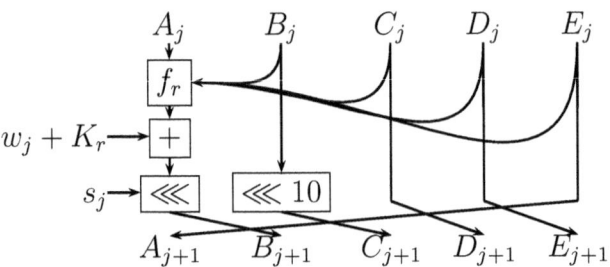

Fig. 1. Step Function in PKC98-Hash

Finally, feed-forward is computed; The initial value $A_0, B_0, \ldots, E_0$ and the output of the last step $A_{96}, B_{96}, \ldots, E_{96}$ are added by wordwise modular addition[2] as an output of the compression function.

2.2 Specification of HAS-V

HAS-V was proposed by Park *et al.* in [12]. It processes 1024-bit message blocks and produces a hash value of $128 + 32k$ ($k = 0, 1, \ldots, 6$) bits. HAS-V also adopts the Merkle-Damgård structure. Again, let $M = M_0 \| M_1 \| \cdots \| M_{t-1}$ be a t-block message (after padding). We refer to the description of the padding method in [12]. Hash value H is computed by the iteration of compression function CF : $\{0,1\}^{320} \times \{0,1\}^{1024} \to \{0,1\}^{320}$ using the recurrence $H_{i+1} \leftarrow \mathrm{CF}(H_i, M_i)$ ($i = 0, 1, \ldots, t-1$) where H_0 is a predefined initial value. Compression function CF updates chaining value H_i in two-parallel lines (left and right) adopting the Davies-Meyer structure where the intermediate values of the two lines are swapped at the end of each round.

Message Schedule. A 1024-bit input M_i is divided into two 512-bit blocks $X \| Y \leftarrow M_i$, and then divided into 2 sets of sixteen words,

$$\begin{cases} X_0 \| X_1 \| \cdots \| X_{15} \leftarrow X \\ Y_0 \| Y_1 \| \cdots \| Y_{15} \leftarrow Y \end{cases}$$

We denote expanded messages to compute step j of the left and right compression functions by $w^{\triangleleft\prime}_j$ and $w^{\triangleright\prime}_j$, respectively. The index of $w^{\triangleleft\prime}_j$ and $w^{\triangleright\prime}_j$ are identical and determined by Table 3, where $w^{\triangleleft\prime}_j$ and $w^{\triangleright\prime}_j$ for $j = 16, 17, 18, 19$ are computed by XOR of 4 message words indexed by Table 4.

---

[2] There is no description of this operation in the original paper. We can interpret it as XOR instead of the modular addition. However, this choice is not important because the best proposed attack can work in both cases. As an independent interest, in Section B.2, we show an attack that works well for addition feed-forward, but not for XOR.

Table 3. Message Ordering in HAS-V (Common between $w^{\vartriangleleft\prime}{}_j$ and $w^{\vartriangleright\prime}{}_j$)

Round	\multicolumn{20}{c}{Step}
	0 1 2 3 4 5 6 7 8 9 10 11 12 13 14 15 116 17 18 19
0	18 0 1 2 3 19 4 5 6 7 16 8 9 10 11 17 12 13 14 15
1	18 3 6 9 12 19 15 2 5 8 16 11 14 1 4 17 7 10 13 0
2	18 12 5 14 7 19 0 9 2 11 16 4 13 6 15 17 8 1 10 3
3	18 7 2 13 8 19 3 14 9 4 16 15 10 5 0 17 11 6 1 12
4	18 15 9 5 3 19 12 8 6 2 16 13 11 7 1 17 14 10 4 0

Table 4. Extra Message Words for HAS-V

j	Round 0	Round 1	Round 2	Round 3	Round 4
16	0,1,2,3	3,6,9,12	12,5,14,7	7,2,13,8	15,9,5,3
17	4,5,6,7	15,2,5,8	0,9,2,11	3,14,9,4	12,8,6,2
18	8,9,10,11	11,14,1,4	4,13,7,15	15,10,5,0	13,11,7,1
19	12,13,14,15	7,10,13,0	8,1,10,3	11,6,1,12	14,10,4,0

Table 5. Amount of Shift for HAS-V Step Function

$j \bmod 20$	0 1 2 3 4 5 6 7 8 9 10 11 12 13 14 15 16 17 18 19
s_j	5 11 7 13 15 6 13 9 5 11 7 12 8 15 13 8 15 6 7 14

State Update Transformation. The state update transformation updates intermediate value $H_i = p^{\vartriangleleft\prime}{}_0 \| p^{\vartriangleright\prime}{}_0$ in two parallel lines using the same step function in both lines:

$$
\begin{cases}
p^{\vartriangleleft\prime}{}_0 \| p^{\vartriangleright\prime}{}_0 & \leftarrow H_i, \\
p^{\vartriangleleft\prime}{}_{j+1} \| p^{\vartriangleright\prime}{}_{j+1} & \leftarrow R_j(p^{\vartriangleleft\prime}{}_j, w^{\vartriangleleft\prime}{}_j) \| R_j(p^{\vartriangleright\prime}{}_j, w^{\vartriangleright\prime}{}_j) \quad (j = 0, 1, \ldots, 99), \\
H_{i+1} & \leftarrow H_i + (p^{\vartriangleleft\prime}{}_{100} \| p^{\vartriangleright\prime}{}_{100}).
\end{cases}
$$

$R_j : \{0,1\}^{160} \times \{0,1\}^{32} \rightarrow \{0,1\}^{160}$ is the step function of HAS-V described below. It is used to update the $p^{\vartriangleleft\prime}{}_j$ and $p^{\vartriangleright\prime}{}_j$ in 5 rounds of 20 steps each using $w^{\vartriangleleft\prime}{}_j$ and $w^{\vartriangleright\prime}{}_j$ computed by the message schedule. $p^{\vartriangleleft\prime}{}_j$ and $p^{\vartriangleright\prime}{}_j$ are swapped after each round to make the two lines dependent on each other.

For each "$\vartriangleleft$" (left side) and "$\vartriangleright$" (right side), step function $R_j(p'_j, w'_j)$ is computed as follows.

$$
\begin{cases}
A_{j+1} \leftarrow
\begin{cases}
A_j^{\lll s_j} + f_r(B_j, C_j, D_j, E_j) + w'_j + K_r & \text{(for left side "$\vartriangleleft$")} \\
A_j^{\lll s_j} + f_{4-r}(B_j, C_j, D_j, E_j) + w'_j + K_r & \text{(for right side "$\vartriangleright$")}
\end{cases} \\
B_{j+1} \leftarrow A_j \\
C_{j+1} \leftarrow B_j^{\ggg 2} \\
D_{j+1} \leftarrow C_j \\
E_{j+1} \leftarrow D_j
\end{cases}
\tag{3}
$$

where r is $\lfloor j/20 \rfloor$, K_r is a constant defined by the specification, and s_j is defined in Table 5.

$f_r(B, C, D, E)$ is defined as follows.

$$\begin{cases} f_0 = BC \oplus (B \oplus 1)D \oplus CE \oplus DE \\ f_1 = BD \oplus C \oplus E \\ f_2 = BC \oplus (B \oplus 1)E \oplus D \end{cases}, \tag{4}$$

where $\mathbf{1}$ represents bit-string, 1^{32}. Moreover, $f_3(B, C, D, E) = f_1(C, B, D, E)$ and $f_4(B, C, D, E) = f_0(B, D, C, E)$.

Finally, feed-forward is computed; The initial values and the output of the last step are added by wordwise modular addition in both lines as an output of the compression function.

When the digest is shorter than 320 bits, output tailoring is applied after the last iteration of the compression function. Table 9 in Appendix A shows the details of the output tailoring of HAS-V.

Alternative Description. For the following analysis, we use the alternative description shown in [14, Section 3]. Let $h_i\|g_i \leftarrow H_i$ and $h(h_i, M_i)$ and $g(g_i, M_i)$ denote the state update function in the left and right stream. CF can be written as

$$\begin{cases} h_{i+1} \leftarrow h_i + g(g_i, M_i) \\ g_{i+1} \leftarrow g_i + h(h_i, M_i) \end{cases}$$

Notations $p^{\triangleleft}{}_j$, $p^{\triangleright}{}_j$, $w^{\triangleleft}{}_j$, and $w^{\triangleright}{}_j$ correspond to the variables indicated with prime notation but the swapping procedure is omitted. That is, $p^{\triangleleft}{}_0\|p^{\triangleright}{}_0 = h_0\|g_0$, $p^{\triangleleft}{}_1 = p^{\triangleleft\prime}{}_1$, and $p^{\triangleleft}{}_{22} = p^{\triangleright\prime}{}_{22}$, for example. Figures 2 and 3 show the graphical explanation of this correspondence.

2.3   Conversion from Pseudo-preimage Attack to Preimage Attack

[5, Fact 9.99] describes a generic conversion from a pseudo-preimage attack to a preimage attack. When the complexity of a pseudo-preimage attack is 2^t, a preimage attack can be constructed in $2^{(n+t)/2+1}$ where n is the number of bits in the internal state length, that is, 160 for PKC98-Hash and 320 for HAS-V. The memory complexity of the converted attack requires the order of $2^{(n-t)/2}$.

When the pseudo-preimage attack satisfies several conditions, more effective conversion can be used such as a tree structure [17], P^3 graph [14,18], and GMTPP [19].

2.4   Pseudo-preimage Attack against HAS-V

Mendel and Rijmen [14, Algorithm 1] proposed an algorithm to invert the compression function of HAS-V as given below.

Input: The final hash value $h_{i+1}\|g_{i+1}$ and an arbitrary intermediate hash value h_i.
Output: Intermediate hash value g_i and message M_i such that $\mathrm{CF}(h_i\|g_i, M_i) = h_{i+1}\|g_{i+1}$.
  1. Guess M_i and calculate: $g_i \leftarrow g_{i+1} - h(h_i, M_i)$.
  2. Check if the following equation holds: $h_{i+1} = h_i + g(g_i, M_i)$.
The time complexity of the attack is expected to be 2^{160}.

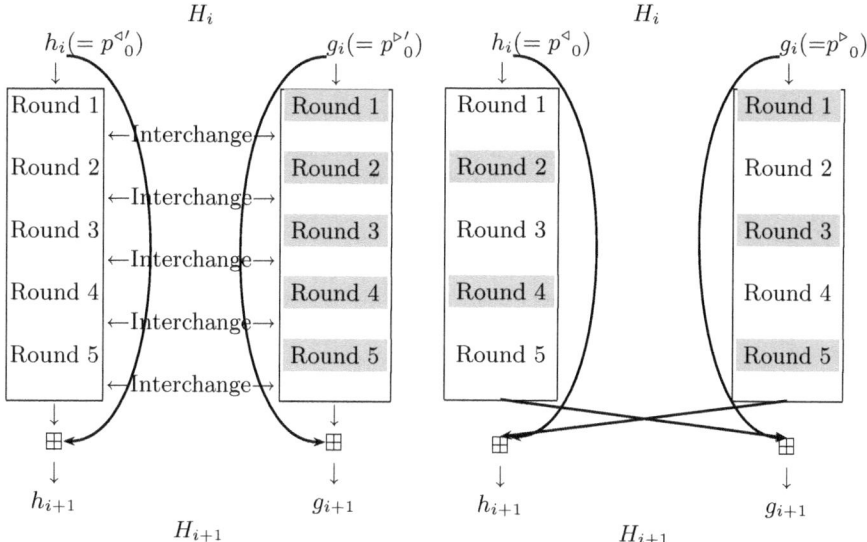

Fig. 2. Original Description of HAS-V Compression Function

Fig. 3. Alternative Description of HAS-V Compression Function

3   Preimage Attacks against PKC98-Hash

This section describes the vulnerability of non-injective step functions. In all proposed attacks, we exploit the following property:

The value of one input chaining variable can be ignored for the remaining computations.

In Section 3.1, we show that this property leads to a simple pseudo-preimage attack with 2^{128} computations. Then, in Section 3.2, we apply this idea to generate preimages of 1-block long with 2^{96} computations. However, due to the effect of the message-dependent rotation, this attack only works with the probability of $1 - e^{-1}$ for a randomly given target.

We also show that non-injective step functions enable attackers to convert pseudo-preimages to a preimage by performing a pre-computation with a lower complexity level. Because the best proposed attack in Section 3.2 directly generates preimages, the impact of this attack is limited, and thus we explain it only in the appendix. However, explaining this conversion will be useful to learn bad behaviors of non-injective step functions. In addition, by combining this conversion with the results of Section 3.1, we can generate preimages for any given target value with 2^{128} computations. In Appendix B.1, we explain this conversion that combines the ideas of the weak property of non-injective step functions with the multi-collision technique [16]. Then, in Appendix B.2, we show another efficient conversion that works well only if the feed-forward operation is modular addition, not XOR.

3.1 Basic Idea and Its Application for Pseudo-preimage Attack

Equation (1) shows that register A is only used for the input of f_r inside the step function. Moreover, for $f_0 = (A \wedge B) \oplus (C \wedge D) \oplus (B \wedge C \wedge D) \oplus E$ in Eq. (2), A does not impact the output when $B = 0$. Therefore, if B_0 is set to 0, $A_{96}\|B_{96}\|C_{96}\|D_{96}\|E_{96}$ can be computed regardless of the value of A_0. By utilizing these properties, we develop a pseudo-preimage attack against compression function CF of PKC98-Hash as follows.

For given hash value $H_A\|H_B\|H_C\|H_D\|H_E$,

1. Let the input of CF be $A_0\|B_0\|C_0\|D_0\|E_0$. Set B_0 to 0 and C_0, D_0, and E_0 to randomly fixed values. Note that $A_{96}\|B_{96}\|C_{96}\|D_{96}\|E_{96}$ can be computed without knowing the value of A_0.
2. Randomly generate message M which is the input of CF, and compute the output. Confirm whether or not the output agrees with the given hash value without H_A, that is $H_B\|H_C\|H_D\|H_E$.
3. If it agrees, set A_0 as $H_A - A_{96}$, and we have a pseudo-preimage;

$$H_A\|H_B\|H_C\|H_D\|H_E = \mathrm{CF}(A_0\|B_0\|C_0\|D_0\|E_0, M)$$

Since the success probability of Step 2 is about 2^{-128} $(= (2^{-32})^4)$, we have a pseudo-preimage of CF with high probability by iterating the above procedure 2^{128} times. Although this attack can be converted to a preimage attack, we only describe it in Appendices because it is not the main purpose of this subsection.

3.2 One-Block Attack Utilizing More Weakness

This attack uses two major improved techniques compared to the basic attack in Section 3.1.

Start from real IV. The attack in Section 3.1 requires that B_0 be fixed to 0 to ignore the impact of the value of A_0. Because register B of the initial value of PKC98-Hash is not 0, the attack cannot generate 1-block long preimages. We solve this problem by using an ignoring property in the latter step of CF. Namely, we set an intermediate variable to the form of $(x\|c_B\|c_C\|c_D\|c_E)$, where x is an unfixed value and c_B, c_C, c_D, and c_E are constant values making the output of the f_r function in this step independent of x. Then, the remaining steps become independent of x and we use the freedom degrees of message words so that a given hash value can be computed from $(x\|c_B\|c_C\|c_D\|c_E)$. Finally, the attack complexity is the one that achieves $(x\|c_B\|c_C\|c_D\|c_E)$, which is 2^{128} and faster than the brute force attack.

Reduce the complexity more. The attack in Section 3.1 only makes one chaining variable (A_0) irrelevant to the remaining computations. This limits the advantage of the attack to at most 2^{32}. We solve this problem by continuing to ignore the intermediate variable in several steps. Remember that the f_r function for the last round is $f_1(A, B, C, D, E) = B \oplus ((D \wedge E) \vee (A \wedge C))$. We observe that $f_1(x, 0, 0, 0, y)$ returns 0 regardless of the values of x and y. At the next step, the input of f_1 function can be written as $(y, c_B, 0, 0, 0)$,

Table 6. Intermediate Variables in the Last 5 Steps

j	A_j	B_j	C_j	D_j	E_j	w_j	s_j
91	x	0	0	0	y	X_{19}	X_{19}
92	y	$A_{96}^{\ggg 10}$	0	0	0	X_8	X_{20}
93	0	$E_{96}^{\ggg 10}$	A_{96}	0	0	X_9	X_{21}
94	0	$D_{96}^{\ggg 10}$	E_{96}	A_{96}	0	X_{11}	X_{22}
95	0	$C_{96}^{\ggg 10}$	D_{96}	E_{96}	A_{96}	X_1	X_{23}
96	A_{96}	B_{96}	C_{96}	D_{96}	E_{96}		

where c_B is a fixed constant. $f_1(y, c_B, 0, 0, 0)$ returns c_B regardless of y. Finally, the attack complexity can be improved to that achieving $(x, 0, 0, 0, y)$ at an intermediate step, which is 2^{96}.

By considering the above two improvements, we set the value of $(A_{91}, B_{91}, C_{91}, D_{91}, E_{91})$ to $(x, 0, 0, 0, y)$ and adjust each variable for a fixed target during the last five steps. Table 6 shows how to control the last five steps, and Eq. (5) shows the message schedule for the last five steps.

$$\begin{cases} X_{19} = (X_3 \oplus X_5 \oplus X_{10} \oplus X_{15})^{\lll 1} \\ X_{20} = (X_4 \oplus X_6 \oplus X_{11} \oplus (X_0 \oplus X_2 \oplus X_7 \oplus X_{12})^{\lll 1})^{\lll 1} \\ X_{21} = (X_5 \oplus X_7 \oplus X_{12} \oplus (X_1 \oplus X_3 \oplus X_8 \oplus X_{13})^{\lll 1})^{\lll 1} \\ X_{22} = (X_6 \oplus X_8 \oplus X_{13} \oplus (X_2 \oplus X_4 \oplus X_9 \oplus X_{14})^{\lll 1})^{\lll 1} \\ X_{23} = (X_7 \oplus X_9 \oplus X_{14} \oplus (X_3 \oplus X_5 \oplus X_{10} \oplus X_{15})^{\lll 1})^{\lll 1} \end{cases} \quad (5)$$

Based on Table 6, we develop the following procedure for given hash value $H_A \| H_B \| H_C \| H_D \| H_E$.

1. Set $A_{96} \leftarrow H_A - A_0$, $B_{96} \leftarrow H_B - B_0$, $C_{96} \leftarrow H_C - C_0$, $D_{96} \leftarrow H_D - D_0$, and $E_{96} \leftarrow H_E - E_0$, where $A_0 \| B_0 \| C_0 \| D_0 \| E_0 (= H_0)$ is the initial value.
2. Set A_j, B_j, C_j, D_j, and E_j as specified in Table 6 for $j = 95, 94, \ldots, 91$, where x and y are unfixed values. Compute $w_j \leftarrow B_{j+1}^{\ggg s_j} - K_3 - f_1(A_j, B_j, C_j, D_j, E_j)$ for $j = 95, 94, 93, 92$ to achieve these values. Remember that $f_1(x, 0, 0, 0, y)$ and $f_1(y, A_{96}^{\ggg 10}, 0, 0, 0)$ are fixed constants regardless of the values of x and y. On s_{92}, s_{93}, s_{94}, and s_{95}, we set arbitrary values.
3. For Step 91, by exhaustively trying the least significant 5 bits of $w_{91} (= s_{91} = X_{19})$, check if there exists X_{19} satisfying the following with respect to the least significant 5 bits. Fix the least significant 5 bits and compute $X_{19} = (A_{96}^{\ggg 10})^{\ggg X_{19} \bmod 2^5} - f_1(A_j, B_j, C_j, D_j, E_j) - K_3$. If the least significant 5 bits of the newly computed X_{19} matches the fixed value, the solution exists. Otherwise, the attack fails.
4. Set X_{13}, X_{14}, and X_{15} as the appropriate padding rule for a 1 block message. Now, we determine the full bits of X_1, X_8, X_9, X_{11}, X_{13}, X_{14}, X_{15}, and X_{19}.
5. We have not yet determined X_0, X_2, X_3, X_4, X_5, X_6, X_7, X_{10}, and X_{12}, but we have the constraint for X_{19}. So, we still have $32 \times 8 = 256$ free bits. Equation (5) specifies 5×4 bits for s_j for $j = 95, 94, 93, 92$. In total, we have $256 - 5 \times 4 = 236$ free bits by solving the linear system in GF(2) of Eq. (5).

Note $X_{19}, X_{20}, \ldots, X_{23}$ include at least one non-determined message word, for example, $X_3, X_4, \ldots, X_7$.

6. For each message candidate, compute $A_{91} \| B_{91} \| C_{91} \| D_{91} \| E_{91}$. This will match the form of $* \| 0 \| 0 \| 0 \| *$ with probability 2^{-96}. Hence, we expect to find a preimage in 2^{96}.

At Step 3, the success probability of the check is approximately $1 - (1 - \frac{1}{32})^{32} \approx 1 - e^{-1}$. In fact, for 2^{32} candidates of A_{96}, we experimentally confirmed the success probability of the check. As a result, 2745131115 points out of 2^{32} points passed this check. This is almost the same, but strictly speaking it is 1.11% better than the original approximation: $2^{32} \cdot (1 - e^{-1}) \approx 2714937127$.

4   Preimage Attack against HAS-V

Mendel and Rijmen [14] proposed a preimage attack on HAS-V-$(128 + 32k)$ $(k \geq 2)$ with 2^{162} in time and 20×2^{160} in memory. The generated preimages are 161 blocks long.

Equation (3) shows that, inside the step function of HAS-V, register E of an intermediate variable is only used for the input of f_r. Hence, the same approach as PKC98-Hash is possible. In Section 4.1, we propose a 1-block preimage attack on HAS-V with a complexity of 2^{256} compression function operations by using an idea similar to that in Section 3.2. Due to its complexity, the attack can be applied to 320-bit and 288-bit outputs. Moreover, if the details of the output tailoring function are carefully analyzed, the complexity for a 288-bit output can be slightly improved, which takes $2^{253.5}$, and we can attack a 256-bit output faster than 2^{256}, which takes $2^{249.2}$. Because the analysis on the output tailoring function is too complicated and too specific for HAS-V, we show these optimization only in Appendix B.3. In Section 4.2, we propose the basic idea to obtain the best pseudo-preimage attack on HAS-V by using the idea in Section 4.1. Finally, in Section 4.3, we further reduce the complexity for tailored output sizes by carefully considering the output tailoring function. Note that these attacks are the first results on HAS-V-160 and HAS-V-128.

4.1   One-Block Attack

As was done in Section 3.2, we set an intermediate variable (input to Step 98) to the form of $(c_A \| c_B \| c_C \| c_D \| *_E)$ for both left and right sides, where $*_E$ is an unfixed value and c_A, c_B, c_C, and c_D are constant values making the output of f_r independent of $*_E$. Note that, for the left and right sides, $f_r(B, C, D, E)$ for the last round is $BC \oplus (B \oplus 1)D \oplus CE \oplus DE$ and $BD \oplus (B \oplus 1)C \oplus CE \oplus DE$, respectively. Both of them are independent of E as long as $C = D$ is satisfied. For a random message, the form $(c_A \| c_B \| c_C \| c_D \| *_E)$ is satisfied with a probability of 2^{-128} on each side. Hence, the attack will succeed in $2^{256} (= (2^{128})^2)$. The detailed attack procedure is as follows.

Table 7. Intermediate Variables in the Last 2 Steps

j	$A^{\triangleleft}_{j}$	$B^{\triangleleft}_{j}$	$C^{\triangleleft}_{j}$	$D^{\triangleleft}_{j}$	$E^{\triangleleft}_{j}$	$A^{\triangleright}_{j}$	$B^{\triangleright}_{j}$	$C^{\triangleright}_{j}$	$D^{\triangleright}_{j}$	$E^{\triangleright}_{j}$	$w^{\triangleleft}_{j}$	$w^{\triangleright}_{j}$
98	$C^{\triangleleft\lll 2}_{100}$	$D^{\triangleleft\lll 2}_{100}$	$E^{\triangleleft}_{100}$	$E^{\triangleleft}_{100}$	x	$C^{\triangleright\lll 2}_{100}$	$D^{\triangleright\lll 2}_{100}$	$E^{\triangleright}_{100}$	$E^{\triangleright}_{100}$	y	X_4	Y_4
99	$B^{\triangleleft}_{100}$	$C^{\triangleleft\lll 2}_{100}$	$D^{\triangleleft}_{100}$	$E^{\triangleleft}_{100}$	$E^{\triangleleft}_{100}$	$B^{\triangleright}_{100}$	$C^{\triangleright\lll 2}_{100}$	$D^{\triangleright}_{100}$	$E^{\triangleright}_{100}$	$E^{\triangleright}_{100}$	X_0	Y_0
100	$A^{\triangleleft}_{100}$	$B^{\triangleleft}_{100}$	$C^{\triangleleft}_{100}$	$D^{\triangleleft}_{100}$	$E^{\triangleleft}_{100}$	$A^{\triangleright}_{100}$	$B^{\triangleright}_{100}$	$C^{\triangleright}_{100}$	$D^{\triangleright}_{100}$	$E^{\triangleright}_{100}$		

1. For a given hash value and initial value H_0, compute $A^{\triangleleft}_{100}, B^{\triangleleft}_{100}, \ldots, E^{\triangleright}_{100}$ by reversely applying the feed-forward operation.
2. Compute $A^{\triangleleft}_{j}, B^{\triangleleft}_{j}, \ldots, E^{\triangleright}_{j}$ for $j = 98, 99$ as specified in Table 7, and message words X_0, Y_0, X_4, and Y_4 by solving the first assignment line for w'_j in Eq. (3).
3. Set Y_{13}, Y_{14}, and Y_{15} as an appropriate padding rule for a 1-block message.
4. Set non-specified message words X_j and Y_j randomly, compute Steps 0 to 97, and confirm whether or not the output of Step 97 $p^{\triangleleft}_{98}\|p^{\triangleright}_{98}$ matches

$$C^{\triangleleft\lll 2}_{100}\|D^{\triangleleft\lll 2}_{100}\|E^{\triangleleft}_{100}\|E^{\triangleleft}_{100}\| * \|C^{\triangleright\lll 2}_{100}\|D^{\triangleright\lll 2}_{100}\|E^{\triangleright}_{100}\|E^{\triangleright}_{100}\| *$$

The match will occur with the probability 2^{-256}, and we expect to find a preimage in 2^{256}.

Since the output tailoring function of HAS-V can easily be inverted, we can also compute 1-block preimages of HAS-V-288 with the complexity of 2^{256}.

4.2   A Pseudo-preimage Attack

Section 2.4 describes a pseudo-preimage attack in 2^{160}. We further reduce the complexity by combining the idea described in Section 4.1. Note that this attack is a pseudo-preimage attack and thus the goal is finding a pair of $h_i\|g_i$ and M_i which generates a given hash value $h_{i+1}\|g_{i+1}$ (g_i can be given instead of being chosen by the attacker). This attack uses the left half of Table 7.

1. For given hash value $h_{i+1}\|g_{i+1}$ and right half of input g_i, compute $A^{\triangleleft}_{100}$, $B^{\triangleleft}_{100}, \ldots, E^{\triangleleft}_{100}$ (only the left side) by reversely applying the feed-forward operation, namely $g_{i+1} - g_i$.
2. Set $A^{\triangleleft}_{j}, B^{\triangleleft}_{j}, \ldots, E^{\triangleleft}_{j}$ for $j = 98, 99$ as specified in Table 7, so that $A^{\triangleleft}_{100}$, $B^{\triangleleft}_{100}, \ldots, E^{\triangleleft}_{100}$ can be achieved for any unfixed value x. Compute X_0 and X_4 using Eq. (3).
3. Set Y_{13}, Y_{14}, and Y_{15} as an appropriate padding rule for a 1-block message.
4. Generate non-specified message words X_j and Y_j randomly, and compute the following.

$$
\begin{aligned}
p^{\triangleright}_{0} &\leftarrow g_i, \\
p^{\triangleright}_{j+1} &\leftarrow R(p^{\triangleright}_{j}, w^{\triangleright}_{j}) &&\text{for } j = 0, 1, \ldots, 99, \\
p^{\triangleleft}_{0} &\leftarrow h_{i+1} - p^{\triangleright}_{100}, \\
p^{\triangleleft}_{j+1} &\leftarrow R(p^{\triangleleft}_{j}, w^{\triangleleft}_{j}) &&\text{for } j = 0, 1, \ldots, 97.
\end{aligned}
$$

Table 8. Two Neutral Words Allocation

j	$A^{\triangleleft}_j$	$B^{\triangleleft}_j$	$C^{\triangleleft}_j$	$D^{\triangleleft}_j$	$E^{\triangleleft}_j$	$w^{\triangleleft}_j$
98	$c_0 \oplus 1$	y	c_0	c_0	x	X_4
99		$c_0 \oplus 1$	$y^{\ggg 2}$	c_0	c_0	X_0
100			$(c_0 \oplus 1)^{\ggg 2}$	$y^{\ggg 2}$	c_0	
	$A^{\triangleright}_j$	$B^{\triangleright}_j$	$C^{\triangleright}_j$	$D^{\triangleright}_j$	$E^{\triangleright}_j$	$w^{\triangleright}_j$
0		c_1	$c_1^{\ggg 2}$	y'	c_1	Y_0
1			$c_1^{\ggg 2}$	$c_1^{\ggg 2}$	y'	Y_1

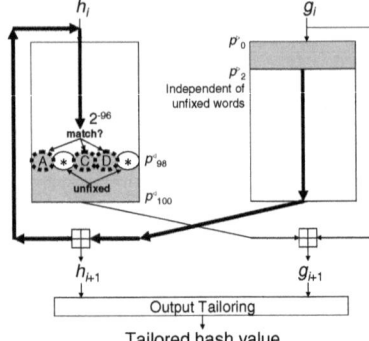

Fig. 4. Pseudo-Preimages on Tailored HAS-V

5. If the newly computed $p^{\triangleleft}_{98}$ matches the form of $C^{\triangleleft \lll 2}_{100} \| D^{\triangleleft \lll 2}_{100} \| E^{\triangleleft}_{100} \| E^{\triangleleft}_{100} \| *$, we obtain a pseudo-preimage. Otherwise, go back to Step 4.

The success probability of Step 5 is 2^{-128}, and we expect to find a pseudo-preimage in 2^{128}.

Since we can easily invert the output tailoring function, we expect to find pseudo-preimages faster than with a brute-force search for all output lengths of HAS-V except for the 128-bit output.

4.3   Pseudo-preimage Attacks with Output Tailoring

Regarding HAS-V with shorter hash lengths, we can further reduce the complexity of the pseudo-preimage attack in Section 4.2 by using the redundancy introduced in the output tailoring function.

Overall Strategy. Similar to Section 4.2, we make unfixed words in the last several steps of the left line ($p^{\triangleleft}_{98}$). However, different from Section 4.2, we make two unfixed words. The overall strategy is as follows, which is also illustrated in Fig. 4.

- Set two unfixed words on registers B and E of $p^{\triangleleft}_{98}$.
- Set intermediate variables in $p^{\triangleleft}_{98}$, $p^{\triangleleft}_{99}$, $p^{\triangleleft}_{100}$, $p^{\triangleright}_0$, and $p^{\triangleright}_1$ so that $p^{\triangleright}_2$ and the following computations can be carried out independently of the values of unfixed words.
- For a randomly chosen message, compute $p^{\triangleright}_2 \rightarrow p^{\triangleright}_{100} \rightarrow p^{\triangleright}_0 \rightarrow p^{\triangleleft}_{98}$. Finally check that the registers of A, C, and D of the computed $p^{\triangleleft}_{98}$ match the values of $p^{\triangleleft}_{98}$ set in advance.

Computations from $p^{\triangleleft}_{98}$ to $p^{\triangleright}_2$ and Use of Output Tailoring Function. Table 8 describes the intermediate variables at the last 3 steps of the left line and the first 2 steps of the right line. c_0 and c_1 are constants to be determined by the method below and x and y are unfixed variables. y' is a variable that changes depending on y. The variables in Table 8 guarantee that two unfixed words (registers B and E of $p^{\triangleleft}{}_{98}$) never impact $p^{\triangleright}{}_2$ regardless of their values. We explain this computation step by step.

- $f_4(B, C, D, E)$ for the left line is $BD \oplus (B \oplus 1)C \oplus DE \oplus CE$.

For $R(p^{\triangleleft}{}_{98}, w^{\triangleleft}{}_{98})$**:** $f_r(y, c_0, c_0, x) = (y \wedge c_0) \oplus ((y \oplus 1) \wedge c_0) \oplus (c_0 \wedge x) \oplus (c_0 \wedge x) = c_0 \oplus 0 = c_0$.

For $R(p^{\triangleleft}{}_{99}, w^{\triangleleft}{}_{99})$**:** $f_r(c_0 \oplus 1, y^{\ggg 2}, c_0, c_0) = ((c_0 \oplus 1) \wedge c_0) \oplus (c_0 \wedge y^{\ggg 2}) \oplus (c_0 \wedge c_0) \oplus (y^{\ggg 2} \wedge c_0) = c_0$.

- Assume that $p^{\triangleright}{}_0$ in Table 8 can be achieved from $p^{\triangleleft}{}_{98}$ in Table 8. We later show that this assumption is reasonable. The Boolean function of the right line for the first round, f_r, is the same as f_4 for the left line, which is $BD \oplus (B \oplus 1)C \oplus DE \oplus CE$.

For $R(p^{\triangleright}{}_0, w^{\triangleright}{}_0)$**:** $f_r(c_1, c_1^{\ggg 2}, y', c_1) = (c_1 \wedge y') \oplus ((c_1 \oplus 1) \wedge c_1^{\ggg 2}) \oplus (y' \wedge c_1) \oplus (c_1^{\ggg 2} \wedge c_1) = c_1^{\ggg 2}$.

For $R(p^{\triangleright}{}_1, w^{\triangleright}{}_1)$**:** $f_r(*, c_1^{\ggg 2}, c_1^{\ggg 2}, y') = (* \wedge c_1^{\ggg 2}) \oplus ((* \oplus 1) \wedge c_1^{\ggg 2}) \oplus (c_1^{\ggg 2} \wedge y') \oplus (c_1^{\ggg 2} \wedge y') = c_1^{\ggg 2}$.

Finally, we can conclude that the two unfixed words of $p^{\triangleleft}{}_{98}$ are always ignored in the f_r function in each step, and thus do not impact $p^{\triangleright}{}_2$.

The remaining problem is how to satisfy the above assumption, namely, how we choose variables c_0 and c_1 so that they are consistent with the transformation from $p^{\triangleleft}{}_{100}$ to $p^{\triangleright}{}_0$. Specifically, we need to make sure that $C^{\triangleleft}{}_{100} + C^{\triangleright}{}_0$, which is $(c_0 \oplus 1)^{\ggg 2} + c_1^{\ggg 2}$, is equal to the third word of g_{i+1} and, at the same time, $E^{\triangleleft}{}_{100} + E^{\triangleright}{}_0$, which is $c_0 + c_1$ is equal to the fifth word of g_{i+1} (We do not have to consider register D because y' is a variable that can change depending on the value of y). This is usually impossible if g_{i+1} is a given and a fixed target value. Therefore, this attack is impossible for HAS-V-320. However, for the shorter outputs, there exist multiple candidates of g_{i+1} that can produce the given short hash values. This redundancy introduced by the output tailoring function enables us to satisfy the constraints on registers C and E simultaneously.

Satisfying Constraints Between $p^{\triangleleft}{}_{100}$ and $p^{\triangleright}{}_0$. Let us denote five words of h_{i+1} and g_{i+1} by $A^{\triangleleft}\|B^{\triangleleft}\|C^{\triangleleft}\|D^{\triangleleft}\|E^{\triangleleft}$ and $A^{\triangleright}\|B^{\triangleright}\|C^{\triangleright}\|D^{\triangleright}\|E^{\triangleright}$, respectively. The output tailoring function is shown in Table 9. The goal here is to find a tuple of values $(c_0, c_1, C^{\triangleright}, E^{\triangleright})$ that can produce the given target hash value of a reduced size and satisfy equations $(c_0 \oplus 1)^{\ggg 2} + c_1^{\ggg 2} = C^{\triangleright}$ and $c_0 + c_1 = E^{\triangleright}$.

We start with an observation: $C^{\triangleright}$ *may not have freedom depending on the output size.* For example, in a 256-bit case, 8 LSBs of $C^{\triangleright}$ can take any value by choosing $E^{\triangleleft}{}_{[15-8]}$ so that $C^{\triangleright}{}_{[7-0]} + E^{\triangleleft}{}_{[15-8]}$ can be equal to the 8 LSBs of the 6th word of the given target value. However, the other 24 bits of $C^{\triangleright}$ must be fixed uniquely. Therefore, in this attack procedure, we regard that $C^{\triangleright}$ is fixed by the given target hash value (Even if $C^{\triangleright}$ has the freedom, we fix $C^{\triangleright}$ to one of the possible values and do not use the freedom). Hence, if we determine the value of c_0 (resp. c_1), the value of c_1 (resp. c_0) will be fixed by $(c_0 \oplus 1)^{\ggg 2} + c_1^{\ggg 2} = C^{\triangleright}$. This also fixes the value of $E^{\triangleright}$ by $c_0 + c_1 = E^{\triangleright}$. Then, we have another observation: *For a given hash value, $E^{\triangleright}$ can take any value for* *any output sizes.* Table 9 shows that for all output sizes, $E^{\triangleright}$ is always added by other variables. Hence, for any value of $E^{\triangleright}$, we can produce the given target hash value by appropriately choosing values of these added variables.

Finally, the constraints on registers C and E can always be satisfied and the variables in Table 9 can always be set.

Attack Procedure. The detailed attack procedure for the tailored HAS-V is as follows.

1. Fix the value of $C^{\triangleright}$ so that it can be consistent with the given target hash value. Then, choose arbitrary value c_0 and compute c_1 by using $(c_0 \oplus 1)^{\ggg 2} + c_1^{\ggg 2} = C^{\triangleright}$ and $E^{\triangleright}$ by using $c_0 + c_1 = E^{\triangleright}$.
2. Compute $A^{\triangleleft}$, $B^{\triangleleft}$, ..., $E^{\triangleleft}$, $A^{\triangleright}$, $B^{\triangleright}$, and $D^{\triangleright}$. We can choose any possible value as long as they produce the target hash value. Moreover, compute X_4 to satisfy $B^{\triangleright}{}_0 = c_1$.
3. Randomly generate message words except for X_4 and intermediate variables between the left line of Step 98 and the right line of Step 1 that are not determined yet, and compute the following:

$$
\begin{array}{ll}
p^{\triangleleft}{}_{j+1} \leftarrow R(p^{\triangleleft}{}_j, w^{\triangleleft}{}_j) & \text{for } j = 99 \\
p^{\triangleright}{}_{j+1} \leftarrow R(p^{\triangleright}{}_j, w^{\triangleright}{}_j) & \text{for } j = 0, 1, \ldots, 99 \\
p^{\triangleleft}{}_0 \quad \leftarrow (A^{\triangleleft}\|B^{\triangleleft}\|C^{\triangleleft}\|D^{\triangleleft}\|E^{\triangleleft}) - p^{\triangleright}{}_{100} & \\
p^{\triangleleft}{}_{j+1} \leftarrow R(p^{\triangleleft}{}_j, w^{\triangleleft}{}_j) & \text{for } j = 0, 1, \ldots, 97
\end{array}
$$

4. If the newly computed $p^{\triangleleft}{}_{98}$ matches the form of $(c_0 \oplus 1 \| * \| c_0 \| c_0 \| *)$, we find a pseudo-preimage. Otherwise, go back to Step 3.

The success probability of Step 4 is 2^{-96}, and we expect to find a pseudo-preimage in 2^{96}.

5   Conclusions

This paper presented attacks against Davies-Meyer hash functions with non-injective "block cipher", including PKC98-Hash and HAS-V. The best attacks in the paper are: a preimage attack on PKC98-Hash in time 2^{96}, a preimage attack on HAS-V in time 2^{256}, and a pseudo-preimage attack on tailored HAS-V in time 2^{96}. All of these attacks only require a negligible amount of memory.

The non-injective property of the step function enables attackers to cancel a part of the input chaining variable of the compression function. This property leads to a simple pseudo-preimage attack and an efficient conversion method from pseudo-preimages to a preimage. Moreover, the attacks work for any number of steps. Therefore, we suggest that hash function designers should avoid the Davies-Meyer construction with non-injective state-update functions.

References

1. Wang, X., Yu, H.: How to break MD5 and other hash functions. In: Cramer, R. (ed.) EUROCRYPT 2005. LNCS, vol. 3494, pp. 19–35. Springer, Heidelberg (2005)
2. Rivest, R.L.: Request for Comments 1321: The MD5 Message Digest Algorithm. The Internet Engineering Task Force (1992),
   http://www.ietf.org/rfc/rfc1321.txt
3. Shin, S.U., Rhee, K.H., Ryu, D.H., Lee, S.J.: A new hash function based on MDx-family and its application to MAC. In: Imai, H., Zheng, Y. (eds.) PKC 1998. LNCS, vol. 1431, pp. 234–246. Springer, Heidelberg (1998)
4. Hong, D., Koo, B., Kim, W.H., Kwon, D.: Preimage attacks on reduced steps of ARIRANG and PKC98-hash. In: Lee, D., Hong, S. (eds.) ICISC 2009. LNCS, vol. 5984, pp. 315–331. Springer, Heidelberg (2010)
5. Menezes, A.J., van Oorschot, P.C., Vanstone, S.A.: Handbook of applied cryptography. CRC Press, Boca Raton (1997)
6. Han, D., Park, S., Chee, S.: Cryptanalysis of the modified version of the hash function proposed at PKC'98. In: Daemen, J., Rijmen, V. (eds.) FSE 2002. LNCS, vol. 2365, pp. 252–262. Springer, Heidelberg (2002)
7. Chang, D., Sung, J., Sung, S., Lee, S., Lim, J.: Full-round differential attack on the original version of the hash function proposed at PKC'98. In: Nyberg, K., Heys, H.M. (eds.) SAC 2002. LNCS, vol. 2595, pp. 160–174. Springer, Heidelberg (2003)
8. Mendel, F., Pramstaller, N., Rechberger, C.: Improved collision attack on the hash function proposed at PKC'98. In: Rhee, M.S., Lee, B. (eds.) ICISC 2006. LNCS, vol. 4296, pp. 8–21. Springer, Heidelberg (2006)
9. Aoki, K., Sasaki, Y.: Meet-in-the-middle preimage attacks against reduced SHA-0 and SHA-1. In: Halevi, S. (ed.) CRYPTO 2009. LNCS, vol. 5677, pp. 70–89. Springer, Heidelberg (2009)
10. Aoki, K., Sasaki, Y.: Preimage attacks on one-block MD4, 63-step MD5 and more. In: Avanzi, R., Keliher, L., Sica, F. (eds.) SAC 2008. LNCS, vol. 5381, pp. 103–119. Springer, Heidelberg (2009)
11. Sasaki, Y., Aoki, K.: Finding preimages in full MD5 faster than exhaustive search. In: Joux, A. (ed.) EUROCRYPT 2009. LNCS, vol. 5479, pp. 134–152. Springer, Heidelberg (2009)
12. Park, N.K., Hwang, J.H., Lee, P.J.: HAS-V: A new hash function with variable output length. In: Stinson, D.R., Tavares, S. (eds.) SAC 2000. LNCS, vol. 2012, pp. 202–216. Springer, Heidelberg (2001)
13. Lim, C.H., Lee, P.J.: A study on the proposed korean digital signature algorithm. In: Ohta, K., Pei, D. (eds.) ASIACRYPT 1998. LNCS, vol. 1514, pp. 175–186. Springer, Heidelberg (1998)
14. Mendel, F., Rijmen, V.: Weaknesses in the HAS-V compression function. In: Nam, K.-H., Rhee, G. (eds.) ICISC 2007. LNCS, vol. 4817, pp. 335–345. Springer, Heidelberg (2007)
15. Mouha, N., Cannière, C.D., Indesteege, S., Preneel, B.: Finding collisions for a 45-step simplified HAS-V. In: Youm, H.Y., Yung, M. (eds.) WISA 2009. LNCS, vol. 5932, pp. 206–225. Springer, Heidelberg (2009)
16. Joux, A.: Multicollisions in iterated hash functions. Application to cascaded constructions. In: Franklin, M. (ed.) CRYPTO 2004. LNCS, vol. 3152, pp. 306–316. Springer, Heidelberg (2004)
17. Leurent, G.: MD4 is not one-way. In: Nyberg, K. (ed.) FSE 2008. LNCS, vol. 5086, pp. 412–428. Springer, Heidelberg (2008)

18. De Cannière, C., Rechberger, C.: Preimages for reduced SHA-0 and SHA-1. In: Wagner, D. (ed.) CRYPTO 2008. LNCS, vol. 5157, pp. 179–202. Springer, Heidelberg (2008); slides on preliminary results presented at ESC 2008 seminar, http://wiki.uni.lu/esc/
19. Guo, J., Ling, S., Rechberger, C., Wang, H.: Advanced meet-in-the-middle preimage attacks: First results on full Tiger, and improved results on MD4 and SHA-2. In: Abe, M. (ed.) ASIACRYPT 2010. LNCS, vol. 6477, pp. 56–75. Springer, Heidelberg (2010); IACR Cryptology ePrint Archive: Report 2010/016, http://eprint.iacr.org/2010/016
20. Lamberger, M., Mendel, F.: Structural attacks on two SHA-3 candidates: Blender-n and DCH-n. In: Samarati, P., Yung, M., Martinelli, F., Ardagna, C.A. (eds.) ISC 2009. LNCS, vol. 5735, pp. 68–78. Springer, Heidelberg (2009)
21. Joux, A., Lucks, S.: Improved generic algorithms for 3-collisions. In: Matsui, M. (ed.) ASIACRYPT 2009. LNCS, vol. 5912, pp. 347–363. Springer, Heidelberg (2009)
22. Kelsey, J., Schneier, B.: Second preimages on n-bit hash functions for much less than 2^n work. In: Cramer, R. (ed.) EUROCRYPT 2005. LNCS, vol. 3494, pp. 474–490. Springer, Heidelberg (2005)

A   Output Tailoring of HAS-V

When the length of output is shorter than 320, the output tailoring shown in Table 9 is applied, where $A^{\triangleleft}\|B^{\triangleleft}\|C^{\triangleleft}\|D^{\triangleleft}\|E^{\triangleleft}\|A^{\triangleright}\|B^{\triangleright}\|C^{\triangleright}\|D^{\triangleright}\|E^{\triangleright}$ is a 320-bit output of the HAS-V compression function, and suffix $[u-v]$ means the bitstring extracted from bit positions v to u and the rightmost bit is the 0th bit.

B   Additional Attacks

This section describes the evidence why the attacks work correctly. The details are very complicated so we did not include these results in the main body of the paper.

B.1   Efficient Pseudo-preimage Conversion Using a Partial-Multi-collision

This section describes an efficient conversion from pseudo-preimages to a preimage by combining the property of non-injective step functions with the idea of multi-collisions proposed by Joux [16]. This method can be used for both of modular-addition feed-forward and XOR feed-forward. Because the attack in Section 3.2 can directly find preimages and has a lower level of complexity, the impact of this attack for PKC98-Hash is limited. However, the attack would be useful to learn the behavior of non-injective functions, and thus we explain the attack.

The attack generates a preimage of 34 blocks. For given H_{34}, compute as follows. The attack is also illustrated in Fig 5.

Table 9. HAS-V Output Tailoring

Output length (in bits)	0th word 3rd word 6th word	1st word 4th word 7th word	2nd word 5th word 8th word
128	$A^\lhd + A^\rhd + E^\lhd_{[31-16]}$	$B^\lhd + B^\rhd + E^\lhd_{[15-0]}$	$C^\lhd + C^\rhd + E^\rhd_{[31-16]}$
	$D^\lhd + D^\rhd + E^\rhd_{[15-0]}$		
160	$A^\lhd + A^\rhd$	$B^\lhd + B^\rhd$	$C^\lhd + C^\rhd$
	$D^\lhd + D^\rhd$	$E^\lhd + E^\rhd$	
192	$A^\lhd+(E^\rhd_{[31-21]}\|D^\rhd_{[20-10]})$	$B^\lhd+(E^\lhd_{[20-10]}\|D^\lhd_{[9-0]})$	$C^\lhd+(E^\rhd_{[9-0]}\|D^\rhd_{[31-21]})$
	$A^\rhd+(E^\lhd_{[31-21]}\|D^\lhd_{[20-10]})$	$B^\rhd+(E^\rhd_{[20-10]}\|D^\rhd_{[9-0]})$	$C^\rhd+(E^\lhd_{[9-0]}\|D^\lhd_{[31-21]})$
224	$A^\lhd+(E^\rhd_{[31-24]}\|D^\rhd_{[23-16]})$	$B^\lhd+(E^\rhd_{[23-16]}\|D^\rhd_{[15-8]})$	$C^\lhd+(E^\rhd_{[15-8]}\|D^\rhd_{[7-0]})$
	$D^\lhd+(E^\lhd_{[7-0]}\|D^\rhd_{[31-24]})$	$A^\rhd + E^\lhd_{[31-21]}$	$B^\rhd + E^\lhd_{[20-10]}$
	$C^\rhd + E^\lhd_{[9-0]}$		
256	$A^\lhd + E^\rhd_{[31-24]}$	$B^\lhd + E^\rhd_{[23-16]}$	$C^\lhd + E^\rhd_{[15-8]}$
	$D^\lhd + E^\rhd_{[7-0]}$	$A^\rhd + E^\lhd_{[31-24]}$	$B^\rhd + E^\lhd_{[23-16]}$
	$C^\rhd + E^\lhd_{[15-8]}$	$D^\rhd + E^\lhd_{[7-0]}$	
288	$A^\lhd + E^\rhd_{[31-25]}$	$B^\lhd + E^\rhd_{[24-18]}$	$C^\lhd + E^\rhd_{[17-12]}$
	$D^\lhd + E^\rhd_{[11-6]}$	$E^\lhd + E^\rhd_{[5-0]}$	$A^\rhd$
	$B^\rhd$	$C^\rhd$	$D^\rhd$

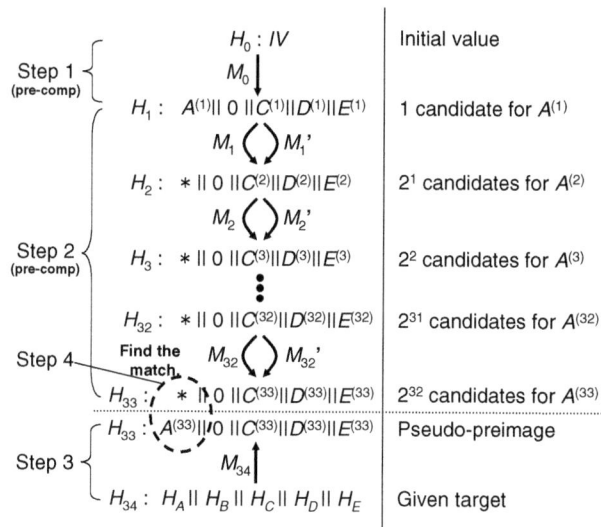

Fig. 5. Efficient Conversion from PPI to PI with Partial Multi-Collision

1. Generate M_0 randomly until $*\|0\| * \| * \|* = \mathrm{CF}(H_0, M_0)$ holds. Let $A^{(1)}\|0 \|C^{(1)}\|D^{(1)}\|E^{(1)} = \mathrm{CF}(H_0, M_0)$.
2. For $i = 1, 2, \ldots, 32$, find (M_i, M'_i) such that the outputs of $\mathrm{CF}(H_i, M_i)$ and $\mathrm{CF}(H_i, M'_i)$ collide with registers C, D, and E and register B is 0. Let the

colliding values be $*\|0\|C^{(i+1)}\|D^{(i+1)}\|E^{(i+1)}$. Store $A_{96}^{(i)}$ and $A_{96}'^{(i)}$, which will be used later. Note that we can compute registers B, C, D, and E of H_{i+1} without knowing the value of $A^{(i)}$ or $A'^{(i)}$.

3. Find a pseudo-preimage for the last block by using the attack in Section 3.1. First, fix registers B, C, D, and E of the input chaining variable to 0, $C^{(33)}$, $D^{(33)}$, and $E^{(33)}$, and fix the target hash value to H_{34}. Then, find $A^{(33)}$ and M_{33} where $H_{34} = \mathrm{CF}(H_{33}, M_{33})$ and M_{33} satisfy the padding for a 34-block message.

4. Determine which of M_i and M_i' $(i = 1, 2, \ldots, 32)$ should be used for a preimage. This is done by the exhaustive search, namely, for all message combinations, check $A^{(33)}$ fixed in Step 3 is achieved or not. This can work with high probability, because the following equation holds.

$$A^{(33)} = A^{(1)} + \sum_{i=0}^{31} \begin{cases} A_{96}^{(i)} & \text{if } M_i \text{ is used.} \\ A_{96}'^{(i)} & \text{if } M_i' \text{ is used.} \end{cases}$$

This can be done with the previous technique such as one described in [20]. Note that we write the case that the feed-forward operation is modular addition. Replace the modular operation with XOR if the feed-operation is XOR.

The complexity of the attack is as follows.

1. 2^{32} because we need to specify that register B is 0.
2. For each i, collision without register A can be found in 2^{64}, and register B of such a collision is 0 with a probability of approximately 2^{-32}. In total, 2^{101} ($= 32 \times 2^{64} \cdot 2^{32}$). The computation can be done with negligible memory, while we can compute this in 2^{85} with 2^{48} blocks of memory.
3. 2^{128} followed by Section 3.1.
4. Less than 2^{32} because 32 additions are expected in less than one CF computation.

In total, $2^{32} + 2^{101} + 2^{128} + 2^{32} \approx 2^{128}$ computations are required to obtain a preimage. Note that Steps 1 and 2 can be pre-computed before the target hash value is given. Also note that if 2^{32} words of memory is allowed, Step 4 can also be precomputed.

The above attack uses 2 collisions. We can shorten the length of the preimage with a slightly higher level of complexity and much more memory. When we use k-collisions, the length of the multi-collision part is $\lceil \log_k(2^{32}) \rceil$ blocks, and the complexity for pre-computation is

$$\lceil \log_k(2^{32}) \rceil \times 2^{32} \times (k! \times 2^{96 \times (k-1)/k})$$

and the memory requirement for the attack is approximately $2^{96 \times (k-1)/k}$ blocks. Thanks to the Joux and Lucks attack [21], we can reduce the memory complexity to 2^{32} for the case of $k = 3$. When $k \leq 6$, the increase of the time complexity is negligible with the use of k-collisions.

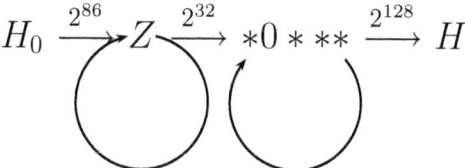

Fig. 6. Intermediate Values

B.2 Efficient Pseudo-preimage Conversion Applicable for Addition Feed-Forward

In this section, we describe another efficient conversion from pseudo-preimages to a preimage. This conversion is worse than the one in Section B.1 with respect to two points: 1) pre-computation requires higher level of complexity, and 2) the generated preimages are longer (2^{33} blocks). However, this attack has a unique characteristic where the attack works well for modular-addition feed-forward but does not work for XOR feed-forward. As far as we know, this is the first case where the attack efficiency is significantly different between these two operations, and thus, we explain the attack.

Note that this attack uses an expandable message. (a, b)-expandable message is a set of messages, where each message outputs the same hash value and the block length of each message covers all values between a and b for a given input H. $(k, k+2^k-1)$-expandable messages can be computed in $2^k + k \times 2^{n/2+1}$ [22].

Followed by Section 3.1, we can find (A', M_*) in 2^{128} that satisfies

$$A\|0\|C\|D\|E = \mathrm{CF}(A'\|0\|C\|D\|E, M_*)$$

with given A, C, D, and E. Let X be $-A_{96}$, then we have

$$A' = A + X.$$

Because X is the constant only depending on C, D, E, and M_*,

$$A\|0\|C\|D\|E = \mathrm{CF}^i((A + iX)\|0\|C\|D\|E, M_*)$$

holds. Utilizing the property, we compute a preimage of $H_A\|H_B\|H_C\|H_D\|H_E$. Figure 6 shows a graphical explanation of this attack.

1. Construct a $(33, 33 + 2^{33} - 1)$-expandable message whose input is H_0, and let Z be its output.
2. Find (A'', C, D, E, M_Z) such that $A''\|0\|C\|D\|E = \mathrm{CF}(Z, M_Z)$, by randomly choosing M_Z.
3. Find (A, M_{LAST}) such that $H_A\|H_B\|H_C\|H_D\|H_E = \mathrm{CF}(A\|0\|C\|D\|E, M_{\mathrm{LAST}})$ using Section 3.1. M_{LAST} should follow the padding rule for a message whose length is $512 \times 2^{33} + 447$ bits.
4. Find (A', M_*) such that $A\|0\|C\|D\|E = \mathrm{CF}(A'\|0\|C\|D\|E, M_*)$ using Section 3.1.

5. Compute i such that $A'' \equiv A + i(A' - A) \pmod{2^{32}}$. If there is no solution, go back to Step 4.
6. With the preceding and appropriate expandable part depending on i to satisfy the padding rule, $M_Z \| (M_*)^i \| M_{\text{LAST}}$ is a preimage for the given hash value.

The complexity of the attack is as follows.

1. $2^{33} + 33 \times 2^{81} \approx 2^{86}$.
2. 2^{32} because we need to specify the second word as 0.
3. 2^{128} followed by Section 3.1.
4. 2^{128} followed by Section 3.1.
5. Negligible but the success probability is about $\frac{2}{3}$, where the success probability is estimated as follows. Let $X = A' - A$ and $Y = A'' - A$. Assume X and Y are uniformly distributed. When X is odd, there is a solution. This happens with probability $\frac{1}{2}$. When X is even and Y is odd, there is no solution. When X is even, Y is even, and $X/2$ is odd, there is a solution. This happens with probability $\frac{1}{2^3}$. We can continue to confirm the parity of each bit to the most significant bit, thus the success probability is $\frac{1}{2} + \frac{1}{2^3} + \frac{1}{2^5} + \cdots \approx \frac{\frac{1}{2}}{1 - \frac{1}{2^2}} = \frac{2}{3}$.
6. The time to output a message whose length is about 2^{33} blocks, and we regard this step as negligible.

In total, the complexity is about

$$2^{86} + 2^{32} + 2^{128} + \left(\frac{2}{3}\right)^{-1} 2^{128} \approx 2^{129.3}.$$

The length of the preimage is about 2^{33} blocks, but the memory requirement of the attack is negligible since most of the message block is a repetition of M_* and thus all we have to remember is a counter which tells the number of repetitions of M_*. Moreover, we can compute the expandable message with a memoryless collision search.

In Step 5, we need to solve modular equation $\text{mod} 2^{32}$. If the feed-forward operation in Davies-Meyer is $\oplus$, the equation hardly has a solution, and the attack described above does not work well. When designing a hash function, the difference in security between modular addition and XOR of feed-forward operation in Davies-Meyer as well as the difference in performance may need to be considered.

B.3   One-Block Attack on PKC98-Hash with Output Tailoring

We can compute 1-block preimages of HAS-V-256 faster than the brute-force search using the redundancy in the output tailoring function, and we can also reduce the complexity of the attack against HAS-V-288 using the same idea.

The outline of the attack is as follows.

Table 10. Intermediate Variables to Compute 1-Block Preimage of HAS-V-256

j	$A^{\lhd}_j$	$B^{\lhd}_j$	$C^{\lhd}_j$	$D^{\lhd}_j$	$E^{\lhd}_j$	$A^{\rhd}_j$	$B^{\rhd}_j$	$C^{\rhd}_j$	$D^{\rhd}_j$	$E^{\rhd}_j$	$w^{\lhd}_j\; w^{\rhd}_j$
97	$D^{\lhd}_{100}\!\lll\!2$	$E^{\lhd}_{100}\!\lll\!2$	$C^{\lhd}_{100}\!\lll\!2$	$C^{\lhd\lhd}_{100}\!\lll\!2$	x						X_{10}
98	$C^{\lhd}_{100}\!\lll\!2$	$D^{\lhd}_{100}\!\lll\!2$	$E^{\lhd}_{100}$	$C^{\lhd\lhd}_{100}\!\lll\!2$	$C^{\lhd\lhd}_{100}\!\lll\!2$	$C^{\rhd}_{100}\!\lll\!2$	$D^{\rhd}_{100}\!\lll\!2$	$E^{\rhd}_{100}$	$E^{\rhd}_{100}$	y	$X_4\;\;Y_4$
99	$B^{\lhd}_{100}$	$C^{\lhd}_{100}\!\lll\!2$	$D^{\lhd}_{100}$	$E^{\lhd}_{100}$	$C^{\lhd\lhd}_{100}\!\lll\!2$	$B^{\rhd}_{100}$	$C^{\rhd}_{100}\!\lll\!2$	$D^{\rhd}_{100}$	$E^{\rhd}_{100}$	$E^{\rhd}_{100}$	$X_0\;\;Y_0$
100	$A^{\lhd}_{100}$	$B^{\lhd}_{100}$	$C^{\lhd}_{100}$	$D^{\lhd}_{100}$	$E^{\lhd}_{100}$	$A^{\rhd}_{100}$	$B^{\rhd}_{100}$	$C^{\rhd}_{100}$	$D^{\rhd}_{100}$	$E^{\rhd}_{100}$	
0	$A^{\rhd}_0$	$B^{\rhd}_0$	$C^{\rhd}_0$	$D^{\rhd}_0$	$E^{\rhd}_0$	$A^{\lhd}_0$	$B^{\lhd}_0$	$C^{\lhd}_0$	$D^{\lhd}_0$	$E^{\lhd}_0$	Fixed
	$A^{\rhd}$	$B^{\rhd}$	$C^{\rhd}$	$D^{\rhd}$	$E^{\rhd}$	$A^{\lhd}$	$B^{\lhd}$	$C^{\lhd}$	$D^{\lhd}$	$E^{\lhd}$	
	$E^{\lhd}_{[31-24]}$	$E^{\lhd}_{[23-16]}$	$E^{\lhd}_{[15-8]}$	$E^{\lhd}_{[7-0]}$		$E^{\rhd}_{[31-24]}$	$E^{\rhd}_{[23-16]}$	$E^{\rhd}_{[15-8]}$	$E^{\rhd}_{[7-0]}$		Adjustable
	$H_{A^{\rhd}}$	$H_{B^{\rhd}}$	$H_{C^{\rhd}}$	$H_{D^{\rhd}}$		$H_{A^{\lhd}}$	$H_{B^{\lhd}}$	$H_{C^{\lhd}}$	$H_{D^{\lhd}}$		Given

1. Setup the intermediate variables similar to Section 4.1. Moreover, the least significant few bits of register B will not influence the following computation for updating register A.
2. Compute the right line, and check the match. The least significant few bits of register B will influence a register that inputs to the output tailoring function.
3. Fortunately, the influenced bits cannot influence the intermediate values for the left line. We can finally absorb the influenced bits by matching in the left line.

Table 10 shows the initial setup for the attack. Let $A^{\lhd}$, $B^{\lhd}$, ..., $E^{\rhd}$ be the input of the output tailoring function and $H_{A^{\lhd}}\|H_{B^{\lhd}}\|H_{C^{\lhd}}\|H_{D^{\lhd}}\|H_{A^{\rhd}}\|H_{B^{\rhd}}\|H_{C^{\rhd}}\|H_{D^{\rhd}}$ be the given hash value for the preimage to be computed. How to construct Table 10 is as follows. Note that we have full redundancy in register E, since $E^{\rhd}_0 + E^{\lhd}_{100} = E^{\rhd}$ and $E^{\lhd}_0 + E^{\rhd}_{100} = E^{\lhd}$ hold. First, we construct the equations using the variables with $E^{\lhd}_{99}$ and $E^{\rhd}_{100}$, then we compute these variables from the fixed and the given values.

- Focus on the right line. For Step 98, we want to ignore $E^{\rhd}_{98}$ and $B^{\rhd}_{98}$. We can ignore the variables under the condition $D^{\rhd}_{98} = C^{\rhd}_{98}\ (= E^{\rhd}_{100})$, since the Boolean function of the step is f_0.
- For Step 99, we want to ignore $C^{\rhd}_{99}$. It can be achieved with $B^{\rhd}_{99} = E^{\rhd}_{99}\ (= D^{\rhd}_{98})$. Using the condition in Step 98, we have

$$C^{\rhd}_{100}\lll 2 = E^{\rhd}_{100} \tag{6}$$

- For the output of Step 99, namely $j = 100$, $D^{\rhd}_{100}$ should be consistent with the output tailoring function and feed-forward operation,

$$D^{\rhd}_{100} = H_{D^{\lhd}} - D^{\lhd}_0 - E^{\rhd}_{[7-0]}. \tag{7}$$

- Now focus on the left line. For Step 97, we want to ignore $E^{\lhd}_{97}$ and $B^{\lhd}_{97}$ $(= E^{\lhd}_{100})$. This can be achieved by $D^{\lhd}_{97} = C^{\lhd}_{97}\ (= E^{\lhd}_{99})$, since the Boolean function of the step is f_4.

- For Step 98, we want to ignore $C^\lhd{}_{98}$ $(= E^\lhd{}_{100})$. This can be achieved by $\mathbf{1} \oplus B^\lhd{}_{98} = E^\lhd{}_{98}$ $(= D^\lhd{}_{97})$. Using the condition in Step 97, we have

$$\mathbf{1} \oplus D^{\lhd \lll 2}{}_{100} = E^\lhd{}_{99} \tag{8}$$

- For Step 99, we need to ignore $D^\lhd{}_{99}$ $(= E^\lhd{}_{100})$. This can be achieved by $B^\lhd{}_{99} = E^\lhd{}_{99}$. That is,

$$C^{\lhd \lll 2}{}_{100} = E^\lhd{}_{99}. \tag{9}$$

Focusing on Eq. (7), for all possible $H_{D^\lhd}$, we can only adjust the least significant 8 bits of $D^\rhd{}_{100}$ at most. The adjustment is achieved by $E^\rhd{}_{[7-0]}$, and its value determines the least significant 8 bits of $E^\lhd{}_{100}$. From Eqs. (8) and (9), we need to satisfy

$$\mathbf{1} \oplus D^{\lhd \lll 2}{}_{100} = C^{\lhd \lll 2}{}_{100}. \tag{10}$$

Rotating two bits, and substituting $\mathbf{1} \oplus D^\lhd{}_{100} = -(1 + D^\lhd{}_{100})$ and $D^\lhd{}_{100} = H_{D^\rhd} - D^\rhd{}_0 - E^\lhd{}_{[7-0]}$ and $C^\lhd{}_{100} = H_{C^\rhd} - C^\rhd{}_0 - E^\lhd{}_{[15-8]}$, we have $E^\lhd{}_{[15-8]} + E^\lhd{}_{[7-0]} = H_{C^\rhd} + H_{D^\rhd} - C^\rhd{}_0 - D^\rhd{}_0 + 1$. Since we can choose an arbitrary value for $E^\lhd{}_{99}$ using X_0, the least significant 9 bits of Eq. (10) can almost always be satisfied using $E^\lhd{}_{[15-8]}$ and $E^\lhd{}_{[7-0]}$. This condition can cancel the influence of the following computation in $D^\lhd{}_{99[10-2]}$ $(= E^\lhd{}_{100[10-2]})$ and $C^\lhd{}_{98[10-2]}$ $(= E^\lhd{}_{100[10-2]})$. The range we want to cancel is $[7-0]$ in $E^\lhd{}_{100}$, the shared bits we can cancel are $E^\lhd{}_{100[7-2]}$. Thus, we choose $E^\lhd{}_{99} \leftarrow C^{\lhd \lll 2}{}_{100}$, and adjust the least significant few bits of $D^\lhd{}_{100}$.

When all bits in $E^\lhd$ and $E^\rhd$ are determined, all variables in Table 10 are determined. We choose $E^\lhd$ and $E^\rhd$ to satisfy the remaining and unsatisfied condition $B^\rhd{}_{99[7-2]} = E^\rhd{}_{99[7-2]}$ and $B^\lhd{}_{98[7-2]} = E^\lhd{}_{98[7-2]}$ which comes from Eqs. (6) and (8) as follows.

1. Fix $E^\rhd{}_{[15-8]}$ randomly.
2. Compute $C^\lhd \leftarrow H_{C^\lhd} - E^\rhd{}_{[15-8]}$, and $C^\rhd{}_{100} \leftarrow C^\lhd - C^\lhd{}_0$.
3. Compute $E^\lhd{}_{100[7-0]} \leftarrow (C^{\rhd \lll 2}{}_{100})_{[7-0]}$ to satisfy the part of Eq. (6), and $E^\lhd{}_{[7-0]} \leftarrow E^\rhd{}_{100[7-0]} + E^\lhd{}_{0[7-9]} \pmod{2^8}$.
4. Compute $D^\rhd \leftarrow H_{D^\rhd} - E^\lhd{}_{[7-0]}$ and $D^\lhd{}_{100} \leftarrow D^\rhd - D^\rhd{}_0$. Compute $C^\lhd{}_{100[7-0]} \leftarrow (\mathbf{1} \oplus D^\lhd{}_{100})_{[7-0]}$ to satisfy the part of Eq. (8).
5. Compute $C^\rhd{}_{[7-0]} \leftarrow C^\lhd{}_{100[7-0]} + C^\rhd{}_{0[7-0]} \pmod{2^8}$ and $E^\lhd{}_{[15-8]} \leftarrow H_{C^\rhd[7-0]} - C^\rhd{}_{[7-0]} \pmod{2^8}$.
6. Choose $E^\lhd{}_{[31-16]}$, $E^\rhd{}_{[31-16]}$, and $E^\rhd{}_{[7-0]}$ randomly, and compute $A^\lhd$, $B^\lhd$, ..., $E^\rhd$. Note that some bits are already determined.
7. Compute the intermediate variables $A^\lhd{}_j, B^\lhd{}_j, \ldots, E^\lhd{}_j$ for $j = 97, 98, 99, 100$, and $A^\rhd{}_j, B^\rhd{}_j, \ldots, E^\rhd{}_j$ for $j = 98, 99, 100$. Note that some bits are already determined.
8. Reversely compute the message words $X_{10}, X_4, X_0, Y_4,$ and Y_0 using Eq. (3).

We are now ready to write the detailed attack procedure.

1. Choose all message words that are not determined yet.

2. Compute $R(p^{\triangleright}{}_j, w^{\triangleright}{}_j)$ for $j = 0, 1, \ldots, 97$ and confirm whether or not the resulting $p^{\triangleright}{}_{98}$ matches the form of

$$C^{\triangleright}{}^{\lll 2}_{100} \| ((D^{\triangleright}{}^{\lll 2}_{100})_{[31-10]} \| *_{[9-4]} \| (D^{\triangleright}{}^{\lll 2}_{100})_{[3-0]}) \| E^{\triangleright}{}_{100} \| E^{\triangleright}{}_{100} \| * .$$

If there is no match, go back to the first step. The match will occur with probability 2^{-122}.

3. Compute $E^{\triangleright}{}_{[7-0]} \leftarrow H_{D^{\triangleleft}[7-0]} - D^{\triangleleft}{}_{0[7-0]} - D^{\triangleright}{}_{100[7-0]} \pmod{2^8}$, and reassign $E^{\triangleleft}{}_{100[7-0]} \leftarrow E^{\triangleright}{}_{[7-0]} - E^{\triangleright}{}_{0[7-0]}$.

4. Compute $R(p^{\triangleleft}{}_j, w^{\triangleleft}{}_j)$ for $j = 0, 1, \ldots, 96$ and confirm whether or not the resulting $p^{\triangleleft}{}_{97}$ matches the form of $D^{\triangleleft}{}^{\lll 2}_{100} \| E^{\triangleleft}{}^{\lll 2}_{100} \| C^{\triangleleft}{}^{\lll 2}_{100} \| C^{\triangleleft}{}^{\lll 2}_{100} \| *$. If there is no match, go back to the first step. The match will occur with probability 2^{-128}.

As a result, we will find a preimage in 2^{250}.

Note for Weak Hash Values. After setting up the intermediate variables in Table 10, $(C^{\triangleleft}{}^{\lll 2}_{100})_{[k]} = (D^{\triangleleft}{}^{\lll 2}_{100})_{[k]}$ for $k = 0, 1$ holds with probability $\frac{1}{2}$. Thus, we can compute a preimage for $\frac{1}{4}$ of the hash value in 2^{248}, $\frac{1}{2}$ of the hash value in 2^{249}, and $\frac{1}{4}$ of the hash value in 2^{250}. On average, the attack complexity is $2^{249.2}$.

An Attack against HAS-V-288. We can attack HAS-V-288 in a similar manner. On the right line, we can cancel the effect of $B^{\triangleright}{}_{98[7-2]}$. On the left line, unfortunately we cannot satisfy Eq. (10) anymore because $E^{\triangleleft}$ is directly used to compute a part of the hash value. However, the condition is sometimes automatically satisfied for a given hash value. For $l = 0, 1, \ldots, 6$, we can compute a preimage of HAS-V-288 in 2^{256-l} for $\frac{\binom{6}{l}}{2^6}$ of the hash values. On average, the attack complexity is $2^{253.5}$.

Passive Cryptanalysis of the UnConditionally Secure Authentication Protocol for RFID Systems

Mohammad Reza Sohizadeh Abyaneh

Department of Informatics, University of Bergen
{reza.sohizadeh@ii.uib.no}

Abstract. Recently, Alomair et al. proposed the first *UnConditionally Secure* mutual authentication protocol for low-cost RFID systems(UCS-RFID). The security of the UCS-RFID relies on five dynamic secret keys which are updated at every protocol run using a fresh random number (nonce) secretly transmitted from a reader to tags.

Our results show that, at the highest security level of the protocol (security parameter= 256), inferring a nonce is feasible with the probability of 0.99 by eavesdropping(observing) about 90 runs of the protocol. Finding a nonce enables a passive attacker to recover all five secret keys of the protocol. To do so, we propose a three-phase probabilistic approach in this paper. Our attack recovers the secret keys with a probability that increases by accessing more protocol runs. We also show that tracing a tag using this protocol is also possible even with less runs of the protocol.

Keywords: RFID, Authentication Protocol, Passive Attack.

1 Introduction

As of today, RFID (Radio Frequency Identification) is referred to as the next technological revolution after the Internet. A typical RFID system involves a *reader*, a number of *tags*, which may range from the battery-powered, to the low-cost ones with even no internal power, and a *database*. RFID systems enable the identification of objects in various environments. They can potentially be applied almost everywhere from electronic passports[20,21], contactless credit cards[19], to supply chain management[22,23,24].

Keeping RFID systems secure is imperative, because they are vulnerable to a number of malicious attacks. For low-cost RFID systems, security problems become much more challenging, as many traditional security mechanisms are inefficient or even impossible due to resource constraints. Some existing solutions utilize traditional cryptographic primitives such as hash or encryption functions, which are often too expensive to be implemented on low-cost RFID tags.

Another method of securing RFID systems has been the lightweight approach. These solutions base themselves on mostly lightweight operations (e.g. bitwise or simple arithmetic operations) instead of more expensive cryptographic

K.-H. Rhee and D. Nyang (Eds.): ICISC 2010, LNCS 6829, pp. 92–103, 2011.

primitives. The HB-family(HB$^+$,HB^{++}, HB*,etc.) [1,2,3,4,5,7,6,8] and the MAP-family(LMAP,EMAP,M2AP,etc)[9,10,11] authentication protocols, are some examples of this kind. However, proposed lightweight protocols so far have been targeted to various successful attacks and therefore, the search for a concrete lightweight solution for authentication in low-cost RFID tags still continues.

Recently, Alomair et al. embarked on the notion of UnConditionally Secure mutual authentication protocol for RFID systems (UCS-RFID)[17]. UCS-RFID's security relies mainly on the freshness of five secret keys rather than the hardness of solving mathematical problems. Freshness in the keys is guaranteed with a *key updating* phase at every protocol run by means of a fresh random number (nonce). This nonce is generated at the reader side due to low-cost tags constraints, and delivered to the tag secretly. This allows the tags to benefit from the functionalities of random numbers without the hardware to generate them.

Our Contribution. In this paper, we present a three-phase probabilistic passive attack against the UCS-RFID protocol to recover all the secret keys in the protocol. Our attack is mainly based on a weakness observed in the protocol(section 3). To put in a nutshell, the weakness implies that the more outputs we have from consecutive runs of the protocol, the more knowledge we will obtain on the nonces in these protocol runs. In other words, having more number of protocol run outputs observed, we are able to determine some of the nonces (*victim* nonces) with higher probability. It should be noted that this weakness has also been tackled by the authors in [17]. Nevertheless we will show that the security margin they expected from the protocol has been overestimated. Finding the victim nonce in the protocol paves the way toward adopting an attacking scenario to achieve all of the five secret keys in the system.

Outline. The remainder of this paper is organized as follows. In section 2, we briefly describe the UCS-RFID protocol. In section 3 the weakness of the protocol is investigated thoroughly. Section 4 and 5 describes our attacking scenario to recover the keys, and trace the tag in the protocol. Finally, section 6 concludes the paper.

2   Description of the UCS-RFID Protocol

The UCS-RFID authentication protocol consists of two phases: the *mutual authentication phase* and the *key updating phase*. The former phase mutually authenticates an RFID reader and a tag. In the latter phase both the reader and the tag update their dynamic secret keys for next protocol runs.

In this protocol, first the security parameter, N, is specified and a $2N$-bit prime integer, p, is chosen. Then, each tag T is loaded with an N-bit long identifier, $A^{(0)}$, and five secret keys, $k_a^{(0)}, k_b^{(0)}, k_c^{(0)}, k_d^{(0)}$ and $k_u^{(0)}$ chosen independently and uniformly from $\mathbb{Z}_{2^N}, \mathbb{Z}_p, \mathbb{Z}_p\backslash\{0\}, \mathbb{Z}_{2^N}$ and $\mathbb{Z}_p\backslash\{0\}$ respectively.

Notation
- N: security parameter.
- p: a prime number in $\mathbb{Z}_{2^N}$

- A^x, B^x, C^x, D^x: observable outputs of xth protocol run
- $n = n_l \| n_r$: random number in $\mathbb{Z}_{2^N}$
- n_l, n_r: left and right *half-nonces*

2.1   Mutual Authentication Phase

Figure 1 shows one instance run of the mutual authentication phase in the UCS-RFID protocol. The reader starts the interrogation with a "Hello" message which is responded by tag's dynamic identifier $A^{(i)}$. The reader then looks up in the database for a set of five keys$(k_a, k_b, k_c, k_d, k_u)$ which corresponds to $A^{(i)}$. If this search is successful, it means that the tag is authentic. Having the tag authenticated, the reader generates a $2N$-bit random nonce $n^{(i)}$ uniformly drawn from $\mathbb{Z}_p^*$, calculates messages $B^{(i)}$, $C^{(i)}$ by (2),(3) and sends them to the tag.

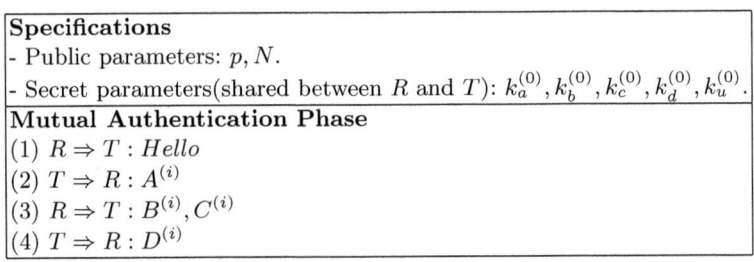

Specifications
- Public parameters: p, N.
- Secret parameters(shared between R and T): $k_a^{(0)}, k_b^{(0)}, k_c^{(0)}, k_d^{(0)}, k_u^{(0)}$.
Mutual Authentication Phase
(1) $R \Rightarrow T : Hello$
(2) $T \Rightarrow R : A^{(i)}$
(3) $R \Rightarrow T : B^{(i)}, C^{(i)}$
(4) $T \Rightarrow R : D^{(i)}$

Fig. 1. i^{th} run of the mutual authentication phase in the UCS-RFID protocol

$$A^{(i)} \equiv n_l^{(i-1)} + k_a^{(i)} \ mod \ 2^N \tag{1}$$

$$B^{(i)} \equiv n^{(i)} + k_b^{(i)} \ mod \ p \tag{2}$$

$$C^{(i)} \equiv n^{(i)} \times k_c^{(i)} \ mod \ p \tag{3}$$

The tag first checks the integrity of the received messages by (4):

$$(B^{(i)} - k_b^{(i)}) \times k_c^{(i)} \equiv C^{(i)} \ mod \ p \tag{4}$$

This check implies the authenticity of the reader as well. Then, the tag extracts the nonce $n^{(i)}$ by (5.)

$$n^{(i)} \equiv (B^{(i)} - k_b^{(i)}) \ mod \ p \tag{5}$$

To conclude the mutual authentication phase, the tag transmits $D^{(i)}$ as a receipt of obtaining $n^{(i)}$.

$$D^{(i)} = n_l^{(i)} \oplus k_d^{(i)} \tag{6}$$

2.2   Key Updating Phase

After a successful mutual authentication, both the reader and the tag update their keys and dynamic identifier $(A^{(i)})$ for the next protocol run.

$$k_a^{(i+1)} = n_r^{(i)} \oplus k_a^{(i)} \tag{7}$$

$$k_b^{(i+1)} \equiv k_u^{(i)} + (n^{(i)} \oplus k_b^{(i)}) \bmod p \tag{8}$$

$$k_c^{(i+1)} \equiv k_u^{(i)} \times (n^{(i)} \oplus k_c^{(i)}) \bmod p \tag{9}$$

$$k_d^{(i+1)} = n_r^{(i)} \oplus k_d^{(i)} \tag{10}$$

$$k_u^{(i+1)} \equiv k_u^{(i)} \times n^{(i)} \bmod p \tag{11}$$

$$A^{(i+1)} \equiv n_l^{(i)} + k_a^{(i+1)} \bmod 2^N \tag{12}$$

It should be noted that the dynamic values have been proved to preserve their properties of independency and uniformity after updating[17].

3   Observation

In this section, we shed more light on a weakness in the UCS-RFID protocol which becomes the origin of our proposed attack presented in the subsequent section.

By xoring (7) and (10), we have:

$$k_a^{i+1} \oplus k_d^{i+1} = k_a^i \oplus k_d^i \tag{13}$$

Equation (13) shows that the difference between k_a and k_d remains the same for two consecutive runs of the protocol. This statement can also be generalized for every r arbitrary run of the protocol the as following:

$$k_a^{r+1} \oplus k_d^{r+1} = k_a^r \oplus k_d^r = \ldots = k_a^0 \oplus k_d^0 = L \tag{14}$$

By using (14), for outputs A and D in m consecutive runs of the protocol, we have:

$$A^{(i)} \equiv n_l^{(i-1)} + k_a^{(i)} \bmod 2^N \tag{15}$$

$$D^{(i)} = n_l^{(i)} \oplus (k_a^{(i)} \oplus L) \tag{16}$$

$$A^{(i+1)} \equiv n_l^{(i)} + (k_a^{(i)} \oplus n_r^{(i)}) \bmod 2^N \tag{17}$$

$$D^{(i+1)} = n_l^{(i+1)} \oplus (k_a^{(i)} \oplus L \oplus n_r^{(i)}) \tag{18}$$

$$\vdots$$

$$A^{(i+m-1)} \equiv n_l^{(i+m-2)} + (k_a^{(i)} \bigoplus_{j=i}^{i+m-2} n_r^{(j)}) \bmod 2^N \tag{19}$$

$$D^{(i+m-1)} = n_l^{(i+m-1)} \oplus (k_a^{(i)} \oplus L \bigoplus_{j=i}^{i+m-2} n_r^{(j)}) \tag{20}$$

It is apparent that we have a set of $2m$ equations with $2m + 2$ variables. These variables can be divided into two groups:

1. $2m$ half-nonces: $n_l^{(i-1)}, \ldots, n_l^{(i+m-1)}, n_r^{(i)}, \ldots, n_r^{(i+m-2)}$
2. L and $k_a^{(i)}$.

So, if we fix the value of variables L and $k_a^{(i)}$, we end up with $2m$ equations and $2m$ half-nonce variables. This implies that the $2m$ half-nonces can not be chosen independently and fulfil the above equations simultaneously. In other words, if we observe the outputs of m consecutive runs of the protocol, it is only necessary to search over all possible sequences of $k_a^{(i)}$ and L, which is 2^{2N}, and then it will be possible to find all $2m$ half-nonces uniquely. As we will see, this weakness is the result of introduction of a tighter bound for the half-nonces while we keep observing more runs of the protocol.

By the randomness nature of the generated half-nonces, the total number of possible sequences for them(2^{2N}) is uniformly distributed over them. This implies that each of the $2m$ half-nonces is expected to have a bound of $\sqrt[2m]{2^{2N}}$ possible values (comparing to its previous bound which was N). Therefore, for m consecutive protocol runs, the total number of possible values distributed over the $2m$ half-nonces is $2m \sqrt[2m]{2^{2N}}$ [17].

Now, if we exclude the value which half-nonces has taken already($2m \sqrt[2m]{2^{2N}} - 2m$), we can calculate the probability that at least one half-nonce does not receive another possible value (remains constant). To do so, we utilize the well-known problem in probability theory(i.e. Given r balls thrown uniformly at random at b bins, the probability that at least one bin remains empty which is calculated by (21))[18]:

$$\Pr(\text{at least one bin remains empty}) = 1 - \frac{\binom{r-1}{b-1}}{\binom{b+r-1}{b-1}} \qquad (21)$$

Now, it only requires to substitute $b = 2m$ and $r = 2m. \sqrt[2m]{2^{2N}} - 2m$ in (21) and then we will have (22). The result is plotted in Figure 2.

$$P_h = \Pr(\text{at least one half-nonce remains constant}) = 1 - \frac{\binom{2m. \sqrt[2m]{2^{2N}} - 2m - 1}{2m-1}}{\binom{2m. \sqrt[2m]{2^{2N}} - 1}{2m-1}}$$
$$(22)$$

Figure 2 shows the probability of inferring at least one half-nonce in terms of the number of consecutive runs of the protocol required to be observed to do so. For example, if we observe 35 runs of the protocol runs with $N=256$, we will be able to determine at least one of the 70 transmitted half-nonces with the probability of more than 0.99.

We will use the term "victim half-nonce" for inferred half-nonce and notation m_h instead of m for the number of consecutive runs of the protocol required to infer one half-nonce hereafter.

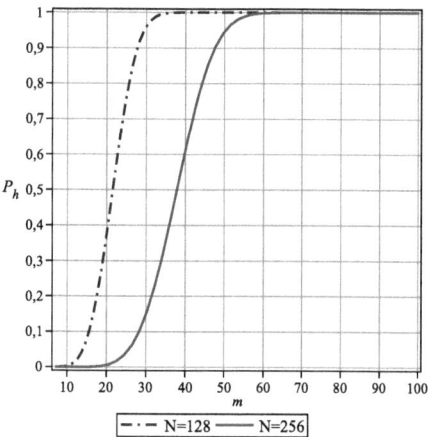

Fig. 2. The number of consecutive protocol runs an adversary must observe(m) in order to infer at least one half-nonce for $N = 128, 256$

4   Our Attack Scenario

In the previous section, we presented a probabilistic approach to find the number of consecutive runs of the protocol to infer one half-nonce. But in our attack, we need to have a complete nonce(left and right corresponding half-nonces) to recover all secret keys. To achieve this goal, we propose an attacking scenario which consists of the three following phases:

1. Finding the total number of necessary consecutive runs of the protocol to find a complete victim nonce (m_t).
2. Finding the victim nonce.
3. Recovering the secret keys.

4.1   Phase I: Finding m_t

In section 3, we proposed a probabilistic way to calculate the number of consecutive runs that must be observed by an adversary to infer a half-nonce(m_h). It is obvious that if we keep observing more runs of the protocol(i.e. more than m_h), after each extra observation, another half-nonce can be inferred. This is simply possible by eliminating the two equations which contain the first victim half-nonce and adding two newly observed equations to the set of equations (15-20) and then, we again have $2m_h$ equations and $2m_h + 2$ variables which yield another half-nonce inference.

If we intend to find a complete nonce, we must continue observing the runs of the protocol until we infer two corresponding victim half-nonces to form a complete nonce. To do so, we should first calculate the probability that the inferred half-nonce at $(m_e + m_h)^{th}$ run matches one of the previously victim half-nonces.

As we know, after m_h runs of the protocol, we accomplish to find one victim half-nonce, after m_e extra runs of the protocol, we have $\beta = 2m_h + 2m_e$ equations and β half-nonces which $m_e + 1$ of them can be inferred. The probability that none of these $m_e + 1$ half-nonces match is:

$$\Pr(\text{Having no pair after } m_h + m_e \text{ runs}) = \frac{(\beta - 1)}{\beta} \times \frac{(\beta - 2)}{\beta} \times \ldots \times \frac{(\beta - m_e)}{\beta}$$

$$= \frac{\prod_{i=1}^{m_e}(\beta - i)}{\beta^{(m_e)}} \tag{23}$$

Consequently, the probability of having at least one pair after observing m_e runs is simply calculated by (24).

$$P_e = \Pr(\text{Having at least one pair of matching half-nonces after } m_h + m_e \text{ runs})$$
$$= 1 - \frac{\prod_{i=1}^{m_e}(\beta - i)}{\beta^{(m_e)}} \tag{24}$$

By using (22) and (24) the total number of protocol runs to have at least one complete victim nonce ($m_t = m_h + m_e$) can be calculated by (25) and is plotted in Figure 3.

$$P_t = \Pr(\text{Having at least one complete nonce after } m_t \text{ runs})$$
$$= (P_e|m_h = h) \times \Pr(m_h = h) = (P_e|m_h = h) \times P_h(h) \tag{25}$$

Remark. The authors of [17] have also calculated m_t by using some other protocol outputs (B and C). Figure 3 compares our results with what the authors "Expected". This comparison has been conducted for two different security parameters $N{=}128, N{=}256$ which are plotted on the left and right respectively. The results show that the security margin of the protocol in terms of the number of consecutive runs that must be observed to infer one nonce is less than what the designers of the protocol expected. In other words, we need less number of protocol runs to infer at least one nonce. For example a passive adversary is able to infer a complete nonce with high probability of 0.99 by eavesdropping less that 60 and 90 runs of the protocol for the key size of 128 and 256 bits respectively. These numbers were expected to be 110 and 200 respectively.

4.2   Phase II: Finding the Constant Nonce

Having m_h consecutive runs of the protocol observed, we have one constant half-nonce or one half-nonce with only one possible value. In order to find this half-nonce, we adopt the following algorithm.

Algorithm Inputs : $A^{(i)}, \ldots, A^{(i+m_t-1)}, D^{(i)}, \ldots, D^{(i+m_t-1)}$

1. Determine a level of confidence(probability) for the final results.
2. Find the m_h, m_t related to the determined probability from Figures 1,2 respectively.

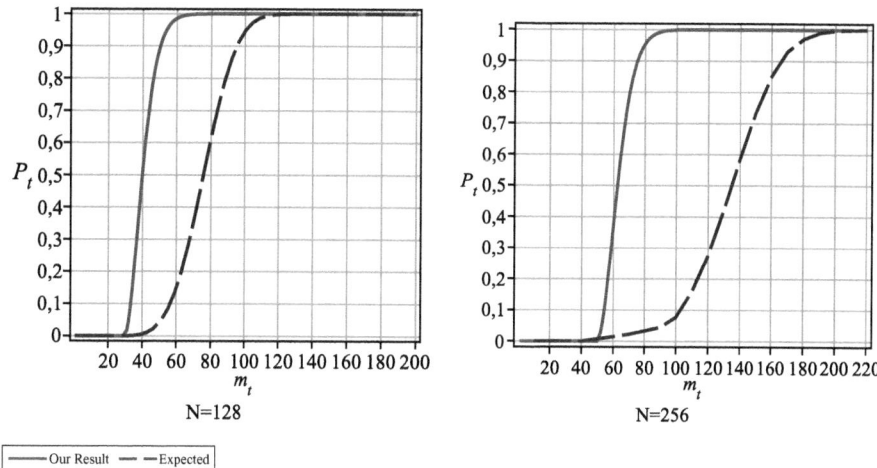

Our Result — — Expected

Fig. 3. Comparison of expected security margin of the UCS-RFID protocol and our results in terms of the number of consecutive protocol runs an adversary must observe in order to infer at least one nonce

3. Calculate $m_e = m_t - m_h$
4. Choose two random numbers from $\mathbb{Z}_{2^N}$ and assign them to $L, k_a^{(i)}$ respectively.
5. Find $2m$ nonces $(n_l^{(i-1)}, \ldots, n_l^{(i+m_h-1)}, n_r^{(i)}, \ldots, n_r^{(i+m_h-2)})$ as follows.
   - Find $n_l^{(i-1)}$ from (15) i.e. $n_l^{(i-1)} \equiv A^{(i)} - k_a^{(i)} \bmod 2^N$.
   - Find $n^{(i)}$ from (16) i.e. $n_l^{(i)} = D^{(i)} \oplus (k_a^{(i)} \oplus L)$.
   - Find $n_r^{(i)}$ from (17) i.e. $n_r^{(i)} \equiv (A^{(i+1)} - n_l^{(i)} \bmod 2^N) \oplus k_a^{(i)}$.

   $\vdots$

   - Find $n_r^{(i+m_h-2)}$ from (19)i.e. $n_r^{(i+m_h-2)} \equiv (A^{(i+m_h-1)} - n_l^{(i+m_h-2)} \bmod 2^N) \oplus (k_a^{(i)} \bigoplus_{j=i}^{i+m_h-2} n_r^{(j)})$.
   - Find $n_l^{(i+m_h-1)}$ from (20) i.e. $n_l^{(i+m_h-1)} = D^{(i+m_h-1)} \oplus (k_a^{(i)} \oplus L) \bigoplus_{j=i}^{i+m_h-2} n_r^{(j)}$.

6. Repeat 4 and 5 as many times as we observe that only one half-nonce keeps its value for all of the repetitions.
7. Save the constant(victim) half-nonce.
8. Observe another run of the protocol.
   - $A^{(i+m_h)} \equiv n_l^{(i+m_h-1)} + (k_a^{(i)} \bigoplus_{j=i}^{i+m_h-1} n_r^{(j)}) \bmod 2^N$
   - $D^{(i+m_h)} = n_l^{(i+m_h)} \oplus (k_a^{(i)} \oplus L \bigoplus_{j=i}^{i+m_h-1} n_r^{(j)})$.
9. Replace the equations corresponding to the found victim half-nonce with two newly observed equations in the equation set (15-20).
10. Repeat 4,5,6,7,8 for m_e times.

11. Match two corresponding victim half-nonces(e.g. $n_l^{(j)}, n_r^{(j)}$).

12. Output the victim nonce $(n^{(j)} = n_l^{(j)} || n_r^{(j)})$.

4.3 Phase III: Key Recovery

In the previous two phases of our attack, we accomplished to find a complete victim nonce $n^{(j)}$,with a certain probability, by observing m_t consecutive runs of the protocol. Now, we present how an adversary is able to recover all five secret keys of the protocol. To find $k_a^{(j)}, k_b^{(j)}, k_c^{(j)}$ and $k_d^{(j)}$, we should follow(26-29).

$$k_a^{(j)} \equiv (A^{(j+1)} - n_l^{(j)}) \oplus n_r^{(j)} \bmod 2^N \tag{26}$$

$$k_b^{(j)} \equiv B^{(j)} - n^{(j)} \bmod p \tag{27}$$

$$k_c^{(j)} \equiv (\frac{1}{n^{(j)}} \bmod p) \times C^{(j)} \bmod p \tag{28}$$

$$k_d^{(j)} = n_l^{(j)} \oplus D^{(j)} \tag{29}$$

To recover $k_u^{(j)}$, we need to find the nonce in the next run $(n^{(j+1)})$, thus we should calculate the updated keys for the $(j+1)^{th}$ run using (7) and (10).

$$k_a^{(j+1)} = k_a^{(j)} \oplus n_r^{(j)} \tag{30}$$

$$k_d^{(j+1)} = k_d^{(j)} \oplus n_r^{(j)} \tag{31}$$

Then we have:

$$n_l^{(j+1)} = D^{(j+1)} \oplus k_d^{(j+1)} \tag{32}$$

$$k_a^{(j+2)} = A^{(j+2)} \oplus n_l^{(j+1)} \tag{33}$$

Using (30) and (33), we can write:

$$n_r^{(j+1)} = k_a^{(j+2)} \oplus k_a^{(j+1)} \tag{34}$$

Finally, by using (27),(32) and,(34) we can find $k_u^{(j)}$.

$$k_u^{(j)} \equiv B^{(j+1)} - n^{(j+1)} - (k_b^{(j)} \oplus n^{(j+1)}) \bmod p \tag{35}$$

The procedure above provides us with our objective to recover all of the secret keys with a certain probability(P_t). This probability can be increased by paying the price of having more protocol run outputs available.

Furthermore, as it can be seen from the (32) and (34), next nonce is also achievable. This implies that the secret keys of the next run can also be calculated by using (26-35) for the next run. This is an ongoing procedure which yields the keys of any arbitrary run of the protocol(r) which $r > j$. Being able to generate the future secret keys, an adversary is capable of either impersonating both the reader and the tag or tracing the tag.

5   On the Traceability of the UCS-RFID

In the previous section, we presented a probabilistic key recovery attack against the UCS-RFID protocol. We mentioned that according to Figure 3, we need to have about 90 runs of the protocol to be almost sure that our found keys are correct. But with less number of protocol run outputs, we still can apply an attack against the traceability of the protocol. In this section, we formally investigate the *untraceability* of the UCS-RFID based on the formal description in [12].

5.1   Adversarial Model

According to [12], the means that are accessible to an attacker are the following: We denote a tag and a reader in i^{th} run of the protocol by $\mathcal{T}_i$ and $\mathcal{R}_i$, respectively.

- Query($\mathcal{T}_i, m_1, m_3$): This query models the attacker $\mathcal{A}$ sending a message m_1 to the tag and sending the m_3 after receiving the response.
- Send($\mathcal{R}_i, m_2$): This query models the attacker $\mathcal{A}$ sending a message m_2 to the Reader and being acknowledged.
- Execute($\mathcal{T}_i, \mathcal{R}_i$): This query models the attacker $\mathcal{A}$ executing a run of protocol between the Tag and Reader to obtain the exchanged messages.
- Reveal($\mathcal{T}_i$): This query models the attacker $\mathcal{A}$ obtaining the information on the Tag's memory.

A *Passive Adversary*, $\mathcal{A}_\mathcal{P}$, is capable of eavesdropping all communications between a tag and a reader and accesses only to the Execute($\mathcal{T}_i, \mathcal{R}_i$): .

5.2   Attacking Untraceability

The result of application of an oracle for a passive attack $\mathcal{O}_\mathcal{P} \subseteq \{Execute(.)\}$ on a tag T in the run i is denoted by $w_i(T)$. Thus, a set of I protocol run outputs, $\Omega_I(T)$, is:
$\Omega_I(T) = \{w_i(T)|i \in I\} ; I \subseteq N;(N$ denotes the total set of protocol runs).
The formal description of attacking scenario against untraceability of a protocol is as following:

1. $\mathcal{A}_\mathcal{P}$ requests the *Challenger* to give her a target T.
2. $\mathcal{A}_\mathcal{P}$ chooses I and calls $Oracle(T, I, \mathcal{O}_\mathcal{P})$ where $|I| \leq l_{ref}$ receives $\Omega_I(T)$.
3. $\mathcal{A}_\mathcal{P}$ requests the *Challenger* thus receiving her challenge T_1, T_2 ,$I_1$and I_2
4. $\mathcal{A}_\mathcal{P}$ calls $Oracle(T_1, I_1, \mathcal{O}_\mathcal{P})$, $Oracle(T_2, I_2, \mathcal{O}_\mathcal{P})$ then receives $\Omega_{I_1}(T_1)$, $\Omega_{I_2}(T_2)$.
5. $\mathcal{A}_\mathcal{P}$ decides which of T_1 or T_2 is T, then outputs her guess T'.

For a security parameter,k, if $Adv_{\mathcal{A}_\mathcal{P}}^{UNT}(k) = 2Pr(T' = T) - 1 > \epsilon$ then we can say that the protocol is traceable.

For UCS-RFID case, as Figure 3 implies, an adversary $\mathcal{A}_\mathcal{P}$ needs only to access to about 40 and 65 consecutive runs of the protocol to be able to determine $n^{(j)}$ with a probability of more than 0.5 (e.g. 0.6) for $k =128$ and 256 respectively and then according to section 4.3, she will be able to recover the keys of subsequent runs. After, key recovery, the adversary can easily distinguish a target tag with any other challenge tag given by the challenger. So we have:

$\forall l_{ref} \geq 40, Adv_{A_P}^{UNT}(128) = 2Pr(T' = T) - 1 = 0.1 > \epsilon.$

$\forall l_{ref} \geq 65, Adv_{A_P}^{UNT}(256) = 2Pr(T' = T) - 1 = 0.1 > \epsilon.$

6   Conclusions

The design of suitable lightweight security protocols for low-cost RFID tags is still a big challenge due to their severe constraints. Despite of interesting proposals in the literature, this field still lacks a concrete solution.

Recently, Alomair et al have proposed the first authentication protocol based on the notion of unconditional security. Regardless of some inefficiencies in UCS-RFID authentication protocol, such as: large key sizes, using modular multiplication ,etc ,which makes this protocol an unsuitable nominate for low-cost RFID tag deployment, we presented a passive attack which showed that even the security margin which was expected to be yielded by UCS-RFID has also been overestimated.

In our attack, we showed that a passive adversary is able to achieve the all secret keys of the system with a high probability of 0.99 by eavesdropping less that 60 and 90 runs of the protocol for the key size of 128 and 256 bits respectively. Tracing the tag in the protocol is also feasible even by less number of runs of the protocol (e.g. 40, 65).

Our results suggest a major rethink in the design of the authentication protocols for RFID systems based on unconditional security notion. Drastic changes are necessary to fulfil both technological constraints and security concerns in RFID systems.

References

1. Hopper, N.J., Blum, M.: Secure Human Identification Protocols. In: Boyd, C. (ed.) ASIACRYPT 2001. LNCS, vol. 2248, pp. 52–66. Springer, Heidelberg (2001)
2. Bringer, J., Chabanne, H., Dottax, E.: HB++: a Lightweight Authentication Protocol Secure Against Some Attacks. In: IEEE International Conference on Pervasive Services, Workshop on Security, Privacy and Trust in Pervasive and Ubiquitous Computing SecPerU (2006)
3. Bringer, J., Chabanne, H.: Trusted-HB: a low-cost version of HB+ secure against man-in-the-middle attacks. CoRR, abs/0802.0603 (2008)
4. Bringer, J., Chabanne, H., Dottax, E.: HB++: a lightweight authentication protocol secure against some attacks. In: Second International Workshop on Security, Privacy and Trust in Pervasive and Ubiquitous Computing (SecPerU 2006), pp. 28–33. IEEE Computer Society, Los Alamitos (2006)
5. Duc, D.N., Kim, K.: Securing HB+ against GRS man-in-the-middle attack. In: Institute of Electronics, Information and Communication Engineers, Symposium on Cryptography and Information Security, Sasebo, Japan, January 23-26, p. 123 (2007)

6. Gilbert, H., Robshaw, M.J.B., Seurin, Y.: HB♯: Increasing the security and effciency of HB+. In: Smart, N.P. (ed.) EUROCRYPT 2008. LNCS, vol. 4965, pp. 361–378. Springer, Heidelberg (2008)
7. Munilla, J., Peinado, A.: HB-MP: A further step in the HB-family of lightweight authentication protocols. Computer Networks (2007)
8. Madhavan, M., Thangaraj, A., Sankarasubramaniam, Y., Viswanathan, K.: NLHB: A Non-Linear Hopper Blum Protocol. In: IEEE National Conference on Communications (NCC) 2010, CoRR abs/1001.2140 (2010)
9. Peris-Lopez, P., Hernandez-Castro, J.C., Estevez Tapiador, J., Ribagorda, A.: LMAP: A Real Lightweight Mutual Authentication Protocol for Low-cost RFID tags. In: RFIDSec 2006 (2006)
10. Peris-Lopez, P., Hernandez-Castro, J.C., Estevez-Tapiador, J., Ribagorda, A.: M2AP: A minimalist mutual-authentication protocol for low-cost RFID tags. In: Ma, J., Jin, H., Yang, L.T., Tsai, J.J.-P. (eds.) UIC 2006. LNCS, vol. 4159, pp. 912–923. Springer, Heidelberg (2006)
11. Peris-Lopez, P., Hernandez-Castro, J.C., Estevez-Tapiador, J., Ribagorda, A.: EMAP: An Efficient Mutual-Authentication Protocol for Low-cost RFID tags. In: OTM Federated Conferences and Workshop: IS Workshop (2006)
12. Avoine, G.: Adversarial Model for Radio Frequency Identification. Cryptology ePrint Archive, Report 2005/049 (2005)
13. Ohkubo, M., Suzuki, K., Kinoshita, S.: Cryptographic Approach to Privacy-Friendly Tags. In: RFID Privacy Workshop (2003)
14. Henrici, D., Muller, P.: Hash-based Enhancement of Location Privacy for Radio Frequency Identification Devices using Varying Identifiers. In: Proceedings of PerSec 2004, IEEE PerCom, pp. 149–153 (2004)
15. Henrici, D., Muller, P.: Providing Security and Privacy in RFID Systems Using Triggered Hash Chains. In: PerCom 2008, pp. 50–59 (2008)
16. Kulseng, L.S.: Lightweight Mutual Authentication, Owner Transfer, and Secure Search Protocols for RFID Systems, Master Thesis, Iowa State University, Ames (2009)
17. Alomair, B., Lazos, L., Poovendran, R.: Securing Low-cost RFID Systems: an Unconditionally Secure Approach. In: RFIDsec 2010, Asia, Singapore (2010)
18. Feller, W.: An Introduction to Probability Theory and its Applications. Wiley India Pvt. Ltd, Chichester (2008)
19. Heydt-Benjamin, T.S., Bailey, D.V., Fu, K., Juels, A., O'Hare, T.: Vulnerabilities in first-generation RFID-enabled credit cards. In: Dietrich, S., Dhamija, R. (eds.) FC 2007 and USEC 2007. LNCS, vol. 4886, pp. 2–14. Springer, Heidelberg (2007)
20. Carluccio, D., Lemke, K., Paar, C.: E-passport: The global traceability or how to feel like a UPS package. In: Lee, J.K., Yi, O., Yung, M. (eds.) WISA 2006. LNCS, vol. 4298, pp. 391–404. Springer, Heidelberg (2007)
21. Hoepman, J.-H., Hubbers, E., Jacobs, B., Oostdijk, M., Schreur, R.W.: Crossing borders: Security and privacy issues of the european e-passport. In: Yoshiura, H., Sakurai, K., Rannenberg, K., Murayama, Y., Kawamura, S.-i. (eds.) IWSEC 2006. LNCS, vol. 4266, pp. 152–167. Springer, Heidelberg (2006)
22. CASPIAN, Boycott Benetton (2007), http://www.boycottbenetton.com
23. Mitsubishi Electric Asia Switches on RFID (2006), http://www.rfidjournal.com/article/articleview/2644/
24. Target, Wal-Mart Share EPC Data (2005), http://www.rfidjournal.com/article/articleview/642/1/1/

Cryptanalysis of RSA with Small Prime Combination*

Xianmeng Meng

School of Statistics and Mathematics
Shandong University of Finance
Jinan 250014, P.R. China
mxmeng@gmail.com

Abstract. Let $N = pq$ be RSA modulus where primes p and q are of the same bit-length. If $|\rho q - p| = N^{\frac{1}{4}+\gamma}$ where ρ is a known constant satisfying $1 \leq \rho \leq 2$ and the constant γ satisfies $0 < \gamma < \frac{1}{4}$, we show the factorization attack on N and weak key attack against RSA modulus N. We present algorithms to find the factorization of N in time $O(N^{\gamma+\varepsilon})$ by some square root attacks, such as the baby-step giant-step method and a more sophisticated square root attack. Using similar techniques of Blömer and May (PKC 2004), we present a weak key attack and find new weak keys over the work of Maitra and Sarkar (ISC 2008).

Keywords: RSA, Factorization, Weak Keys.

1 Introduction

RSA is a public key cryptosystem, which was proposed by Rivest, Shamir and Adleman [12] in 1978. The public exponent e and the private exponent d are chosen to be inverse of each other modulo $\phi(N) = (p-1)(q-1)$.

It is well-known that small prime difference makes RSA insecure. In [14], Weger showed that if $p - q$ is small, RSA system with small exponent is much more vulnerable. In [8] [9], Maitra and Sarker revisited Wiener's continued fraction method and showed that given ρ ($1 \leq \rho \leq 2$) is known to the attacker the RSA keys are weak when $d = N^\delta$ and $\delta < \frac{1}{2} - \frac{\gamma}{2}$, where the RSA modulus $N = pq$ with $|\rho q - p| \leq \frac{N^\gamma}{16}$. And using similar techniques they also present new result over the work of Blömer and May [1]. In fact, a similar result was also obtained independently by Han and Xu [5].

In this paper, we investigate the square root attack on factoring $N = pq$ with small $|\rho q - p|$ where ρ is a known constant satisfying $1 \leq \rho \leq 2$. We show that for $|\rho q - p| = N^{\frac{1}{4}+\gamma}$ and $0 < \gamma < \frac{1}{4}$, one can find the factorization of N in time $O(N^{\gamma+\varepsilon})$. Furthermore, we present a weak key attack against RSA with small $|\rho q - p|$. We find new weak keys over the work of Maitra and Sarkar [8]. And we are

---

* This research is partially supported by Project 973 (no: 2007CB807902) and the science and technology foundation of the ministry of education (no: 210123) and the natural science foundation in Shandong province (no: Y2008A22) in China.

K.-H. Rhee and D. Nyang (Eds.): ICISC 2010, LNCS 6829, pp. 104–112, 2011.

able to show that for $|\rho q - p| = N^{\frac{1}{4}+\gamma}$ there are at least $N^{1-\gamma-\varepsilon}$ weak RSA-keys. From this, we get a probabilistic factorization algorithm with expected running time $O(N^{\gamma+\varepsilon})$. This result perfectly matches our factoring algorithm result.

On the other hand, one can see that the weak key attack has a nice interpolation property towards our factoring algorithm: As $|\rho q - p|$ decreases, the number of weak public keys increases. For γ approaching zero almost all keys are weak, corresponding to the fact that N can be factored without any hint that is encoded in e.

The remainder of this paper is organized as follows. In section 2, we investigate algorithms of factorization. In section 3, we present new result on weak RSA keys. We conclude this paper in section 4.

2   Factorization Attack

Let $N = pq$ be an RSA-modulus, where p and q are primes of equal bit-size (wlog $q < p < 2q$). Let $p - q = N^\theta$. If $0 < \theta \leq \frac{1}{4}$, we have that $p - \sqrt{N} < p - q \leq N^{\frac{1}{4}}$. Then it is known that the factorization of N can be recovered by the following Coppersmith's theorem [2].

Theorem 1 (Coppersmith). *Let $N = pq$ where primes p and q are of the same bit-size. Suppose we given an approximation of p with additive error at most $N^{\frac{1}{4}}$. Then N can be factored in time polynomial in $\log N$.*

If $\frac{1}{4} < \theta < \frac{1}{2}$, Weger [14] showed that the Fermat factoring method can factor N in time $O(N^{2\theta-\frac{1}{2}})$. Here we show a square root attack by applying the baby-step giant-step method of Shanks [13] which deals with the discrete logarithm problem.

Theorem 2. *Let $N = pq$ where primes p and q satisfy $p - q = N^\theta$ with $\frac{1}{4} < \theta < \frac{1}{2}$, then N can be factored in time $O(N^{\theta-\frac{1}{4}+\varepsilon})$.*

Proof. We have
$$(p - q)^2 = (p + q)^2 - 4N = (p + q - 2\sqrt{N})(p + q + 2\sqrt{N}).$$
Hence
$$p + q - 2\sqrt{N} = \frac{(p-q)^2}{p+q+2\sqrt{N}} < \frac{(p-q)^2}{4\sqrt{N}} = \frac{1}{4}N^{2\theta-\frac{1}{2}}. \tag{1}$$
The exponent of the multiplicative group modulo N is $\mathrm{lcm}(p - 1, q - 1)$. Let $\lambda(N) = \mathrm{lcm}(p - 1, q - 1)$. Here as usual, we assume $\gcd(p - 1, q - 1) = 2$ and
$$\lambda(N) = \frac{(p-1)(q-1)}{2} = \frac{N+1}{2} - \frac{p+q}{2}. \tag{2}$$
Choose M to be an integer no less than $\frac{1}{2\sqrt{2}}N^{\theta-\frac{1}{4}}$. By (1) and (2), we have that there exist integers u_0 and v_0 such that
$$\lambda(N) = \frac{N+1}{2} - \left\lfloor \sqrt{N} \right\rfloor - u_0 - v_0 \cdot M$$
with $0 \leq u_0, v_0 < M$.

We choose an integer a such that $\gcd(a, N) = 1$. We have that $a^{\lambda(N)} \equiv 1 \bmod N$. Let $b = a^{\frac{N+1}{2} - \lfloor \sqrt{N} \rfloor}$. We construct two lists as follows,

$$L_1 = \left\{ a^{v \cdot M} \bmod N \,\middle|\, 0 \leq v < M \right\}$$

and

$$L_2 = \left\{ b \cdot a^{-u} \bmod N \,\middle|\, 0 \leq u < M \right\}.$$

According to Algorithm 7.5.1 in [3], we can sort these lists and find a common value $a^{v_0 \cdot M} = b \cdot a^{-u_0} \bmod N$ in time $O(M \log M)$. A low-storage alternative is to use Pollard's λ method [11]. Then u_0 and v_0 are known and $p + q$ can be recovered by $p + q = 2 \left(\lfloor \sqrt{N} \rfloor + u_0 + v_0 \cdot M \right)$. Finally we can get p and q from the solutions of the equation $y^2 - (p + q)y + N = 0$. Thus the factorization of N follows.

Instead of considering $p - q = N^\theta$, now we consider the case $2q - p = N^\theta$ to get an additional result. We present an algorithm to factor $N = pq$ with small $2q - p$ by using a more sophisticated square root attack which applied to RSA with small CRT secret exponent. One can see [10] for details.

Theorem 3. *Let $N = pq$ where primes p and q satisfy $2q - p = N^\theta$ with $\frac{1}{4} < \theta < \frac{1}{2}$, then N can be factored in time $O(N^{\theta - \frac{1}{4} + \varepsilon})$.*

Proof. We have

$$(2q - p)^2 = (p + 2q)^2 - 8N = (p + 2q - 2\sqrt{2N})(p + 2q + 2\sqrt{2N}).$$

If $2q - p = N^\theta$, then

$$p + 2q - 2\sqrt{2N} = \frac{(2q - p)^2}{p + 2q + 2\sqrt{2N}} < \frac{(2q - p)^2}{4\sqrt{2N}} \leq \frac{1}{4} N^{2\theta - \frac{1}{2}}.$$

Hence

$$\phi(N) = N + 1 - p - q = N + 2 - (p + 2q) + (q - 1)$$
$$= N + 2 - \lfloor 2\sqrt{2N} \rfloor - w + (q - 1) \tag{3}$$

with the unknown w satisfying $0 \leq w < \frac{1}{4} N^{2\theta - \frac{1}{2}}$.

By (3), we have that

$$N + 2 - \lfloor 2\sqrt{2N} \rfloor - w \equiv 0 \bmod (q - 1), \tag{4}$$

and

$$w - (N + 2 - \lfloor 2\sqrt{2N} \rfloor) \equiv 0 \bmod (q - 1). \tag{5}$$

Since $w < q - 1$ and q is a prime, we can recover w from equation (5). Choose an integer a such that $\gcd(a, N) = 1$. We have $a^{q-1} \equiv 1 \bmod q$. Thus by (5), we have

$$a^{w - (N + 2 - \lfloor 2\sqrt{2N} \rfloor)} = 1 \bmod q.$$

Choose M to be an integer no less than $\frac{1}{2}N^{\theta-\frac{1}{4}}$. Since $w \leq \frac{1}{4}N^{2\theta-\frac{1}{2}}$, we have that w can be written as $w = u_0 + v_0 \cdot M$ with the unknown integers u_0 and v_0 satisfying $0 \leq u_0, v_0 \leq M$.

Let $\bar{b} = a^{-(N+2-\lfloor 2\sqrt{2N}\rfloor)} \bmod N$, and

$$G(x) = \prod_{v=0}^{M} (\bar{b} \cdot a^{M \cdot v}x - 1) \bmod N.$$

Computing the polynomial $G(x)$ takes time $O(M)$ and storing $G(x)$ requires space $O(M)$. Note that $G(a^{u_0}) \equiv 0 \bmod q$ since $w = u_0 + v_0 \cdot M$. Evaluate $G(x) \bmod N$ at a^u for all $0 \leq u \leq M$. This gives a list of M numbers, one of which has a non-trivial gcd with N. Then the factorization of N is obtained.

This method requires the evaluation of $G(x)$ at M points. By the Fast Fourier Transform (FFT), one can evaluate a polynomial of degree M at M points in time $O(M \log M)$ operations (see Chapter 10 in [4]). Since $G(x)$ is a polynomial of degree $M+1$, these M numbers can be obtained in time $O(M \log M)$ operations. One of these values has a non-trivial gcd with N, which can be calculated in time $O(M \log N)$ operations by the Euclidean algorithm. This completes the proof of Theorem 3.

Now we consider the case $|\rho q - p| = N^\theta$ with a known constant ρ satisfying $1 < \rho < 2$.

Theorem 4. *Let* $N = pq$ *be an integer where primes* p *and* q *satisfy* $|\rho q - p| = N^\theta$ *with* $\frac{1}{4} < \theta < \frac{1}{2}$. *If* ρ *is a known simple fraction between* 1 *and* 2, *then* N *can be factored in time* $O(N^{\theta-\frac{1}{4}+\varepsilon})$.

Proof. Since ρ is a known simple fraction between 1 and 2, we write $\rho = \frac{s}{t}$ with the known integers s and t satisfying $s > t > 0$ and $\gcd(s,t) = 1$. We have $|sq - tp| = tN^\theta$, and

$$sq + tp - 2\sqrt{stN} = \frac{(sq - tp)^2}{sq + tp + 2\sqrt{stN}}$$
$$< \frac{t^2 N^{2\theta}}{4\sqrt{stN}} < \frac{t}{4}N^{2\theta-\frac{1}{2}}. \tag{6}$$

Since

$$t(N - p) - s(q - 1) = 0 \bmod (q - 1),$$

we have

$$tN + s - (tp + sq) = 0 \bmod (q - 1). \tag{7}$$

By (6), we have that $tp + sq$ can be written as

$$tp + sq = \left\lfloor 2\sqrt{stN} \right\rfloor + w,$$

with $0 < w \leq \frac{t}{4}N^{2\theta-\frac{1}{2}}$. Thus by (7) we have

$$tN + s - \left\lfloor 2\sqrt{stN} \right\rfloor - w = 0 \bmod (q-1). \tag{8}$$

This is similar to equation (4). Then similarly as proving Theorem 3, we have that Theorem 4 holds.

3   Weak Key Attack and New Weak Keys

In [1], Blömer and May proved that p and q can be found in polynomial time for every N and e satisfying $ex + y = 0 \bmod \phi(N)$, with $0 < x \leq \frac{1}{3}\sqrt{\frac{\phi(N)}{e}} \frac{N^{\frac{3}{4}}}{p-q}$ and $|y| \leq \frac{p-q}{\phi(N)N^{\frac{1}{4}}}ex$. And if $p - q = N^{\frac{1}{4}+\gamma}$, it is proved that the number of these *weak keys* is at least $N^{1-\gamma-\varepsilon}$. In [8], Maitra and Sarkar showed that the same result holds for $2q - p = N^{\frac{1}{4}+\gamma}$ instead of $p - q$. Later in [9], Maitra and Sarkar considered the case $|\rho q - p| = N^{\frac{1}{4}+\gamma}$, and proved that the factorization of N can be recovered in polynomial time for every N and e satisfying $ex + y = 0 \bmod \phi(N)$, with $0 < x \leq \frac{1}{6}\sqrt{\frac{\phi(N)}{e}}N^{\frac{3-4\gamma}{8}}$ and $|y| \leq \frac{|\rho q-p|}{\phi(N)N^{\frac{1}{4}}}ex$. And it is obtained that the number of these *weak keys* is at least $N^{\frac{1}{2}-\varepsilon}$. In this section, we revisit the weak key attack and find new weak keys. As usually, we formalize the notion of weak keys as follows.

Definition 1. *Let C be a class of RSA public keys (N, e). The size of the class C is defined by*

$$Size_C(N) = \left|\{e \in \mathbb{Z}^*_{\phi(N)}|(N, e) \in C\}\right|.$$

A class C is called weak if:
1. $Size_C(N) = \Omega(N^\tau)$ for some $\tau > 0$;
2. There exists a probabilistic algorithm A that on every input $(N, e) \in C$ outputs the factorization of N in time polynomial in $\log N$.

Now we revisit the weak key attack and then present new weak keys over the work of Maitra and Sarkar [9].

Theorem 5. *Let $N = pq$ be an integer with primes satisfying $11N^{\frac{1}{4}} \leq |\rho q-p| = N^\theta < \sqrt{N}$ where ρ is a known constant satisfying $1 < \rho \leq 2$. Denote $\rho - 1$ by $\bar{\rho}$. Define $[\bar{\rho}q] = \lfloor\bar{\rho}q\rfloor$ if $\lfloor\bar{\rho}q\rfloor$ is even, otherwise $[\bar{\rho}q] = \lceil\bar{\rho}q\rceil$. Suppose that e satisfies the equation*

$$ex + y + k[\bar{\rho}q] = k\phi(N) \tag{9}$$

for $k > 0$. Then N can be factored in polynomial time in $\log N$ when $0 < x \leq \frac{1}{3}\sqrt{\frac{\phi(N)}{e}}\frac{N^{\frac{3}{4}}}{|\rho q-p|}$ and $|y| \leq \frac{|\rho q-p|}{\phi(N)N^{\frac{1}{4}}}ex$.

We will apply the following classical theorem on diophantine approximations (see Corollary 2, [1, § 2] in [6]).

Lemma 1 (Legendre). *Let ξ be a real number. If the coprime integers X and Y satisfy*

$$\left| \xi - \frac{X}{Y} \right| < \frac{1}{2Y^2},$$

then $\frac{X}{Y}$ is a convergent of ξ.

Now we show the proof of Theorem 5, which is similar to the proof of Theorem 4 in [1].

Proof. We have $(\rho q - p)^2 = N^{2\theta} = (p + \rho q)^2 - 4\rho N = (p + \rho q - 2\sqrt{\rho N})(p + \rho q + 2\sqrt{\rho N})$, hence

$$p + \rho q - 2\sqrt{\rho N} = \frac{N^{2\theta}}{(p + \rho q + 2\sqrt{\rho N})} < \frac{N^{2\theta}}{4\sqrt{N}}.$$

Rearrange the terms in (9), we have

$$ex + y = k(N + 1 - p - \lceil \rho q \rceil).$$

Let $z = 2\lfloor \sqrt{\rho N} \rfloor - p - \lceil \rho q \rceil$, and $M = N + 1 - 2\lfloor \sqrt{\rho N} \rfloor$. Then by the above, we have that

$$ex + y = k(M + z), \tag{10}$$

and

$$\frac{e}{M} - \frac{k}{x} = \frac{kz - y}{Mx}. \tag{11}$$

Here we can assume $\gcd(k, x) = 1$. Since if $\gcd(k, x) = t$, we have that t divides y, which gives us an equation $ex' + y' = 0 \pmod{M + z}$ with even smaller parameters x' and y'. Hence we can assume that $\frac{k}{x}$ is a fraction in its lowest terms.

By Lemma 1, the fraction $\frac{k}{x}$ is among the convergents of the continued fraction expansion of $\frac{e}{M}$ if $|\frac{e}{M} - \frac{k}{x}| < \frac{1}{2x^2}$ is satisfied. Thus it remains to show that

$$\left| \frac{kz - y}{Mx} \right| < \frac{1}{2x^2}.$$

We have $|y| \leq \frac{1}{4}ex$ and

$$\frac{3}{4}\frac{ex}{\phi(N)} \leq k = \frac{ex + y}{M} \leq \frac{6}{4}\frac{ex}{\phi(N)},$$

where the last inequality holds for $N > 288$. Wlog we assume $N > 288$ during the proof of the theorem. We have $|z| \leq \frac{1}{4}N^{2\theta - \frac{1}{2}}$ and

$$|kz - y| \leq |kz| + |y| \leq \frac{6}{4}\frac{ex}{\phi(N)}\frac{1}{4}N^{2\theta - \frac{1}{2}} + N^{\theta - \frac{1}{4}}x \leq \frac{ex}{\phi(N)}N^{2\theta - \frac{1}{2}}.$$

Therefore we have to satisfy

$$\frac{ex}{\phi(N)}N^{2\theta - \frac{1}{2}}\frac{1}{Mx} < \frac{1}{2x^2}.$$

This holds by our upper bound $x \leq \frac{1}{3}\sqrt{\frac{\phi(N)}{e}}\frac{N^{\frac{3}{4}}}{|\rho q - p|}$ and $N > 288$.

Hence the fraction $\frac{k}{x}$ must be among the convergents of the continued fraction expansion of $\frac{e}{M}$. There are only $O(\log N)$ many convergents, thus we can apply the following process to each candidate for k and x until our algorithm succeeds.

We now show that the correct k and x yield the factorization of N. Let us write (10) as

$$\frac{ex}{k} - M = z - \frac{y}{k}. \tag{12}$$

Since the left is know to us, we can compute an approximation of z up to some unknown error term $\frac{y}{k}$, which can be bounded by $|\frac{y}{k}| \le \frac{4}{3}N^{\theta - \frac{1}{4}}$ using (11). Let $S = 2\lfloor\sqrt{\rho N}\rfloor + M - \frac{ex}{k}$. Since $p + [\rho q] = 2\lfloor\sqrt{\rho N}\rfloor - z$, we have that S is an approximation of $p + [\rho q]$ with additive error at most $\frac{4}{3}N^{\theta - \frac{1}{4}}$ using (12). Let $T = \sqrt{S^2 - 4\rho N}$. Now we show that T is well defined by proving $S^2 - 4\rho N \ge 0$. Since $S = p + [\rho q] - \frac{y}{k}$, we write $S = p + \rho q + \varpi$ with $|\varpi| \le |\frac{y}{k}| + 1 \le \frac{4}{3}N^{\theta - \frac{1}{4}} + 1$. We have that $S^2 - 4\rho N = (p + \rho q + \varpi)^2 - 4\rho N = (\rho q - p)^2 + 2\varpi(p + \rho q) + \varpi^2$. Since $p + \rho q \le (2 + \sqrt{2})\sqrt{N}$, we have that $2\varpi(p + \rho q) \le 2(\frac{4}{3}N^{\theta - \frac{1}{4}} + 1) \cdot (2 + \sqrt{2})\sqrt{N} < 11N^{\theta + \frac{1}{4}} \le N^{2\theta}$, which holds by our lower bound $N^\theta \ge 11N^{\frac{1}{4}}$. This implies $S^2 - 4\rho N \ge 0$.

We now show that T is an approximation of $|\rho q - p|$ with an additive error that can be bounded by $N^{\frac{1}{4}}$. Since $|\frac{y}{k}| \le \frac{4}{3}N^{\theta - \frac{1}{4}} < \frac{1}{2}\sqrt{N} \le \frac{1}{4}(p + \rho q)$, we have that $S = p + [\rho q] + \frac{y}{k} \le \frac{5}{4}(p + \rho q)$. Then we have

$$\begin{aligned}
T - |\rho q - p| &= \sqrt{S^2 - 4\rho N} - |\rho q - p| \\
&= \frac{(S - (p + \rho q))(S + (p + \rho q))}{\sqrt{S^2 - 4\rho N} + |\rho q - p|} \\
&\le \frac{\frac{4}{3}N^{\theta - \frac{1}{4}} \cdot \frac{9(2 + \sqrt{2})}{4}\sqrt{N}}{N^\theta} \le 12N^{\frac{1}{4}}.
\end{aligned}$$

We suppose that $\rho q > p$. Thus the term $\frac{1}{2}(S - T)$ is an approximation of p with error at most

$$\begin{aligned}
\left|\frac{1}{2}(S - T) - p\right| &= \frac{1}{2}\left(|S - p - \rho q - T - p + \rho q|\right) \\
&\le \frac{1}{2}\left(|S - p - \rho q| + |T - \rho q + p|\right) \\
&\le \frac{2}{3}N^{\theta - \frac{1}{4}} + \frac{12}{2}N^{\frac{1}{4}} \le 7N^{\frac{1}{4}}.
\end{aligned}$$

Let $\bar{p} = \frac{1}{2}(S - T)$. Then one of the seven values $\bar{p} + 2kN^{\frac{1}{4}}, k = -3, -2, -1, 0, 1, 2, 3$ is an approximation of p up to an error of at most $N^{\frac{1}{4}}$ in absolute value. We apply Coppersmith's algorithm (Theorem 1) to all these values. The correct term will then lead to the factorization of N.

If $\rho q < p$, then the term $\frac{1}{2}(S + T)$ is an approximation of p, and the factorization of N can also be recovered. This completes the proof of Theorem 5.

Every key tuple (N, e) that satisfies (9) with $0 < x \leq \frac{1}{3}\sqrt{\frac{\phi(N)}{e}} \frac{N^{\frac{3}{4}}}{|\rho q - p|}$ and $|y| \leq \frac{|\rho q - p|}{\phi(N) N^{\frac{1}{4}}} ex$ yields the factorization of N in polynomial time. These tuples (N, e) are weak keys that should not be used in the design of a crypto-system.

Our attack in Theorem 5 defines a weak class C. We are going to discuss how large this weak class is. We are interested in a lower bound of C.

Theorem 6. *Let p and q be safety prime numbers with $|\rho q - p| = N^{\frac{1}{4}+\gamma}$ and $0 < \gamma < \frac{1}{4}$. Further, let C be the weak key class that is given by the public key tuples (N, e) defined in Theorem 5 with the additional restriction that $e \in \mathbb{Z}^*_{\phi(N)}$ and $e > \frac{\phi(N)}{4}$, then*

$$Size_C(N) = \Omega\left(\frac{N^{1-\gamma}}{\log\log^2(N^2)}\right).$$

Proof. Denote $\Theta(N) = \phi(N) - [\bar{\rho}q]$. Let $\bar{C}$ be the weak keys that is given by the public key tuples (N, e) defined in Theorem 5 with the additional restriction that $e \in \mathbb{Z}^*_{\Theta(N)}$ and $e > \frac{\Theta(N)}{4}$. Similarly as the proof of Theorem 7 in [1], we have

$$Size_{\bar{C}}(N) = \Omega\left(\frac{N^{1-\gamma}}{\log\log^2(N^2)}\right).$$

If p and q are safety prime numbers, then $p = 2p' + 1$ and $q = 2q' + 1$, and $\phi(N) = 4p'q'$ with p' and q' being primes. Thus we have

$$\sum_{\substack{i=\phi(N)/4 \\ \gcd(i,\phi(N)/4)\neq 1}}^{\phi(N)} 1 \leq 3\sqrt{N}.$$

We have

$$Size_C(N) = \left|\{e \in \mathbb{Z}^*_{\phi(N)} | (N, e) \in C\}\right|$$

$$\geq \left|\{e \in \mathbb{Z}^*_{\Theta(N)} | (N, e) \in C, \gcd(e, \phi(N)) = 1\}\right|$$

$$\geq Size_{\bar{C}}(N) - 3\sqrt{N}$$

$$= \Omega\left(\frac{N^{1-\gamma}}{\log\log^2(N^2)}\right).$$

This completes the proof of Theorem 6.

Thus when p and q are safety prime numbers, the result of Theorem 5 presents new weak keys other than those $N^{\frac{1}{2}-\varepsilon}$ weak keys presented in [8].

4   Conclusion

In this paper, we investigate the cryptoanalysis of RSA modulus $N = pq$ with small prime combination $|\rho q - p|$ where ρ is a known constant $(1 \leq \rho \leq 2)$. We show some square root attacks on the factorization of N, and also present a weak key attack against RSA modulus N. When p and q are safety prime numbers, we find new weak keys over the work of Maitra and Sarkar [8].

References

1. Blömer, J., May, A.: A generalized wiener attack on RSA. In: Bao, F., Deng, R., Zhou, J. (eds.) PKC 2004. LNCS, vol. 2947, pp. 1–13. Springer, Heidelberg (2004)
2. Coppersmith, D.: Small solutions to polynomial equations and low exponent RSA vulnerabilities. Journal of Cryptology 10(4), 223–260 (1997)
3. Crandall, R., Pomerance, C.: Prime Numbers, 2nd edn. Springer, Heidelberg (2005)
4. von zur Gathen, J., Gerhard, J.: Modern Computer Algebra, 2nd edn. Cambridge University Press, Cambridge (2003)
5. Han, L.D., Xu, G.W.: Generalization of Some Attacks on RSA with Small Prime Combination and Small Private Exponent. In: 2009 Asia-Pacific Conference on Information Processing, vol. 1, pp. 445–449 (2009)
6. Lang, S.: Introduction to diophantine approximations. Addison-Wesley Pub. Co., Reading (1966)
7. Lewis, D.J. (ed.): Number Theory Institute 1969. Proceedings of Symposia in Pure Mathematics, vol. 20. American Mathematical Society, Providence RI (1971)
8. Maitra, S., Sarkar, S.: Revisiting Wiener's attack - new weak keys in RSA. In: Wu, T.-C., Lei, C.-L., Rijmen, V., Lee, D.-T. (eds.) ISC 2008. LNCS, vol. 5222, pp. 228–243. Springer, Heidelberg (2008)
9. Maitra, S., Sarkar, S.: Revisiting Wiener's attack - new weak keys in RSA, http://eprint.iacr.org/2008/228.pdf
10. Nguyen, P.Q.: Recent Trends in Cryptography. In: Luengo, I. (ed.) Public-Key Cryptanalysis. Contemporary Mathematics series, vol. 477, AMS-RSME (2009)
11. Pollard, J.M.: Monte Carlo methods for index computation (mod p). Math. Comp. 32, 918–924 (1978)
12. Rivest, R.L., Shamir, A., Adleman, L.: A method for obtaining digital signatures and public-key cryptosystems. Commun. of the ACM 21, 120–126 (1978)
13. Shanks, D.: Class number, a theory of factorization and genera. In: Lewis [7], pp. 415–440 (1971)
14. de Weger, B.: Cryptanalysis of RSA with small prime difference. Applicable Algebra in Engineering, Communication and Computing 13, 17–28 (2002)

The Twin Bilinear Diffie-Hellman Inversion Problem and Applications

Yu Chen[1] and Liqun Chen[2]

[1] School of Electronics Engineering and Computer Science, Peking University, China
Key Laboratory of High Confidence Software Technologies, Ministry of Education
`chenyu@infosec.pku.edu.cn`
[2] Hewlett-Packard Laboratories, Bristol, United Kingdom
`liqun.chen@hp.com`

Abstract. We propose a new computational problem and call it the *twin bilinear Diffie-Hellman inversion (BDHI) problem*. Inspired by the technique proposed by Cash, Kiltz and Shoup, we have developed a new trapdoor test which enables us to prove that the twin BDHI problem is at least as hard as the ordinary BDHI problem even in the presence of a decision oracle that recognizes a solution to the problem. The relation between the two problems implies that many of the cryptographic constructions based on ordinary BDHI problem can be improved with a tighter security reduction. As one such application, we present a new variant of Sakai-Kasahara Identity-Based Encryption (SK-IBE) with a simple and efficient security proof in the random oracle model, under the computational BDHI problem. We also present a new Identity-Based Key Encapsulation Mechanism (ID-KEM) based on SK-IBE, which has a better security analysis than previous results.

Keywords: bilinear Diffie-Hellman inversion problem, twin bilinear Diffie-Hellman inversion problem, trapdoor test, identity-based encryption, identity-based key encapsulation.

1 Introduction

The bilinear Diffie-Hellman inversion (BDHI) problem has found its applications in many cryptographic constructions, such as identity-based encryption (IBE) [6, 11, 22], identity-based key encapsulation mechanism [12], identity-based signatures (IBS) [2], identity-based signcryption [2], identity-based key agreement (IBKA) [14, 21], and verifiable random function [18] etc.

In this paper, we introduce a new problem based on the original BDHI problem, named the *twin bilinear Diffie-Hellman inversion problem*, which has the following properties:

- The twin BDHI problem can easily be employed in many cryptographic constructions where one would usually use the ordinary BDHI problem; as a result, it will improve the security analysis of these constructions.
- The twin BDHI problem is hard, even given access to the corresponding decision oracle, assuming the ordinary BDHI problem is hard.

K.-H. Rhee and D. Nyang (Eds.): ICISC 2010, LNCS 6829, pp. 113–132, 2011.

1.1   Related Work

In EuroCrypt 2008, Cash, Kiltz and Shoup [10] proposed a new computational problem called the *twin Diffie-Hellman (DH) problem*, i.e. given a random triple of the form $(X_1, X_2, Y) \in \mathbb{G}^3$ for a cyclic group $\mathbb{G}$, compute $\text{dh}(X_1, Y)$ and $\text{dh}(X_2, Y)$, where dh is the DH function. They also proposed the *strong twin DH problem*, which is the twin DH problem under the condition that an adversary is given access to a corresponding decision twin DH oracle. They developed an ingenious trapdoor test, which enables them to prove that the strong twin Diffie-Hellman problem is as hard as the original Diffie-Hellman problem. They also extended this technique to the bilinear Diffie-Hellman (BDH) problem.

Trapdoor test is the main contribution of [10]. Concretely speaking, when a DH adversary $\mathcal{B}$ is given a challenge $(g, X, Y) = (g, g^x, g^y)$, it creates a twin DH challenge with a trapdoor by setting $X_1 = X$ and $X_2 = g^s / X_1^r$. In this way, a linear relation with *two degrees of freedom*: $x_2 = s - rx_1$ is embedded into the twin DH challenge. Based on the observation that the solution exponents $(x_1 \hat{y}, x_2 \hat{y})$ are linear to (x_1, x_2), $\mathcal{B}$ can answer the twin DH decision queries by testing if the linear relationship holds accordingly between $\hat{Z}_1$ and $\hat{Z}_2$. This trapdoor test can be extended to BDH problem in an analogous way without any difficulty, since the BDH problem is a natural extension of the DH problem in groups equipped with a bilinear map. Using the trapdoor test as a tool, they proved that the strong twin DH/BDH problem is as hard as the ordinary DH/BDH problem. Benefiting from this result, they improved a bunch of cryptographic schemes by tailoring them to fit the twin-type problem, such as Diffie and Hellman non-interactive key exchange protocol [17], Cramer-Shoup encryption [15], Boneh-Franklin identity based encryption [7], etc. Particularly, we call the tailoring method as *twinning technique*.

For clarity of further discussion, we take a close look at how to apply the twinning technique to Boneh-Franklin IBE (BF-IBE for short) scheme. The twin BF-IBE uses two hash functions, K (outputs symmetric keys) and H (hashes identities to group elements), and a symmetric cipher (Enc, Dec). The master public key is (X_1, X_2), where $X_i = g^{x_i}$ for $i = 1, 2$. The master private key is (x_1, x_2). The private key for an identity $\text{ID} \in \{0, 1\}^*$ is $(S_1, S_2) = (H(\text{ID})^{x_1}, H(\text{ID})^{x_2})$. To encrypt a message M for identity ID, one chooses $y \in \mathbb{Z}_p$ at random and sets $Y = g^y$, $Z_1 = e(H(\text{ID}, X_1))^y$, $Z_2 = e(H(\text{ID}, X_2))^y$, derives $k = K(\text{ID}, Y, Z_1, Z_2)$, encrypts $C = \text{Enc}(k, M)$. To decrypt using the private key (S_1, S_2) for ID, one computes $Z_1 = e(S_1, Y)$, $Z_2 = e(S_2, Y)$, $k = K(\text{ID}, Y, Z_1, Z_2)$, then decrypts $M = \text{Dec}(k, C)$. We highlight that this is essentially a KEM-DEM construction.

We study the possible improvements brought by the twinning technique as follows, with an emphasis on public key encryption schemes based on the DH-type problems in the random oracle model [4].

1. Tighten the security reductions to the computational DH-type problems. Public key encryption schemes based on the DH-type problems can be divided into two groups, one is based on the decisional problem [6, 16, 19], the other one is based on the computational problems [1, 7, 22]. In the provable

security paradigm, the standard security notions (CPA and CCA) of public key encryption systems are formally defined via an interactive decisional problem: it is hard for an adversary to distinguish the challenge ciphertext is the encryption of M_0 or M_1. When the underlying hard problem is also a decisional problem, the simulator always outputs its solution based on the adversary's output. When the underlying hard problem is a computational problem, a common proving technique in the random oracle model is the simulator extracting the solution from a randomly picked entry in the associated random oracle query list, whose correctness is based on the assertion that if the adversary can answer the decision problem with advantage ϵ, then it must issue the associated query related to the challenge with probability at least 2ϵ. However, the "random picking" step loses a factor of Q_h in the security reduction, where Q_h is the maximum number of random oracle queries an adversary can make. If applying the twin technique to the original scheme, the simulator can identify the wanted query precisely with the help of a corresponding decision oracle. Thereby the security reduction can be tighten by a multiplicative factor Q_h immediately.

2. Facilitate a kind of redundancy free KEM construction without making a stronger assumption.

For the public key encryption schemes based on the DH/BDH problem, there exists a simple and elegant KEM construction which derives a symmetric key from a DH/BDH tuple. For example, in DHIES [1] (based on the strong DH assumption), Abdalla, Bellare and Rogaway constructed the KEM as $k := K(y^r)$, where $y = g^x$ is the public key. The associated ciphertext C is g^r. In [20], Libert and Quisquater proposed a redundancy free variant of BF-IBE (based on the Gap-BDH assumption). Their scheme essentially adopts the KEM-DEM methodology, and the ID-KEM is constructed as $k := K(e(g^a, g^b)^r)$ [20], while g^a is the master public key and g^b is the public key of ID. The associated ciphertext C is g^r. It is easy to see that (g^x, g^r, y^r) is a DH tuple, and $(g^a, g^b, g^r, e(g^a, g^b)^r)$ is a BDH tuple. Compare to other KEM constructions, the advantage of this special construction is that it is redundancy free. So it is not very surprisingly that the associated security reduction requires a decisional DH/BDH oracle available to distinguish "valid" decapsulation queries. This explains why such a kind of KEM construction has to resort to the strong DH/BDH assumptions. Via applying twinning technique to this kind of KEM construction, the security can be reduced to the ordinary assumption at a minor cost of the ciphertext size and the computation overhead. For example, Cash, Kiltz and Shoup [10] improved ElGamal encryption and BF-IBE in this way. The resulting schemes are redundancy free.

1.2   Our Contributions

Cash, Kiltz and Shoup [10] mentioned that their ideas can also be applied to the Sakai-Kasahara IBE scheme (SK-IBE for short) [22] based on the BDHI assumption in an analogous way as they did to BF-IBE. However, after investigating their approach, it appears to us that building such a trapdoor test for the BDHI problem and employing it to the SK-IBE or the SK-ID-KEM is in fact not trivial.

Our motivation of this work is to figure out: 1) how to construct a trapdoor test for the BDHI problem, and 2) how to apply the twinning technique to the schemes based on the BDHI problem.

The contributions of this paper can be summarized as follows:

1. We first define the *twin* BDHI problem as

$$2\text{bdhi}(g_1, g_1^x, \ldots, g_1^{x^q}; g_2, g_2^y, \ldots, g_2^{y^q}) \mapsto (e(g_1, g_1)^{1/x}, e(g_2, g_2)^{1/y})$$

then devise a trapdoor test which allow a BDHI adversary to implement an effective decision oracle for the twin bdhi predicate:

$$2\text{bdhip}(g_1, g_1^x, \ldots, g_1^{x^q}, T_x; g_2, g_2^y, \ldots, g_2^{y^q}, T_y) :=$$
$$(e(g_1, g_1)^{1/x}, e(g_2, g_2)^{1/y}) \stackrel{?}{=} (T_x, T_y)$$

Note that the formalization of the BDHI problem is quite different from that of the DH/BDH problem, due to the exponentiation $1/x$ of the solution not being linear to the exponentiations x^i appearing in the problem. So it is not straightforward to construct a trapdoor test for the BDHI problem in a similar way as Cash, Kiltz and Shoup did for the DH/BDH problem. We overcome this obstacle by embedding two relations in the twin BDHI instances simultaneously: one is between the generators by setting $g_2 = g_1^{ax+b}$; the other one is between the exponentiations by setting $y = cx$. Exploiting the bilinearity of the pairing, the two relations are verifiable without knowing the corresponding discrete logarithms x^i. In this way we build an efficient trapdoor test for the BDHI problem. With the trapdoor test as a basic tool, we prove that *strong* twin BDHI problem is at least as hard as the ordinary BDHI problem. Thus a decisional twin BDHI oracle is available in the security reduction without relying on a strong assumption.

2. SK-IBE [22], as a representative of the IBE family featured by *exponent inversion* [9], is proven to be secure based on the computational BDHI problem [11]. If we follow the strategy that Cash, Kiltz and Shoup used in twin BF-IBE [10] to improve SK-IBE, we need to construct an ID-KEM as $k = K(y^r, y, e(g, g)^r)$, where the public key of ID is y and the ciphertext is y^r [13]. As the preceding remark indicates, a special oracle named the bilinear inversion DH predicate (bidhp) [13] is needed to keep the simulation coherent in the security reduction (bidhp$(g, g^a, g^b, T) := e(g, g)^{a/b} \stackrel{?}{=} T$). Apparently such an oracle cannot be built from the trapdoor test for the BDHI problem, since the bidhp input is incompatible with the bdhip input. So their strategy is not applicable for SK-IBE. However, we can still apply the twinning technique to SK-IBE in a common way to enjoy a tighter security reduction, just as we analyzed before. We show how to tailor SK-IBE [11]

and SK-ID-KEM [12] to fit the twin BDHI problem, yielding twin SK-IBE and twin SK-ID-KEM, which outperform the original schemes in terms of efficiency especially for security analysis. A comparison to prior schemes is shown in Table 1.

Table 1. Comparison to prior schemes

Scheme	Type	Ciphertext size	Key size	Enc	Dec	Reduction factor
Chen et al.[11]	IBE	$|\mathbb{G}| + 2\ell$	$|\mathbb{G}|$	2E	1E+1P	$1/Q_{h_1}Q_{h_2}(Q_{h_3} + Q_{h_4})$
Ours	IBE	$2|\mathbb{G}| + 2\ell$	$2|\mathbb{G}|$	4E	2E+2P	$1/Q_{h_1}$
Chen et al.[12]	ID-KEM	$|\mathbb{G}| + \ell$	$|\mathbb{G}|$	2E	1E+1P	$1/Q_{h_1}Q_{h_2}(Q_{h_3} + Q_{h_4})$
Ours	ID-KEM	$2|\mathbb{G}| + \ell$	$2|\mathbb{G}|$	4E	2E+2P	$1/Q_{h_1}$

P denotes a pairing operation, and E a group exponentiation in $\mathbb{G}$ or $\mathbb{G}_T$. A common estimate used here are $Q_{h_i} = 2^{60}$ for $1 \le i \le 4$ (suggested by Bellare and Rogaway [4]).

Judging from appearances our schemes seem to be less efficient because our schemes double the number of group exponentiations and pairing operations in computation overheads and increase one group element in the ciphertext, one group element in the private key. Before illustrating why our twin schemes are more efficient than prior schemes at the same security level, we first review the concept of the tightness of security reduction [3, 19]. Consider a scheme with a security reduction showing that an adversary attacking the scheme in time t with advantage ϵ implies another adversary breaking some intractable problem in time $t + \omega_1$ with advantage $\epsilon' \ge \epsilon/\omega_2$. The tightness of security reduction refers to the values of ω_1 and ω_2. If the underlying assumptions are the same, the scheme with a looser security reduction needs to increase the size of the groups to obtain the same security level of the one with a tighter security reduction. Using the general method, exponentiations and pairings in a group whose elements can be represented in r bits takes roughly $\mathcal{O}(r^3)$ time. As a concrete example, performing two 128-bit group exponentiations/pairings can be significantly faster than a single 256-bit group exponentiation/pairing. From the comparison in Table 1, we learn that to obtain the same security guarantee, the group size of [11] and [12] is roughly 120 bits larger than the size of our schemes. Therefore at the same security level, our twin schemes are more efficient than prior schemes while the ciphertext size and the key size are roughly the same.

2   The Twin Bilinear Diffie-Hellman Inversion Assumption

In this section, we introduce the twin bilinear Diffie-Hellman inversion problem. Let us first review some necessary facts about bilinear maps. Suppose $\mathbb{G}$ and $\mathbb{G}_T$ are two multiplicative groups of prime order p, and g is the generator of $\mathbb{G}$. The groups are equipped with a map $e : \mathbb{G} \times \mathbb{G} \to \mathbb{G}_T$ which satisfies the following properties:

- bilinear: for all $u, v \in \mathbb{G}$ and all $a, b \in \mathbb{Z}_p$, we have $e(u^a, v^b) = e(u, v)^{ab}$.
- non-degeneracy: $e(g, g) \neq 1_{\mathbb{G}_T}$.
- computable: there is an efficient algorithm to compute $e(u, v)$ for all $u, v \in \mathbb{G}$.

2.1 The BDHI Assumption

We recall that the bilinear Diffie-Hellman inversion (BDHI) problem, denoted by q-BDHI, is defined [6] as follows: given the $q + 1$ tuple $(g, g^x, \ldots, g^{x^q}) \in (\mathbb{G}^*)^{q+1}$ as input, compute $e(g, g)^{1/x} \in \mathbb{G}_T^*$. More formally, we explain this problem by using the following BDHI function:

$$\text{bdhi:} \quad \mathbb{G}^{q+1} \to \mathbb{G}_T$$
$$(g, g^x, \ldots, g^{x^q}) \mapsto e(g, g)^{1/x}$$

We also use a predicate

$$\text{bdhip}(g, g^x, \ldots, g^{x^q}, \hat{T}) := e(g, g)^{1/x} \stackrel{?}{=} \hat{T}$$

to denote the corresponding *decisional* BDHI problem.

Definition 2.1. *We say that the (decisional) (t, q, ϵ)-BDHI assumption holds in $\mathbb{G}$ if no t-time algorithm has advantage at least ϵ in solving the (decisional) q-BDHI problem in $\mathbb{G}$.*

2.2 The Twin BDHI Assumption

We define the *twin q-BDHI function* as

$$\text{2bdhi:} \quad (\mathbb{G}^*)^{2(q+1)} \to \mathbb{G}_T^2$$
$$(g_1, g_1^x, \ldots, g_1^{x^q}; g_2, g_2^y, \ldots, g_2^{y^q}) \mapsto (\text{bdhi}(g_1, g_1^x, \ldots, g_1^{x^q}), \text{bdhi}(g_2, g_2^y, \ldots, g_2^{y^q})).$$

We also define a corresponding *twin q-BDHI predicate*:

$$\text{2bdhip}(g_1, g_1^x, \ldots, g_1^{x^q}, \hat{T}_x; g_2, g_2^y, \ldots, g_2^{y^q}, \hat{T}_y) :=$$
$$\text{2bdhi}(g_1, g_1^x, \ldots, g_1^{x^q}; g_2, g_2^y, \ldots, g_2^{y^q}) \stackrel{?}{=} (\hat{T}_x, \hat{T}_y).$$

The *twin q-BDHI assumption* states that given random $g_1, g_2 \in \mathbb{G}^*$ and $x, y \in \mathbb{Z}_p^*$ it is hard to compute $\text{2bdhi}(g_1, g_1^x, \ldots, g_1^{x^q}; g_2, g_2^y, \ldots, g_2^{y^q})$. The *strong twin q-BDHI assumption* states that given random $g_1, g_2 \in \mathbb{G}^*$ and $x, y \in \mathbb{Z}_p^*$ it is hard to compute $\text{2bdhi}(g_1, g_1^x, \ldots, g_1^{x^q}, g_2, g_2^y, \ldots, g_2^{y^q})$, along with access to a decision oracle for the predicate $\text{2bdhip}(g_1, g_1^x, \ldots, g_1^{x^q}, \cdot; g_2, g_2^y, \ldots, g_2^{y^q}, \cdot)$, which on input $(\hat{T}_x, \hat{T}_y)$, returns $\text{2bdhip}(g_1, g_1^x, \ldots, g_1^{x^q}, \hat{T}_x; g_2, g_2^y, \ldots, g_2^{y^q}, \hat{T}_y)$.

We have the following result to address the relation between the BDHI assumption and the strong twin BDHI assumption:

Theorem 2.2 *The $(q+1)$-BDHI assumption holds if the strong twin q-BDHI assumption holds.*

To prove this theorem we first create a trapdoor test as follows.

Theorem 2.3 (Trapdoor Test for the Twin BDHI Problem) *Let $\mathbb{G}$ be a cyclic group of prime order p, generated by $g \in \mathbb{G}$. Suppose $(g, g^x, \ldots, g^{x^{q+1}})$ is given as a BDHI instance. Pick three random values a, b and c from $\mathbb{Z}_p$. Let $g_1 = g$, then define $g_2 = g_1^{ax+b}$, $y = xc$ and form $(g_2, g_2^y, \ldots, g_2^{y^q})$ accordingly. Then we have:*

1. *y and g_2 are uniformly distributed over $\mathbb{Z}_p$ and $\mathbb{G}^*$, respectively;*
2. *$(g_1, g_1^x, \ldots, g_1^{x^q})$ and $(g_2, g_2^y, \ldots, g_2^{y^q})$ are independent;*
3. *if $T_x = e(g_1, g_1)^{1/x}$ and $T_y = e(g_2, g_2)^{1/y}$, then the probability that the truth value of*

$$(T_x)^{b^2/c} \cdot e(g_1, g_1)^{2ab/c} \cdot e(g_1^x, g_1)^{a^2/c} = T_y \tag{1}$$

does not agree with the truth value of

$$T_x = e(g_1, g_1)^{1/x} \wedge T_y = e(g_2, g_2)^{1/y} \tag{2}$$

is at most $1/p$; moreover, if (2) holds, then (1) certainly holds.

Proof. Observe that the items 1 and 2 holds because a, b, c are randomly chosen from $\mathbb{Z}_p$. Now we begin to prove item 3. In the conditional probability space on fixed values of g_1, g_2, x and y (c is implicitly fixed), b is uniformly distributed over $\mathbb{Z}_p$. If (2) holds, by substituting the two equations in (2) into (1), we see that (2) certainly holds. Conversely, if (2) does not holds, we show that (1) holds with probability at most $1/p$. Observe that (1) is equivalent to

$$\left(\frac{T_x}{e(g_1, g_1)^{1/x}}\right)^{b^2} = \left(\frac{T_y}{e(g_2, g_2)^{1/y}}\right)^c. \tag{3}$$

It is easy to see if $T_x = e(g_1, g_1)^{1/x}$ and $T_y \neq e(g_2, g_2)^{1/y}$, then (3) certainly does not hold, since $\mathbb{G}_T$ is a group of prime order. This leaves us with the case $T_x \neq e(g_1, g_1)^{1/x}$ and $T_y = e(g_2, g_2)^{1/y}$. In this case, the right hand side of (3) is 1. There are two values or no value for b satisfying (3). Since b is uniformly distributed over $\mathbb{Z}_p$, (3) holds with probability at most $2/p$ in this case.   □

Using the trapdoor test, we can prove Theorem 2.2. For an adversary $\mathcal{B}$ against the BDHI problem, denote Adv-BDHI$_\mathcal{B}$ to be the possibility that $\mathcal{B}$ solves the BDHI problem. For an adversary $\mathcal{A}$ against the strong twin q-BDHI problem, denote Adv-2BDHI$_\mathcal{A}$ to be the possibility that $\mathcal{A}$ solves the strong twin BDHI problem. We have:

Theorem 2.4 *Suppose $\mathcal{A}$ is a strong twin q-BDHI adversary that makes at most Q_d queries to its decision oracle, and runs in time at most τ. Then there exists* a $(q+1)$-BDHI *adversary* $\mathcal{B}$ *with the following properties:* $\mathcal{B}$ *runs in time at most* τ, *plus the time to perform* $\mathcal{O}(Q_d \log p)$ *group operations and some minor bookkeeping; moreover,*

$$\text{Adv-BDHI}_{\mathcal{B}} \geq \left(1 - \frac{2Q_d}{p}\right) \text{Adv-2BDHI}_{\mathcal{A}}.$$

Proof. The $(q+1)$-BDHI adversary $\mathcal{B}$ works as follows, given a $(q+1)$-BDHI challenge instance, $\mathcal{B}$ randomly chooses $a, b, c \in \mathbb{Z}_p$ at random and generate the twin BDHI challenge instance as showed in Theorem 2.3. Second, $\mathcal{B}$ processes each decision query using the trapdoor test. Finally, when $\mathcal{A}$ outputs (T_x, T_y) to $\mathcal{B}$, $\mathcal{B}$ outputs T_x as its solution to the BDHI challenge. Provided the oracle simulation is perfect, and adversary $\mathcal{A}$'s view is identical to its view in the real environment. It remains to calculate the accuracy of the trapdoor test. Note that the probability of the trapdoor test returning a wrong decision result for a query is at most $2/p$, and this happens at most Q_d times. Therefore the trapdoor test can simulate the decision oracle perfectly with probability at least $1 - 2Q_d/p$. This proves the result we desire. $\qquad\square$

We remark that the above result also holds when the q-BDHI problem and the twin q-BDHI problem are constructed in asymmetric pairing groups. In the next two sections, we present a variant of SK-IBE [11, 22] and a variant of SK-ID-KEM [12] respectively. The related security notions of IBE and ID-KEM will be shown in Appendix A.

3   Twin SK-IBE

In this section, we apply the twinning technique to SK-IBE [11], to yield the twin SK-IBE scheme, which works as follows:

Setup. The system parameters are generated as follows:

1. Pick two random generators $g_1, g_2 \in \mathbb{G}^*$.
2. Pick two random element $s_1, s_2 \in \mathbb{Z}_p^*$ and set $u_1 = g_1^{s_1}$ and $u_2 = g_2^{s_2}$.
3. Pick four cryptographic hash functions $H_1 : \{0,1\}^* \to \mathbb{Z}_p^*$, $H_2 : \mathbb{G}_T \times \mathbb{G}_T \to \{0,1\}^n$ for some for some integer $n > 0$, $H_3 : \{0,1\}^n \times \{0,1\}^n \to \mathbb{Z}_p^* \times \mathbb{Z}_p^*$ and $H_4 : \{0,1\}^n \to \{0,1\}^n$.

The message space is $\mathcal{M} = \{0,1\}^n$. The ciphertext space is $\mathcal{C} = \mathbb{G}^2 \times \{0,1\}^n \times \{0,1\}^n$. The master public key mpk and the master secret key msk are given by

$$mpk = (g_1, g_2, u_1, u_2), \ msk = (s_1, s_2).$$

Extract. The private key for identity $\text{ID} \in \{0,1\}^*$ is generated as follows:

$$d_{\text{ID}} = (d_1, d_2) = \left(g_1^{\frac{1}{s_1 + H_1(\text{ID})}}, g_2^{\frac{1}{s_2 + H_1(\text{ID})}}\right)$$

Encrypt. A message $M \in \mathcal{M}$ is encrypted for an identity ID as follows:

1. Compute $t_1 = u_1 g_1^{H_1(\text{ID})} = g_1^{s_1 + H_1(\text{ID})}$ and $t_2 = u_2 g_2^{H_1(\text{ID})} = g_2^{s_2 + H_1(\text{ID})}$.

2. Choose a random $\sigma \in \{0,1\}^n$ and compute $(r_1, r_2) = H_3(\sigma, M)$.

3. Set the ciphertext to be $C = (t_1^{r_1}, t_2^{r_2}, \sigma \oplus H_2(e(g_1, g_1)^{r_1}, e(g_2, g_2)^{r_2}), M \oplus H_4(\sigma))$.

Decrypt. Let $C = \langle U_1, U_2, V, W \rangle \in \mathcal{C}$ be a ciphertext encrypted using ID. Then C can be decrypted by $d_{\text{ID}} = (d_1, d_2)$ as:

1. Compute $V \oplus H_2(e(U_1, d_1), e(U_2, d_2)) = \sigma$.

2. Compute $W \oplus H_4(\sigma) = M$.

3. Set $(r_1, r_2) = H_3(\sigma, M)$. Test whether $U_1 = t_1^{r_1}$ and $U_2 = t_2^{r_2}$. If not, reject the ciphertext; otherwise output M.

Theorem 3.1 *Twin SK-IBE is chosen ciphertext secure* (IND-ID-CCA) *provided that H_i ($1 \le i \le 2$) are random oracles and the $(Q_{h_1} + 1)$-BDHI assumption holds. Specifically, suppose there exists an* IND-ID-CCA *adversary $\mathcal{A}$ against twin SK-IBE that has advantage ϵ. Suppose during the attack $\mathcal{A}$ makes at most Q_{h_i} queries to H_i for $1 \le i \le 2$ respectively. Then there exists an algorithm $\mathcal{B}'$ to solve the $(Q_{h_1} + 1)$-BDHI problem with advantage*

$$\text{Adv-BDHI}_{\mathcal{B}'} \ge 2\epsilon \cdot \frac{1}{Q_{h_1}}\left(1 - \frac{2Q_{h_1}}{p}\right)$$

and a running time $\mathcal{O}(time(\mathcal{A}))$.

Proof. We first build an algorithm $\mathcal{B}$ that uses $\mathcal{A}$ to solve the strong twin Q_{h_1}-BDHI problem in $\mathbb{G}$. $\mathcal{B}$ is given as input a random strong twin Q_{h_1}-BDHI instance $(\hat{g}_1, \hat{g}_1^{x}, \ldots, \hat{g}_1^{x^q}; \hat{g}_2, \hat{g}_2^{y}, \ldots, \hat{g}_2^{y^q})$. $\mathcal{B}$'s goal is to output $Z_1 = e(\hat{g}_1, \hat{g}_1)^{1/x}$ and $Z_2 = e(\hat{g}_2, \hat{g}_2)^{1/y}$. $\mathcal{B}$ works by interacting with $\mathcal{A}$ in an IND-ID-CCA game, specified in Appendix A.1, as follows:

Preparation. Let $q = Q_{h_1}$, $\mathcal{B}$ builds generators $g_1, g_2 \in \mathbb{G}^*$ for which it knows $q - 1$ triples of the form $(w_0 + w_i, g_1^{1/(x+w_i)}, g_2^{1/y+w_i})$ for random $w_1, \ldots, w_{q-1} \in \mathbb{Z}_p^*$. This is done as follows:

1. Pick random $w_0, \ldots, w_{q-1} \in \mathbb{Z}_p^*$ and let $f(z)$ be the polynomial $f(z) = \prod_{i=1}^{q-1}(z + w_i)$. Reformulate f to get $f(z) = \sum_{i=0}^{q-1} c_i z^i$. The constant term c_0 is non-zero because $w_i \ne 0$. The c_i are computable from w_i.

2. Compute

$$g_1 = \prod_{i=0}^{q-1}(\hat{g}_1^{x^i})^{c_i} = \hat{g}_1^{f(x)}; \quad u_1 = g_1^{-w_0}\prod_{i=0}^{q-1}(\hat{g}_1^{x^{i+1}})^{c_i} = g_1^{-w_0}\hat{g}_1^{xf(x)} = g_1^{x-w_0};$$

$$g_2 = \prod_{i=0}^{q-1}(\hat{g}_2^{y^i})^{c_i} = \hat{g}_2^{f(y)}; \quad u_2 = g_2^{-w_0}\prod_{i=0}^{q-1}(\hat{g}_2^{y^{i+1}})^{c_i} = g_2^{-w_0}\hat{g}_2^{yf(y)} = g_2^{y-w_0}.$$

3. Check that $g_1 \in \mathbb{G}^*$. The case $g_1 = 1_\mathbb{G}$ means that $w_j = -x$ for some easily identifiable w_j, at which point $\mathcal{B}$ would be able to solve the challenge directly. Similarly, check that $g_2 \in \mathbb{G}^*$. The case $g_2 = 1_\mathbb{G}$ means that $w_j = -y$ for some easily identifiable w_j, at which point $\mathcal{B}$ would be able to solve the challenge directly. We thus assume that all $w_j \neq -x$, $w_j \neq -y$.

4. For any $i = 1, \ldots, q - 1$, it is easy for $\mathcal{B}$ to construct the triple $(w_0 + w_i, g_1^{1/(x+w_i)}, g_2^{1/(y+w_i)})$. To see this, write $f_i(z) = f(z)/(z+w_i) = \sum_{i=0}^{q-2} d_i z^i$. Then $g_1^{\frac{1}{x+w_i}} = \hat{g}_1^{f_i(x)} = \prod_{i=0}^{q-2}(\hat{g}_1^{x^i})^{d_i}$ and $g_2^{\frac{1}{y+w_i}} = \hat{g}_2^{f_i(y)} = \prod_{i=0}^{q-2}(\hat{g}_2^{y^i})^{d_i}$.

5. $\mathcal{B}$ computes

$$T_1' = \prod_{i=0}^{q-2}(\hat{g}_1^{x^i})^{c_{i+1}} = \hat{g}_1^{\frac{f(x)-c_0}{x}} \ , \ T_1 = e(T_1', \hat{g}_1^{f(x)+c_0}) = e(\hat{g}_1, \hat{g}_1)^{\frac{f(x)^2-c_0^2}{x}},$$

$$T_2' = \prod_{i=0}^{q-2}(\hat{g}_2^{y^i})^{c_{i+1}} = \hat{g}_2^{\frac{f(y)-c_0}{y}} \ , \ T_2 = e(T_2', \hat{g}_2^{f(y)+c_0}) = e(\hat{g}_2, \hat{g}_2)^{\frac{f(y)^2-c_0^2}{y}}.$$

We will use these values throughout the simulation.

Setup. $\mathcal{B}$ sets the master public key $mpk = (g_1, g_2, u_1, u_2)$, implicitly sets the master secret key $msk = (x - w_0, y - w_0)$, which is unknown to $\mathcal{B}$. $\mathcal{B}$ generates a set S containing w_0, and $(w_0 + w_i)$ for $i = 1, \ldots, q - 1$.

H_1-**queries.** At any time algorithm $\mathcal{A}$ can query the random oracle H_1. To respond to these queries $\mathcal{B}$ maintains a list of tuples $\langle \mathsf{ID}_i, W_i \rangle$ indexed by ID_i (as explained below). We refer to this list as the L_1 list which is initially empty. When $\mathcal{A}$ queries the oracle H_1 at a point ID_i algorithm $\mathcal{B}$ responds as follows:

1. If ID_i already appears on L_1 in a tuple $\langle \mathsf{ID}_i, W_i \rangle$ then $\mathcal{B}$ responds with $H_1(\mathsf{ID}_i) = W_i \in \mathbb{Z}_p^*$.

2. Otherwise, $\mathcal{B}$ randomly picks an element from S,
   - If the element has the form $w_0 + w_j$, $\mathcal{B}$ adds the pair $\langle \mathsf{ID}_i, W_i = w_0 + w_j \rangle$ into L_1 and answers $\mathcal{A}$ with $H_1(\mathsf{ID}_i) = w_0 + w_j$.
   - If the element is w_0, $\mathcal{B}$ adds the pair $\langle \mathsf{ID}_i, w_0 \rangle$ into L_1 and responds $\mathcal{A}$ with $H_1(\mathsf{ID}_i) = w_0$.
   - Delete this element from S.

Note that either way $H_1(\mathsf{ID}_i)$ is uniform in $\mathbb{Z}_p^*$ and independent of $\mathcal{A}$'s current view as required.

H_2-**queries.** To respond the queries to H_2 oracle, $\mathcal{B}$ maintains a list of tuples $\langle v_{i,1}, v_{i,2}, \theta_i \rangle$ indexed by $(v_{i,1}, v_{i,2})$ (as explain below). We refer to this list as the L_2 list which is initially empty. To respond to a query on $(v_{i,1}, v_{i,2})$, $\mathcal{B}$ carries out the following operations:

1. If there is a tuple indexed by $(v_{i,1}, v_{i,2})$ on L_2, $\mathcal{B}$ responds with θ_i.

2. Otherwise, $\mathcal{B}$ randomly picks a $\theta_i \in \{0, 1\}^n$, runs the self-decryption function (as described below in the simulation algorithm of decryption oracle) with composing $\langle v_{i,1}, v_{i,2}, \theta_i \rangle$ and each tuple $\langle \hat{\mathsf{ID}}, \hat{C} \rangle$ in the current R list

(as explained below) as input. If the self-decryption function returns false, $\mathcal{B}$ responds to $\mathcal{A}$ with θ_i and inserts $\langle v_{i,1}, v_{i,2}, \theta_i \rangle$ in L_2. Otherwise, $\mathcal{B}$ picks another $\theta_i \in \{0,1\}^n$ until the self-decryption function returns false on input $\langle v_{i,1}, v_{i,2}, \theta_i \rangle$ and each tuple $\langle \hat{ID}, \hat{C}_i \rangle$. For a randomly chosen θ_i, the probability that $\langle v_{i,1}, v_{i,2}, \theta_i \rangle$ enables at least one $\langle \hat{ID}, \hat{C} \rangle$ in the R list to be valid is at most $|R|/p$. Thus for a random chosen θ_i the self-decryption function returns false with probability at least $1 - |R|/p$. Hence $\mathcal{B}$ can find such a θ_i efficiently. The intuition of running the self-decryption function is to keep the coherence of the decryption oracle (as explained below): an "invalid" ciphertext is always invalid.

Phase 1: Private key queries. To respond to the private key query on ID_i, $\mathcal{B}$ does the following:

1. If the tuple in L_1 indexed by ID_i has the form of $\langle ID_i, w_0 + w_j \rangle$, then $\mathcal{B}$ answers $\mathcal{A}$ with $d_{ID_i} = (g_1^{\frac{1}{x+w_j}}, g_2^{\frac{1}{y+w_j}})$.

2. If the tuple in L_1 indexed by ID_i has the form of $\langle ID_i, w_0 \rangle$, then $\mathcal{B}$ cannot answer it and aborts the game.

Phase 1: Decryption queries. In order to simulate the decryption oracle in coherence with H_2 oracle, $\mathcal{B}$ maintains a list of tuples $\langle ID_j, C_j \rangle$. We refer to this list as the R list, which is initially empty. We remark that the R list stores all the decryption queries which are rejected by $\mathcal{B}$. Let $\langle ID_i, C_i \rangle$ be a decryption query issued by algorithm $\mathcal{A}$, where $C_i = \langle U_{i,1}, U_{i,2}, V_i, W_i \rangle$. $\mathcal{B}$ simulates the decryption oracle to answer this query as follows:

- If $\mathcal{B}$ can extract the private key $d_{ID_i} = (d_{i,1}, d_{i,2})$ of ID_i, $\mathcal{B}$ uses the private key to processes the decryption query normally.

- If $\mathcal{B}$ can not extract the corresponding private key, $\mathcal{B}$ runs the following test with composing $\langle ID_i, U_{i,1}, U_{i,2}, V_i, W_i \rangle$ and each tuple $\langle \hat{v}_1, \hat{v}_2, \hat{\theta} \rangle$ on the L_2 list as input:

  1. Compute $\hat{\sigma} = V_i \oplus \hat{\theta}$, $\hat{M} = W_i \oplus H_4(\hat{\sigma})$, $(\hat{r}_1, \hat{r}_2) = H_3(\hat{M}, \hat{\sigma})$;
  2. Check if $U_{i,1} = t_{i,1}^{\hat{r}_1}$, $U_{i,2} = t_{i,2}^{\hat{r}_2}$, $\hat{v}_1 = e(g_1, g_1)^{\hat{r}_1}$, and $\hat{v}_2 = e(g_2, g_2)^{\hat{r}_2}$ hold simultaneously, where $t_{i,1} = u_1 g_1^{H_1(ID_i)}$, $t_{i,2} = u_2 g_2^{H_1(ID_i)}$.
  3. If so, return true. Else, continue the test with next input. Finally, if no input can go through this test, return false.

If the test return true, $\mathcal{B}$ returns the associated $\hat{M}$. Otherwise, $\mathcal{B}$ returns $\perp$ to $\mathcal{A}$ and inserts $\langle ID_i, U_{i,1}, U_{i,2}, V_i, W_i \rangle$ into the R list. Particularly, we refer to the above test as *the self-decryption function*. The correctness of the decryption oracle simulation is based on the fact that an adversary does not have the capbility to generate a valid ciphertext without making the associated random oracle queries, which is implied by the chosen ciphertext security of twin SK-IBE (the hypothesis we use in the proof).

Challenge. Once $\mathcal{A}$ decides that Phase 1 is over it outputs two messages M_0, M_1 and a target identity ID^* on which it wishes to be challenged. $\mathcal{B}$ operates as follows:

1. If $H_1(\mathsf{ID}^*) \neq w_0$, then $\mathcal{B}$ aborts.
2. Otherwise, $\mathcal{B}$ randomly picks $r_1', r_2' \in \mathbb{Z}_p^*$, a string $\sigma^* \in \{0,1\}^n$, and a random bit $\beta \in \{0,1\}$. $\mathcal{B}$ sets the ciphertext $C^* = \langle U_1^*, U_2^*, V^*, W^* \rangle$, where $U_1^* = g_1^{r'}$, $U_2^* = g_2^{r'}$, V^*, W^* are two random strings from $\{0,1\}^n$. Notice that $g_1^{r_1'} = (g_1^{x-w_0+w_0})^{r_1'/x}$, $g_2^{r_2'} = (g_2^{y-w_0+w_0})^{r_2'/y}$, thus the real randomness factors are $r_1^* = r_1'/x$, $r_2^* = r_2'/y$. $\mathcal{B}$ responds to $\mathcal{A}$ with C^*.

We remark that C^* is a valid ciphertext with probability at most $1/p$. However, two claims below show that this does not affect the validity of the security reduction at all.

Phase 2. $\mathcal{B}$ proceeds in the same way as it did in Phase 1.

Guess. $\mathcal{A}$ outputs its guess β' for β.

When the IND-ID-CCA game finishes, for each tuple $\langle v_1, v_2, \theta \rangle$ on the L_2 list, $\mathcal{B}$ computes $Z_1 = v_1^{1/r_1'}$, $Z_2 = v_2^{1/r_2'}$, $T_x = (Z_1/T_1)^{1/c_0^2}$, $T_y = (Z_2/T_2)^{1/c_0^2}$, then submits (T_x, T_y) to its twin q-BDHI decision oracle until the decision oracle returns true. $\mathcal{B}$ outputs the associated (T_x, T_y) as its answer to the twin q-BDHI challenge.

It is easy to see that if $\mathcal{A}$ queries H_2 at (v_1^*, v_2^*), where $v_1^* = e(g_1, g_1)^{r_1'/x}$ and $v_2^* = e(g_2, g_2)^{r_2'/y}$, then $Z_1 = e(g_1, g_1)^{1/x} = e(\hat{g}_1, \hat{g}_1)^{f(x)^2/x}$, $Z_2 = e(g_2, g_2)^{1/y} = e(\hat{g}_2, \hat{g}_2)^{f(y)^2/y}$; $Z_1/T_1 = e(\hat{g}_1, \hat{g}_1)^{c_0^2/x}$, $Z_2/T_2 = e(\hat{g}_2, \hat{g}_2)^{c_0^2/y}$. Therefore, the associated $T_x = (Z_1/T_1)^{1/c_0^2} = e(\hat{g}_1, \hat{g}_1)^{1/x}$ and $T_y = (Z_2/T_2)^{1/c_0^2} = e(\hat{g}_2, \hat{g}_2)^{1/y}$ are the desired answer. We borrow the proving technique from [7] to show that if algorithm $\mathcal{B}$ does not abort, it outputs the correct answer (T_x, T_y) with probability at least 2ϵ.

Let $\mathtt{abort}$ be the event that $\mathcal{B}$ aborts during the simulation above, F be the event that algorithm $\mathcal{A}$ issues a query for $H_2(v_1^*, v_2^*)$ during the simulation above. Define event $F' = F|\overline{\mathtt{abort}}$, which means the probability that at the end of the simulation (v_1^*, v_2^*) appears in some entry on L_2 if $\mathcal{B}$ does not abort. We show that $\Pr[F'] \geq 2\epsilon$. This will prove that algorithm $\mathcal{B}$ outputs (T_x, T_y) with probability at least 2ϵ if it does not abort. We also study the event F in the real attack game, namely the event that $\mathcal{A}$ issues a query for $H_2(v_1^*, v_2^*)$ when communicating with a real challenger and a real random oracle for H_2.

Claim 1: $\Pr[F']$ in the simulation above is equal to $\Pr[F]$ in the real attack.

Proof. Let F_ℓ' be the event that $\mathcal{A}$ makes a query for $H_2(v_1^*, v_2^*)$ in one of its first ℓ queries to the H_2 oracle in the simulation that $\mathcal{B}$ does not abort. Let F_ℓ be the event that $\mathcal{A}$ makes a query for $H_2(v_1^*, v_2^*)$ in one of its first ℓ queries to the H_2 oracle in the real attack. We prove by induction on ℓ that $\Pr[F_\ell]$ is equal to $\Pr[F_\ell']$ in the simulation for all $\ell \geq 0$. Clearly $\Pr[F_0'] = \Pr[F_0] = 0$. Now suppose that for some $\ell > 0$ we have that $\Pr[F_{\ell-1}']$ is equal to $\Pr[F_{\ell-1}]$. We show that the same holds for F_ℓ' and F_ℓ. We know that:

$$\Pr[F_\ell'] = \Pr[F_\ell'|F_{\ell-1}']\Pr[F_{\ell-1}'] + \Pr[F_\ell'|\neg F_{\ell-1}']\Pr[\neg F_{\ell-1}']$$
$$= \Pr[F_{\ell-1}'] + \Pr[F_\ell'|\neg F_{\ell-1}']\Pr[\neg F_{\ell-1}'] \qquad (4)$$

We argue that $\Pr[F'_\ell|\neg F'_\ell]$ is equal to $\Pr[F_\ell|\neg F_{\ell-1}]$ in the real attack. To see this observe that as long as $\mathcal{A}$ does not issue a query for $H_2(v_1^*, v_2^*)$, its view during the simulation is identical to its view in the real attack (against a real challenger and a real random oracle for H_2), because all responses to H_1, H_2 queries and the challenge are distributed as in the real attack. We also remark that if $\mathcal{A}$ issues the query for $H_2(v_1^*, v_2^*)$, $\mathcal{A}$ will distinguish the simulation from the real attack by detecting the challenge is not a genuine one, however, this does not affect the security reduction. Therefore, $\Pr[F'_\ell|\neg F'_{\ell-1}]$ is equal to $\Pr[F_\ell|\neg F_{\ell-1}]$ in the real attack. It follows by (4) and the inductive hypothesis that $\Pr[F_\ell]$ in the real attack is equal to $\Pr[F'_\ell]$ in the simulation. By induction on ℓ we obtain that $\Pr[F']$ is equal to $\Pr[F]$ in the real attack. $\qquad\square$

Claim 2: In the real attack we have $\Pr[F] \geq 2\epsilon$.

Proof. In the real attack, if $\mathcal{A}$ never issues a query for $H_2(v_1^*, v_2^*)$ then the decryption of C is independent of $\mathcal{A}$'s view (since $H_2(v_1^*, v_2^*)$ is independent of $\mathcal{A}$'s view). Therefore, in the real attack $\Pr[\beta = \beta'|\neg F] = 1/2$. By definition of $\mathcal{A}$, we know that in the real attack $|\Pr[\beta = \beta'] - 1/2| \geq \epsilon$. We show that these two facts imply that $\Pr[F] \geq 2\epsilon$. To do so we derive upper and lower bounds on $\Pr[\beta = \beta']$:

$$\Pr[\beta = \beta'] = \Pr[\beta = \beta'|\neg F]\Pr[\neg F] + \Pr[\beta = \beta'|F]\Pr[F]$$

$$\leq \Pr[\beta = \beta'|\neg F]\Pr[\neg F] + \Pr[F] = \frac{1}{2}\Pr[\neg F] + \Pr[F] = \frac{1}{2} + \frac{1}{2}\Pr[F],$$

$$\Pr[\beta = \beta'] \geq \Pr[\beta = \beta'| \neq F]\Pr[\neg F] = \frac{1}{2} - \frac{1}{2}\Pr[F].$$

It follows that $\epsilon \leq |\Pr[\beta = \beta'] - 1/2| \leq \frac{1}{2}\Pr[F]$. Therefore, in the real attack $\Pr[F] \geq 2\epsilon$. $\qquad\square$

To complete the proof we still have to bound the probability that $\mathcal{B}$ aborts during the game. Throughout the simulation, $\mathcal{B}$ will abort the game for the following two reasons:

1. $\mathcal{A}$ issues the private query on ID^*, the probability of which is Q_e/Q_{h_1}.
2. $\mathcal{A}$ does not choose ID^* as the challenge identity, the possibility of which is $1 - 1/(Q_{h_1} - Q_e)$.

So the probability that $\mathcal{B}$ does not abort during the simulation is

$$\Pr[\overline{\text{abort}}] = \left(1 - \frac{Q_e}{Q_{h_1}}\right)\left(\frac{1}{Q_{h_1} - Q_e}\right) = \frac{1}{Q_{h_1}}. \tag{5}$$

Therefore the advantage of $\mathcal{B}$ against strong twin BDHI problem is

$$\text{Adv-2BDHI}_\mathcal{B} \geq \Pr[F|\overline{\text{abort}}]\Pr[\overline{\text{abort}}] = 2\epsilon \cdot \frac{1}{Q_{h_1}}. \tag{6}$$

Combined with Theorem 2.2, we have

$$\text{Adv-BDHI}_{\mathcal{B}'} \geq 2\epsilon \cdot \frac{1}{Q_{h_1}} \left(1 - \frac{2Q_{h_1}}{p}\right).$$

The running time of $\mathcal{B}'$ is easy to be verified.   □

4   Twin SK-ID-KEM

In this section, we apply the twinning technique to a KEM scheme from SK-IBE [12], to yield the twin SK-ID-KEM scheme. We present the scheme by describing the four algorithms: Setup, Extract, Encaps, Decaps.

Setup. To generate system parameters, the algorithm works as follows:

1. Pick two random generators $g_1, g_2 \in \mathbb{G}^*$.
2. Pick two random $s_1, s_2 \in \mathbb{Z}_p^*$, compute $u_1 = g_1^{s_1}$ and $u_2 = g_2^{s_2}$.
3. Pick four cryptographic hash functions $H_1 : \{0,1\}^* \to \mathbb{Z}_p^*$, $H_2 : \mathbb{G}_T \times \mathbb{G}_T \to \{0,1\}^n$ for some integer n, $H_3 : \{0,1\}^n \to \mathbb{Z}_p^* \times \mathbb{Z}_p^*$ and $H_4 : \{0,1\}^n \to \{0,1\}^\lambda$ for some integer λ.

The key space is $\mathcal{K} = \{0,1\}^n$. The ciphertext space is $\mathcal{C} = \mathbb{G}^* \times \mathbb{G}^* \times \{0,1\}^n$. The master public key mpk and the master secret key msk are given by

$$mpk = (g_1, g_2, u_1, u_2), \ msk = (s_1, s_2)$$

Extract. Given an identity $\mathsf{ID} \in \{0,1\}^*$, the algorithm sets the private key to be

$$d_{\mathsf{ID}} = (d_1, d_2) = \left(g_1^{\frac{1}{s_1 + H_1(\mathsf{ID})}}, g_2^{\frac{1}{s_2 + H_1(\mathsf{ID})}}\right)$$

Encaps. To encapsulate a key of ID do the following:

1. Pick a random $\sigma \in \{0,1\}^n$ and compute $(r_1, r_2) = H_3(\sigma)$.
2. Compute $t_1 = u_1 g_1^{H_1(\mathsf{ID})} = g_1^{s_1 + H_1(\mathsf{ID})}$ and $t_2 = u_2 h_1^{H_1(\mathsf{ID})} = g_2^{s_2 + H_1(\mathsf{ID})}$.
3. Set the ciphertext to be $C = (t_1^{r_1}, t_2^{r_2}, \sigma \oplus H_2(e(g_1, g_1)^{r_1}, e(g_2, g_2)^{r_2}))$.
4. Encapsulate the key $k = H_4(\sigma)$.

Decaps. Let $C = (U_1, U_2, V) \in \mathcal{C}$ be a ciphertext encrypted using ID. To decrypt C using the private key $d_{\mathsf{ID}} = (d_1, d_2)$ compute:

1. Compute $V \oplus H_2(e(U_1, d_1), e(U_2, d_2)) = \sigma$.
2. Compute $(r_1, r_2) = H_3(\sigma)$. Test whether $U_1 = t_1^{r_1}$ and $U_2 = t_2^{r_2}$. If not, reject the ciphertext.
3. Output $k = H_4(\sigma)$.

Theorem 4.1 *Twin SK-ID-KEM is* IND-ID-CCA *secure provided that* H_i ($1 \leq i \leq 2$) *are random oracles and the* $(Q_{h_1} + 1)$-BDHI *assumption holds. Specially, suppose there exists an* IND-ID-CCA *adversary* $\mathcal{A}$ *against twin SK-ID-KEM that has advantage* ϵ. *Suppose during the attack* $\mathcal{A}$ *makes at most* Q_e

extraction queries, Q_d decapsulation queries, and at most Q_{h_i} queries on H_i for $1 \leq i \leq 2$ respectively. Then there exists an algorithm $\mathcal{B}$ to solve the $(Q_{h_1}+1)$-BDHI problem with advantage

$$\text{Adv-BDHI}_{\mathcal{B}'} \geq 2\epsilon \cdot \frac{1}{Q_{h_1}} \left(1 - \frac{2Q_{h_1}}{p}\right)$$

and running time $\mathcal{O}(time(\mathcal{A}))$.

Proof. We first build an algorithm $\mathcal{B}$ that uses $\mathcal{A}$ to solve the strong twin q-BDHI problem in $\mathbb{G}$. $\mathcal{B}$ is given as input a strong twin q-BDHI instance $(\hat{g}_1, \hat{g}_1^x, \ldots, \hat{g}_1^{x^q}; \hat{g}_2, \hat{g}_2^y, \ldots, \hat{g}_2^{y^q})$ and is expected to output $Z_1 = e(\hat{g}_1, \hat{g}_1)^{1/x}$ and $Z_2 = e(\hat{g}_2, \hat{g}_2)^{1/y}$. $\mathcal{B}$ works by interacting with $\mathcal{A}$ in an IND-ID-CCA game as follows:

Preparation. Same as in the proof of twin SK-IBE scheme in Section 3.

Setup. Same as in the proof of twin SK-IBE scheme Section 3.

H_1-queries. Same as in the proof of twin SK-IBE scheme Section 3.

H_2-queries. Same as in the proof of twin SK-IBE scheme Section 3, except that we replace the self-decryption funtion with the corresponding self-decapsulation function (as explained below).

Phase 1: Private key queries. Same as in the proof of twin SK-IBE scheme 3.

Phase 1: Decapsulation queries. In order to simulate the decapsulation oracle coherently with a H_2 oracle, $\mathcal{B}$ maintains a list of tuples $\langle ID_j, C_j \rangle$. We refer to this list as the R list, which is initially empty. The R list is used to store the invalid ciphertexts issued by $\mathcal{A}$. Let $\langle ID, C_i \rangle$ be a decapsulation query issued by algorithm $\mathcal{A}$, where $C_i = \langle U_{i,1}, U_{i,2}, V_i \rangle$. $\mathcal{B}$ simulates the decapsulation oracle to answer this query as follows:

– If $\mathcal{B}$ can extract the private key $d_{ID_i} = (d_{i,1}, d_{i,2})$ of ID_i, $\mathcal{B}$ uses the private key to process the decapsulation query normally.

– If $\mathcal{B}$ can not extract the corresponding private key, for every tuple $(\hat{v}_1, \hat{v}_2, \hat{\theta})$ on L_2, $\mathcal{B}$ runs the following test with composing $\langle ID_i, U_{i,1}, U_{i,2}, V_i \rangle$ and each tuple $\langle \hat{v}_1, \hat{v}_2, \hat{\theta} \rangle$ on the L_2 list as input,

  1. Compute $t_{i,1} = u_1 g_1^{H_1(ID_i)}$, $t_{i,2} = u_2 g_2^{H_1(ID_i)}$;
  2. Compute $\hat{\sigma} = V_i \oplus \hat{\theta}$, $(\hat{r}_1, \hat{r}_2) = H_3(\hat{\sigma})$.
  3. Check if $U_{i,1} = t_{i,1}^{\hat{r}_1}$, $U_{i,2} = t_{i,2}^{\hat{r}_2}$, $\hat{v}_1 = e(g_1, g_1)^{\hat{r}_1}$, and $\hat{v}_2 = e(g_2, g_2)^{\hat{r}_2}$ simultaneously.
  4. If so, return true. Else, continue the test with the next input.
  5. Finally, if no input can go through this test, return false.

  If the test returns true, $\mathcal{B}$ returns the associated $\hat{M}$. Otherwise, $\mathcal{B}$ returns $\perp$ to $\mathcal{A}$ and inserts $\langle ID_i, U_{i,1}, U_{i,2}, V_i \rangle$ into the R list. Particularly, we refer to the above test as the *self-decapsulation function*.

Challenge. Once $\mathcal{A}$ decides that Phase 1 is over it outputs a target identity ID^* on which it wishes to be challenged. $\mathcal{B}$ operates as follows:

1. If the $H_1(\mathsf{ID}^*) \neq w_0$, then $\mathcal{B}$ aborts.
2. Otherwise, $\mathcal{B}$ randomly picks $r_1', r_2' \in \mathbb{Z}_p^*$, a string $\sigma^* \in \{0,1\}^n$. $\mathcal{B}$ sets the ciphertext $C^* = \langle U_1^*, U_2^*, V^* \rangle$, where $U_1^* = g_1^{r_1'}$, $U_2^* = g_2^{r_2'}$, V^* is a random string from $\{0,1\}^n$. Note that $g_1^{r_1'} = (g_1^{x-w_0+w_0})^{r_1'/x}$, $g_2^{r_2'} = (g_2^{y-w_0+w_0})^{r_2'/y}$, so the real random factors are $r_1^* = r_1'/x$, $r_2^* = r_2'/y$. $\mathcal{B}$ picks a random bit $\beta \in \{0,1\}$, if $\beta = 0$, $\mathcal{B}$ picks a random string $k_0 \in \{0,1\}^\lambda$ and returns (C^*, k_0) to $\mathcal{A}$. If $\beta = 1$, $\mathcal{B}$ computes $k_1 = H_2(\sigma^*)$ and responds to $\mathcal{A}$ with (C^*, k_1).

We remark that C^* is a valid ciphertext with probability at most $1/p$. However, this does not affect the final result of the security reduction.

Phase 2. $\mathcal{B}$ proceeds the same way as it did in Phase 1.

Guess. $\mathcal{A}$ outputs its guess β' for β.

The remaining part of the proof is almost identical to the proof of the twin SK-IBE scheme in Section 3. Due to lack of space, we include the complete proof in the full version. We give the final result as follows. The advantage of $\mathcal{B}$ against the twin strong BDHI problem is

$$\text{Adv-2BDHI}_\mathcal{B} \geq 2\epsilon \cdot \frac{1}{Q_{h_1}} \tag{7}$$

Combined with Theorem 2.2, we have

$$\text{Adv-BDHI}_{\mathcal{B}'} \geq 2\epsilon \cdot \frac{1}{Q_{h_1}} \left(1 - \frac{2Q_{h_1}}{p} \right)$$

The running time of $\mathcal{B}'$ is easy to be verified. $\qquad\square$

5   Conclusion

In this paper we propose a new computational problem, named the twin bilinear Diffie-Hellman inversion assumption. We construct a new trapdoor test which enables us to prove that the strong twin Diffie-Hellman inversion problem is at least as hard as the original bilinear Diffie-Hellman inversion problem. Based on this result, we show how to apply the twinning technique to SK-IBE and SK-ID-KEM, respectively. It is worth to point out that the improvement on the tightness of security reductions of SK-IBE [11] and SK-ID-KEM [12] comes from two aspects, one which benefits from the twinning technique, the other one which benefits from the using of self-decryption/self-decapsulation function.

Acknowledgements. We would like to thank Eike Kiltz for helpful discussions, and to the anonymous reviewers for their valuable suggestions. We also thank Michael Scott for careful comments that improved the presentation of our results. Yu Chen was supported by the China Scholarship Council.

References

1. Abdalla, M., Bellare, M., Rogaway, P.: The Oracle Diffie-Hellman Assumptions and an Analysis of DHIES. In: Naccache, D. (ed.) CT-RSA 2001. LNCS, vol. 2020, pp. 143–158. Springer, Heidelberg (2001)
2. Barreto, P.S.L.M., Libert, B., McCullagh, N., Quisquater, J.-J.: Efficient and provably-secure identity-based signatures and signcryption from bilinear maps. In: Roy, B. (ed.) ASIACRYPT 2005. LNCS, vol. 3788, pp. 515–532. Springer, Heidelberg (2005)
3. Bellare, M., Ristenpart, T.: Simulation without the Artificial Abort: Simplified Proof and Improved Concrete Security for Waters' IBE Scheme. In: Joux, A. (ed.) EUROCRYPT 2009. LNCS, vol. 5479, pp. 407–424. Springer, Heidelberg (2009)
4. Bellare, M., Rogaway, P.: The Exact Security of Digital Signatures - How to Sign with RSA and Rabin. In: Maurer, U.M. (ed.) EUROCRYPT 1996. LNCS, vol. 1070, pp. 399–416. Springer, Heidelberg (1996)
5. Bentahar, K., Farshim, P., Malone-Lee, J., Smart, N.P.: Generic Constructions of Identity-Based and Certificateless KEMs. Cryptology ePrint Archive, Report 2005/058 (2005), http://eprint.iacr.org/
6. Boneh, D., Boyen, X.: Efficient Selective-ID Secure Identity-Based Encryption Without Random Oracles. In: Cachin, C., Camenisch, J.L. (eds.) EUROCRYPT 2004. LNCS, vol. 3027, pp. 223–238. Springer, Heidelberg (2004)
7. Boneh, D., Franklin, M.: Identity-Based Encryption from the Weil Pairing. In: Kilian, J. (ed.) CRYPTO 2001. LNCS, vol. 2139, pp. 213–229. Springer, Heidelberg (2001)
8. Boneh, D., Franklin, M.K.: Identity-Based Encryption from theWeil Pairing. SIAM Journal on Computation 32, 586–615 (2003)
9. Boyen, X.: General ad hoc encryption from exponent inversion ibe. In: Naor, M. (ed.) EUROCRYPT 2007. LNCS, vol. 4515, pp. 394–411. Springer, Heidelberg (2007)
10. Cash, D., Kiltz, E., Shoup, V.: The Twin Diffie-Hellman Problem and Applications. In: Smart, N.P. (ed.) EUROCRYPT 2008. LNCS, vol. 4965, pp. 127–145. Springer, Heidelberg (2008)
11. Chen, L., Cheng, Z.: Security Proof of Sakai-Kasahara's Identity-Based Encryption Scheme. In: Smart, N.P. (ed.) Cryptography and Coding 2005. LNCS, vol. 3796, pp. 442–459. Springer, Heidelberg (2005)
12. Chen, L., Cheng, Z., Malone-Lee, J., Smart, N.P.: An Efficient ID-KEM Based On The Sakai-Kasahara Key Construction. In: IEE Proceedings of Information Security, pp. 19–26 (2006)
13. Cheng, Z.: Simple SK-ID-KEM unpublished notes (2005), http://www.cs.mdx.ac.uk/staffpages/m_cheng/link/simple-sk-kem.pdf
14. Cheng, Z., Chen, L.: On security proof of McCullagh-Barreto's key agreement protocol and its variants. IJSN 2(3/4), 251–259 (2007)
15. Cramer, R., Shoup, V.: A Practical Public Key Cryptosystem Provably Secure Against Adaptive Chosen Ciphertext Attack. In: Krawczyk, H. (ed.) CRYPTO 1998. LNCS, vol. 1462, pp. 13–25. Springer, Heidelberg (1998)
16. Cramer, R., Shoup, V.: Design and Analysis of Practical Public-Key Encryption Schemes Secure against Adaptive Chosen Ciphertext Attack. SIAM Journal on Computing 33, 167–226 (2001)
17. Diffie, W., Hellman, M.E.: New Directions in Cryptograpgy. IEEE Transactions on Infomation Theory 22(6), 644–654 (1976)

18. Dodis, Y., Yampolskiy, A.: A Verifiable Random Function with Short Proofs and Keys. In: Vaudenay, S. (ed.) PKC 2005. LNCS, vol. 3386, pp. 416–431. Springer, Heidelberg (2005)
19. Gentry, C.: Practical Identity-Based Encryption Without Random Oracles. In: Vaudenay, S. (ed.) EUROCRYPT 2006. LNCS, vol. 4004, pp. 445–464. Springer, Heidelberg (2006)
20. Libert, B., Quisquater, J.J.: Identity based encryption without redundancy. In: Ioannidis, J., Keromytis, A.D., Yung, M. (eds.) ACNS 2005. LNCS, vol. 3531, pp. 285–300. Springer, Heidelberg (2005)
21. McCullagh, N., Barreto, P.S.L.M.: A New Two-Party Identity-Based Authenticated Key Agreement. In: Menezes, A. (ed.) CT-RSA 2005. LNCS, vol. 3376, pp. 262–274. Springer, Heidelberg (2005)
22. Sakai, R., Kasahara, M.: ID based Cryptosystems with Pairing on Elliptic Curve. Cryptology ePrint Archive, Report 2003/054 (2003), http://eprint.iacr.org/

A   Security Notions

In this section, we briefly review the security notions for IBE and ID-KEM.

A.1   Chosen Ciphertext Security for IBE

Recall that an IBE system consists of four algorithms [7]: Setup, KeyGen, Encrypt, Decrypt. Via $(mpk, msk) = $ Setup(1^κ) the PKG generate the master key pair (mpk, msk). Via $sk \leftarrow$ KeyGen(msk, ID) the PKG uses the master secret key msk to generate the private key sk corresponding to ID. Via $C \leftarrow$ Enc(mpk, M, ID) the encryption algorithm encrypts messages for a given identity and the decryption algorithm decrypts ciphertexts using the private key via $M \leftarrow$ Dec(sk, C). The definition of adaptive chosen ciphertext security for IBE was first formalized by Boneh and Franlkin in [7, 8]. An IBE scheme $\mathcal{E}$ is said to be secure against an adaptively chosen ciphertext attack (IND-ID-CCA) if no probabilistic polynomial time (PPT) algorithm $\mathcal{A}$ has a non-negligible advantage against the challenger in the following game:

Setup. The challenger run the Setup on security parameter κ to generate the public parameters mpk and the master secret msk, gives the adversary the public parameters, and keeps the master secret to itself.

Phase 1. The adversary issues queries $q_1, \ldots, q_m$ where query q_i is one of:

- Extraction query $\langle \mathsf{ID}_i \rangle$. The challenger responds by running algorithm Extract to generate the private key d_i corresponding to ID_i. It sends d_i to the adversary $\mathcal{A}$.
- Decryption query $\langle \mathsf{ID}_i, C_i \rangle$. The challenger responds by running algorithm Extract to generate the private key d_i corresponding to ID_i. It then runs algorithm Decrypt to decrypt the ciphertext C_i using the private key d_i. It sends the resulting plaintext to the adversary $\mathcal{A}$.

These queries may be asked adaptively, that is, each query q_i may depend on the replies to $q_1, \ldots, q_{i-1}$.

Challenge. Once the adversary decides that Phase 1 is over it outputs two equal length plaintexts $M_0, M_1 \in \mathcal{M}$ and an identity ID on which it wishes to be challenged. The only constraint is that ID did not appear in any private key extraction query in Phase 1. The challenger picks a random bit $\beta \in \{0,1\}$ and sets $C = \mathsf{Encrypt}(mpk, \mathsf{ID}, M_\beta)$. It sends C as the challenge to the adversary.

Phase 2. The adversary issues more queries $q_{m+1}, \ldots, q_r$ where q_i is one of:
- Extraction query $\langle \mathsf{ID}_i \rangle \neq \mathsf{ID}$. Challenger responds as in Phase 1.
- Decryption query $\langle \mathsf{ID}_i, C_i \rangle \neq \langle \mathsf{ID}, C \rangle$. Challenger responds as in Phase 1.

These queries may be asked adaptively as in Phase 1.

Guess. Finally, the adversary outputs a guess $\beta' \in \{0,1\}$ and wins the game if $\beta = \beta'$.

We refer to such an adversary $\mathcal{A}$ as an IND-ID-CCA adversary. We define adversary $\mathcal{A}$'s advantage over the scheme $\mathcal{E}$ by $\mathrm{Adv}_{\mathcal{E},\mathcal{A}}^{\mathsf{CCA}}(\kappa) = \left| \Pr[\beta = \beta'] - \frac{1}{2} \right|$, where κ is the security parameter. The probability is over the random bits used by the challenger and the adversary.

Definition A.1. *We say that an IBE scheme $\mathcal{E}$ is* IND-ID-CCA *secure if for any probabilistic polynomial time* IND-ID-CCA *adversary $\mathcal{A}$ the advantage* $\mathrm{Adv}_{\mathcal{E},\mathcal{A}}^{\mathsf{CCA}}(\kappa)$ *is negligible.*

A.2   Chosen Ciphertext Security for ID-KEM

The natural way to perform public key encryption for large messages is to use *hybrid encryption*. A hybrid encryption consists of two parts: one part uses public key techniques to encrypt a one-time symmetric key, known as the *key encapsulation mechanism* (KEM), while the other part uses the symmetric key to encrypt the actual message, known as the *data encapsulation mechanism* (DEM). Cramer and Shoup first formalized the concept of KEM/DEM in [16]. Bentahar et al. [5] extended the KEM concept to the identity based setting. An identity based key encapsulation mechanism consists of four polynomial-time algorithms: (Setup, KeyGen, Encap, Decap). Algorithms Setup and KeyGen behave exactly as those of IBE. Via $(C, k) \leftarrow \mathsf{Encap}(mpk, \mathsf{ID})$ the randomized encapsulation algorithm creates an uniformly distributed symmetric key $k \in \{0,1\}^\lambda$, together with a ciphertext C; via $k \leftarrow \mathsf{Decap}(sk, C)$ the possessor of private key sk decrypts ciphertext C to get back a key $k \in \{0,1\}^\lambda$ or a special symbol $\bot$.

The notion of adaptive chosen ciphertext security for identity based key encapsulation mechanism is similar to that for IBE, except that there are no challenge messages to encrypt. Instead, in the challenge phase the challenger flips a coin $\beta \in \{0,1\}$, and the adversary is given a ciphertext C^* and a string k^*, which will be a session key if $\beta = 1$, or a random string if $\beta = 0$.

Setup. The challenger takes a security parameter κ and runs the KeyGen algorithm. It gives the adversary the resulting system parameters. It keeps the master key to itself.

Phase 1. The adversary issues queries $q_1, \ldots, q_m$ where query q_i is one of:

- Extraction query $\langle \mathsf{ID}_i \rangle$. The challenger responds by running the algorithm Extract to generate the private key d_i corresponding to the identity ID_i. It sends d_i to the adversary.
- Decapsulation query $\langle \mathsf{ID}_i, C_i \rangle$. The challenger responds by running algorithm Extract to generate the private key d_i corresponding to ID_i. It then runs algorithm Decapsulate to decapsulate the encapsulation C_i using the private key d_i. It sends the resulting session key to the adversary.

Challenge. Once the adversary decides that Phase 1 is over it outputs an identity ID^* on which it wishes to be challenged. The only constraint is that ID^* did not appear in any private key extraction query in Phase 1. The challenger computes $(C^*, k_1^*) = \mathsf{Encapsulate}(mpk, \mathsf{ID}^*)$, then picks a random bit $\beta \in \{0, 1\}$. If $\beta = 1$, it sends (C^*, k_1^*) as the challenge to the adversary, where k_1^* is the real session key. Otherwise, it sends (C^*, k_1^*) as the challenge to the adversary, where k_0^* is randomly chosen from $\mathcal{K}$.

Phase 2. The adversary issues more queries $q_{m+1}, \ldots, q_r$ where query q_i is one of:

- Extraction query $\langle \mathsf{ID}_i \rangle$ where $\mathsf{ID}_i \neq \mathsf{ID}^*$. Challenger responds as in Phase 1.
- Decapsulation query $\langle \mathsf{ID}_i, C_i \rangle \neq \langle \mathsf{ID}^*, C^* \rangle$. Challenger responds as in Phase 1.

Thess queries may be asked adaptively as in Phase 1.

Guess. Finally, the adversary outputs a guess $\beta' \in \{0, 1\}$ and wins the game if $\beta = \beta'$.

We refer to such an adversary $\mathcal{A}$ as an IND-ID-CCA adversary. We define adversary $\mathcal{A}$'s advantage over the ID-KEM scheme $\mathcal{E}$ by $\mathrm{Adv}_{\mathcal{E},\mathcal{A}}^{\mathsf{CCA}}(\kappa) = \left| \Pr[\beta = \beta'] - \frac{1}{2} \right|$, where κ is the security parameter. The probability is over the random bits used by the challenger and the adversary.

Definition A.2. *We say that an ID-KEM scheme $\mathcal{E}$ is* IND-ID-CCA *secure if for any probabilistic polynomial time* IND-ID-CCA *adversary $\mathcal{A}$ the advantage* $\mathrm{Adv}_{\mathcal{E},\mathcal{A}}^{\mathsf{CCA}}(\kappa)$ *is negligible.*

Group Signatures are Suitable for Constrained Devices

Sébastien Canard[1,*], Iwen Coisel[2,**],
Giacomo De Meulenaer[2,***], and Olivier Pereira[2]

[1] Orange Labs - 42 rue des Coutures - BP6234 - F-14066 Caen Cedex - France
sebastien.canard@orange-ftgroup.com
[2] Université Catholique de Louvain - B-1348 Louvain-la-Neuve - Belgium
{iwen.coisel,giacomo.demeulenaer}@uclouvain.be

Abstract. In a group signature scheme, group members are able to sign messages on behalf of the group. Moreover, resulting signatures are anonymous and unlinkable for every verifier except for a given authority. In this paper, we mainly focus on one of the most secure and efficient group signature scheme, namely XSGS proposed by Delerablée and Pointcheval at Vietcrypt 2006. We show that it can efficiently be implemented in a sensor node or an RFID tag, even if it requires 13 elliptic curve point multiplications, 2 modular exponentiations and one pairing evaluation to produce a group signature. This is done by securely outsourcing part of the computation to an untrusted powerful intermediary. The result is that XSGS can be executed in the MICAz (8-bit 7.37MHz ATmega128 microprocessor) and the TelosB (16-bit 4MHz MSP430 processor) sensor nodes in less than 200 ms.

Keywords: Constrained devices, server-aided computation, group signature, anonymity.

1   Introduction

Group signatures have been introduced by Chaum and van Heyst [9], and showed to be extremely useful in various applications such as anonymous credentials, e-cash, e-vote and identity management. These signatures allow any member of a group to sign a document and any verifier to confirm that the signature has been computed by a group member. Moreover, group signatures are anonymous and unlinkable for every verifier except, when needed, for a given authority. While being very appealing, implementing these signature schemes on low-power devices, like sensor nodes or RFID tags, appears to be a particularly challenging task, as the computation of a signature typically requires numerous modular exponentiations or pairing evaluations. For instance, it is necessary to compute 13 elliptic curve point multiplications, 2 modular exponentiations and 1 pairing

---

* Supported by French ANR PACE project.
** Supported by Walloon Region project SEE.
*** Supported by Walloon Region project Nanotic.

K.-H. Rhee and D. Nyang (Eds.): ICISC 2010, LNCS 6829, pp. 133–150, 2011.
© Springer-Verlag Berlin Heidelberg 2011

to produce a DLIN based eXtremely Short Group Signature (XSGS) [14], one of the most efficient and powerful schemes available today.

In this paper, we design a cooperative variant of the XSGS group signature scheme [14]. Our result is efficient enough to permit the group member to be associated to a constrained device which interacts with an intermediary (which can belong to the group member, e.g. a personal computer).

Related Work. The way to embed cryptography into low-power devices has been largely studied by the cryptographic community. One solution is to make pre-computations but this has the drawback of consuming a lot of memory space and thus simply shifts the problem. Another possibility is to modify the crypto-graphic mechanism to fit the device restrictions. This has already been done in the RFID case [19,15] or when considering the integration of group signatures in a smart card [8,36]. This may necessitates important modifications of the initial algorithm, and may imply some stronger (and questionable) assumptions such as, e.g. , tamper-resistance.

In this paper, we focus on a second approach, which consists in studying how a more powerful entity can help a small device to provide a group signature, as introduced in [27] and later used for DAA in *e.g.* [6]. Another approach has also been taken in the CAFE project [10,11], which consists in designing schemes where a powerful prover interacts with a non-trusted smart card to perform some computations in such a way that the prover is unlinkable w.r.t. the smart card.

Our Contributions. The introduction of an intermediary device in a signature scheme must be carefully understood if one wants to avoid introducing severe security flaws in the system. Our contribution in this direction is threefold.

To begin with, we propose the first complete security model for coopera-tive group signatures. Our model allows clarifying the exact level of trust that is placed into the intermediary, this trust directly impacting the amount of compu-tation that can be outsourced by the tag. The trust we place in the intermediary is quite limited: even compromised, it is not able to impersonate the signer. With this property, the security of standard group signature systems can be improved: the group members' secrets can be stored in well-protected embedded devices like contactless smart cards instead of being present in their personal computer, which may be unsecure (e.g. infected with a trojan).

Then, we propose a new cooperative group signature scheme, based on the XSGS protocol [14], and prove its security in our model. Our scheme is efficient enough to be implemented on small embedded devices.

We demonstrate this by documenting implementation results on two common wireless sensor nodes, the MICAz and the TelosB sensor nodes, the processor of the TelosB node being also used in contactless secure government electronic ID chips. The on-line phase of the protocol can be completed in less than 200 ms. The off-line phase requires four to six seconds to be completed, but this can be mitigated with a coupon mode (from [29] and the EU project CAFE ESPRIT 7023), which allows a node to pre-compute or pre-load up to 5000 coupons in advance, while satisfying the memory constraints of our devices. We also show that, in our case, new coupons can be added at any time (at home, for instance), without any problem.

This article thus provides fundamental building blocks for the deployment of low-power privacy preserving applications, in contexts where the nodes involved in the applications cannot perform heavy computations, which is the case most of the time for the moment.

Outline. After a brief reminder on group signatures (Section 2), we extend the standard security properties to the cooperative setting (Section 3). Next, we present our cooperative XSGS protocol and prove its security with respect to our new security definitions (Section 4). We eventually demonstrate that our protocol can be executed on small devices, by presenting and discussing the performance of its implementation on small wireless sensor devices, of which the controllers are also sometimes included in RFID tags (Section 5).

2   Definition of Group Signature Schemes

In [9], Chaum and van Heyst introduce the notion of *group signature* schemes [1,3,17,14,5,22] where members of the group can sign documents and any verifier can confirm that the signature comes from a group member. Moreover, group signatures are anonymous and unlinkable for every verifier except for a given authority.

2.1   Generic Description of Group Signatures

Formally speaking, a group signature scheme $\mathcal{GS}$ is described by the following polynomial-time procedures, where λ is a security parameter.

- GENPARAM is a probabilistic algorithm which takes as input 1^λ and which outputs the public parameters of the system $\mathcal{PP} = (\mathsf{Gpk}, \mathsf{Rpk}, \mathsf{params})$ where Gpk is the group's public key, Rpk is the public key of the opening manager and params are public parameters (e.g. mathematical groups, generators,...), it also outputs the group manager's secret key gmsk and the opening manager's secret key rsk;
- USERKEYGEN is a probabilistic algorithm which attributes to a user a pair of secret/public key (usk, Upk) respecting a PKI.
- JOIN is a probabilistic protocol between the group manager and a new group member U_i to provide the latter with his group secret key $\mathsf{gsk}[i]$. The group manager makes an entry $\mathsf{Tab}[i]$ in the registration table Tab, with the entire transcript of the process.
- SIGN is a probabilistic algorithm which takes as input a secret signing key $\mathsf{gsk}[i]$ and a message m and returns the group signature σ on m;
- VERIF is a deterministic algorithm which takes as input a message m, and a signature σ and returns either 1 if the signature is valid or 0 otherwise;

- OPEN is a deterministic algorithm which takes as input the opening manager secret key rsk, the registration table Tab, a message m and a signature σ and returns either an identity i or the symbol $\perp$ to indicate a failure, together with a proof τ of this claim;
- JUDGE is a deterministic algorithm which takes as input the registration table Tab, a message m, a signature σ, an identity i and a proof τ and returns 1 if the proof τ is a valid proof that user i has produced the signature σ and 0 otherwise.

2.2   Security Properties

We outline the formal security properties from the BSZ model, introduced by Bellare et al. [2], that are expected for (dynamic) group signature schemes.

- **Correctness:** a signature produced by a valid user U_i must be accepted by a verifier. Furthermore, the opening of this signature must return the identity of U_i and the judge must validate this opening.
- **Anonymity:** given several signatures of a user (randomly chosen among two users), it is infeasible to distinguish which of these two users have produced this set of signatures.
- **Traceability:** it is infeasible to produce a valid signature which cannot be opened or where the proof outputted by OPEN cannot be verified. This property must be verified even if several users and the group manager collude.
- **Non-Frameability:** it is infeasible, even for the opening and the group manager, to claim falsely that a signature has been produced by a user.

To prove that a scheme ensures these properties, Bellare et al. [2] define for each of these properties an experiment played by an adversary. Depending on the concerned property, the adversary has several possibilities to interact with the system. For example, an adversary can corrupt some users and thus obtains their group secret keys. In some cases, the group manager can be corrupted and thus the adversary can play his role during a JOIN procedure. All the possible interactions are realized through oracles which are listed below. Moreover, a list of honest users $\mathcal{HU}$ and one of corrupted users $\mathcal{CU}$ are needed.

- $\mathcal{O}^{\mathsf{CreateU}}$: this oracle generates a new user i using USERKEYGEN.
- $\mathcal{O}^{\mathsf{AddU}}(i)$: this oracle adds a new user i in the group using USERKEYGEN and the interactive protocol JOIN. The identity of this new user is added to the list $\mathcal{HU}$. The new public key Upk_i is outputted.
- $\mathcal{O}^{\mathsf{SJoin}}(i)$: during the request to this oracle, the adversary will play the role of the group manager during a JOIN protocol with a new honest user. First of all, the oracle generates a new user i with USERKEYGEN and simulates him during the protocol with the adversary. This new user is added to $\mathcal{HU}$.
- $\mathcal{O}^{\mathsf{UJoin}}$: this oracle simulates the group manager during a JOIN protocol where the adversary plays the role of the user.

- $\mathcal{O}^{\mathsf{CrptU}}(i)$: this oracle gives the total control of the user i to the adversary. In other words, the adversary obtains all the information related to this user (secret keys, random values, ...). The member i is moved from $\mathcal{HU}$ to $\mathcal{CU}$.
- $\mathcal{O}^{\mathsf{Reveal}}(i)$: this oracle outputs the secret keys $(\mathsf{usk}[i], \mathsf{gsk}[i])$ of the member i.
- $\mathcal{O}^{\mathsf{SignU}}(i, m)$: this oracle outputs the signature σ on m of the member i and adds the tuple (m, σ, i) in Set (initially empty).
- $\mathcal{O}^{\mathsf{Open}}(m, \sigma)$: this oracle outputs the identity of the user which produced σ.
- $\mathcal{O}^{\mathsf{Choose}_b}(m, i_0, i_1)$: if i_0 or i_1 have not been given as input to $\mathcal{O}^{\mathsf{CrptU}}$ (i.e. $i_0, i_1 \notin \mathcal{HU}$), this oracle outputs the signature σ on the message m of the member i_b (where b is a bit). σ cannot be given in input to the $\mathcal{O}^{\mathsf{Open}}$ oracle.

2.3 Some Group Signature Constructions

In this paper we focus on group signatures based on the use of pairings [3,17,14] since they are relatively efficient (compared to standard model based group signatures [22]) and does not need the manipulation of big integers (contrary to [1,7]). The BBS scheme [3] only considers static group while others [17,14] are secure in the dynamic case [2]. We here base our study on a variant of the XSGS protocol from [14], which one is described in Appendix A. In fact, our study is also relevant for the scheme in [17] but we have chosen the XSGS one as it includes a complete security study[1]. To prevent the use of the XDH assumption, which may be seen as a too strong assumption, we adapt XSGS (as suggested in [14]) by replacing the El Gamal encryption scheme [18] by the Linear encryption [3], at the cost of a slightly bigger group signature. In a nutshell, a user owns a group secret key gsk and a certificate (A, x) such that $(x + \mathsf{gmsk}).A = G_1 + \mathsf{gsk}.\mathsf{Rpk}_1$, where G_1 is a parameter, Rpk_1 the opening manager public key, and gmsk is the group manager secret key. To sign a message on behalf of the group, a member produces a double encryption of A and a signature of knowledge of m which must prove that the double encryption contains a part of a valid certificate (and is thus linked to the group master secret key).

The main drawback of XSGS is that it needs one pairing evaluation, many elliptic curve point multiplications and modular exponentiations (13 and 2 respectively) to produce a group signature. But this is the case for many other group signature schemes. In fact, this complexity places XSGS as one of the most efficient group signature scheme which is today available. Our purpose in this paper is now to propose a secure (see next section) and efficient (see Sections 4 and 5) cooperative version of XSGS.

3 Security of Cooperative Group Signatures

A cooperative group signature [27] allows a group member, with constrained resources, to be helped by some powerful entity, called an intermediary, in the

---

[1] In order to reach the full anonymity property described in [2], the proposal in [14] uses the Naor-Yung methodology [30], and thus twice the same encryption scheme with the same message together with a proof. The scheme in [17] does not totally uses this method and the resulting security is not discussed in the paper.

production of the group signature. The group member is here the constrained device, while the role of the intermediary is played by a more powerful entity, e.g. a personal computer. Note that the cooperative system also requires verifiers, which have the same role as in standard group signatures. The problem on which we focus is that the intermediary may have some more knowledge that is traditionally not available for the adversary. We thus give all the assumptions about the intermediary and we next adapt the security properties of a group signature scheme in this context. This work has not been totally done in [27,6].

3.1   Concept and Assumptions

We first assume that the intermediary does not know any secret information and thus, at the beginning of a protocol, the signer may transfer some data to the intermediary in order to decrease its computation complexity. Consequently, an adversary may obtain more information to break the security of the scheme, e.g. by eavesdropping. It is thus obvious that we must model all her new abilities.

In the cooperative setting, the "standard" adversary (meaning the adversary of the original group signature scheme) can be improved in three different manners. Firstly, the adversary can obtain from the intermediary all data that have been sent by the device. Secondly, she can eavesdrop all communications during a signature protocol (at least the shared data but potentially more information). Finally, she can impersonate the intermediary, and thus obtain all the exchanged information. Moreover, she can learn all the choices made by the intermediary during the protocol (e.g. random values). It is clear that the last adversary is more powerful than the two others. Consequently, we only formally model this one in the cooperative setting and thus introduce the new following oracle.

- $\mathcal{O}^{\mathsf{PartialSign}(i,m)}$: this oracle simulates for the adversary the behaviour of the user i realizing the cooperative signature of a message m. Several exchanges between the oracle and the adversary can be done as it simulates a real cooperative protocol execution between a constrained device and the adversary playing the role of the intermediary.

Based on this new adversary's ability, we must adapt the security properties of group signature schemes to the cooperative setting, based on the formal model of Bellare et al. [2].

3.2   Adaptation of the Correctness

The first security property concerns the correctness and focus on the signature, the opening and the judge verification. In our cooperative context, we decide that the intermediary realizes the connection with the "outside world". Thus, if it decides to send a false signature, the signer cannot do anything to rectify this. Consequently, it seems impossible to ensure such correctness when the adversary can actively participate during the experiment. Nevertheless, the cooperative protocol should at least ensure the "standard" correctness property.

Definition 1. *The* correctness predicate *of a group signature scheme, denoted* $\mathcal{E}_{corr}^{GSS}$, *is verified for a user i and a message m if and only if the following conditions are verified:*

$$\text{VERIF}(m, \text{SIGN}(\textit{gsk}[i], m)) = 1 \wedge \text{OPEN}(\textit{rsk}, \textit{Tab}, m, \text{SIGN}(\textit{gsk}[i], m)) = (i, \tau)$$

$$\wedge \text{JUDGE}(\textit{Tab}, m, \text{SIGN}(\textit{gsk}[i], m), i, \tau) = 1$$

We denote $\mathcal{E}_{corr}^{GSS}(i, m) = 1$ *if this predicate is true, and 0 otherwise.*

A cooperative scheme *ensures the* correctness *property if there exists a negligible function $\epsilon(\lambda)$ such that:*

$$\forall (i, m): \quad Pr[\mathcal{E}_{corr}^{GSS}(i, m) = 0] < \epsilon(\lambda)$$

Remark 1. In case the intermediary is "behind" the signer and has no link with the verifier, it is possible to define a strong cooperative completeness where the intermediary can be corrupted. Since this is not the practical case we are studying, we will not consider it.

We say that a protocol is the *cooperative version* of another one (called the standard one) if their outputs are constructed identically. Then, a cooperative version of a protocol ensures the correctness property if the standard is also correct in the BSZ model [2].

3.3 Adaptation of the Anonymity

From the anonymity point of view, it is possible to assume that the signer and the intermediary live in a personal environment. In fact, as the intermediary can most of the time recognize the signer by some other means, allowing it to know the user identity does not introduce a threat. In this case, the cooperative scheme should only verifies the "standard" anonymity property. More precisely, if the initial group signature scheme provides anonymity (in the BSZ sense), then a cooperative version necessarily verifies this "weak" anonymity property. By doing this assumption, it is generally possible to transfer more data to the intermediary and thus to reduce the signer's complexity by a better factor.

Definition 2 (Anonymity Property). *A cooperative scheme ensures the* anonymity *property if there exists a negligible function $\epsilon(\lambda)$ such that:*

$$\left| Pr[\mathcal{A}(\textit{gmsk}) \to 1 \mid b = 1] - Pr[\mathcal{A}(\textit{gmsk}) \to 1 \mid b = 0] \right| < \epsilon(\lambda)$$

for any polynomial adversary $\mathcal{A}$, who have access to $\mathcal{O}^{CreateU}$, $\mathcal{O}^{AddU}$, $\mathcal{O}^{SJoin}$, $\mathcal{O}^{UJoin}$, $\mathcal{O}^{CrptU}$, $\mathcal{O}^{Reveal}$, $\mathcal{O}^{SignU}$, $\mathcal{O}^{Open}$ and $\mathcal{O}^{Choose_b}$.

Remark 2. In some cases, being unlinkable *w.r.t.* the intermediary is a really important issue and needs to be studied. For the completeness of the model, it is consequently possible to provide a stronger definition for the cooperative anonymity property. The adversary is thus additionally given access to the $\mathcal{O}^{PartialSign}$ oracle in the above experiment.

3.4 Adaptation of the Traceability

Concerning the two remaining security properties, it is necessary to give access to the adversary the $\mathcal{O}^{\mathsf{PartialSign}(i,m)}$ oracle in the cooperative versions of the related experiments.

Definition 3 (Traceability Property). *The* traceability predicate *of a group signature scheme, denoted* $\mathcal{E}^{GSS}_{trac}$*, is verified for* (m,σ) *if and only if the following conditions are verified:*

$$\mathrm{VERIF}(m,\sigma) = 1 \wedge \Big[\mathrm{OPEN}(m,\sigma,\mathsf{rsk}) = \perp \vee$$

$$\Big(\mathrm{OPEN}(m,\sigma,\mathsf{rsk}) = (\mathsf{Upk},\tau) \wedge \mathrm{JUDGE}(\sigma,m,\tau,\mathsf{Upk}) = \perp\Big)\Big]$$

We denote $\mathcal{E}^{GSS}_{trac}(m,\sigma) = 1$ *if this predicate is true, and 0 otherwise.*

A cooperative scheme ensures the traceability *property if there exists a negligible function* $\epsilon(\lambda)$ *such that for any polynomial adversary* $\mathcal{A}$*, who have access to* $\mathcal{O}^{\mathsf{CreateU}}$, $\mathcal{O}^{\mathsf{AddU}}$, $\mathcal{O}^{\mathsf{SJoin}}$, $\mathcal{O}^{\mathsf{UJoin}}$, $\mathcal{O}^{\mathsf{CrptU}}$, $\mathcal{O}^{\mathsf{Reveal}}$, $\mathcal{O}^{\mathsf{SignU}}$, $\mathcal{O}^{\mathsf{Open}}$, $\mathcal{O}^{\mathsf{PartialSign}}$:

$$Pr\Big[\mathcal{A}(gmsk) \rightarrow (m,\sigma) : \mathcal{E}^{GSS}_{trac}(m,\sigma) = 1\Big] < \epsilon(\lambda).$$

Note that the traceability predicate is verified even when the user, which possess Upk, is corrupted, as in the standard security definition [2].

3.5 Adaptation of the Non-frameability

We next study the non-frameability property, for which we introduce a list Set which contains all valid signatures outputted during the experiment (i.e. realized by the $\mathcal{O}^{\mathsf{SignU}}$ oracle).

Definition 4 (Non-Frameability Property). *The* non-frameability predicate *of a group signature scheme, denoted* $\mathcal{E}^{GSS}_{NonFra}$*, is verified for* (m,σ) *if and only if the following conditions are verified, where* $(\mathsf{Upk}_i,\tau) = \mathrm{OPEN}(m,\sigma,\mathsf{rsk},\mathsf{Tab})$:

$$\mathrm{VERIF}(m,\sigma) = 1 \wedge (m,\sigma,i) \notin \mathsf{Set} \wedge i \in \mathcal{HU} \wedge \mathrm{JUDGE}(m,\sigma,\tau,\mathsf{Upk}_i,\mathsf{Tab}) = 1.$$

We denote $\mathcal{E}^{GSS}_{NonFra}(m,\sigma) = 1$ *if this predicate is true, and 0 otherwise.*

A cooperative scheme ensures the non-frameability *property if there exists a negligible function* $\epsilon(\lambda)$ *such that for any polynomial adversary* $\mathcal{A}$*, who have access to* $\mathcal{O}^{\mathsf{AddU}}$, $\mathcal{O}^{\mathsf{CrptU}}$, $\mathcal{O}^{\mathsf{Reveal}}$, $\mathcal{O}^{\mathsf{SignU}}$, $\mathcal{O}^{\mathsf{Open}}$, $\mathcal{O}^{\mathsf{PartialSign}}$:

$$Pr\Big[\mathcal{A}(gmsk,\mathsf{rsk}) \rightarrow (m,\sigma) : \mathcal{E}^{GSS}_{NonFra}(m,\sigma) = 1\Big] < \epsilon(\lambda).$$

4   The Cooperative Version of XSGS

Our aim is now to adapt the XSGS protocol [14] (described in Appendix A) in a secure cooperative manner such that it can be embedded in a RFID tag. For this reason, we consider that the tag is not anonymous w.r.t. the reader. We thus describe a cooperative version of the XSGS scheme and prove its security.

4.1   Protocol Description

To obtain the best possible gain in terms of efficiency, we adapt the XSGS proto-col such that the user will not be anonymous for the intermediary. However, we will prove that all the other security properties remain. Thus, at the beginning of a signature protocol, the user transmits its certificate (A, x), which is part of its group secret key (see Appendix A for details) to the intermediary which will performs all the computations related to this certificate. The user keeps secret his group secret key and computes all values based on it. The obtained cooper-ative version of the protocol is described in Figure 1. The efficiency gain from the user's point of view is huge as there only remains one point multiplication to compute instead of one pairing, 13 point multiplications and 2 modular expo-nentiations in the DLIN based XSGS protocol. In this section we use notations introduced in Section 2 and Appendix A. In a nutshell, a group member owns a group secret key gsk and a certificate (A, x) obtained during the JOIN protocol. The revocation manager has two couples of key $(\mathsf{rsk}_1, \mathsf{Rpk}_1), (\mathsf{rsk}_2, \mathsf{Rpk}_2)$. The group public key is denoted GMpk. Finally, $(q, \mathbb{G}_1, \mathbb{G}_2, \mathbb{G}_T, e, \psi)$ is a bilinear en-vironment (see Appendix A.1) and $\mathcal{H}$ denotes a cryptographically secure hash function.

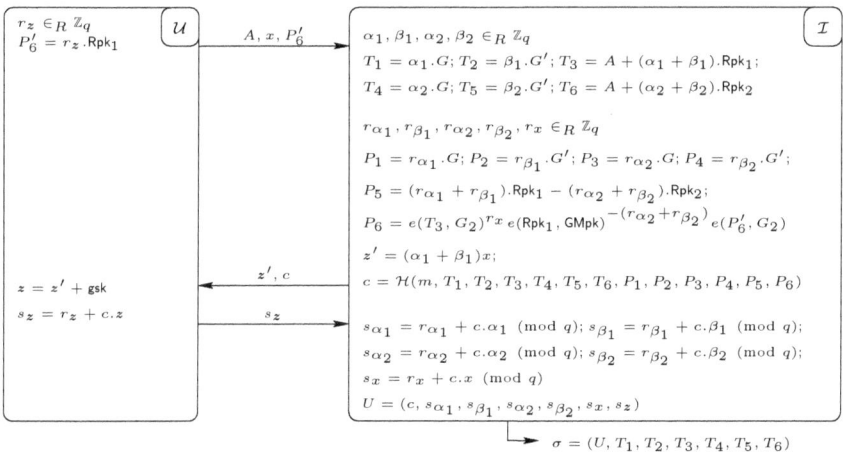

Fig. 1. Sign procedure of the cooperative XSGS Protocol

4.2   Security Analysis

Intuitively, the transmission of the certificate does not introduce any security flaw since both **Traceability** and **Non-Frameability** assume that even the group manager cannot break these properties. As this entity knows all users' certificates, it should be also hard for an active adversary to break them. Nev-ertheless we will formally prove these results. Note that our proof are in the random oracle model as in the original paper [14]. First of all, we prove that our cooperative protocol perfectly realizes a XS group signature, and thus that it verifies the correctness property.

Theorem 1. *The Coop-XSGS protocol ensures the Correctness property.*

Proof. Recall that for this property we assume that the intermediary has a honest behaviour and executes perfectly his part of the protocol. We remark that only the signature of knowledge U slightly deviates from the standard one and we need to verify its correctness. More precisely, the deviation is on the P_6 value. Based on the pairing property (see Section A.1) it is obvious that:

$$P_6 = e(T_3, G_2)^{r_x} e(\mathsf{Rpk}_1, \mathsf{GMpk})^{-(r_{\alpha_1} + r_{\beta_1})} e(P_6', G_2)$$
$$= e(T_3, G_2)^{r_x} e(\mathsf{Rpk}_1, \mathsf{GMpk})^{-(r_{\alpha_1} + r_{\beta_1})} e(\mathsf{Rpk}_1, G_2)^{r_z}$$

Thus, the whole group signature is computed identically as in the standard protocol. Consequently, the cooperative protocol is correct.   □

Theorem 2. *The Coop-XSGS protocol ensures the Anonymity property.*

Proof. As explained in Section 3.3 and since the scheme ensures correctness, the proposed cooperative scheme ensures the anonymity.   □

Theorem 3. *The Coop-XSGS protocol ensures the Traceability property.*

Proof. This proof is obvious since the adversary has no more information than the adversary in the standard model. Indeed, this property is verified even when the adversary represents a collusion of members (thus knowing their group secret keys and certificates). Thus, the cooperative version of the XSGS protocol trivially verifies the traceability property.   □

Theorem 4. *The Coop-XSGS protocol ensures the Non-Frameabilty property.*

Proof. In the original proof (see proof of Theorem 11 in [14]), the authors use the "unforgeability techniques" to retrieve the certificate and the group secret key used in a signature outputted by an adversary. Thus, they build an algorithm which interacts with this adversary in order to break the discrete logarithm either in $\mathbb{G}_1$ or in $\mathbb{G}_2$ (depending of the retrieved group secret key). This proof only works if the adversary does not know the group secret key of the targeted user. As the JOIN procedure does not leak any information about it, their proof is correct. For the cooperative protocol, this proof can also be applied if we prove that an active adversary cannot learn any information about this secret key.

For this purpose, we first highlight the fact that the protocol between the constrained device and the intermediary can be interpreted as a Schnorr protocol (see [33] for further details) which has been proven to be a zero-knowledge proof of knowledge. As a consequence, the values P_6' and s_z do not reveal any information about gsk. It is next obvious that the intermediary has no information about the value r_z under the discrete logarithm assumption. Consequently, given a fixed value of s_z, whatever the value gsk is equal to, there exists one value r_z such that $s_z = r_z + ((\alpha_1 + \beta_1)x + \mathsf{gsk})c$. As a result, a perfect simulation of $\langle P_6', c, s_z \rangle$ can be realized as for the zero-knowledge property of Schnorr's protocol (see [33]). Then, if r_z is uniformly chosen in $\mathbb{Z}_p$, the group secret key of the user is perfectly hidden in s_z. $\square$

Theoretically speaking, this protocol appears to be really efficient. However, in order to demonstrate this efficiency in practice, we describe in the next section an implementation of this new cooperative protocol.

5 Implementation of Coop-XSGS in a RFID Tag

To assess the suitability of the cooperative XSGS protocol for small portable devices, we have implemented it using a wireless sensor node for the prover and a laptop for the powerful helping entity. Wireless sensor nodes are small autonomous devices equipped with a small microcontroller and a transceiver. In this work, we studied the performances of the protocol on two representative sensor node platforms, the MICAz [12] equipped with an 8-bit 7.37MHz ATmega128L microprocessor and the TelosB [13], based on the 16-bit 4MHz MSP430 processor. These devices are conceptually quite close to contactless smart cards. For instance, the TI RF360 chip for contactless secure government electronic ID embeds the same MSP430 processor as the TelosB node [34]. Therefore, although we consider an active device, the results of our implementation can easily be extended to platforms such as contactless smart cards.

The protocol implemented follows the cooperative sign procedure described in Figure 1. The most costly operation for the prover is the point multiplication $P_6' = r_z.\mathsf{Rpk}_1$. It can be computed prior to the interactions with the intermediary, either during idle time (in case of an active device) or precomputed and preloaded on the tag. The latter case corresponds to the *coupon* mode, as in [20], where a *coupon* is a pair of $(r_z, r_z.\mathsf{Rpk}_1)$ loaded on the device. In the following, the operation leading to P_6' is denoted as the off-line phase, although it might still be computed on-line in the case of a passive device avoiding coupons.

Concerning the pairing parameters, we chose an asymmetric pairing, as it allows to use small-length inputs on the tag, reducing therefore the storage and bandwidth costs. For the elliptic curves on which the pairing is applied, we selected the so-called type D curves (following the classification of [25]), i.e., the ordinary curves with embedding degree 3, 4 or 6 known as MNT curves [28]. This type of curve ensures a small input length (around 170 bits for an embedding degree 6) together with an efficient pairing computation [25].

The prover computations were coded in TinyOS [35] on both the MICAz and TelosB sensors. For the point multiplication, we extended the TinyECC library [24] to support the MNT curves. The parameters of the used curve were taken from the PBC library [26] written by B. Lynn. They are labeled as the d159 parameters, where 159 is the size of the base field of the curve. Their security level is equivalent to the hardness of the discrete log problem on $6 \cdot 159 = 954$ bits. Parameters for a higher security could be selected if required. On the

Table 1. Running times of the various phases of the cooperative XSGS protocol. The on-line phase of the sign procedure takes less than 200 ms.

Phase	Detail	Time (ms)
Off-line Sign		
	Prover (MICAz)	3800
	Prover (TelosB)	6400
	Intermediary	195
On-line Sign		
	Prover (MICAz)	7
	Prover (TelosB)	9
	Intermediary	65
	Communication	120
	Total	**< 200**
Verify	–	55
Open	–	35

intermediary side, the computations were implemented in C using the GMP [21] library and the PBC library to achieve the pairings. They took place on a laptop equipped with a 64-bit 1.4 GHz Intel Core 2 Duo. The pairing computations were rather fast in this setting: 11 ms were sufficient to perform a Tate pairing.

The running times for the various parts of the sign procedure are given in Table 1. The on-line phase of the protocol is performed in less than 200 ms with both sensor nodes. The communication delay and the intermediary computations are the main components of the on-line phase latency. The time required by the prover computations, i.e. , the calculation of s_z, is marginal (<10 ms).

In the off-line phase, the computation of the point multiplication giving P_6' lasts about 4 and 6 seconds on the MICAz and TelosB respectively. While reasonably efficient, the TinyECC library, on which our implementation is based, is however significantly slower than recently proposed ECC implementations on sensor nodes. We therefore expect the point multiplication to be significantly faster using the same techniques as for instance the implementation proposed in [23], which performs a fixed-base point multiplication in less than a second on a 192-bit elliptic curve group. As a result, even if the prover device has to compute the point multiplication P_6' during the on-line phase, the whole procedure can be done within a few seconds (much less if a dedicated hardware ECC engine is available on the tag).

The fast on-line phase of the cooperative protocol makes group signatures of practical interest for small devices like contactless smart cards. By contrast, signing with the original XSGS protocol would require a much longer interaction between the passive device and the reader. To give a rough idea, a pairing evaluation takes about 5.5 second on the MICAz (with the TinyPBC library [31]) and an ECC point multiplication requires 0.71 second with the implementation from [23] (there are 13 of these in XSGS): the whole protocol would take on the order of 15 seconds with the MICAz.

Table 2. Storage requirements (in bytes) of our implementation on the sensor nodes. In coupon mode, 21 B must be added per stored coupon.

Mode	Memory	MICAz	TelosB
Coupon	RAM	1067 (26%)	1071 (10%)
	ROM	23764 (18%)	14470 (29%)
No Coupon	RAM	1699 (41%)	1739 (17%)
	ROM	36576 (28%)	19164 (39%)

The original XSGS protocol would also consume a considerable amount of memory. On the other hand, the memory usage of the cooperative protocol is relatively modest (Table 2), even when the ECC point multiplication is performed on the node. The tiny operating system already consumes a significant fraction of the used memory (about 700 B RAM and a little more than 10 kB ROM on both nodes). A coupon $(r_z, r_z.\mathsf{Rpk}_1)$ requires normally 60 B, i.e. three 20-B field elements, but it can be reduced to a little more than 20 B using point compression and a PRNG sequence for storing all the r_z, as done in [20]. As the coupons can be placed in RAM or ROM, their storage in both the MICAz and TelosB is not a problem. Considering the available memory resources, the MICAz and the TelosB could be filled with more than 5000 and 2500 coupons respectively, which is more than practical for many applications.

References

1. Ateniese, G., Camenisch, J., Joye, M., Tsudik, G.: A practical and provably secure coalition-resistant group signature scheme. In: Bellare, M. (ed.) CRYPTO 2000. LNCS, vol. 1880, pp. 255–270. Springer, Heidelberg (2000)
2. Bellare, M., Shi, H., Zhang, C.: Foundations of group signatures: The case of dynamic groups. In: Menezes, A. (ed.) CT-RSA 2005. LNCS, vol. 3376, pp. 136–153. Springer, Heidelberg (2005)
3. Boneh, D., Boyen, X., Shacham, H.: Short group signatures. In: Franklin, M.K. (ed.) CRYPTO 2004. LNCS, vol. 3152, pp. 41–55. Springer, Heidelberg (2004)
4. Boneh, D., Boyen, X.: Short signatures without random oracles and the sdh assumption in bilinear groups. J. Cryptology 21(2), 149–177 (2008)
5. Boyen, X., Waters, B.: Full-domain subgroup hiding and constant-size group signatures. In: Okamoto, T., Wang, X. (eds.) PKC 2007. LNCS, vol. 4450, pp. 1–15. Springer, Heidelberg (2007)
6. Brickell, E.F., Camenisch, J., Chen, L.: Direct anonymous attestation. In: ACM Conference on Computer and Communications Security 2004, pp. 132–145 (2004)
7. Camenisch, J., Groth, J.: Group signatures: Better efficiency and new theoretical aspects. In: Blundo, C., Cimato, S. (eds.) SCN 2004. LNCS, vol. 3352, pp. 120–133. Springer, Heidelberg (2005)
8. Canard, S., Girault, M.: Implementing group signature schemes with smart cards. In: CARDIS 2002, pp. 1–10. USENIX (2002)
9. Chaum, D., van Heyst, E.: Group signatures. In: Davies, D.W. (ed.) EUROCRYPT 1991. LNCS, vol. 547, pp. 257–265. Springer, Heidelberg (1991)

10. Chaum, D., Pedersen, T.P.: Wallet databases with observers. In: Brickell, E.F. (ed.) CRYPTO 1992. LNCS, vol. 740, pp. 89–105. Springer, Heidelberg (1993)
11. Cramer, R., Pedersen, T.P.: Improved privacy in wallets with observers (extended abstract). In: Helleseth, T. (ed.) EUROCRYPT 1993. LNCS, vol. 765, pp. 329–343. Springer, Heidelberg (1994)
12. CrossBow. MICAz low-power wireless sensor module (April 2010), http://www.xbow.com/-Products/Product_pdf_files/Wireless_pdf/MICAz_Datasheet.pdf
13. CrossBow. TelosB low-power wireless sensor module (April 2010), http://www.xbow.com/-Products/Product_pdf_files/Wireless_pdf/TELOSB_Datasheet.pdf
14. Delerablée, C., Pointcheval, D.: Dynamic fully anonymous short group signatures. In: Nguyên, P.Q. (ed.) VIETCRYPT 2006. LNCS, vol. 4341, pp. 193–210. Springer, Heidelberg (2006)
15. Feldhofer, M., Dominikus, S., Wolkerstorfer, J.: Strong authentication for rfid systems using the aes algorithm. In: Joye, M., Quisquater, J.-J. (eds.) CHES 2004. LNCS, vol. 3156, pp. 357–370. Springer, Heidelberg (2004)
16. Fiat, A., Shamir, A.: How to prove yourself: Practical solutions to identification and signature problems. In: Odlyzko, A.M. (ed.) CRYPTO 1986. LNCS, vol. 263, pp. 186–194. Springer, Heidelberg (1987)
17. Furukawa, J., Imai, H.: An efficient group signature scheme from bilinear maps. In: Boyd, C., González Nieto, J.M. (eds.) ACISP 2005. LNCS, vol. 3574, pp. 455–467. Springer, Heidelberg (2005)
18. Gamal, T.E.: A public key cryptosystem and a signature scheme based on discrete logarithms. IEEE Transactions on Information Theory 31(4), 469–472 (1985)
19. Girault, M., Lefranc, D.: Public key authentication with one (online) single addition. In: Joye, M., Quisquater, J.-J. (eds.) CHES 2004. LNCS, vol. 3156, pp. 413–427. Springer, Heidelberg (2004)
20. Girault, M., Juniot, L., Robshaw, M.: The Feasibility of On-the-Tag Public Key Cryptography. In: Workshop on RFID Security – RFIDSec 2007, Malaga, Spain (July 2007)
21. GMP. The GNU Multiple Precision Arithmetic Library (April 2010), http://gmplib.org/
22. Groth, J.: Fully anonymous group signatures without random oracles. In: Kurosawa, K. (ed.) ASIACRYPT 2007. LNCS, vol. 4833, pp. 164–180. Springer, Heidelberg (2007)
23. Lederer, C., Mader, R., Koschuch, M., Großschädl, J., Szekely, A., Tillich, S.: Energy-Efficient Implementation of ECDH Key Exchange for Wireless Sensor Networks. In: Markowitch, O., Bilas, A., Hoepman, J.-H., Mitchell, C.J., Quisquater, J.-J. (eds.) WISTP 2009. LNCS, vol. 5746, pp. 112–127. Springer, Heidelberg (2009), http://www.cs.bris.ac.uk/Publications/Papers/2001061.pdf
24. Liu, A., Ning, P.: TinyECC: A configurable library for elliptic curve cryptography in wireless sensor networks. In: IPSN, pp. 245–256 (April 2008)
25. Lynn, B.: On the implementation of pairing-based cryptosystems. PhD thesis. Stanford University (2007)
26. Lynn, B.: PBC, the Pairing-Based Cryptography Library (April 2010), http://crypto.stanford.edu/pbc/
27. Maitland, G., Boyd, C.: Co-operatively formed group signatures. In: Preneel, B. (ed.) CT-RSA 2002. LNCS, vol. 2271, pp. 218–235. Springer, Heidelberg (2002)
28. Miyaji, A., Nakabayashi, M., Takano, S.: New explicit conditions of elliptic curve traces for fr-reduction (2001)

29. Naccache, D., M'Raïhi, D., Vaudenay, S., Raphaeli, D.: Can D.S.A. Be improved? In: De Santis, A. (ed.) EUROCRYPT 1994. LNCS, vol. 950, pp. 77–85. Springer, Heidelberg (1995)
30. Naor, M., Yung, M.: Universal one-way hash functions and their cryptographic applications. In: STOC, pp. 33–43. ACM, New York (1989)
31. Oliveira, L.B., Scott, M., Lopez, J., Dahab, R.: Tinypbc: Pairings for authenticated identity-based non-interactive key distribution in sensor networks. In: 5th International Conference on Networked Sensing Systems, INSS 2008, pp. 173–180 (June 2008)
32. Paillier, P.: Public-key cryptosystems based on composite degree residuosity classes. In: Stern, J. (ed.) EUROCRYPT 1999. LNCS, vol. 1592, pp. 223–238. Springer, Heidelberg (1999)
33. Schnorr, C.-P.: Efficient identification and signatures for smart cards. In: Brassard, G. (ed.) CRYPTO 1989. LNCS, vol. 435, pp. 239–252. Springer, Heidelberg (1990)
34. TexasInstruments. Texas Instruments Government Electronic Identification (April 2010), http://www.ti.com/rfid/docs/manuals/brochures/govid_trifold.pdf
35. TinyOS. An open-source operating system designed for wireless embedded sensor networks (April 2010), http://www.tinyos.net/
36. Xu, S., Yung, M.: Accountable ring signatures: A smart card approach. In: CARDIS 2004, pp. 271–286. Kluwer, Dordrecht (2004)

A   XSGS Group Signature

In this section, we give some useful tools and next focus on the XSGS group signature scheme, using additive notations.

A.1   Some Notations and Tools

Bilinear Groups. Let $\mathbb{G}_1$, $\mathbb{G}_2$ and $\mathbb{G}_T$ be multiplicative cyclic groups of prime order q and let ψ be an isomorphism from $\mathbb{G}_2$ to $\mathbb{G}_1$. G_1 (resp. G_2) is a generator of $\mathbb{G}_1$ (resp. $\mathbb{G}_2$). Finally, let e be a computable bilinear map $\mathbb{G}_1 \times \mathbb{G}_2 \longrightarrow \mathbb{G}_T$ such that $e(G_1, G_2) \neq 1$ and for all $P_1 \in \mathbb{G}_1$, $P_2 \in \mathbb{G}_2$ and $a, b \in \mathbb{Z}$, $e(a.P_1, b.P_2) = e(P_1, P_2)^{ab}$. $(q, \mathbb{G}_1, \mathbb{G}_2, \mathbb{G}_T, G_1, G_2, e, \psi)$ is called a bilinear environment.

Zero-Knowledge Proofs of Knowledge. Roughly speaking, a Zero Knowledge Proof of Knowledge (ZKPK) is an interactive protocol during which an entity $\mathcal{P}$ proves to a verifier $\mathcal{V}$ that he knows a set of secret values $\alpha_1, \ldots, \alpha_q$ verifying a given relation R without revealing anything else. These protocols are also used to prove that some public values are well-formed from secret values known by the prover. It is possible to transform these protocol into non-interactive proof of knowledge, generally called signature of knowledge, using the Fiat-Shamir heuristic [16].

In the sequel, we denote by $\text{SoK}(\alpha_1, \ldots, \alpha_q : \text{R}(\alpha_1, \ldots, \alpha_q))$ a signature of knowledge of the secrets $\alpha_1, \ldots, \alpha_q$ verifying the relation R. We also define π as the interactive protocol between a prover $\mathcal{P}$, on input $\alpha_1, \ldots, \alpha_q$ and R and a verifier $\mathcal{V}$ on input R and which allows $\mathcal{P}$ to prove that she knows the secrets in a zero-knowledge manner. The output of $\mathcal{V}$ is either 1 if the prover is accepted and 0 otherwise.

A.2   Decision Linear Problem and Encryption

The Decision Linear Problem has been introduced in [3] and is defined as follows.

Definition 5. *Given $G, G', H, \alpha.G, \beta.G', \gamma.H \in \mathbb{G}$ as input, the Decision Linear Problem consists to decide if $\gamma = \alpha + \beta$ or not.*

A great advantage of this problem is that it is still a hard problem even in bilinear groups where the DDH problem is easy. Based on this problem, the authors of [3] introduced a new encryption scheme called Linear Encryption:

- GENPARAM(1^λ): let $\mathbb{G}$ be a group of prime order q. Select three generators G, G' and Rpk such that there exists $\mathsf{rsk}_1, \mathsf{rsk}_2 \in \mathbb{Z}_q$ which verify $\mathsf{Rpk} = \mathsf{rsk}_1.G = \mathsf{rsk}_2.G'$. The public-key of the system is the tuple (G, G', Rpk) while the secret key is $(\mathsf{rsk}_1, \mathsf{rsk}_2)$.
- ENC(m): to encrypt the message $m \in \mathbb{G}$, this algorithm selects two random values $\alpha, \beta \in \mathbb{Z}_q$ and computes $T_1 = \alpha.G, T_2 = \beta.G', T_3 = m + (\alpha + \beta)\mathsf{Rpk}$. The encrypted message is (T_1, T_2, T_3).
- DEC(T_1, T_2, T_3): to decrypt a message, this algorithm computes $m = T_3 - \mathsf{rsk}_1.T_1 - \mathsf{rsk}_2.T_2$.

To define the parameters of this scheme verifying is $\mathsf{Rpk} = \mathsf{rsk}_1.G = \mathsf{rsk}_2.G'$, a solution is described by the next steps:

- choose a random generator $G \in \mathbb{G}$;
- choose a random value $\mathsf{rsk} \in_R \mathbb{Z}_q$ and compute $G' = \mathsf{rsk}.G$;
- choose a first secret key $\mathsf{rsk}_1 \in_R \mathbb{Z}_q$ and compute $\mathsf{Rpk} = \mathsf{rsk}_1.G$;
- compute $\mathsf{rsk}_2 = \mathsf{rsk}_1/\mathsf{rsk} \pmod q$.

A.3   The XSGS Group Signature Scheme

We now focus on the XSGS protocol, introduced by Delerablée and Pointcheval in [14]. For security reasons (see Section 8.1 of the extended version of [3] for more details), we use the XSGS scheme without the XDH assumption. We thus use the double linear encryption scheme, introduced by Boneh et al. in [3] and described in Appendix A.2, instead of a double ElGamal encryption, as suggested in [14]. The group signature scheme is described by the following procedures, where λ is a security parameter.

- GENPARAM(1^λ): this procedure generates the public parameters of the system and also the keys of the different entities as follows:
  - a bilinear environment $(q, \mathbb{G}_1, \mathbb{G}_2, \mathbb{G}_T, e, \psi)$;
  - the parameters for the double linear encryption, i.e. a generator $G \in_R \mathbb{G}_1$ and another generator $G' = \mathsf{rsk}.G$ where $\mathsf{rsk} \in_R \mathbb{Z}_q$.
  - the secret keys of the opening judge $(\mathsf{rsk}_1, \mathsf{rsk}_3) \in_R \mathbb{Z}_q^2$, $\mathsf{rsk}_2 = \mathsf{rsk}_1/\mathsf{rsk}$, $\mathsf{rsk}_4 = \mathsf{rsk}_3/\mathsf{rsk}$ and the associated public keys $\mathsf{Rpk}_1 = \mathsf{rsk}_1.G = \mathsf{rsk}_2.G'$ and $\mathsf{Rpk}_2 = \mathsf{rsk}_3.G = \mathsf{rsk}_4.G'$;

- the secret key $\mathsf{gmsk} \in_R \mathbb{Z}_q$ of the group manager $\mathcal{GM}$ and the associated public key $\mathsf{GMpk} = \mathsf{gmsk}.G_2$;
- the parameters of Paillier's encryption scheme [32] for EXTCOMMIT;

The public parameters of the system are $\mathcal{PP} = \{\lambda, q, \mathbb{G}_1, \mathbb{G}_2, \mathbb{G}_T, e, \psi, G_1, G_2, G, G', \mathsf{GMpk}, \mathsf{Rpk}_1, \mathsf{Rpk}_2\}$.

- USERKEYGEN(1^λ): before that a user, denoted U_i, can join a group, he has to be registered in a PKI. This procedure permits to ensure the unlinkability and the non-repudiation of the system. At the end of this procedure, the user obtain a couple of key $(\mathsf{usk}_i, \mathsf{Upk}_i)$. The value Upk_i is added in a table UPK which is supposed public.
- JOIN[$U_i(\mathsf{usk}_i, \mathsf{Upk}_i) \leftrightarrow \mathcal{GM}(\mathsf{UPK}, \mathsf{gmsk})$]: this interactive protocol between a user U_i and the group manager results by the adhesion of the new user to the group. Consequently, the user obtain a group certificate $\mathsf{cert}_i = (A_i, x_i)$, and his group secret key gsk_i. The group manager add an entry $(\mathsf{Upk}_i, A_i, x_i, S)$ in Tab, where S is a signature of A_i made by the user U_i with his secret key usk_i. This interactive protocol is presented in Figure 2 where EXTCOMMIT is an extractable commitment done with the Paillier's encryption scheme [32].

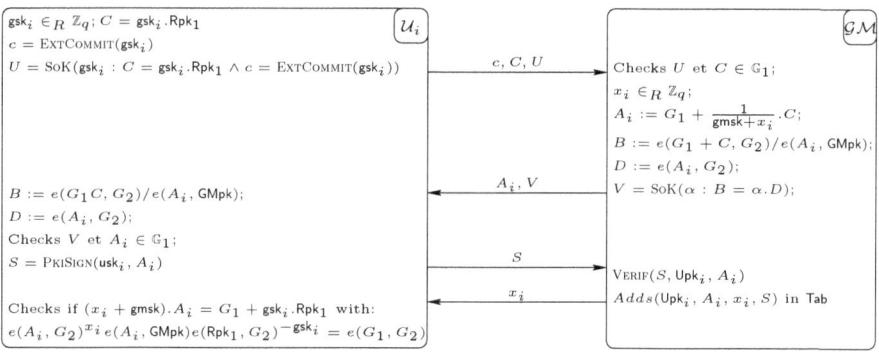

Fig. 2. XSGS JOIN protocol

- SIGN($m, \mathsf{gsk}_i, \mathsf{cert}_i$): the signature of m is composed of two steps
  - a double linear encryption, namely, the user randomly chooses $(\alpha_1, \beta_1, \alpha_2, \beta_2) \in_R \mathbb{Z}_q$ and computes the fours values

$$T_1 = \alpha_1.G; \quad T_2 = \beta_1.G'; \quad T_3 = A + (\alpha_1 + \beta_1)\mathsf{Rpk}_1;$$
$$T_4 = \alpha_2.G; \quad T_5 = \beta_2.G'; \quad T_6 = A + (\alpha_2 + \beta_2)\mathsf{Rpk}_2;$$

  - a signature of knowledge U, where $z = (\alpha_1 + \beta_1).x + \mathsf{gsk}_i$:

$$U = \mathrm{SoK}\Big(\alpha_1, \beta_1, \alpha_2, \beta_2, x, z : T_1 = \alpha_1.G \wedge T_2 = \beta_1.G' \wedge T_4 = \alpha_2.G$$
$$\wedge T_5 = \beta_2.G' \wedge T_3 - T_6 = (\alpha_1 + \beta_1)\mathsf{Rpk}_1 - (\alpha_2 + \beta_2)\mathsf{Rpk}_2 \wedge$$
$$e(T_3, G_2)^x e(\mathsf{Rpk}_1, \mathsf{GMpk})^{-(\alpha_1+\beta_1)} e(\mathsf{Rpk}_1, G_2)^{-z} = \frac{e(G_1, G_2)}{e(T_3, \mathsf{GMpk})}\Big)(m).$$

The user outputs the signature $\sigma = (T_1, T_2, T_3, T_4, T_5, T_6, U)$.

- $\text{VERIF}(m, \sigma)$ this procedure simply verifies the correctness of the signature of knowledge U, as detailed in Section A.4.
- $\text{OPEN}(m, \sigma, (\mathsf{rsk}_1, \mathsf{rsk}_2, \mathsf{rsk}_3, \mathsf{rsk}_4), \mathsf{Tab})$ if σ is valid, the opening judge computes $A = T_3 - (\mathsf{rsk}_1.T_1 + \mathsf{rsk}_2.T_2)$ and realizes the signature of knowledge $\tau = \text{SoK}(\mathsf{rsk}_1, \mathsf{rsk}_2 : A = T_3 - (\mathsf{rsk}_1.T_1 + \mathsf{rsk}_2.T_2) \wedge \mathsf{Rpk}_1 = \mathsf{rsk}_1.G \wedge \mathsf{Rpk}_1 = \mathsf{rsk}_2.G')$. By using Tab, the judge can retrieve the key Upk_i associated to the user certificate A. Then he outputs Upk, $S(= \text{PKISIGN}_\mathsf{usk}(A))$, A and τ.
- $\text{JUDGE}(m, \sigma, A, \tau, \mathsf{Upk}, \mathsf{Tab})$ this procedure verifies the correctness of the signature of knowledge τ. The signature S, stored in Tab, permits to check that the certificate A is the one which have been given to the user during the JOIN procedure. If both signatures (τ and S) are valid, the procedure outputs 1 else it outputs 0.

This protocol ensures all the security requirements of a group signature scheme under the q-SDH [4] and the decision linear assumption (see Section A.2). We refer the interested reader to [14] and [3] for the security aspects of this scheme.

A.4   Focus on the Signature of Knowledge

During the signature of a message, a user must produce the signature of knowledge U. We detailed here how this should be done.

- Choose $r_{\alpha_1}, r_{\beta_1}, r_{\alpha_2}, r_{\beta_2} \in_R \mathbb{Z}_p$; $r_x, r_z \in_R \mathbb{Z}_q$
- Compute

$$P_1 = r_{\alpha_1}.G; P_2 = r_{\beta_1}.G'; P_3 = r_{\alpha_2}.G; P_4 = r_{\beta_2}.G';$$
$$P_5 = (r_{\alpha_1} + r_{\beta_1})\mathsf{Rpk}_1 - (r_{\alpha_2} + r_{\beta_2})\mathsf{Rpk}_2;$$
$$P_6 = e(T_3, G_2)^{r_x} e(\mathsf{Rpk}_1, \mathsf{GMpk})^{-(r_{\alpha_1} + r_{\beta_1})} e(\mathsf{Rpk}_1, G_2)^{-r_z}$$

- Compute $c = \mathcal{H}(m, T_1, T_2, T_3, T_4, T_5, T_6, P_1, P_2, P_3, P_4, P_5, P_6)$
- Compute

$$s_{\alpha_1} = r_{\alpha_1} + c.\alpha_1 \pmod q; s_{\beta_1} = r_{\beta_1} + c.\beta_1 \pmod q;$$
$$s_{\alpha_2} = r_{\alpha_2} + c.\alpha_2 \pmod q; s_{\beta_2} = r_{\beta_2} + c.\beta_2 \pmod q;$$
$$s_x = r_x + c.x \pmod q; s_z = r_z + c.z \pmod q.$$

The signature is the tuple $U = (c, s_{\alpha_1}, s_{\beta_1}, s_{\alpha_2}, s_{\beta_2}, s_x, s_z)$.

The verification of this signature of knowledge is done as follow. The verifier first computes:

- $P_1 = s_{\alpha_1}.G - c.T_1, P_2 = s_{\beta_1}.G' - c.T_2, P_3 = s_{\alpha_2}.G - c.T_4, P_4 = s_{\beta_2}.G' - c.T_5$ and $P_5 = (s_{\alpha_1} + s_{\beta_1})\mathsf{Rpk}_1 - (s_{\alpha_2} + s_{\beta_2})\mathsf{Rpk}_2 - c.(T_3 - T_6)$
- $P_6 = e(T_3, G_2)^{s_x} e(\mathsf{Rpk}_1, \mathsf{GMpk})^{-(s_{\alpha_1} + s_{\beta_1})} e(\mathsf{Rpk}_1, G_2)^{-s_z} \left(\frac{e(G_1, G_2)}{e(T_3, \mathsf{GMpk})} \right)^{-c}$

Finally the verifier validates the signature of knowledge if:

$$c = \mathcal{H}(m, T_1, T_2, T_3, T_4, T_5, T_6, P_1, P_2, P_3, P_4, P_5, P_6).$$

A Lightweight 256-Bit Hash Function for Hardware and Low-End Devices: Lesamnta-LW

Shoichi Hirose[1], Kota Ideguchi[2], Hidenori Kuwakado[3], Toru Owada[2],
Bart Preneel[4], and Hirotaka Yoshida[2,4]

[1] Graduate School of Engineering, University of Fukui
3-9-1, Bunkyo, Fukui 910-8507, Japan
[2] Systems Development Laboratory, Hitachi, Ltd.
292 Yoshida-cho, Totsuka-ku, Yokohama, Kanagawa 244-0817, Japan
[3] Graduate School of Engineering, Kobe University
1-1 Rokkodai, Nada, kobe 657-8501, Japan
[4] Department of Electrical Engineering ESAT/SCD-COSIC,
Katholieke Universiteit Leuven
Kasteelpark Arenberg 10, B–3001 Heverlee, Belgium

Abstract. This paper proposes a new lightweight 256-bit hash function Lesamnta-LW with claimed security levels of at least 2^{120} with respect to collision, preimage, and second preimage attacks. We adopt the Merkle-Damgård domain extension; the compression function is constructed from a dedicated AES-based block cipher using the LW1 mode, for which a security reduction can be proven. In terms of lightweight implementations, Lesamnta-LW offers a competitive advantage over other 256-bit hash functions. Our size-optimized hardware implementation of Lesamnta-LW requires only 8.24 Kgates on 90 nm technology. Our software implementation of Lesamnta-LW requires only 50 bytes of RAM and runs fast on short messages on 8-bit CPUs.

Keywords: Hash functions, lightweight cryptography, security reduction proofs.

1 Introduction

The next decade will witness an ever growing demand for applications using small electronic devices such as sensor nodes, RFID tags and smart devices. These devices have to cope with security problems such as confidentiality, and more importantly, authentication and privacy. The key tools to develop these applications are lightweight cryptographic algorithms which can be implemented under restricted resources, such as low-cost, low-energy, or low-power environments. Lightweight cryptographic algorithms such as PRESENT [11], H-PRESENT [12], KATAN [14], MAME [45], SQUASH [41] and QUARK [2] have been proposed to target these environments. The lightweight symmetric-key encryption algorithms attract users for providing very compact authentication using MACs.

Among these cryptographic techniques, we argue that cryptographic hash functions are of particular importance because hash functions are needed in

K.-H. Rhee and D. Nyang (Eds.): ICISC 2010, LNCS 6829, pp. 151–168, 2011.

any serious library: they can be used in applications such as digital signatures, certificates, MAC algorithms, randomness extraction, and public key encryption (e.g., RSA-OAEP), etc. During the last years, there has been substantial progress in cryptanalysis [46,47] of widely-used hash functions such as MD5 [39] and SHA-1 [37]. In response, NIST started the SHA-3 competition in 2007 [38], and selected 14 hash functions as Round-2 candidates in last July. Thus, lightweight software/hardware implementations could use SHA-256 [37] or the SHA-3 Round-2 candidates. However, most of these hash functions could be too expensive for small devices since they are designed for generic purpose; they are fast on high-end 32/64-bit CPUs and have in general a large internal state for the resistance against multi-collision-type of attacks [29,30].

We argue that there is an increasing demand for lightweight hash functions providing a high security level. A reasonable application would be code signing for small but highly sensitive devices which can be targeted at medical applications or car electronics. Code signing requires hashing and public key cryptography (PKC). Some recent works [3,22] have shown that implementations of elliptic curve cryptography (ECC) can be so compact that implementations of ECC are targeted at wireless sensor networks (WSN). Therefore it would be a nice challenge to fit ECC and hashing in a small area such as 25 Kgates.

In addition, applications using small portable electronic devices employing low-cost 8-bit CPUs have gained increasing attention. It has been reported that about 55 % of all CPUs sold in the world are 8-bit microcontrollers and microprocessors and over 4 billion 8-bit controllers were sold in 2006 [48,40]. Since the memory size of devices employing low-cost CPUs are often very small, RAM/ROM requirements are an important factor for implementations.

This paper proposes a 256-bit lightweight hash function, *Lesamnta-LW*. Its domain extension is the strengthened Merkle-Damgård construction and its underlying component is an AES-based block cipher taking a 256-bit plaintext and a 128-bit key. Note that Lesamnta-LW is somewhat a lightweight version of Lesamnta [25] that was submitted to the SHA-3 competition. The feature of Lesamnta-LW is summarized below.

1. Lesamnta-LW can be implemented efficiently on both of a dedicated hardware and 8-bit CPUs. In hardware, it only requires 8.24 Kgates on 90 nm technology, which is substantially smaller than those of most of Round-2 SHA-3 candidates. In software, it gains clear advantages over SHA-256 with respect to speed on short messages and RAM requirements for 8-bit CPUs.
2. The compression function is a new mode of a block cipher, called the *LW1 mode*. Notice that the PGV mode cannot be used because Lesamnta-LW uses the block cipher such that the key size is smaller than the block size in order to achieve the efficient implementation. Unlike the DM mode and the MMO mode, the LW1 mode does not have the feedforward of inputs, which contributes to reduction of the size of required memory. This structure enables us to provide proofs reducing the security of Lesamnta-LW to that of the underlying block cipher which has also been designed to offer adequate security against all relevant known attacks.

3. Lesamnta-LW is designed to provide at least 2^{120} security levels against both collision and (second-)preimage attacks, where $2^{120} = 2^n/(n+1)$ with $n = 128$. Actually, it is easy to see that a meet-in-the-middle attack can find a preimage of Lesamnta-LW with complexity at most 2^{128}.

For the security levels, an ideal 256-bit hash function would provide the 2^{256} security level against preimage attacks. However, the 2^{120} security level is sufficient for most applications, especially on small devices. We give preference to the hardware cost over the preimage resistance in the design of Lesamnta-LW. The security and the cost do not go together generally.

As an important application of Lesamnta-LW, we consider the key-prefix (KP) mode which is similar to HMAC but more efficient. The KP mode of a hash function is required in PPP Challenge Handshake Authentication Protocol [42]. We give a security reduction for this mode.

The outline of this paper is as follows. In Sect. 2, we explain our design strategy. In Sect. 3, we give the specification of the Lesamnta-LW hash function. In Sect. 4, we discuss the provable security of Lesamnta-LW. In Sect. 5, we evaluate the security of Lesamnta-LW against all relevant attacks. Section 6 presents implementation results. Section 7 concludes the paper.

2   Design Principle

Our main design goal is to develop a secure 256-bit hash function which achieves small hardware/software implementations. More specifically, the most important aspects are to have security proofs, to have a small footprint for hardware, and to have low working memory (RAM) requirement for software. Our next target is to achieve fairly fast speed, considering the ways hash functions are used: the processing message length and the modes of operation, etc. This is because the required efficiency could include speed on very short messages such as IDs or speed of the pseudorandom function from a hash function such as HMAC or Key-Prefix mode as discussed in this paper.

2.1   Padding Method

For the padding method of Lesamnta-LW, the last block does not contain any part of the message input. It only contains the length of the message input. This property is required to guarantee preimage resistance of Lesamnta-LW.

2.2   LW1 Mode

Sophisticated designs and attacks on block ciphers were presented in the AES competition. Knowledge on block ciphers is useful in designing secure hash functions. This is why Lesamnta-LW is designed as a block-cipher-based hash function. A few reasons for choosing the LW1 mode are also listed below. First, from the viewpoint of attacks on a block cipher, recent collision attacks use the fact that an attacker can directly control the key of a block cipher. In contrast, the LW1 mode does not allow attackers to control the key of the block cipher directly. Second, the LW1 mode is theoretically analyzed. It enables us to reduce the security of Lesamnta-LW to that of the underlying block cipher to a greater extent than the popular DM mode used by the SHA family.

2.3 Block Cipher

The block cipher is designed to meet the following requirements:

- The security analysis should be simple to have confidence in the design.
- It should be compact in software/hardware.
- It should offer a reasonable speed on high-end/low-end CPUs.

For this purpose, the block cipher is an AES-based design such that Lesamnta-LW can gain certain clear advantages over know block-cipher based designs such as SHA-256 and MAME. The key scheduling function ensures a strong non-linearity and an excellent diffusion property by re-using the 32-bit permutation of the mixing function; this reduces the hardware complexity since a part of the hardware can be reused. The round constants sequentially generated from a linear feedback shift register introduce randomness and asymmetry into the key scheduling function.

3   Specification

3.1 Message Padding

The first step of the hash computation is the padding of the message. The purpose of the padding is to ensure that the input consists of a multiple of 128 bits. Suppose that the length of a message M is l bits. Append the bit "1" to the end of the message, followed by $k + 63$ zero bits, where k is the smallest non-negative integer such that $l + k \equiv 0 \pmod{128}$. Then, append a 64-bit block equal to the number l as expressed in binary representation. The length of the padded message should now be a multiple of 128 bits.

3.2 Compression Function and Domain Extension

Lesamnta-LW is a Merkle-Damgård iterated hash function [19,36] using the compression function operates as follows on the 128-bit words $H_0^{(i-1)}$, $H_1^{(i-1)}$, and $M^{(i)}$:

$$h(H^{(i-1)}, M^{(i)}) = E(H_0^{(i-1)}, M^{(i)} \parallel H_1^{(i-1)}) \ ,$$

where $H^{(i-1)} = H_0^{(i-1)} \parallel H_1^{(i-1)}$ and $E(K, \cdot)$ is a 256-bit block cipher with a 128-bit key K. We call this method to construct a compression function the LW1 mode, which is illustrated in Fig. 1.

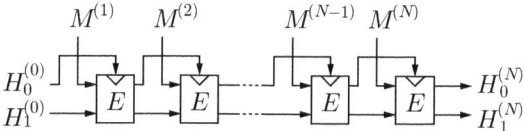

Fig. 1. The domain extension using the LW1 mode

3.3   Block Cipher

Lesamnta-LW uses a 64-round block cipher E that takes as input a 128-bit key and a 256-bit plaintext. The block cipher consists of two parts: the key scheduling function mapping the key to the round keys and the mixing function taking as input a plaintext and the round keys to produce a ciphertext. Both of them use a type-1 4-branch generalized Feistel network (GFN) (cf. Zheng et al. [50]). One round of the block cipher is illustrated in Fig. 2. The input variables to round r for the mixing function and the key scheduling function are denoted by $(x_0^{(r)}, x_1^{(r)}, x_2^{(r)}, x_3^{(r)})$ and $(k_0^{(r)}, k_1^{(r)}, k_2^{(r)}, k_3^{(r)})$ respectively. Each $x_i^{(r)}$ is a 64-bit word and each $k_i^{(r)}$ is a 32-bit word.

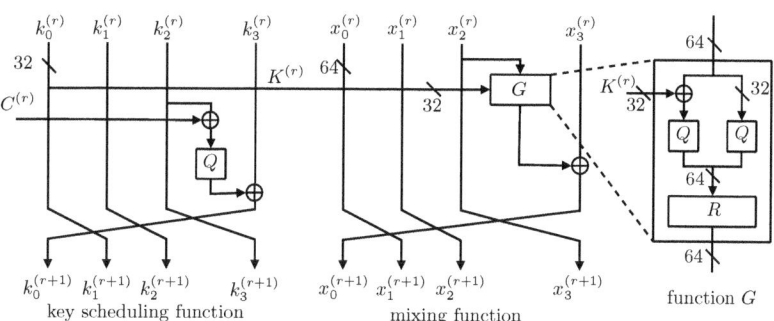

Fig. 2. The round function

The mixing function consists of XORs, a wordwise permutation, and a non-linear function G. Taking as input a 32-bit round key $K^{(r)}$, the mixing function updates its intermediate state in the following manner:

$$x_0^{(r+1)} = x_3^{(r)} \oplus G(x_2^{(r)}, K^{(r)}), \quad x_1^{(r+1)} = x_0^{(r)}, \quad x_2^{(r+1)} = x_1^{(r)}, \quad x_3^{(r+1)} = x_2^{(r)}.$$

The function G consists of XOR operations, a 32-bit non-linear permutation Q, and a function R. For a 64-bit input $y = y_0 \parallel y_1$ and a 32-bit round key $K^{(r)}$, $G(y, K^{(r)})$ is defined as follows:

$$G(y, K^{(r)}) = R(Q(y_0 \oplus K^{(r)}) \parallel Q(y_1)).$$

Using the AES components [18], the function Q is defined as follows:

$$Q = \textbf{MixColumns} \circ \textbf{SubBytes}.$$

The **SubBytes** transformation is a non-linear byte substitution that takes 4 bytes s_0, s_1, s_2, s_3 as input and operates independently on each byte by using the AES S-box. It proceeds as follows:

$$s_i' = \text{S-box}(s_i) \quad \text{for } 0 \leq i < 4.$$

The **MixColumns** step is a bytewise operation that takes 4 bytes s_0, s_1, s_2, s_3 as input. The **MixColumns** step is given by the AES MDS matrix multiplication defined over $\text{GF}(2^8)$ as follows:

$$\begin{bmatrix} s_0' \\ s_1' \\ s_2' \\ s_3' \end{bmatrix} = \begin{bmatrix} 02 & 03 & 01 & 01 \\ 01 & 02 & 03 & 01 \\ 01 & 01 & 02 & 03 \\ 03 & 01 & 01 & 02 \end{bmatrix} \begin{bmatrix} s_0 \\ s_1 \\ s_2 \\ s_3 \end{bmatrix}.$$

For a 64-bit input $s = s_0 \parallel s_1 \parallel s_2 \parallel s_3 \parallel s_4 \parallel s_5 \parallel s_6 \parallel s_7$, the function $R(s)$ is defined as follows: $R(s) = s_4 \parallel s_5 \parallel s_2 \parallel s_3 \parallel s_0 \parallel s_1 \parallel s_6 \parallel s_7$.

One round of the key scheduling function consists of the following two steps:

Firstly, it generates the r-th round-key $K^{(r)} = k_0^{(r)}$.

Secondly, it updates the intermediate state in the following manner:

$$k_0^{(r+1)} = k_3^{(r)} \oplus Q(C^{(r)} \oplus k_2^{(r)}), \quad k_1^{(r+1)} = k_0^{(r)}, \quad k_2^{(r+1)} = k_1^{(r)}, \quad k_3^{(r+1)} = k_2^{(r)},$$

where the 32-bit round constants $C^{(r)}$ are defined in Appendix A.

4   Provable Security

In this section, it is assumed that Lesamnta-LW consists of a block cipher with its key length n and its block length $2n$; specifically, $n = 128$.

4.1   Collision Resistance

The collision resistance of Lesamnta-LW can be proved in the ideal cipher model using the technique by Black et al. in [10]. Lesamnta-LW has a claimed security level of at least 2^{120} block-cipher operations against collision attacks.

Definition. Let $\mathcal{BC}(\kappa, \nu)$ be the set of all (κ, ν) block ciphers, where κ and ν represents their key size and block size, respectively. Let $H[E]$ be a hash function using a block cipher E. Let A be an adversary trying to find a collision for $H[E]$. The col-advantage of A against $H[E]$, $\text{Adv}_{H[E]}^{\text{col}}(A)$, is given by

$$\Pr\left[A^E = (M, M') \wedge M \neq M' \wedge H[E](M) = H[E](M') \mid E \xleftarrow{\$} \mathcal{BC}(\kappa, \nu) \right],$$

where the probabilities are taken over the coin tosses by A and the uniform distribution on $\mathcal{BC}(\kappa, \nu)$. $H[E]$ is said to be collision-resistant if $\text{Adv}_{H[E]}^{\text{col}}(A)$ is negligible for any efficient A.

Result. The following theorem gives an upper bound on the probability of finding a collision of Lesamnta-LW in the ideal cipher model. A proof is given in Appendix B.

Theorem 1. *Let A be a col-adversary against Lesamnta-LW asking at most q queries to E. Then, for $n \geq 4$ and $q \geq 1$,*

$$\mathrm{Adv}_{\mathrm{LW}}^{\mathrm{col}}(A) \leq \frac{2^n n q}{2^{2n} - q} + \frac{q^2}{2^{2n} - q} + \frac{q}{n! \cdot 2^n} \ .$$

4.2   (Second) Preimage Resistance

The preimage resistance of Lesamnta-LW can also be proved in the ideal cipher model using the same technique. Since the compression function is invertible, the Lesamnta-LW hash function also has a claimed security level of at least 2^{120} block-cipher operations against (second-)preimage attacks. On the other hand, Lesamnta-LW cannot provide security larger than 2^{128} because of the LW1 mode.

We note that a second-preimage attack such as the Kelsey-Schneier attack [30] is ineffective in attacking Lesamnta-LW because of the designed security level. For example, consider the Kelsey-Schneier attack using a $(2^{64} - 1)$-bit message, which is the maximum acceptable length of a message. Then, the complexity of the attack is about 2^{192}. However, Lesamnta-LW provides at most the 2^{128} security level against preimage-finding attacks.

4.3   Keyed Hashing Mode

Key-Prefix Mode. The key-prefix (KP) mode is a method to construct a pseudorandom function (PRF) from a given hash function. It simply feeds $K\|M$ to the hash function as an input, where K is a secret key and M is a message input. This mode uses a hash function as a black box. In this sense, it is similar to HMAC.

The KP mode of Lesamnta-LW with the first half of the output chopped off resists any distinguishing attack that requires much fewer than 2^{128} queries if the underlying block cipher is a pseudorandom permutation (PRP) and it also has a mild security property given later.

Definition. Let $\mathcal{F}(\mathcal{X}, \mathcal{Y})$ be a set of all functions from $\mathcal{X}$ to $\mathcal{Y}$. Let $F : \mathcal{K} \times \mathcal{X} \to \mathcal{Y}$ be a keyed function from $\mathcal{X}$ to $\mathcal{Y}$, where $\mathcal{K}$ is its key space. Let A be a probabilistic algorithm which has oracle access to a function from $\mathcal{X}$ to $\mathcal{Y}$ and outputs 0 or 1. The prf-advantage of A against F is given by

$$\mathrm{Adv}_F^{\mathrm{prf}}(A) = \left| \Pr[A^{F_K} = 1 \mid K \xleftarrow{\$} \mathcal{K}] - \Pr[A^{\rho} = 1 \mid \rho \xleftarrow{\$} \mathcal{F}(\mathcal{X}, \mathcal{Y})] \right| \ ,$$

where the probabilities are taken over the coin tosses by A and the uniform distributions on $\mathcal{K}$ and $\mathcal{F}(\mathcal{X}, \mathcal{Y})$. F is called a PRF if $\mathrm{Adv}_F^{\mathrm{prf}}(A)$ is negligible for any efficient A.

Let $\mathcal{P}(\mathcal{X})$ be a set of all permutations on $\mathcal{X}$. Let $F : \mathcal{K} \times \mathcal{X} \to \mathcal{X}$ be a keyed permutation on $\mathcal{X}$. Then, the prp-advantage of A against F is given by

$$\mathrm{Adv}_F^{\mathrm{prp}}(A) = \left| \Pr[A^{F_K} = 1 \mid K \xleftarrow{\$} \mathcal{K}] - \Pr[A^\rho = 1 \mid \rho \xleftarrow{\$} \mathcal{P}(\mathcal{X})] \right| .$$

F is called a PRP if $\mathrm{Adv}_F^{\mathrm{prp}}(A)$ is negligible for any efficient A.

Result. Let h be the Lesamnta-LW compression function: $h(H^{(i-1)}, M^{(i)}) = E(H_0^{(i-1)}, M^{(i)} \| H_1^{(i-1)})$. Let $\mathcal{B} = \{0,1\}^n$. Let $G_1[E] : \mathcal{B} \times \mathcal{B} \to \mathcal{B}^2$ be a keyed function such that $G_1[E](K, M) = h(h(IV, K), M)$, where $K \in \mathcal{B}$ and $M \in \mathcal{B}$. Let $G_2[E] : \mathcal{B}^2 \times \mathcal{B} \to \mathcal{B}^2$ be a keyed function such that $G_2[E](K', M) = h(K', M)$, where $K' \in \mathcal{B}^2$ and $M \in \mathcal{B}$.

Theorem 2. *Let A be a prf-adversary against the KP mode of Lesamnta-LW. Suppose that A runs in time at most t, and makes at most q queries, and each query has at most ℓ message blocks. Then, there exist a prp-adversary B_1 against E and an adversary B_2 against $G_1[E]$ such that*

$$\mathrm{Adv}_{\mathrm{KP\text{-}LW}}^{\mathrm{prf}}(A) \le \ell q \cdot \mathrm{Adv}_E^{\mathrm{prp}}(B_1) + \mathrm{Adv}_{G_1[E]}^{G_2[E]}(B_2) + \frac{\ell q(q-1)}{2^{2n+1}} .$$

where

$$\mathrm{Adv}_{G_1[E]}^{G_2[E]}(B_2) = \left| \Pr[B_2^{G_1[E](K,\cdot)} = 1 \mid K \xleftarrow{\$} \mathcal{B}] - \Pr[B_2^{G_2[E](K,\cdot)} = 1 \mid K \xleftarrow{\$} \mathcal{B}2] \right| .$$

Both B_1 and B_2 make at most q queries and run in time at most $t + O(\ell q T_E)$, where T_E represents the time required to compute E.

$G_2[E]$ is a PRF if E is a PRP. Thus, $\mathrm{Adv}_{G_1[E]}^{G_2[E]}(B_2)$ is negligible for any efficient B_2 if E is a PRP and $G_1[E]$ is a PRF.

The proof of Theorem 2 is omitted due to the page limit.

HMAC. Lesamnta-LW supports HMAC specified in FIPS PUB 198-1:

$$\mathrm{HMAC}(K, M) = H((K \oplus \mathsf{opad}) \| H((K \oplus \mathsf{ipad}) \| M)) ,$$

where H represents Lesamnta-LW and K is a secret key. The security of HMAC using Lesamnta-LW can also be reduced to the security of the underlying block cipher. HMAC using Lesamnta-LW resists any distinguishing attack that requires much fewer than 2^{128} queries if the underlying block cipher E is a PRP, and $G_1[E]((K \oplus \mathsf{opad}) \| \cdot)$ and $G_1[E]((K \oplus \mathsf{ipad}) \| \cdot)$ are independent PRFs.

5   Preliminary Analysis

We evaluate the security of Lesamnta-LW and the underlying block cipher against all relevant attacks. In the analysis of the block cipher, the attacker can have at most 2^{128} complexity because of the key length (128 bits) of the cipher rather than the plaintext length (256 bits). The descriptions of our analysis regarding higher order differential attack, interpolation attack, slide attack, and related-key attack are given in Appendix C.

5.1 Differential and Linear Attacks on Block Ciphers

We explain our method of evaluating the security against differential cryptanalysis [8]. We can apply a similar method to linear cryptanalysis [35] because of its duality to differential cryptanalysis [16]. In order to do this, we compute upper bounds on the probabilities of differential characteristics with the following method:

- Make abstraction of the exact differences used in these characteristics and then just consider patterns of active S-boxes.
- Perform experiments with the Viterbi algorithm to compute lower bounds on the minimum number of the active S-boxes, considering the MDS property.

We can observe that the minimum number of the active S-boxes for 24 rounds is 24. Therefore the probabilities of differential characteristics of 24 rounds of Lesamnta-LW are upperbounded by 2^{-144} because the maximum differential probability of AES S-box is 2^{-6}. As a result, it is very unlikely to apply differential/linear attacks to the full Lesamnta-LW.

5.2 Collision Attacks Using the Message Modification

Wang et al. [46,47] showed methods for finding collisions for widely used hash functions such as SHA-1. Their approach is based on the differential cryptanalysis and the message modification technique which can be used to reduce the attack complexity by using degrees of freedom in the message block space.

In the case of attacks on Lesamnta-LW using the above differential collision finding methods, the attacker has to use messages of at least two blocks because the message block is shorter than the chaining variable. Using multiple block messages, he has some control over 384 bits of the input to the compression function. However, out of these 384 bits, the only input bits over which he can have control in a deterministic way are 128 bits, which correspond to the message block input. He can have control over the remaining 256 bits corresponding to the chaining variable input only in a probabilistic way. On the other hand, we can show that the maximum differential characteristic probabilities for 44 rounds of the mixing function and for 24 rounds of the key scheduling function are less than 2^{-256} and 2^{-128} in the same way we did in Sect 5.1. Their methods for finding collisions require a differential characteristic with a large probability and a large degree of freedom in the message block space. Considering the limited number of bits he can use in a deterministic way and differential characteristics he can find which have a very small probability, we expect that it is very unlikely to apply the differential collision finding methods to Lesamnta-LW.

5.3 Attacks on the Compression Function of Lesamnta

Recently, attacks [13] on the compression function of Lesamnta [25] have been reported. Their main idea is to find some structure in round constants. More specifically, the following properties in Lesamnta are exploited:

- The round constants consisting of two words are word-symmetric up to the least significant bit.
- In one round of the key scheduling function, if the two halves of the input to non-linear function are swapped, then the output is also swapped.

In the case of Lesamnta-LW, the round constants are generated from an LFSR. Therefore, it is difficult to find in them a structure useful for these attacks. The non-linear function Q in the key scheduling function does not have this symmetry due to the linear diffusion layer which uses a single MDS matrix rather than two. On the other hand, the non-linear function G in the mixing function has this symmetry but this is destroyed by the asymmetry introduced from the fact that only 32 bits of round key are added to the 64-bit input to G. Hence, the second property does not hold for Lesamnta-LW. We conclude from the above discussions that these attacks are not relevant to Lesamnta-LW.

5.4 Collision Attack on the 11-Round Lesamnta-LW

We show how to construct collisions for 11-rounds of Lesamnta-LW by applying the method used in collision attacks on Lesamnta in [25]. Now suppose that we can find 2^{96} messages m^* such that all message blocks produce the same value in the lowest 64-bit word of the chaining variable, H_3. Then, we know that due to the birthday paradox two of these message blocks also lead to the same values H_0, H_1, and H_2. In other words, we have constructed a collision for the compression function.

Our attack uses the characteristic for 11 rounds given in Table 1. Note that the symbol ? denotes an arbitrary difference and Δ denotes a message block difference.

Table 1. Characteristic for the collision attack

Round	message block	0	1	2	3	4	5	6	7	8	9	10
	Δ	–	–	?	Δ	–	?	?	Δ	?	?	?
Inputs	–	Δ	–	–	?	Δ	–	?	?	Δ	?	?
	–	–	Δ	–	–	?	Δ	–	?	?	Δ	?
	–	–	–	Δ	–	–	?	Δ	–	?	?	Δ

It is easy to see that this characteristic can be used to fix 64 bits of the output of the compression function. It can be summarized as follows.

1. Select an arbitrary 64-bit value d.
2. Choose a random message block $m = M_0\|M_1\|M_2\|M_3$ and compute $H = H_0\|H_1\|H_2\|H_3$. Check if $H_3 = d$ for a predefined value d.
3. If $H_3 \neq d$, then adjust $\Delta = H_3 \oplus d$ accordingly.
4. Now we have to construct m^* by adjusting m such that $H_3 = d$ as follows:

$$m^* = (M_0 \oplus \Delta)\|M_1\|M_2\|M_3 \ .$$

Hence, we can find a message block m^* such that $H_3 = d$ for an arbitrary value of d with a complexity of about 2 compression function evaluations. Therefore, we can find a collision for the 11 rounds of the compression function (and the hash function) of Lesamnta-LW with a complexity of about 2^{97} compression function evaluations.

6   Implementation Results

6.1   Low-Area ASIC Implementation Results

We have estimated speed and gate count of a hardware architecture of Lesamnta-LW, MAME, and SHA-256. In Table 2, our results[1] are compared to known results on other hash functions such as BLAKE-32 [1], ECHO-224/256 [4], Fugue-256 [24], Grøstl-224/256 [23], Luffa-224/256 [15], and Skein [21]. It is clear that Lesamnta-LW achieves very small implementation and it is substantially smaller than most SHA-3 Round-2 candidates. For designs with a comparable size, Lesamnta-LW offers a reasonably fast speed for the same clock frequency.

Table 2. Our ASIC implementation estimates of Lesamnta-LW, MAME, and SHA-256 with known results on other hash functions. For Imple. Scope, Full means a fully autonomous implementation including the complete functionality of a hash function while Core means an implementation of core functionality comprising only important parts of a hash function such as the compression function. The digest size of SHA-3 candidates is omitted.

Algorithm	Impl. Scope	Logic Process	Area (Kgates)	Throughput (Mbit/s)	Clock (MHz)	Throughput/ Area	Throughput @30MHz (Mbit/s)
BLAKE [44]	Full	0.35 μm	25.57	15.4	31.25	0.60	14.78
ECHO [34]	Full	0.13 μm	82.8	373	66.6	4.50	168.02
Fugue [24]	Full	90 nm	59.22	2000	500	33.77	120.00
Grøstl[44]	Full	0.35 μm	14.62	145.9	55.87	99.79	78.34
Luffa [31]	Full	0.13 μm	18.26	2461	250	134.78	295.32
Skein [44]	Full	0.35 μm	12.89	19.8	80	1.54	7.43
MAME [45]	Core	0.18 μm	8.1	440	333	54.32	39.64
SHA-256 [20]	Full	0.35 μm	10.9	22.5	50	2.06	13.5
T-Quark×16 [2]	Full	0.18 μm	4.64	0.05	0.1	0.01	15
Lesamnta-LW	Full	90 nm	8.24	125.55	188.3	15.31	20.00
MAME	Full	90 nm	12.95	1164.48	436.68	89.9	80.0
SHA-256	Full	90 nm	14.6	1766	220.8	120.0	239.9

---

[1] Note that the values in Throughput/Area and Throughput@30MHz could considerably vary, depending on the technology (logic process). Hence implementation comparisons between these algorithms using the same technology will be welcome.

6.2   Software Implementation Results

For software, Lesamnta-LW is targeted at RAM requirement on the 8-bit CPU employed in smart devices. In low-cost 8-bit CPU applications, hash functions should require limited resources, memory and computation time. We argue that the most important constraint for hash functions is basically the limited RAM which could be critical in many cases.

8-bit CPU. We have estimated RAM requirements of SHA-3 candidates, SHA-256, and Lesamnta-LW. Our results are shown in Table 3. It is clear that Lesamnta-LW achieves very small implementation and substantially smaller than most SHA-3 Round-2 candidates.

We have estimated speed of Lesamnta-LW and SHA-256 on an 8-bit CPU Renesas H8 in assembly language. Our results show that it takes 66434 cycles for Lesamnta-LW to process a one-block message of 128-bit length, which is about 20% faster than SHA-256. However, as for the bulk-speed, our Lesamnta-LW implementation requiring 50 bytes of RAM runs at 1650.9 cycles/byte which, 60% slower than our SHA-256 implementation requiring 330 bytes of RAM.

Table 3. Our estimates of RAM requirements on low-cost 8-bit CPUs

Algorithm	RAM(bytes)	Algorithm	RAM(bytes)
BLAKE[1]	96	JH[49]	128
BMW[28]	192	Keccak[6]	200
CubeHash[5]	128	Luffa[15]	96
ECHO[4]	320	Shabal[17]	176
Fugue[24]	120	SHAvite-3[7]	128
Grøstl[23]	128	SIMD[33]	384
Hamsi[26]	96	Skein[21]	96
Lesamnta-LW	50	MAME[45]	64

32-bit CPU. We have estimated speed of Lesamnta-LW and SHA-256 on the Intel Core i5 processor which offers instructions for fast encryption of AES. Lesamnta-LW runs at 39.5 cycles/byte in assembly language while SHA-256 runs at 26.9 cycles/byte in ANSI C. Our results show that Lesamnta-LW is reasonably fast on this platform.

7   Conclusion

A new lightweight 256-bit hash function Lesamnta-LW has been proposed in this paper. The security of Lesamnta-LW is reduced to that of the underlying block cipher. Lesamnta-LW achieves a very small hardware/software implementation. Moreover, Lesamnta-LW is efficient on short messages on 8-bit CPUs.

Acknowledgments. We would like to mention the people who gave us feedback and important comments on this work: Yasuko Fukuzawa, Kazuo Ota, and Kazuo Sakiyama. This work was partially supported by the National Institute of Information and Communications Technology, Japan.

References

1. Aumasson, J.P., Henzen, L., Meier, W., Phan, R.C.-W.: SHA-3 proposal BLAKE, `http://131002.net/blake/`
2. Aumasson, J.P., Henzen, L., Meier, W., Naya-Plasencia, M.: QUARK: A Lightweight Hash. In: Mangard, S., Standaert, F.-X. (eds.) CHES 2010. LNCS, vol. 6225, pp. 1–15. Springer, Heidelberg (2010)
3. Batina, L., Mentens, N., Sakiyama, K., Preneel, B., Verbauwhede, I.: Low-Cost Elliptic Curve Cryptography for Wireless Sensor Networks. In: Buttyán, L., Gligor, V.D., Westhoff, D. (eds.) ESAS 2006. LNCS, vol. 4357, pp. 6–17. Springer, Heidelberg (2006)
4. Benadjila, R., Billet, O., Gilbert, H., Macario-Rat, G., Peyrin, T., Robshaw, M., Seurin, Y.: SHA-3 Proposal: ECHO, `http://crypto.rd.francetelecom.com/`
5. Bernstein, D.J.: CubeHash Specification (2.B.1), `http://cubehash.cr.yp.to/`
6. Bertoni, G., Daemen, J., Peeters, M., Assche, G.V.: Keccak specifications, `http://keccak.noekeon.org/`
7. Biham, E., Dunkelman, O.: The SHAvite-3 Hash Function, `http://www.cs.technion.ac.il/~orrd/SHAvite-3/`
8. Biham, E., Shamir, A.: Differential Cryptanalysis of the Data Encryption Standard. Springer, Heidelberg (1993)
9. Biryukov, A., Wagner, D.: Advanced slide attacks. In: Preneel, B. (ed.) EUROCRYPT 2000. LNCS, vol. 1807, pp. 589–606. Springer, Heidelberg (2000)
10. Black, J., Rogaway, P., Shrimpton, T.: Black-box analysis of the block-cipherbased hash-function constructions from PGV. In: Yung, M. (ed.) CRYPTO 2002. LNCS, vol. 2442, pp. 320–335. Springer, Heidelberg (2002)
11. Bogdanov, A., Knudsen, L.R., Leander, G., Paar, C., Poschmann, A., Robshaw, M.J.B., Seurin, Y., Vikkelsoe, C.: PRESENT: An Ultra-Lightweight Block Cipher. In: Paillier, P., Verbauwhede, I. (eds.) CHES 2007. LNCS, vol. 4727, pp. 450–466. Springer, Heidelberg (2007)
12. Bogdanov, A., Leander, G., Paar, C., Poschmann, A., Robshaw, M.J.B., Seurin, Y.: Hash Functions and RFID Tags: Mind the Gap. In: Oswald, E., Rohatgi, P. (eds.) CHES 2008. LNCS, vol. 5154, pp. 283–299. Springer, Heidelberg (2008)
13. Bouillaguet, C., Dunkelman, O., Leurent, G., Fouque, P.A.: Another look at complementation properties. In: Preproceedings of Fast Software Encryption 2010 Workshop, pp. 350–367 (2010)
14. De Cannière, C., Dunkelman, O., Knežević, M.: KATAN and KTANTAN a family of small and efficient hardware-oriented block ciphers. In: Clavier, C., Gaj, K. (eds.) CHES 2009. LNCS, vol. 5747, pp. 272–288. Springer, Heidelberg (2009)
15. Canniére, C.D., Sato, H., Watanabe, D.: Hash Function Luffa Specification, `http://www.sdl.hitachi.co.jp/crypto/luffa/`
16. Chabaud, F., Vaudenay, S.: Links between differential and linear cryptanalysis. In: De Santis, A. (ed.) EUROCRYPT 1994. LNCS, vol. 950, pp. 356–365. Springer, Heidelberg (1995)

17. Canteaut, A., Chevallier-Mames, B., Gouget, A., Paillier, P., Pornin, T., Bresson, E., Clavier, C., Fuhr, T., Icart, T., Misarsky, J.-F., Naya-Plasencia, M., Reinhard, J.-R., Thuillet, C., Videau, M.: Shabal, a Submission to NIST's Cryptographic Hash Algorithm Competition, http://www.shabal.com/
18. Daemen, J., Rijmen, V.: The Design of Rijndael: AES -Advanced Encryption Standard. Springer, Heidelberg (2002)
19. Damgård, I.B.: A Design Principle for Hash Functions. In: Brassard, G. (ed.) CRYPTO 1989. LNCS, vol. 435, pp. 416–427. Springer, Heidelberg (1990)
20. Feldhofer, M., Rechberger, C.: A case against currently used hash functions in RFID protocols. In: Meersman, R., Tari, Z., Herrero, P. (eds.) OTM 2006 Workshops. LNCS, vol. 4277, pp. 372–381. Springer, Heidelberg (2006)
21. Ferguson, N., Lucks, S., Schneier, B., Whiting, D., Bellare, M., Kohno, T., Callas, J., Walker, J.: The Skein Hash Function Family, http://www.schneier.com/skein.html
22. Gaubatz, G., Kaps, J.P., Ozturk, E., Sunar, B.: State of the Art in Ultra- Low Power Public Key Cryptography for Wireless Sensor Networks. In: Workshop on Pervasive Computing and Communication Security PerSec (2005)
23. Gauravaram, P., Knudsen, L.R., Matusiewicz, K., Mendel, F., Rechberger, C., Schläffer, M., Thomsen, S.S.: Grøstl - a SHA-3 candidate, http://www.groestl.info/
24. Halevi, S., Hall, W.E., Jutla, C.S.: The Hash Function Fugue, http://domino.research.ibm.com/comm/research_projects.nsf/ pages/fugue.index.html
25. Hirose, S., Kuwakado, H., Yoshida, H.: SHA-3 proposal: Lesamnta, http://csrc.nist.gov/groups/ST/hash/sha-3/Round1/documents/Lesamnta.zip (October 2008) latest version, http://www.sdl.hitachi.co.jp/crypto/lesamnta/
26. Küçük, Ö.: The Hash Function Hamsi okucuk/hamsi/, http://homes.esat.kuleuven.be/
27. Jakobsen, T., Knudsen, L.R.: The interpolation attack on block ciphers. In: Biham, E. (ed.) FSE 1997. LNCS, vol. 1267, pp. 28–40. Springer, Heidelberg (1997)
28. Gligoroski, D., Klima, V., Knapskog, S.J., El-Hadedy, M., Amundsen, J., Mjølsnes, S.F.: Cryptographic Hash Function BLUE MIDNIGHT WISH, http://people.item.ntnu.no/~danilog/Hash/BMW/
29. Joux, A.: Multicollisions in iterated hash functions. Application to cascaded constructions. In: Franklin, M. (ed.) CRYPTO 2004. LNCS, vol. 3152, pp. 306–316. Springer, Heidelberg (2004)
30. Kelsey, J., Schneier, B.: Second preimages on n-bit hash functions for much less than 2^n work. In: Cramer, R. (ed.) EUROCRYPT 2005. LNCS, vol. 3494, pp. 474–490. Springer, Heidelberg (2005)
31. Knezevic, M., Verbauwhede, I.: Hardware evaluation of the Luffa hash family, ftp://ftp.esat.kuleuven.ac.be/cosic/knudsen/trunc.ps.Z
32. Knudsen, L.R.: Truncated and higher order differentials. In: Preneel, B. (ed.) FSE 1994. LNCS, vol. 1008, pp. 196–211. Springer, Heidelberg (1995)
33. Leurent, G., Bouillaguet, C., Fouque, P.-A.: SIMD Is a Message Digest, http://www.di.ens.fr/~leurent/simd.html
34. Lu, L., O'Neill, M., Swartzlander, E.: Hardware Evaluation of SHA- 3 Hash Function Candidate ECHO, http://www.ucc.ie/en/crypto/CodingandCryptographyWorkshop/
35. Matsui, M.: Linear cryptanalysis method for DES cipher. In: Helleseth, T. (ed.) EUROCRYPT 1993. LNCS, vol. 765, pp. 386–397. Springer, Heidelberg (1994)

36. Merkle, R.C.: One way hash functions and DES. In: Brassard, G. (ed.) CRYPTO 1989. LNCS, vol. 435, pp. 428–446. Springer, Heidelberg (1990)
37. National Institute of Standards and Technology, Secure hash standard, Federal Information Processing Standards Publication 180-2 (August 2002), http://csrc.nist.gov/publications/fips/fips180-2/fips180-2.pdf
38. National Institute of Standards and Technology, Announcing request for candidate algorithm nominations for a new cryptographic hash algorithm (SHA-3) family (November 2007), http://csrc.nist.gov/groups/ST/hash/documents/
39. Rivest, R.: The MD5 message-digest algorithm, Request for Comments, no. 1321 (April 1992), ftp://ftp.rfc-editor.org/in-notes/rfc1321.txt
40. http://www.semico.com
41. Shamir, A.: SQUASH – A New MAC with Provable Security Properties for Highly Constrained Devices Such as RFID Tags. In: Nyberg, K. (ed.) FSE 2008. LNCS, vol. 5086, pp. 144–157. Springer, Heidelberg (2008)
42. Simpson, W.: PPP Challenge Handshake Authentication Protocol (CHAP), Request for Comments, no. 1994 (1996), http://www.ietf.org/rfc/rfc1994.txt
43. Suzuki, K., Tonien, D., Kurosawa, K., Toyota, K.: Birthday paradox for multicollisions. IEICE Trans. on Fundamentals E91-A(1), 39–45 (2008)
44. Tillich, S., Feldhofer, M., Issovits, W., Kern, T., Kureck, H., Muhlberghuber, M., Neubauer, G., Reiter, A., Kofler, A., Mayrhofer, M.: Compact hardware implementations of the SHA-3 candidates ARIRANG, BLAKE,Grøstl, and Skein, eprint archive: http://eprint.iacr.org/2009/349.pdf
45. Yoshida, H., Watanabe, D., Okeya, K., Kitahara, J., Wu, H., Küçük, Ö., Preneel, B.: MAME: A compression function with reduced hardware requirements. In: Paillier, P., Verbauwhede, I. (eds.) CHES 2007. LNCS, vol. 4727, pp. 148–165. Springer, Heidelberg (2007)
46. Wang, X., Lai, X., Feng, D., Chen, H., Yu, X.: Cryptanalysis of the hash functions MD4 and RIPEMD. In: Cramer, R. (ed.) EUROCRYPT 2005. LNCS, vol. 3494, pp. 1–18. Springer, Heidelberg (2005)
47. Wang, X., Yin, Y.L., Yu, H.: Finding collisions in the full SHA-1. In: Shoup, V. (ed.) CRYPTO 2005. LNCS, vol. 3621, pp. 17–36. Springer, Heidelberg (2005)
48. Wikipedia, Microprocessor, ch. Market statistics, http://en.wikipedia.org/wiki/Microprocessor
49. Wu, H.: The Hash Function JH , http://www3.ntu.edu.sg/home/wuhj/research/jh/
50. Zheng, Y., Matsumoto, T., Imai, H.: On the construction of block ciphers provably secure and not relying on any unproved hypotheses. In: Brassard, G. (ed.) CRYPTO 1989. LNCS, vol. 435, pp. 461–480. Springer, Heidelberg (1990)

A Lesamnta-LW Example

Initial Hash Value and Round Constants. For Lesamnta-LW, the initial hash value $H^{(0)}$ is $H_0^{(0)} \| H_1^{(0)} \| H_2^{(0)} \| H_3^{(0)} \| H_4^{(0)} \| H_5^{(0)} \| H_6^{(0)} \| H_7^{(0)}$, where each $H_i^{(0)}$ is a 32-bit word 00000256 in hex.

The round constants of sixty-four 32-bit words and the algorithm to generate them are presented in Fig. 3. The algorithm is based on the linear feedback shift register (LFSR) of the following primitive polynomial:

$$g(x) = x^{32} + x^{31} + x^{29} + x^{28} + x^{26} + x^{25} + x^{24} + x^{23} + x^{20}$$
$$+ x^{19} + x^{17} + x^{16} + x^{15} + x^{12} + x^{11} + x^8 + 1.$$

```
a432337f 945e1f8f 92539a11 24b90062      ConstantGenerator(word C[64])
6971c64c d6e3f449 2c2f0da9 33769295      begin
eb506df2 708cebfe b83ab7bf 97df0f17         word c;
9223b802 7fa29140 0ff45228 01fe8a45         c = ffffffff; /*in hexadecimal*/
ed016ee8 1da02ddd ee8aba1b 46c4c223         for i = 0 to (64 * 3) - 1
53cd0d24 d1b46d24 c1fb4124 c3f2a4a4            /* Galois LFSR */
c3b39814 c3bbbf82 759191b0 0eb23236            if c & 00000001 == 00000001
b7fd6c86 a0d48750 141a90ea 6f65b45d               c = (c >> 1) ^ dbcdcc80;
e0d2092b 470fd445 e5df4528 1cbbe8a5            else
eea9c2b4 c618f4d6 aee8345a 783be0cb               c = c >> 1;
5412e979 3c712e0f 87567c21 2619bca4            end if
df0efb14 c02c13e2 75e3643c d571a007            if i mod 3 == 0
9a766de0 134ecdbc d9a41537 9becdb46               C[i/3] = c;
a556b1a8 14aad635 efabe566 abde566c            end if
ceb6064d f4e87f69 286e7ccd e8337039         end for
2bf51d27 85a6fa44 cb7913c8 196f2279      end
```

Fig. 3. The round constants $C^{(i)}$s (in hex) and a pseudocode for generating them

Let the message M be the 24-bit ($l = 24$) ASCII string "abc", which is equivalent to the following binary string: 01100001 01100010 01100011.
    Then the resulting 256-bit message digest is

$$2558c1d3\ 7f9f307b\ e3cddad4\ a23c8654$$
$$518f6079\ 7eb491e7\ 3758727d\ fc83de65\ .$$

B   Proof of Theorem 1

Our analysis uses the following result by Suzuki et al. [43] on multi-collisions.

Proposition 1. *Suppose that there are q balls and t buckets and that the balls are thrown one by one at random into the buckets. For $2 \le s \le q$, let $\mathtt{Col}(t, q, s)$ be the event that there exists at least one bucket that contains at least s balls. Then,*

$$\Pr[\mathtt{Col}(t, q, s)] \le \frac{1}{t^{s-1}} \binom{q}{s} \ .$$

Corollary 1. *Let $t = 2^s$ and $q = 2^{s-2}$. Then, for $s \ge 4$,*

$$\Pr[\mathtt{Col}(2^s, 2^{s-2}, s)] \le \frac{1}{s! \cdot 2^s} \ .$$

Proof. If $s \ge 4$, then $2 \le s \le 2^{s-2}$. From Proposition 1,

$$\Pr[\mathtt{Col}(2^s, 2^{s-2}, s)] \le \frac{1}{(2^s)^{s-1}} \binom{2^{s-2}}{s} \le \frac{1}{2^{s(s-1)}} \frac{(2^{s-2})^s}{s!} = \frac{1}{s! \cdot 2^s} \ .$$

□

Proof of Theorem 1. For $1 \leq i \leq q$, let $(t_i, k_i, w_i \| x_i, y_i \| z_i)$ be a tuple such that $E(k_i, w_i \| x_i) = y_i \| z_i$ and $t_i \in \{\mathsf{e}, \mathsf{d}\}$ obtained by the i-th query. t_i represents the type of the i-th query: encryption (e) or decryption (d). Let $G_1, G_2, \ldots, G_q$ be a sequence of directed graphs such that $G_i = (V_i, L_i)$, where

- $V_1 = \{k_1 \| x_1, y_1 \| z_1\}$, $L_1 = L_i \cup \{(k_1 \| x_1, y_1 \| z_1)\}$, and
- $V_i = V_{i-1} \cup \{k_i \| x_i, y_i \| z_i\}$, $L_i = L_{i-1} \cup \{(k_i \| x_i, y_i \| z_i)\}$ for $2 \leq i \leq q$.

Each edge $(k_i \| x_i, y_i \| z_i)$ is labeled by (t_i, w_i). Notice that $y_i \| z_i = h(k_i \| x_i, w_i)$, where h is the Lesamnta-LW compression funcuion specified in Sect. 2.2.

Suppose that the adversary A first finds a collision of Lesamnta-LW with the i-th query. Then, there must be a path in G_i from $IV_0 \| IV_1$ to some colliding output, which does not exist in $G_1, \ldots, G_{i-1}$. This path also contains the nodes $k_i \| x_i$ and $y_i \| z_i$, and the edge (t_i, w_i).

If $t_i = \mathsf{e}$, that is, the i-th query is an encryption query, then there must be an event such that $y_i \| z_i \in \{y_j \| z_j \mid 1 \leq j \leq i-1\} \cup \{k_j \| x_j \mid 1 \leq j \leq i-1\} \cup \{IV_0 \| IV_1\}$. If $t_i = \mathsf{d}$, then there must be an event such that $k_i \| x_i \in \{y_j \| z_j \mid 1 \leq j \leq i-1\} \cup \{IV_0 \| IV_1\}$.

For the case where $t_i = \mathsf{d}$ and $k_i \| x_i \in \{y_j \| z_j \mid 1 \leq j \leq i-1\}$, let us look into the new path in G_i mentioned above. Let $IV_0 \| IV_1 \xrightarrow{(t_{j_1}, M_{j_1})} v_{j_1} \xrightarrow{(t_{j_2}, M_{j_2})} \cdots \xrightarrow{(t_{j_{l-1}}, M_{j_{l-1}})} v_{j_{l-1}} \xrightarrow{(t_{j_l}, M_{j_l})} v_{j_l}$ be the prefix of the path, where $v_{j_{l-1}} = k_i \| x_i$, $(t_{j_l}, M_{j_l}) = (\mathsf{d}, w_i)$ and $v_{j_l} = y_i \| z_i$. We start from v_{j_l} and go back toward $IV_0 \| IV_1$ until we first find an edge (e, M_{j_k}) or reach the node $IV_0 \| IV_1$ without finding such an edge. Suppose that we reach $IV_0 \| IV_1$. Then, it implies that there is an event such that $t_{i'} = \mathsf{d}$ and $k_{i'} \| x_{i'} = IV_0 \| IV_1$ for some i' such that $1 \leq i' < i$. On the other hand, suppose that we find an edge (e, M_{j_k}). Then, it implies that there is an event such that $t_{i'} = \mathsf{e}$ and $y_{i'} \| z_{i'} \in \{k_j \| x_j \mid 1 \leq j < i'\}$ for some i' such that $1 < i' < i$, or an event such that $t_{i'} = \mathsf{d}$ and $k_{i'} \| x_{i'} \in \{y_j \| z_j \mid 1 \leq j < i' \wedge t_j = \mathsf{e}\}$ for some i' such that $1 < i' \leq i$.

From the discussions above, if A finds a collision with at most q queries, then it implies that there must be at least one of the following events for some i such that $1 \leq i \leq q$:

Ea_i $t_i = \mathsf{e}$ and $y_i \| z_i = IV_0 \| IV_1$,
Eb_i $t_i = \mathsf{e}$ and $y_i \| z_i \in \{y_j \| z_j \mid 1 \leq j \leq i-1\} \cup \{k_j \| x_j \mid 1 \leq j \leq i-1\}$,
Ec_i $t_i = \mathsf{d}$ and $k_i \| x_i = IV_0 \| IV_1$,
Ed_i $t_i = \mathsf{d}$ and $k_i \| x_i \in \{y_j \| z_j \mid 1 \leq j < i \wedge t_j = \mathsf{e}\}$.

It is easy to see that

$$\Pr[\mathsf{Ea}_i] \leq \frac{1}{2^{2n} - (i-1)}, \quad \Pr[\mathsf{Eb}_i] \leq \frac{2(i-1)}{2^{2n} - (i-1)}, \quad \Pr[\mathsf{Ec}_i] \leq \frac{2^n}{2^{2n} - (i-1)}.$$

For Ed_i, the probability of multicollision on y_j should be taken into consideration. From Corollary 1, for $1 \leq q \leq 2^{n-2}$,

$$\Pr[\mathsf{Ed}_i] \leq \frac{(n-1)2^n}{2^{2n} - (i-1)} + \frac{1}{n! \cdot 2^n}.$$

Precisely speaking, the distribution of $y_j \| z_j$ is not uniform on $\{0,1\}^{2n}$ since E is a keyed permutation. However, since $\Pr[y_j \in \{y_1, \ldots, y_{j-1}\}] \leq \Pr[y_j \notin \{y_1, \ldots, y_{j-1}\}]$, the probability of multicollision is smaller in this case.

Thus, for $1 \leq q \leq 2^{n-2}$,

$$\mathrm{Adv}_{C[E]^+}^{\mathrm{col}}(A) \leq \sum_{i=1}^{q} (\Pr[\mathbf{Ea}_i] + \Pr[\mathbf{Eb}_i] + \Pr[\mathbf{Ec}_i] + \Pr[\mathbf{Ed}_i])$$

$$\leq \frac{2^n n q}{2^{2n} - q} + \frac{q^2}{2^{2n} - q} + \frac{q}{n! \cdot 2^n} \ .$$

The upper bound exceeds 1 for $n \geq 4$ and $q > 2^{n-2}$.   □

C   Other Attacks on Block Ciphers

C.1   Higher Order Differential and Interpolation Attack

The higher order differential attack [32] can be mounted if the bits in the intermediate state of the cipher are expressed by Boolean polynomials of degree most d which is a reasonably small value. In the case of Lesamnta-LW, we found that every output bit of the S-box can be expressed as a Boolean polynomial of degree 7 in terms of input bits. Our experiments confirmed that the degree of such polynomials with 19 rounds reaches to the required degree 256. Therefore, we expect that Lesamnta-LW is secure against higher order differential attacks.

The interpolation attack [27] can be mounted if the number of terms in a polynomial expression for a cipher over some field is reasonably small. Lesamnta-LW uses the AES S-box which can be expressed as a polynomial of degree 254 over $\mathrm{GF}(2^8)$. Our experiments have confirmed that after the 16th round, each byte in the intermediate state of the mixing function depends on all the 32 variables while this is not the case just after the 15 rounds. We expect that the number of coefficients grows fast after the 16th round due to the high degree of the S-box and that the full Lesamnta-LW is secure against interpolation attacks.

C.2   Slide and Related-Key Attacks

The round constants introducing randomness into the key scheduling function preclude slide attacks [9] which exploit the similarity between rounds.

Regarding the related-key attacks, we can show that the maximum differential characteristic probabilities for 24 rounds of the key scheduling function are less than 2^{-128} as we did in Sect 5.1. Hence, we expect that it is unlikely to apply related-key attacks to Lesamnta-LW.

Trademarks

- Renesas® and H8® are registered trademarks of Renesas Technology Corporation.
- Intel® is a registered trademark of Intel Corporation in the United States and/or other countries.

Efficient Pairing Computation on Elliptic Curves in Hessian Form⋆

Haihua Gu[1,2], Dawu Gu[2], and WenLu Xie[1]

[1] Computer Science and Engineering Department, Shanghai Jiao Tong University,
200240 Shanghai, P.R. China
[2] Shanghai Huahong Integrated Circuit Co.,Ltd., 201203 Shanghai,P.R. China
guhaihua@shhic.com

Abstract. Pairings in elliptic curve cryptography are functions which map a pair of elliptic curve points to a non-zero element of a finite field. In recent years, many useful cryptographic protocols based on pairings have been proposed. The fast implementations of pairings have become a subject of active research areas in cryptology.

In this paper, we give the geometric interpretation of the group law on Hessian curves. Furthermore, we propose the first algorithm for computing the Tate pairing on elliptic curves in Hessian form. Analysis indicates that it is faster than all algorithms of Tate pairing computation known so far for Weierstrass and Edwards curves excepted for the very special elliptic curves with $a_4 = 0$, $a_6 = b^2$.

Keywords: Elliptic curve, Tate Paring, Hessian form.

1 Introduction

Pairings on elliptic curves are currently of great interest due to their applications in a number of cryptographic protocols such as identity-based encryption [3], group signatures [4], short signatures [12] and the tripartite Diffie Hellman [13]. For implementing such protocols, it is essential to have an efficient algorithm of pairing computation.

Miller [16,17] proposed the first algorithm for iteratively computing the Weil and Tate pairings. Various improvements were published in [5,6,8,11]. In 2009, Arène et.al. [1] first use the geometric interpretation of the group law to show how to compute the Tate pairing on twisted Edwards curves. Their algorithm is faster than all previously proposed formulas for pairings on Edwards curves and competitive with all published formulas for pairing computation on Weierstrass curves. However, to the best of our knowledge, no formulas for pairing computation on Hessian curves are proposed.

The use of Hessian curves in cryptology are explained by [7], [14] and [18]. Smart [18] showed that some sample curves from IEEE, SECG standards can be

---

⋆ This work was supported by Specialized Research Fund for the Doctoral Program of Higher Education (No. 200802480019) and National Natural Science Foundation of China (No. 61073150).

K.-H. Rhee and D. Nyang (Eds.): ICISC 2010, LNCS 6829, pp. 169–176, 2011.

transformed into Hessian form and the point operation for Hessian curves can be implemented in a highly parallel way. As a result, the Hessian curves can provide around a 40 percent performance improvement over the Weierstrass curves [18]. Furthermore, Joye and Quisquater [14] suggest that the unified formula for the addition of points on Hessian curves can be used as a means for preventing side channel attack.

In this paper, we consider the problems on Hessian curves. One of our contribution is giving the geometric interpretation of the group law on Hessian curves. As far as we know, this is the first geometric interpretation. Another contribution is that we develop explicit formulas for computing pairings on Hessian curves. Excepted for the very special elliptic curves with $a_4 = 0$, $a_6 = b^2$, our formulas are fastest for Tate pairing computation up to date.

2 Preliminaries

Let $\mathbb{F}_p$ be a finite field with p elements where $p > 2$ is prime. Consider positive integer r such that r is relative prime to the characteristic of the field $\mathbb{F}_p$. Denote the embedding degree by k, i.e. the smallest positive integer such that r divides $p^k - 1$. The elliptic curve in Weierstrass form is defined as

$$E : y^2 = x^3 + ax + b,$$

where $a, b \in \mathbb{F}_p$, $4a^3 + 27b^2 \neq 0 \in \mathbb{F}_p$. Let P be a point in $E(\mathbb{F}_{p^k})[r]$ and $Q \in E(\mathbb{F}_{p^k})$. Define $f_{i,P}$ to be a function on the elliptic curve with its divisor $\mathrm{div}(f_{i,P}) = i(P) - (iP) - (i-1)(\mathcal{O})$, $i \in \mathbb{Z}$. Consider the divisor $D = (Q+R) - (R)$ with R being a random point in $E(\mathbb{F}_{p^k})$ such that D is coprime with $(P) - (\mathcal{O})$. Then the reduced Tate pairing [9] is a map

$$e_r : E(\mathbb{F}_{p^k})[r] \times E(\mathbb{F}_{p^k})/rE(\mathbb{F}_{p^k}) \to \mathbb{F}_{p^k}^* / (\mathbb{F}_{p^k}^*)^r;$$

$$(P, Q) \mapsto (f_{r,P}(D))^{(p^k-1)/r}.$$

If the function $f_{r,P}$ in the definition is normalized, then one can simply work with the point Q, i.e. the reduced Tate pairing is:

$$e_r(P, Q) = (f_{r,P}(Q))^{(p^k-1)/r}.$$

Let $g_{iP,jP}$ be the function such that $\mathrm{div}(g_{iP,jP}) = (iP) + (jP) - ((i+j)P) - (\mathcal{O})$, then $g_{iP,jP} = l_{iP,jP}/v_{(i+j)P}$, where $l_{iP,jP}$ is the equation of the line through iP, jP and $v_{(i+j)P}$ is the equation of the vertical line through $(i + j)P$. From the definition of $f_{r,P}$, we can see that

$$
\begin{aligned}
\mathrm{div}(f_{i+j,P}) &= (i + j)(P) - ((i + j)P) - (i + j - 1)(\mathcal{O}) \\
&= i(P) - (iP) - (i - 1)(\mathcal{O}) \\
&\quad + j(P) - (jP) - (j - 1)(\mathcal{O}) \\
&\quad + (iP) + (jP) - ((i + j)P) - (\mathcal{O}) \\
&= \mathrm{div}(f_{i,P}) + \mathrm{div}(f_{j,P}) + \mathrm{div}(g_{iP,jP}).
\end{aligned}
\tag{1}
$$

Therefore $f_{i+j,P} = f_{i,P} \cdot f_{j,P} \cdot g_{iP,jP}$. This leads to the following algorithm which computes the Tate pairing [16,17].

Algorithm 1. Miller's algorithm.
Input: integer $r = \sum_{i=0}^{l} b_i 2^i$ with $b_i \in \{0,1\}$, $b_l = 1$
and $P \in E(F_p)[r], Q \in E(F_{p^k})[r]$.
Output: $f = f_{r,P}^{(p^k-1)/r}$.
1. $f \leftarrow 1$, $R \leftarrow P$;
2. for $i \leftarrow l-1$ down to 0 do
    $f \leftarrow f^2 \cdot g_{R,R}(Q)$, $R \leftarrow 2R$;
    if $b_i = 1$ then $f \leftarrow f \cdot g_{R,P}(Q)$, $R \leftarrow R + P$;
3. Return $f \leftarrow f^{(p^k-1)/r}$.

3 Geometric Interpretation of the Group Law on Hessian Curves

An elliptic curve in Hessian form is defined by

$$\mathcal{H} : X^3 + Y^3 + Z^3 = 3dXYZ,$$

where $d \in K$ with $d^3 \neq 1$. The identity element is represented by $(-1 : 1 : 0)$. The negative of $(X : Y : Z)$ is $(Y : X : Z)$. Birational maps between Weierstrass and Hessian curves can be found in [2]. The addition formulas are given by $(X_3 : Y_3 : Z_3) = (X_1 : Y_1 : Z_1) + (X_2 : Y_2 : Z_2)$ where

$$X_3 = 2Y_1^2 X_2 Z_2 - 2X_1 Z_1 Y_2^2,$$
$$Y_3 = 2X_1^2 Y_2 Z_2 - 2Y_1 Z_1 X_2^2,$$
$$Z_3 = 2Z_1^2 X_2 Y_2 - 2X_1 Y_1 Z_2^2.$$

The doubling formulas are given by

$$X_3 = (2X_1 Y_1 - 2Y_1 Z_1)(2X_1 Z_1 + 2(X_1^2 + Z_1^2)),$$
$$Y_3 = (2X_1 Z_1 - 2X_1 Y_1)(2Y_1 Z_1 + 2(Y_1^2 + Z_1^2)),$$
$$Z_3 = (2Y_1 Z_1 - 2X_1 Z_1)(2X_1 Y_1 + 2(X_1^2 + Y_1^2)).$$

Suppose the computational costs of a multiplication, a squaring in the field K are denoted by m,s. The point addition algorithms in [7] and [14] require 12m, while point doubling costs 3m+6s [10].

Theorem 1. *Let $P_1 = (X_1 : Y_1 : Z_1)$ and $P_2 = (X_2 : Y_2 : Z_2)$ be two points on $\mathcal{H}(K)$. Define $P_3 = P_1 + P_2$. Let l_1 be the line passing through P_1 and P_2 while l_2 be the line passing through P_3 and $-P_3$. Then we have*

$$div(l_1/l_2) = (P_1) + (P_2) - (P_3) - (\mathcal{O}).$$

Proof. Let $l_1 : c_X X + c_Y Y + c_Z Z = 0$ be the line passing through P_1 and P_2, where $c_X, c_Y, c_Z \in K$.

If $P_1 \neq P_2$, then we obtain two linear equations in c_X, c_Y and c_Z

$$c_X X_1 + c_Y Y_1 + c_Z Z_1 = 0,$$
$$c_X X_2 + c_Y Y_2 + c_Z Z_2 = 0.$$

It follows that

$$c_X = \begin{vmatrix} Y_1 & Z_1 \\ Y_2 & Z_2 \end{vmatrix} = Y_1 Z_2 - Z_1 Y_2,$$

$$c_Y = \begin{vmatrix} Z_1 & X_1 \\ Z_2 & X_2 \end{vmatrix} = Z_1 X_2 - X_1 Z_2,$$

$$c_Z = \begin{vmatrix} X_1 & Y_1 \\ X_2 & Y_2 \end{vmatrix} = X_1 Y_2 - Y_1 X_2.$$

Recall that the negative point of $P_3 = (X_3 : Y_3 : Z_3)$ is $-P_3 = (Y_3 : X_3 : Z_3)$. Consider the equation

$$c_X Y_3 + c_Y X_3 + c_Z Z_3$$
$$= (Y_1 Z_2 - Z_1 Y_2)Y_3 + (Z_1 X_2 - X_1 Z_2)X_3 + (X_1 Y_2 - Y_1 X_2)Z_3$$
$$= (Y_1 Z_2 - Z_1 Y_2)(X_1^2 Y_2 Z_2 - Y_1 Z_1 X_2^2) + (Z_1 X_2 - X_1 Z_2)(Y_1^2 X_2 Z_2 - X_1 Z_1 Y_2^2)$$
$$\quad + (X_1 Y_2 - Y_1 X_2)(Z_1^2 X_2 Y_2 - X_1 Y_1 Z_2^2)$$
$$= 0.$$

This implies $-P_3$ lies on the line l_1. Therefore

$$\mathrm{div}(l_1) = (P_1) + (P_2) + (-P_3) - 3(\mathcal{O}). \tag{2}$$

Let l_2 be the line passing through $P_3 = (X_3 : Y_3 : Z_3)$ and $-P_3 = (Y_3 : X_3 : Z_3)$. The equation of l_2 can be easily get that

$$(Y_3 Z_3 - X_3 Z_3)(X + Y) + (X_3^2 - Y_3^2)Z = 0.$$

Since the identity element is $\mathcal{O} = (-1 : 1 : 0)$, it follows that the point $\mathcal{O}$ also lies on the line l_2. Therefore

$$\mathrm{div}(l_2) = (P_3) + (-P_3) - 2(\mathcal{O}). \tag{3}$$

Combine Eq.(2) and Eq.(3), we have

$$\mathrm{div}(l_1/l_2) = (P_1) + (P_2) - (P_3) - (\mathcal{O}).$$

In the case of $P_1 = P_2$. Let l_1 be the line tangent to $\mathcal{H}(K)$ at P_1 and it can get that $l_1 : \frac{\partial E}{\partial X_1} X + \frac{\partial E}{\partial Y_1} Y + \frac{\partial E}{\partial Z_1} Z = 0$. Using the similar method, we can also get $\mathrm{div}(l_1/l_2) = (P_1) + (P_2) - (P_3) - (\mathcal{O})$. $\qquad\square$

By Theorem 1, we can know the geometric interpretation of the group law on Hessian curves.

Suppose P, Q are two points on a Hessian curve. The sum R, of P and Q, is defined as follows. First draw a line through P and Q which intersects the Hessian curve at a third point. Then R is the reflection of this point about the line $y = x$. Fig. 1 depicts the addition on $x^3 + y^3 + 1 = 7xy$ over real field.

The double R, of P, is defined as follows. First draw the tangent line to the curve at P. The line intersects the curve at a second point. Then R is the reflection of this point about the line $y = x$.

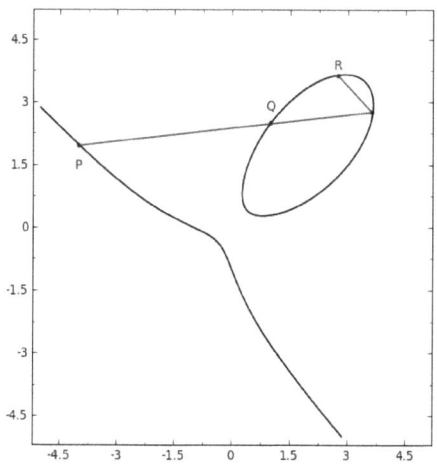

Fig. 1. Point addition on $x^3 + y^3 + 1 = 7xy$ over $\mathbb{R}$

4   Formulas for Pairings on Hessian Curves

Similarly to Eq. (1), using Theorem 1, we can prove the relation

$$f_{i+j,P} = f_{i,P} \cdot f_{j,P} \cdot g_{iP,jP}$$

on Hessian curves. This means pairings on Hessian curves can be computed by Miller's algorithm [17].

In most cryptographic protocols, P can be chosen such that $< P >$ is the unique subgroup of order r in $\mathcal{H}(\mathbb{F}_p)$. Suppose the embedding degree k is even, then the field extension $\mathbb{F}_{p^k}$ is usually constructed via a quadratic subfield as $\mathbb{F}_{p^k} = \mathbb{F}_{p^{k/2}}(\alpha)$, with $\alpha^2 = \delta \in \mathbb{F}_{p^{k/2}}$. Since k is the smallest positive integer such that $r|(p^k - 1)$, it follows that $(p^k - 1)/r$ is a multiple of $p^u - 1$ for some proper divisor u of k. So all elements of $\mathbb{F}_{p^u}^*$ map to 1 when raised to the power $(p^k - 1)/r$.

Without loss of generality, we suppose $d \in \mathbb{F}_{p^{k/2}}$, and select $a \in \mathbb{F}_{p^{k/2}}$ randomly. Repeat $a = a + 1$ until $\exists b \in \mathbb{F}_{p^{k/2}}$ such that $(a + b\alpha)^3 + (a - b\alpha)^3 + 1 =$

$3d(a^2 - b^2\alpha^2)$ and r dividing the order of $Q = (a + b\alpha : a - b\alpha : 1) \in \mathcal{H}(\mathbb{F}_{p^k})$. Note that $\forall u | k$, $Q \notin \mathcal{H}(\mathbb{F}_{p^u})$. The point Q constructed in this way can be applied to identity based cryptosystems [3]. Even if we restrict the point Q as $(a + b\alpha : a - b\alpha : 1)$, it does not affect the security of the ID-based protocol since the point Q is a public point.

To describe clearly, in Algorithm 1, we call $f \leftarrow f \cdot \frac{l_1(Q)}{l_2(Q)}$, $R \leftarrow R + P$ addition and $f \leftarrow f^2 \cdot \frac{l_1(Q)}{l_2(Q)}$, $R \leftarrow 2R$ is referred as doubling. Since $l_2(Q) = (Y_3 Z_3 - X_3 Z_3)(X + Y) + (X_3^2 - Y_3^2)Z = (Y_3 Z_3 - X_3 Z_3)2a + X_3^2 - Y_3^2 \in \mathbb{F}_{p^{k/2}}$, it follows that $l_2(Q)^{(p^k-1)/r} = 1$. Therefore we only need to consider numerator $l_1(Q)$ in additions and doublings.

4.1 Addition

Let $R = (X_1 : Y_1 : Z_1)$ and $P = (X_2 : Y_2 : Z_2)$ be two points on $\mathcal{H}(\mathbb{F}_p)$.

$$f \cdot l_1(Q)$$
$$= f \cdot [c_X(a + b\alpha) + c_Y(a - b\alpha) + c_Z]$$
$$= f \cdot [(c_X + c_Y)a + (c_X - c_Y)b\alpha + c_Z]$$

where $c_X = Y_1 Z_2 - Z_1 Y_2$, $c_Y = Z_1 X_2 - X_1 Z_2$, $c_Z = X_1 Y_2 - Y_1 X_2$. So the explicit formulas for computing $f \leftarrow f \cdot \frac{l_1(Q)}{l_2(Q)}$, $R \leftarrow R + P$ are given as follows:

$A = Y_1 \cdot Z_2$, $B = Z_1 \cdot Y_2$, $C = Z_1 \cdot X_2$, $D = X_1 \cdot Z_2$, $E = X_1 \cdot Y_2$, $F = Y_1 \cdot X_2$, $c_X = A - B$, $c_Y = C - D$, $c_Z = E - F$, $f = f \cdot [(c_X + c_Y)a + (c_X - c_Y)b\alpha + c_Z]$, $X_3 = A \cdot F - B \cdot E$, $Y_3 = D \cdot E - C \cdot F$, $Z_3 = B \cdot C - A \cdot D$.

The cost of these formulas is 1M+km+12m, where M denotes the cost of a multiplication in $\mathbb{F}_{p^k}$ while m is the cost of a multiplication in $\mathbb{F}_p$. Multiplications by $a, b \in \mathbb{F}_{p^{k/2}}$ need $(k/2)m$ each. If the base point P has $Z_2 = 1$, the above costs reduce to 1M+km+10m.

4.2 Doubling

Let $R = (X_1 : Y_1 : Z_1)$ be a point on $\mathcal{H}(\mathbb{F}_p)$. Since the result of $f^2 \cdot \frac{l_1(Q)}{l_2(Q)}$ will not change if the numerator and the denominator are multiplied by 2 simultaneously, we can consider:

$$f^2 \cdot 2l_1(Q)$$
$$= f^2 \cdot 2[(c_X + c_Y)a + (c_X - c_Y)b\alpha + c_Z]$$

where $c_X = 3X_1^2 - 3dY_1 Z_1$, $c_Y = 3Y_1^2 - 3dX_1 Z_1$, $c_Z = 3Z_1^2 - 3dX_1 Y_1$. Hence the explicit formulas for computing $f \leftarrow f^2 \cdot l_1(Q)$, $R \leftarrow 2R$ are given as follows:

$A = X_1^2$, $B = Y_1^2$, $C = Z_1^2$, $D = (Y_1 + Z_1)^2 - B - C$, $E = (X_1 + Z_1)^2 - A - C$, $F = (X_1 + Y_1)^2 - A - B$, $2c_X = 6A - 3dD$, $2c_Y = 6B - 3dE$, $2c_Z = 6C - 3dF$, $f = f^2 \cdot [(2c_X + 2c_Y)a + (2c_X - 2c_Y)b\alpha + 2c_Z]$, $X_3 = (F - D)(E + 2A + 2C)$, $Y_3 = (E - F)(D + 2B + 2C)$, $Z_3 = (D - E)(F + 2A + 2B)$.

The cost of these formulas is 1M+km+1S+3m+6s, where S is the cost of a squaring in $\mathbb{F}_{p^k}$. Note that we have ignored the cost of multiplication by d for the reason that d can be selected specially.

We give an overview of the best formulas in the literature for pairing computation on Edwards curves and for the different forms of Weierstrass curves in Jacobian coordinates. Their performance is summarized in Table 1. We compare the results with our new pairing formulas for Hessian curves. We find that our addition algorithm is fastest and our doubling algorithm is only slower than the one in [8]. However, the curves considered in [8] are extremely special: for $p \equiv 2 \bmod 3$ these curves are supersingular and thus have $k = 2$ and for $p \equiv 1 \bmod 3$ a total of 3 isomorphism classes is covered by this curve shape. Therefore, our new formulas for Tate pairings on Hessian curves are faster than all formulas excepted for the very special curves with $a_4 = 0$, $a_6 = b^2$.

Table 1. Costs of pairing computation

	DBL	mADD
$\mathcal{J}$,[5],[11]	1M+km+1S+1m+11s+1m$_{a_4}$	1M+km+9m+3s
$\mathcal{J}$,[1],[11]	1M+km+1S+1m+11s+1m$_{a_4}$	1M+km+6m+6s
$\mathcal{J}$,$a_4 = -3$,[5]	1M+km+1S+7m+4s	1M+km+9m+3s
$\mathcal{J}$,$a_4 = -3$,[1]	1M+km+1S+6m+5s	1M+km+6m+6s
$\mathcal{J}$,$a_4 = 0$,[5],[6]	1M+km+1S+6m+5s	1M+km+9m+3s
$\mathcal{J}$,$a_4 = 0$,[1]	1M+km+1S+3m+8s	1M+km+6m+6s
$\mathcal{P}$,$a_4 = 0$, $a_6 = b^2$,[8]	1M+km+1S+3m+5s	1M+km+10m+2s+1m$_b$
$\mathcal{E}$,[11]	1M+km+1S+8m+4s+1m$_d$	1M+km+14m+4s+1m$_d$
$\mathcal{E}$,[1]	1M+km+1S+6m+5s	1M+km+12m
$\mathcal{H}$, this paper	1M+km+1S+3m+6s	1M+km+10m

5   Conclusion

In this paper, we first give the geometric interpretation of the group law on Hessian curves, and then propose a new algorithm to compute Tate pairings on elliptic curves in Hessian form. Compared with the methods for Weierstrass curves and Edwards curves, our algorithm is fastest for pairing computation excepted for the very special curves with $a_4 = 0$, $a_6 = b^2$.

References

1. Arène, C., Lange, T., Naehrig, M., Ritzenthaler, C.: Faster Computation of Tate Pairings, Cryptology ePrint Archive, Report 2009/155,
   http://eprint.iacr.org/2009/155.pdf
2. Bernstein, D.J., Lange, T.: Analysis and optimization of elliptic-curve single-scalar multiplication. In: Mullen, G.L., Panario, D., Shparlinski, I.E. (eds.) Finite fields and applications, Contemp. Math., vol. 461, pp. 1–19. American Mathematical Society, Providence (2008)

3. Boneh, D., Franklin, M.K.: Identity-based encryption from the Weil pairing. SIAM J. Comput. 32(3), 586–615 (2003)

4. Boneh, D., Shacham, H.: Group signatures with verifier-local revocation. In: Atluri, V., Pfitzmann, B., McDaniel, P. (eds.) ACM CCS 2004, pp. 168–177. ACM Press, New York (2004)

5. Chatterjee, S., Sarkar, P., Barua, R.: Efficient computation of Tate pairing in projective coordinate over general characteristic fields. In: Park, C., Chee, S. (eds.) ICISC 2004. LNCS, vol. 3506, pp. 168–181. Springer, Heidelberg (2005)

6. Cheng, Z., Nistazakis, M.: Implementing pairing-based cryptosystems. In: 3rd International Workshop on Wireless Security Technologies IWWST 2005, London, UK (April 2005)

7. Chudnovsky, D.V., Chudnovsky, G.V.: Sequences of numbers generated by addition in formal groups and new primality and factorization tests. Adv. Appl. Math. 7(4), 385–434 (1986)

8. Costello, C., Hisil, H., Boyd, C., Nieto, J.M.G., Wong, K.K.H.: Faster pairings on special weierstrass curves. In: Shacham, H., Waters, B. (eds.) Pairing 2009. LNCS, vol. 5671, pp. 89–101. Springer, Heidelberg (2009)

9. Frey, G., Rück, H.G.: A remark concerning m-divisibility and the discrete logarithm in the divisor class group of curves. Math. Comp. 62, 865–874 (1994)

10. Hisil, H., Carter, G., Dawson, E.: New formulae for efficient elliptic curve arithmetic. In: Srinathan, K., Rangan, C.P., Yung, M. (eds.) INDOCRYPT 2007. LNCS, vol. 4859, pp. 138–151. Springer, Heidelberg (2007)

11. Ionica, S., Joux, A.: Another approach to pairing computation in edwards coordinates. In: Chowdhury, D.R., Rijmen, V., Das, A. (eds.) INDOCRYPT 2008. LNCS, vol. 5365, pp. 400–413. Springer, Heidelberg (2008)

12. Joux, A.: The Weil and Tate Pairings as Building Blocks for Public Key Cryptosystems. In: Fieker, C., Kohel, D.R. (eds.) ANTS 2002, Part V. LNCS, vol. 2369, pp. 18–20. Springer, Heidelberg (2002)

13. Joux, A.: A one round protocol for tripartite Diffie-Hellman. J. Cryptol. 17(4), 263–276 (2004)

14. Joye, M., Quisquater, J.-J.: Hessian elliptic curves and side-channel attacks. In: Koç, Ç.K., Naccache, D., Paar, C. (eds.) CHES 2001. LNCS, vol. 2162, pp. 402–410. Springer, Heidelberg (2001)

15. Koblitz, N., Menezes, A.: Pairing-based cryptography at high security levels. In: Smart, P. (ed.) Cryptography and Coding 2005. LNCS, vol. 3796, pp. 13–36. Springer, Heidelberg (2005)

16. Miller, V.S.: Short Programs for Functions on Curves, IBM Watson, T.J. Research Center (1986), http://crypto.stanford.edu/miller/miller.ps

17. Miller, V.S.: The Weil pairing and its efficient calculation. J. Cryptol. 17(4), 235–261 (2004)

18. Smart, N.P.: The Hessian form of an elliptic curve. In: Koç, Ç.K., Naccache, D., Paar, C. (eds.) CHES 2001. LNCS, vol. 2162, pp. 118–125. Springer, Heidelberg (2001)

FPGA Implementation of an Improved Attack against the DECT Standard Cipher

Michael Weiner, Erik Tews, Benedikt Heinz, and Johann Heyszl

Fraunhofer Institute for Secure Information Technology, Munich, Germany
TU Darmstadt, Germany
michaelweiner@mytum.de, e_tews@cdc.informatik.tu-darmstadt.de,
{benedikt.heinz,johann.heyszl}@aisec.fraunhofer.de

Abstract. The DECT Standard Cipher (DSC) is a proprietary stream cipher used for enciphering payload of DECT transmissions such as cordless telephone calls. The algorithm was kept secret, but a team of cryptologists reverse-engineered it and published a way to reduce the key space when enough known keystreams are available [4]. The attack consists of two phases: At first, the keystreams are analyzed to build up an underdetermined linear equation system. In the second phase, a brute-force attack is performed where the equation system limits the number of potentially valid keys. In this paper, we present an improved variant of the first phase of the attack as well as an optimized FPGA implementation of the second phase, which can be used with our improved variant or with the original attack. Our improvement to the first phase of the attack is able to more than double the success probability of the attack, depending of the number of available keystreams. Our FPGA implementation of the second phase of the attack is currently the most cost-efficient way to execute the second phase of the attack.

Keywords: DECT, DECT Standard Cipher, DSC, Stream Cipher, FPGA, Hardware-Accelerated Cryptanalysis.

1 Introduction

Digital Enhanced Cordless Telecommunications (DECT) is a standard for short range cordless communication. DECT is mostly used for phones, however other applications like wireless payment terminals, traffic control and room monitoring are possible. With more than 800 million DECT devices sold[1], it is one of the most commonly used systems for cordless phones besides GSM, UMTS and CDMA. The DECT standard provides mutual authentication of devices and encryption of the payload, however both features are optional and need not be implemented on a device. DECT uses the DECT Standard Authentication Algorithm (DSAA) for authentication and key exchange and the DECT Standard Cipher (DSC) for encryption.

---

[1] http://www.etsi.org/WebSite/NewsandEvents/201004_CATIQ.aspx

K.-H. Rhee and D. Nyang (Eds.): ICISC 2010, LNCS 6829, pp. 177–188, 2011.

First attacks on DECT [2,3] showed that some devices do not use encryption and authentication at all and can easily be eavesdropped on. Even if encryption is used and long-term and session keys are generated in a secure manner, it is still possible to decipher phone calls. In 2009, the DECT Standard Cipher was reverse-engineered and a correlation attack on the cipher was published [4] by Nohl, Tews and Weinmann (NTW-attack). With 2^{15} available keystreams generated with different initialization vectors (IVs), it is possible to recover the session key within minutes to hours on a fast PC or Server. Different tradeoffs are possible. This allows decryption of the call recorded, but does not reveal the long-term keys or keys for the previous or next call.

In this paper, we present an optimized NTW-attack, which reduces the time to recover the key or the number of keystreams required. The optimizations are of general nature and can be used in conjunction with optimized implementations of the attack for CUDA graphics cards or the PS3 cell processor [4] or any other kind of parallel processing hardware. In the second part of the paper we present an optimized FPGA implementation of our optimized NTW-attack, which is currently the most cost-efficient way of searching through the remaining key space the NTW-attack determines.

In Section 2 we describe the attack scenario and point out where our work can be applied. In Section 3, we give an introduction to DSC and the original attack on DSC developed by Nohl, Tews, and Weinmann. Knowledge of the structure of the original attack is essential to understand our improvements. In Section 4, we present our improvements of the first phase of the NTW attack. In a nutshell we introduce a key ranking method making the correct key more likely to be found earlier in the second phase of the attack. In Section 5 we present an FPGA implementation which can be used in conjunction with our improvements from Section 4 to execute the second phase of the attack in the most cost-efficient way currently known. Section 6 concludes our work.

2   Attack Scenario

In this paper, we show that an attacker who is able to eavesdrop on DECT communication can decrypt the encrypted payload faster and more efficiently than previously known. In contrast to some other attack scenarios [2], our attack is passive, i.e. no data needs to be sent by an attacker. Therefore, a victim is not able to detect the presence of an attacker.

At first, the attacker needs to record the raw DECT data being sent over the wireless interface. He can do so, for example, by using a DECT PC-Card using a modified firmware [2] or a generic software radio like USRP[3].

Using the recorded data, the attacker has several options depending on the type of communication and the security services being applied. If the attacker is able to listen to the pairing process between the base station and the handset, he needs at most $10^4 \approx 2^{13.3}$ tries to recover the resulting long-term key (UAK).

---

[2] https://dedected.org/trac/attachment/wiki/25C3/talk-25c3

[3] http://www.ettus.com/

Further decryption is trivial as all other keys are derived from the UAK. However, regular pairing only takes place once when a handset is being installed to a base station and only if the handset is not pre-paired to the station by the manufacturer.

Therefore, we assume that an attacker is only able to eavesdrop on a regular DECT call. In this case, if encryption is enabled, he can either attack the key derivation scheme of DSAA [2,3] that generates the session keys, or he can attack the payload encryption algorithm DSC. Attacking DSAA is especially suitable if the attacked devices have a weak PRNG.

When attacking DSC, the attacker must be able to extract valid DSC keystreams from the recorded data. This is possible because some messages can be predicted – for example, the call duration counter is implemented on the base station for several DECT phones, and the counter value is sent to the handset once per second using a control message. An attacker can predict messages of that type when he knows the start time of the call.

An attack against DSC requires a relatively large number of known keystreams for a reasonable success probability. In this paper, we introduce two means to increase the performance of a DSC attack, which can be applied independent from each other: On the one hand, we provide an algorithmic improvement, and on the other hand, we provide a very efficient implementation on an FPGA.

3   Cryptanalysis of the DECT Standard Cipher

The DECT Standard Cipher is a proprietary stream cipher designed for DECT. It takes a 64 bit key and a 35 bit initialization vector (IV) and generates a keystream of variable length. DECT supports frames of different lengths and formats. For common voice calls, a keystream of 720 bits is generated and split into two keystream segments. The first 360 bits of the output of DSC are used to encrypt traffic from the base station (Fixed Part, FP) to the phone (Portable Part, PP). The first 40 bits can be used to encrypt control traffic (C-channel traffic). If a frame contains no C-channel data, the first 40 bits are discarded. The remaining 320 bits are used to encrypt the actual voice data (B-field). The second part of the keystream is used to encrypt frames sent from the PP to the FP. Again, the first 40 bits are used to encrypt C-channel traffic if present. The remaining 320 bits are used to encrypt the voice data.

The internal design of DSC consists of 4 linear feedback shift registers R1, R2, R3, and R4 of length 17, 19, 21, and 23 bits. Three of them are irregularly clocked, the last one with a length of 23 bits is regularly clocked. A non-linear output combiner is used to generate the output using six bits from the three irregularly clocked registers. Initially, the 35 bit IV is zero-extended to 64 bit and prepended to the 64 bit cipher key resulting in an 128 bit input to the cipher. The input is then clocked into the most significant bit of each register using regular clocking. After the key loading, every bit of every register is just a linear combination of key and IV bits. After key loading, 40 blank rounds are performed using irregular clocking.

To attack DSC, Nohl, Tews, and Weinmann used the following approach: If DSC would be regularly clocked, one could easily recover the secret key. Of course DSC is not regularly clocked, but the probability that register R1 has been clocked i times, R2 has been clocked j times, and R3 has been clocked k times when the lth bit of output is produced is:

$$p_{i,j,k,l} = \binom{40+l}{i-(80+2l)}\binom{40+l}{j-(80+2l)}\binom{40+l}{k-(80+2l)}2^{-(40+l)3}$$

Let $s = x_{1,0}^{(i)}, x_{1,1}^{(i)}, x_{2,0}^{(j)}, x_{2,1}^{(j)}, x_{3,0}^{(k)}, x_{3,1}^{(k)}$ be the six bits of registers R1, R2, and R3, which contributes to the keystream generated by DSC at this moment. To eliminate some variables we may write $x_{1,0}^{(i+1)}$ instead of $x_{1,1}^{(i)}$ because the bit is simply shifted with the next clock. $x_{2,0}^{(j+1)} = x_{2,1}^{(j)}$ and $x_{3,0}^{(k+1)} = x_{3,1}^{(k)}$ also holds. Let z_l be the bit of output produced by DSC and z_{l-1} be the previous bit of output which is now stored in the memory bit of the output combiner. Because s is just a linear combination of key and IV bits, we may split it into a key and IV part $s = s_{\text{key}} + s_{\text{iv}}$. The linear combination of the IV part s_{iv} is known by the attacker for every keystream and the recovery of s_{key} would reveal 6 bit of information about the secret key. If $\mathcal{O}(s, z_{l-1}) = z_l$ holds for a value of s, it can bee seen as an indication that $s_{\text{key}} = s + s_{\text{iv}}$ for a higher probability than guessing ($\frac{1}{64}$).

To execute the attack, a clocking interval C $= [102, 137]$ of length 35 was chosen. This leads to $35^3 = 42875$ possible combinations for the number of clocks i, j, k for the registers R1, R2, and R3 which reveal information about the state variables $x_{\{1,2,3\},0}^{(102)} \cdots x_{\{1,2,3\},0}^{(138)}$. For every choice i, j, k of clocking combinations in this interval a frequency table for the $2^6 = 64$ choices for the key-part key of s is used. For every consecutive pair of bits z_l, z_{l-1} from the keystream where the clocking combination has a none negligible probability and for every choice of s,

$$p = \sum_l p_{i,j,k,l} * [\mathcal{O}(s, z_{l-1}) = z_l] + \frac{1}{2}\left(1 - \sum_l p_{i,j,k,l}\right)$$

is computed and $\ln\frac{p}{1-p}$ is added to the frequency table entry $s_{\text{key}} = s + s_{\text{iv}}$. Instead of representing the equations in the frequency table directly as linear combinations of key-bits, a short form is used where all equations have the form $x_{\{1,2,3\},0}^{(\cdot)} = \{0,1\}$. Every entry in the frequency table contains six of those equations.

After all keystreams have been analyzed, we take every variable v and examine all frequency tables which contain equations of the form $v = b_i, b_i \in \{0,1\}$. We take the top-voted entry from these tables and compute $p_v = \sum_i (2b_i - 1) * p_i$ where p_i is the number of votes for the top voted entry in the table. If p_v is negative, we assume that $v = 0$ holds, 1 otherwise.

In total there are $36 * 3 = 108$ different equations. All of them are sorted according to $|p_v|$. The original attack suggests using the topmost equations (for example 30 equations) to build an equation system of the form $Ak = b$ for the key

k. All possible solutions of the system (using 30 equations leads to $2^{64-30} = 2^{34}$ possible solutions) are then checked against some reference keystreams to check if one of them generates the reference keystream. If so, it can be assumed that this solution is in fact the correct key for the cipher.

4   Key Ranking

To improve the original NTW attack, we introduce a key ranking procedure. The original NTW attack generates equations of the form $\sum_i a_i k_i = \{0, 1\}$ where k_i is a bit of the key and a_i is either 0 or 1. The left part of the equation only depends on the feedback polynomials of the registers. The right part of the equation is either 0 or 1, determined by a voting system. The difference between the number of votes for 0 and 1 is denoted by $|p_v|$. In the original attack, the equations are sorted by $|p_v|$ and the topmost equations are assumed to be correct. Using many equations results only in a small remaining key space which needs to be searched, but increases the probability that at least one equation is incorrect and the key is not found in the set of solutions of the linear equation system.

To improve the attack, we first checked, with which probability the individual equations are correct. We ran 100 experiments against randomly chosen keys and counted in how many times the first, second, third... equation in A was correct. The results are shown in figure 1. The first 10 equations in A (see Section 3) are correct with a probability of at least 99%. This makes it highly unlikely that one of the first 10 equations is incorrect. Starting from equation 30, the probability that the equation is correct drops down to 70-60% for equation 55. This makes these equations only of minor use for the attack and one can assume that at least one of these equations is incorrect with high certainty.

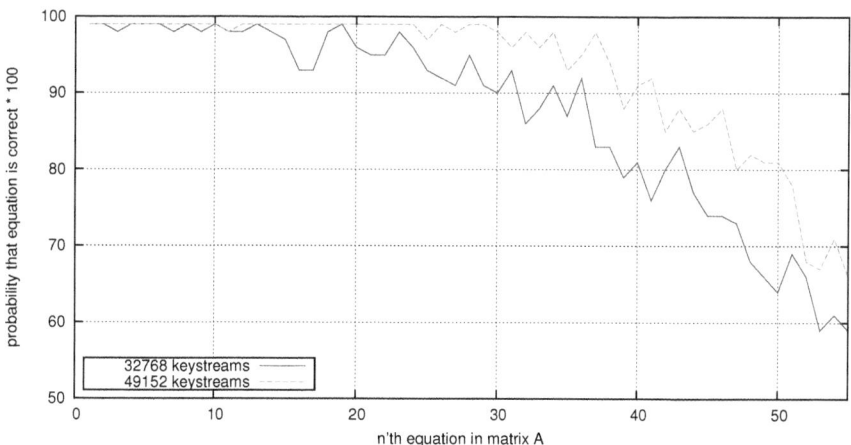

Fig. 1. Success probabilities of the individual equations in matrix A

We decided to look for a strategy to generate highly likely sub key spaces in an order, so that the key spaces which are most likely to contain the correct key are generated first. The key spaces should still be described by a linear equation system and should contain many (at least 2^{26} or more) keys, so that high parallel implementations which can search through such a key space as developed for the original attack can still be used, communication overhead is minimized and pipeline stalls due to too small key spaces are avoided. As a result, at most 36 equations from matrix A should be used.

For the original NTW attack, it is never necessary to compute the success probability of an equation explicitly. Instead, one can just sort all equations by $|p_v|$, assuming that equations with a higher difference have a higher success probability. We decided to compute the explicit probability for an equation from $|p_v|$. First, one can simulate the attack against 100 random keys (using the same number of keystreams) and collect all generated equations with their voting difference and correctness. It is now possible to compute the success probability $P(|p_v|)$ of an equation using this data and a nearest neighbor smoother or similar methods (e.g. kernel smoother). We used a k-nearest-neighbor smoother for this paper.

We did not decide to compute the success probability for an equation from the line number in the matrix A and the number of keystreams. If a systematic problem in the keystream recovery method used would exist, this could decrease the success probability of some equations. Using $|p_v|$ for computing the success probability of each equation seems to be more appropriate.

We can formulate our key ranking approach as a best-first-search over a directed graph: Assuming that we have a set of equations e_i with respective individual success probabilities $P(|p_{x_i}|)$ and that the success probabilities are independent, we can run a best-first-search for the correct key (if we use 64 equations) or for the most promising sub key space (if less than 64 equations are used). We assume that the set of possible keys or sub key spaces is a directed graph $G = (V, E)$. A node v consists of a vector c that indicates which equation e_i is correct ($c_i = 0$) and which of the equations is incorrect ($c_i = 1$). The probability that this node represents the correct sub key space is $\prod_i (|c_i - P(|p_{x_i}|)|)$. The node with the highest probability is the node with $c = (0, \ldots, 0)$ where all equations are assumed to be correct. An edge (v_1, v_2) exists if v_1 and v_2 differ only in a single equation, which is assumed to be correct in v_1 but assumed to be incorrect in v_2.

We can now run a best-first search for the correct sub key space on this graph starting at the node with the highest success probability. Using 64 equations would guarantee that all keys are visited in the exact order of probability, however we think that the number of equations should be limited so that not too much time is spent for generating the keys to check and highly parallel hardware like CUDA graphics cards or FPGAs can be used in an efficient way. Using some kind of data structure for the queue in the best-first-search which allows inserts, searches and removals in $O(\log(n))$ makes generating the sub key spaces very time efficient. However, memory consumption increases because up to $m \cdot g$

solutions need to be tracked in parallel, when m equations are used and g sub key spaces have been generated.

4.1   Performance Results

Executing the old attack against 100 randomly chosen keys only resulted in 71% success rate with 2^{15} keystreams available and 2^{42} keys checked. Using our new key ranking method allowed us to recover the key in 90% of all tests, with also 2^{42} keys checked in total. We used 35 instead of 22 equations, but checked the 8192 most likely sub key spaces. Figure 2 includes more details.

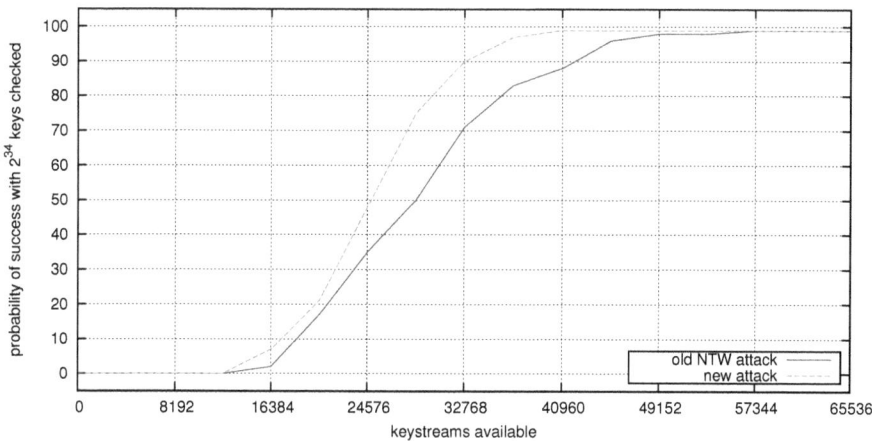

Fig. 2. Success rate of the improved attack

Another advantage of our key ranking strategy is, that the attack time doesn't need to be fixed at the beginning of the attack. Using just a single equation system has the disadvantage that all solutions are checked in an order not depending on their probability. Checking an equation system with 2^n solutions will give the correct key after having checked 2^{n-1} solutions in average (if all equations in the system are correct). Using our approach makes it possible to start the attack with some reasonable parameters and then just wait for the correct key. If a lot of equations in A are correct, the correct key will be found much faster in average than with the original approach. If A contains a lot of incorrect equations, the attack will take longer, but one can decide to continue or cancel the attack at any point of time (assuming that enough main memory for the best-first-search is available).

To speed up the final search through all generated sub key spaces, we present an FPGA implementation of the final search in the next part of this paper.

5   FPGA Implementation

FPGAs are very well-suited for an implementation of the final search phase of a sub key space. Linear Feedback Shift Registers form the main part of the DSC algorithm, and they can be implemented much more efficiently on an FPGA than on a CPU or GPU platform.

5.1   Basic Implementation Idea

Our improved DSC attack requires the knowledge of a valid reference *(IV, Keystream)* pair and an underdetermined equation system

$$A \cdot k = b \tag{1}$$

that constrains the key space. A and b are determined by the first part of the attack (see section 3), k denotes the cipher key.

The FPGA design must iterate over all potentially valid cipher keys according to equation (1), compute the keystream and compare it to the reference keystream. Therefore, a cipher key generator, a DSC keystream generator and a compare unit comparing the keystream output to the reference keystream is necessary for the FPGA implementation. The design shall report cipher keys that produce an identical keystream as the reference.

The most convincing way to implement the key generator is using a counter or full-cycle LFSR that generates "independent" bits and a combinatorial function generating "dependent" bits that use the "independent" bits as an input. The equation systems must be transformed beforehand for this purpose, such that the dependent bits are described as a function of the independent bits. The DSC keystream generator can be implemented straight-forward as described in [4].

5.2   Optimizations

Optimizations are possible on several levels compared to a straight-forward implementation. A list of all matrices used for describing the optimizations is given in Table for clarity reasons.

Table 1. Matrices describing the Key Loading and the Equation System

Matrix	Dimension	Description
k	64×1	Cipher Key
sk	128×1	Session Key
$sk_{i,j}$	$len(R_j) \times 1$	Vector that loads the i-th Bit of sk into Register j
ck	128×1	Zero-extended Cipher Key k
iv	128×1	Zero-extended Initialization Vector
d_i	80×1	DSC State after i clocked in bits, without Output Combiner
$d_{i,j}$	$len(R_j) \times 1$	State of R_j after i clocked in bits
R_j	$len(R_j) \times len(R_j)$	Clock Matrix of Register j
L	80×128	Load Matrix (Session Key to Initial State)
A'	$128 \times len(x)$	Equation System Matrix
b'	128×1	Equation System Offset Vector
x	$len(x) \times 1$	Key Generator Counter Value

Simple Improvements: The key generator can be shared among several DSC units, as it generates one key per cycle whereas the compare units need multiple clock cycles for verifying one key. Unnecessary control signals may be removed and logic delays shall be kept short by inserting registers on critical paths.

DSC Speedup: The fundamental DSC implementation as described in [4] requires three clock cycles per bit of keystream output. This can be reduced to one clock cycle by multiplexing and re-arranging the feedback taps. The corresponding feedback taps can be determined from the feedback matrices R_j^2 and R_j^3. [1]

Key Loading: [4] suggests to load the session key in 128 clock cycles by clocking in one bit per cycle. This can be represented as iterating 128 times over the linear transformation $d_{i,j} = R_j \cdot d_{i-1,j} + sk_{i,j}$ for all four registers j, where $d_0 = (0, ..., 0)$ and $sk_{i,j}$ is a vector with the size of register j in which the most significant position is set to bit i of the session key and all other positions are zero.

The key can be loaded in one cycle by summarizing the four matrices $R_{1,2,3,4}$ into one load matrix L such that

$$d_{128} = L \cdot sk \tag{2}$$

holds. A similar optimization is described in [1], but they only propose to load 16 bits per clock cycle.

As a second step of improvement, the calculation of the full cipher key can be skipped: As described before, the "dependent" part of the cipher key is a combinational function of the "independent" cipher key bits. A matrix A' and a vector b' transforming an independent value x into the cipher key ck can be derived from A and b, such that the equation

$$ck = A' \cdot x + b' \tag{3}$$

generates one key candidate compliant to equation (1) for each value of x. As the session key is the sum of cipher key and initialization vector,

$$sk = ck + iv \tag{4}$$

the whole initial state can be expressed as a function of the independent cipher key bits by inserting equation (4) into equation (2) and then equation (2) into equation (3):

$$d_{128} = \underbrace{LA'x}_{dynamic} + \underbrace{L(b' + iv)}_{static} \tag{5}$$

Hard-Coding: Where the plain NTW attack proposed *one* equation system $A \cdot k = b$, our key ranking allows us to reuse the matrix A and just invert one or more equations, i.e. modify b, if no key has been found for a particular sub key space. Hence, only the b vector needs to be loaded into the FPGA at run time, while A can be hard-coded into the design by a VHDL preprocessor. This saves hardware resources on the FPGA, reduces the complexity and eliminates potentially critical paths.

The reference keystream can be hard-coded as well.

Early Abort: A cipher key can be considered invalid as one bit from the generated keystream differs from the reference. In such a case, the comparison to the reference keystream can be aborted early such that the unit can immediately continue with the next key candidate.

The probability that k subsequent bits of the keystream are correct for a wrong key is 2^{-k}. On average, the comparison for a wrong key already fails after two keystream bits. Therefore, $n - 2$ cycles can be saved in comparison to a deterministic unit that always compares n bits.

We compare at most 32 bits and thus save 30 cycles on average.

Pre-ciphering Pipeline: With the Early Abort optimization, several DSC units are competing to be loaded with a new initial state. As the arbitration logic complexity rises with the number of competing units, this number is to be kept low. A good way to do this is outsourcing the pre-ciphering phase into a strictly sequential, deterministic pipeline. With this optimization, the state *after pre-ciphering* is directly loaded into the computing DSC units.

Input Buffering: Idle time of the FPGA has a negative impact on the effective performance. Therefore, an input buffer is used such that the PC can enqueue multiple tasks and the FPGA can immediately load the next task as soon as the previous one is finished.

5.3   Implementation

For our implementation, a Xilinx Spartan-3E 1200 (XC3S1200E) FPGA on a Digilent Nexys 2 board was used. The PC communication was implemented via the on-board RS-232 interface.

Our final implementation includes all optimizations as described in section 5.2. The runtime of the design is not entirely deterministic, as – for a specific keystream – the position of the first failing comparison is unknown. Therefore, the key generator was given the ability to be paused, which is necessary when all available DSC units are busy.

Figure 3 shows the structure of the key search unit, which forms the essential part of our hardware design. The dotted lines in the diagram denote the hard-coded data. The "State Offset" is sent to the FPGA at run time for each sub key space. It is determined by the attacked IV and the vector b'.

One pipelined key generator (see 5.2) was chosen to serve four DSC units – this is the maximum number implementable on *one* Look-Up Table.

The key search unit consumes about 30% of the FPGA resources in total, such that three instances can be created on our device. This enables searching three sub key spaces at the same time.

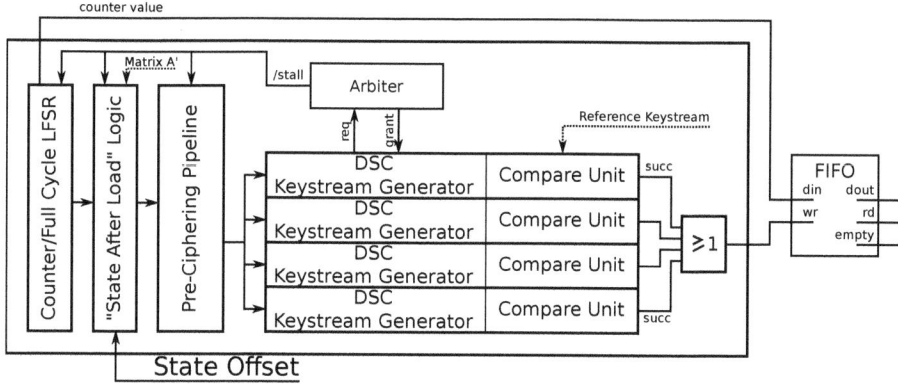

Fig. 3. Block Diagram of Key Search Unit

5.4   Performance Evaluation

This section compares the performance achieved by our FPGA implementation with the CUDA performance published in [4].

We used five different, randomly generated equation systems for evaluating the maximum frequency by synthesizing the design for each of the equation systems. Table 2 shows the achieved results.

Table 2. Performance Evaluation (using 2^{32} equations)

	Max Frequency	Performance $[\frac{keys}{s}]$	Cost $[US\$]$	Cost-Performance
FPGA	140 MHz	$408.8 \cdot 10^6$	169	$2.42 \cdot 10^6 \frac{keys}{US\$\cdot s}$
[4] CUDA / GTX 260	unknown	$148 \cdot 10^6$	190	$0.78 \cdot 10^6 \frac{keys}{US\$\cdot s}$

6   Summary

The final attack could be applied as follows: In the first phase of the attack, the adversary recovers keystreams by eavesdropping on a DECT call. If a phone is used which displays a call duration counter that is implemented on the base station, the adversary might be able to recover about 5 known keystreams per second. After nearly two hours, the adversary has collected 2^{15} known keystreams, which can be processed in the next phase of the attack.

In the second phase of the attack, the adversary needs to generate frequency tables from the known keystreams. We did not modify this step in our paper. In the original attack, Nohl, Tews, and Weinmann used a SUN X4440 using 4 Quad-Core AMD Opteron CPUs running at 2.3 GHz to generate the tables in 20 minutes. This process is highly CPU bound, so that a single Opteron CPU could accomplish the task in about 80 minutes. Because this can be started while the first phase is still running, phases one and two need only two hours to complete.

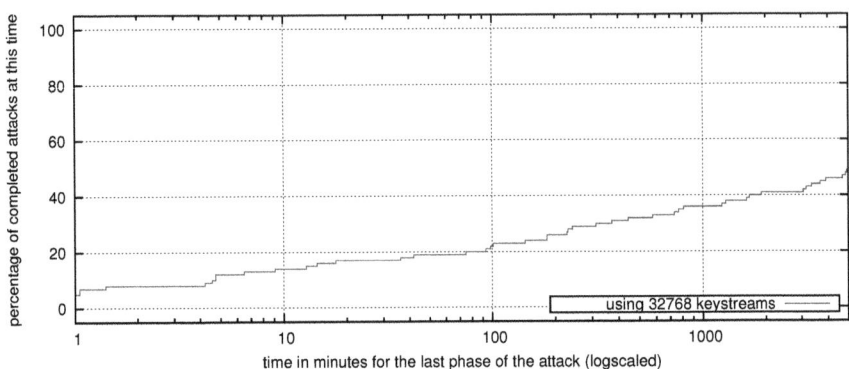

Fig. 4. Time to completion of the attack using a single FPGA and 2^{32} equations

The runtime of these two phases is only affected by the rate of the keystream recovery process and the computing power available.

In the third and last phase, the adversary uses the frequency tables generated in phase two to search for the correct key. He uses a PC which generates the most likely sub key spaces as described in Section 4 and transfers them to a single or multiple FPGAs connected via a serial line or other interfaces. The time for generating the sub key spaces is negligible compared to the time consumed by the FPGAs to check the sub key space, so that many FPGAs can be supplied by a single PC.

Figure 4 shows the time to the completion of the last attack phase, using just a single Xilinx Spartan-3E 1200 (XC3S1200E) FPGA using 2^{15} keystreams. About 20% of our experiments completed within one hour. The next 20% of our experiments needed up to one day to complete. The remaining 60% needed more than a day to complete. Please note that doubling the number of FPGAs used reduces the total time for the last phase by half and the attack scales almost perfectly when the number of available FPGAs is increased.

References

1. Alcatel. Data ciphering device. U.S. Patent 5,608,802 (1994)
2. Lucks, S., Schuler, A., Tews, E., Weinmann, R.P., Wenzel, M.: Attacks on the DECT authentication mechanisms. In: Topics in Cryptology–CT-RSA 2009, pp. 48–65 (2009)
3. Molter, H.G., Ogata, K., Tews, E., Weinmann, R.P.: An Efficient FPGA Implementation for an DECT Brute-Force Attacking Scenario. In: 2009 Fifth International Conference on Wireless and Mobile Communications, pp. 82–86. IEEE, Los Alamitos (2009)
4. Nohl, K., Tews, E., Weinmann, R.P.: Cryptanalysis of the DECT Standard Cipher (February 2010)

Chameleon: A Versatile Emulator for Contactless Smartcards*

Timo Kasper, Ingo von Maurich, David Oswald, and Christof Paar

Horst Görtz Institute for IT Security, Ruhr-University Bochum, Germany
{timo.kasper,ingo.vonmaurich,david.oswald,christof.paar}@rub.de

Abstract. We develop a new, custom-built hardware for emulating contactless smartcards compliant to ISO 14443. The device is based on a modern low-cost microcontroller and can support basically all relevant (cryptographic) protocols used by contactless smartcards today, e.g., those based on AES or Triple-DES. As a proof of concept, we present a full emulation of Mifare Classic cards on the basis of our highly optimized implementation of the stream cipher Crypto1. The implementation enables the creation of exact clones of such cards, including the UID. We furthermore reverse-engineered the protocol of DESFire EV1 and realize the first emulation of DESFire and DESFire EV1 cards in the literature. We practically demonstrate the capabilities of our emulator by spoofing several real-world systems, e.g., creating a contactless payment card which allows an attacker to set the stored credit balance as desired and hence make an infinite amount of payments.

Keywords: RFID, contactless smartcards, payment systems, access control, efficient implementation.

1 Introduction

Radio Frequency Identification (RFID) devices are deployed in a wide range of transportation and access control systems world-wide. If high privacy or security demands have to be met, typically contactless smartcards according to the ISO 14443 standard [13] are employed, as they offer sufficient computational power for cryptographic purposes. Moreover, a growing number of payment systems incorporates secure RFID cards [16], as they offer additional benefits in terms of flexibility and convenience over their contact-based counterpart. State-of-the-art contactless cards, such as the electronic passport ePass [8], provide a high level of security by means of various cryptographic primitives.

In general, RFID technology implies new threats compared to contact-based systems, for instance, a card residing in a pocket or wallet could be read out or modified without the owner taking note of it. Due to the cost sensitivity of such

* The work described in this paper has been supported in part by the European Commission through the ICT programme under contract ICT-2007-216676 ECRYPT II.

K.-H. Rhee and D. Nyang (Eds.): ICISC 2010, LNCS 6829, pp. 189–206, 2011.

high-volume applications, card manufacturers are tempted to use outdated but "cheap" cryptographic components, e.g., in Mifare Classic products.

Since the reverse-engineering of the Crypto1 cipher used in Mifare Classic cards and the subsequently published attacks (cf. Sect. 3.1), the cards have to be regarded as insecure, as the secret keys can be extracted in seconds by means of card-only attacks. Once all keys of the card are known to an attacker, cards can be modified or duplicated. As many systems in the real world still rely on these weak cards, severe security threats may arise.

Accordingly, recently installed contactless systems, especially those with high security demands, are based on the DESFire variant of the Mifare family, and system integrators upgrade the old Mifare Classic technology to these newer cards wherever possible. While the 3DES cipher employed in these cards is secure from the mathematical point of view, the implementation on the card is vulnerable to side-channel analysis, so that it is again possible to extract the secret keys of a card[1], as detailed in Sect. 3.2. Hence, emulating these modern cards is also practical and renders various attacks in real-world scenarios possible.

The resulting security weaknesses can become very costly – one example is a widespread contactless payment system based on Mifare Classic cards as analyzed in [16], where the credit value on the cards can be modified by an adversary with minimal efforts. For many of these systems, the read-only Unique Identifier (UID) of each card constitutes the only means to detect fraud in the backend, as there are no cards available on the market where the UID can be altered. In this paper, we exhibit the possibility of emulating and cloning RFID-enabled smartcards compliant to ISO 14443, including their UID.

1.1 Background and Related Work

Several research groups have proposed custom devices to emulate and counterfeit RFID devices. However, virtually all emulators presented so far suffer from certain drawbacks, e.g., insufficient computational resources, high cost, or impractical dimensions, limiting the threat they pose in the context of attacking real-world systems.

A custom RFID emulation hardware called Ghost is presented in [24]. The Ghost is able to emulate Mifare Ultralight cards which do not use any encryption. Emulating contactless cards employing secure cryptography seems to be impossible using this device due to computational limitations. The OpenPICC project [20] is mainly an RFID sniffing device. There was an approach to offer support for ISO 14443A, but the project seems to be discontinued. The Proxmark III [21] enables sniffing, reading and cloning of RFID tags. Since the device is based on a Field Programmable Gate Array (FPGA), it is also capable of emulating Mifare Classic cards, but at a comparably high cost of $399. The "HF Demo tag" [12] is based on an Atmel ATMega128 microcontroller which is not

---

[1] Note that the effort for extracting secret keys from Mifare DESfire cards by means of side-channel analysis is much higher compared to the Mifare Classic attacks.

computationally powerful enough to perform encryptions with state-of-the-art ciphers in the time window given by the relevant protocols. An embedded system for analyzing the security of contactless smartcards was introduced in [14]. The attack hardware consists of a so-called Fake Tag and an RFID reader and can be used for, e.g., practical relay attacks. The device is based on a Atmel AT-Mega32 [1] processor with a constrained performance and is designed such that all important functionality is provided by the RFID reader. Hence, in addition to the lack of computational power, the Fake Tag cannot operate independently from the reader, which can be a major drawback for practical attacks. The authors also implemented an emulation of Mifare Classic, but similar to the HF Demo tag, the encryption runs too slow so that timing constraints of the protocol cannot be met. We used this work as a starting point for the development of our new stand-alone RFID emulator.

1.2   Contribution of This Paper

We built a freely programmable low-cost device that is capable of emulating various types of contactless smartcards, including those employing secure cryptography. The device operates autonomously without the need of a PC, can be powered from a battery, and possesses an Electronically Erasable Programmable Read-only Memory (EEPROM) for storing received bitstreams or other non-volatile information. An attacker using the presented hardware, which can be built for less than $25, is in full control over all data stored on the emulated card, including its UID and the secret keys.

In order to demonstrate the capabilities of our emulator in the context of real-world attacks, we implemented optimized versions of the Crypto1 stream cipher, the Data Encryption Standard (DES), Triple-DES (3DES) and the Advanced Encryption Standard (AES), as required for emulating the widespread Mifare Classic, Mifare DESFire and Mifare DESFire EV1 cards. With the developed software, it is possible to simulate the presence of one of these cards with an arbitrarily chosen content and identifier, and hence spoof real-world systems in various manners. For example, the emulator can behave as a card that automatically restores its credit value after a payment, or that possesses a new UID and card number on each payment, which impedes the detection of fraud. Besides the simulation of cards, our hardware allows for sniffing, e.g., reverse-engineering of protocols, relay attacks, and testing the vulnerability of RFID readers towards a behavior of the card that does not conform to the specifications, for instance, with respect to timing, intentionally wrong calculation of parity bits, or buffer overflows.

The remainder of this paper is structured as follows: in Sect. 2, we present our custom RFID hardware that serves as a basis for card emulations and attacks. After giving a brief summary of the relevant characteristics and protocols of Mifare Classic, Mifare DESFire and Mifare DESFire EV1 cards in Sect. 3, we detail on our implementations of the respective emulations in Sect. 4. Finally, practical real-world analyses performed with our hardware are described in Sect. 5.

2   Hardware Setup

In the following, we give a brief introduction to the physical characteristics of the RFID technology employed in contactless smartcards. Then, our freely programmable emulator for contactless smartcards is presented.

2.1   RFID Technology

In a typical setup for contactless smartcards, a reading device generates a strong Electro-Magnetic (EM) field at a frequency of 13.56 MHz for supplying the card with energy for its operation. The reader acts as master, while the card serves as slave, thus only the reader can start a communication and issue commands to the card. The ISO 14443 standard specifies the physical characteristics, the data modulation and other characteristics of contactless smartcards. For data transmission, the reader encodes the bits using a pulsed Miller code and transmits it by switching off the EM field for short periods of time. The data to be sent by the card is encoded using a Manchester-code and is afterwards transmitted via the EM field using load-modulation with a 847.5 kHz sub-carrier.

2.2   Our Emulator

For the security analyses in this paper, we developed a custom, freely programmable device termed "Chameleon", which can emulate contactless smartcards compliant to the ISO 14443 standard in a stand-alone manner. Our emulation device consists of off-the-shelf hardware and can be built for less than $25. It is based on an Atmel ATxmega192A3 microcontroller [2,3] which provides 192 kB of program memory, 16 kB SRAM and 4 kB EEPROM memory. Using an FTDI FT245RL chip [9], the ATxmega is able to communicate with a PC via the Universal Serial Bus (USB). This communication link can be used for debugging purposes and data manipulation at runtime. Figure 1 shows the first version of our RFID emulation device.

We chose the ATxmega because it features a hardware acceleration of both DES and AES-128. After loading the key and the data to the corresponding registers, the ATxmega is able to perform a DES en- or decryption in 16 clock

Fig. 1. Our stand-alone RFID emulation device

cycles, i.e., one DES round per clock cycle, whereas the AES engine runs concurrently to the CPU and requires 375 clock cycles until an en- or decryption of one block is finished. The microcontroller is clocked by an external 13.56 MHz crystal, which is internally doubled using a high frequency Phase Locked Loop (PLL).

The coupling to the reader is established by a rectangular coil on the Printed Circuit Board (PCB). Variable capacitors are placed in parallel to form a parallel resonant circuit that is tuned to the carrier frequency. Analog circuitry assists the microcontroller in extracting the encoded data from the EM field and transmitting bitstreams. The design is similar to [14] and mainly shapes the signals according to the ISO 14443 standard and converts them to the appropriate voltage levels. Our emulation device can either be powered via the USB interface or run on battery. As all functionality is directly provided by the microcontroller, the Chameleon operates autonomously without the support of a PC. The full schematics of the developed hardware are given in the Appendix B.

3 Mifare Cards

This section covers the details of Mifare Classic, DESFire and DESFire EV1 cards. We present important facts required for the emulation of the cards and detail on the different authentication protocols, as implemented in Sect. 4.1 and Sect. 4.2. For reference, the complete protocols including the command codes and the low-level format are provided in Appendix A.

3.1 Mifare Classic

Since its introduction more than a decade ago, allegedly over 1 billion Mifare Classic ICs and 7 million reader components have been sold [18]. The cards provide data encryption and entity authentication based on the proprietary stream cipher Crypto1 for preventing from attacks like eavesdropping, cloning, replay and unauthorized reading or modification of the data stored on the card. Crypto1 is based on a Linear Feedback Shift Register (LFSR) with a length of 48 bit.

Basically, a Mifare Classic card can be regarded as a secured EEPROM memory with an RFID communication interface. In this work, we focus on the by far most widely employed Mifare Classic 1K version with 1024 byte EEPROM. All Mifare Classic variants comply to Parts 1-3 of ISO 14443A [13]. While the standard also allows for higher data rates, the cards communicate at a fixed data rate of 106 kBit/s. In addition, they feature a proprietary high-level protocol that diverges from Part 4 of ISO 14443A.

The memory of a Mifare Classic card is divided into sectors, whereas each sector consists of four blocks, as illustrated in Fig. 2. Each sector can be secured by means of two cryptographic keys A and B that are stored along with a set of access conditions in the last block of each sector. Before a sector can be accessed, a proprietary mutual authentication protocol with the appropriate secret key has to be carried out, cf. Protocol 1. The access conditions determine the commands

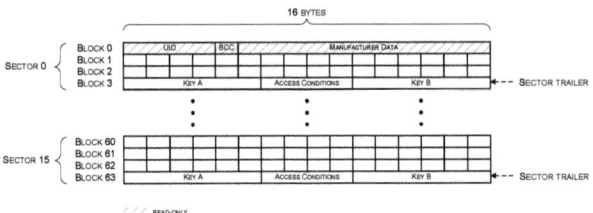

Fig. 2. The memory structure of a Mifare Classic 1K card

that are allowed for each block of the sector (read, write, increment, decrement) and define the role of the keys [19]. The other blocks of each sector can be used for data storage. Note that the first block of the first sector differs from this scheme: it always contains a UID, along with some other manufacturer-specific data. The first block is written to the chip at manufacturing time, making it impossible to change the UID.

When a card is placed close to a reader, the anticollision and select procedure as defined in ISO 14443A is carried out. Then, an authentication command is issued by the reader that specifies for which sector the authentication is performed. The card replies with a 32-bit nonce n_C generated by its internal Pseudo-Random Number Generator (PRNG). The reader replies with an encrypted nonce n_R and an answer a_R, which is generated by loading n_C into the PRNG and clocking it 64 times. For the encryption, the keystream generated by the Crypto1 cipher is used in groups ks_1, ks_2, ... of 32 bit each. After the card has sent the encrypted answer a_C, both parties are mutually authenticated. From that point onwards, the reader can read, write or modify blocks in the chosen sector. If another sector has to be accessed, the authentication procedure must be repeated with a slightly modified protocol.

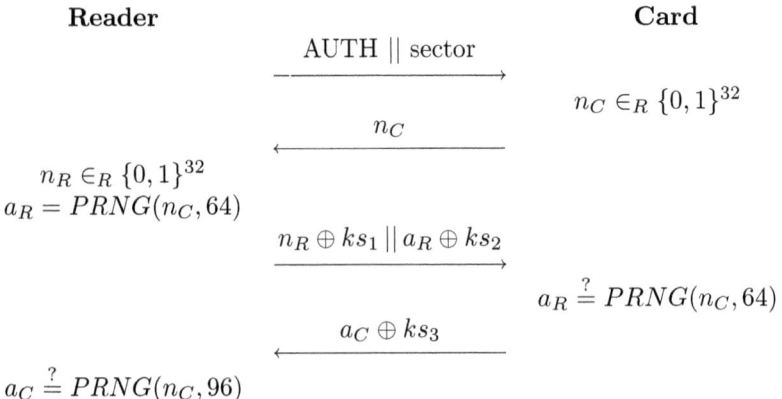

Protocol 1. The Mifare Classic authentication protocol

Security of Mifare Classic. Since its invention, the internal structure of Crypto1 was kept secret and no open review process was performed. The cipher and its PRNG were later recovered by [17] using low-cost hardware reverse-engineering techniques. The authors pointed out several design flaws, i.e., the short key length of 48 bit, mathematical weaknesses in the feedback functions of the LFSR, the weak 16-bit PRNG and the fact that the nonce generated by the PRNG depends on the time elapsed between power-up of the card and the authentication command. Subsequently, strong attacks on Mifare Classic were published: an attack described in [7] utilizes a fixed timing to generate the same nonces for repeated authentications and obtain parts of the keystream. A method to recover a secret sector key is proposed in [10], requiring two recorded genuine authentications to one sector. The most powerful attacks are card-only attacks as presented in [11] and [5]. They exploit amongst others the weakness that a card sends an encrypted NACK (`0x5`) each time the parity bits of the message $n_R \oplus ks_1 \,\|\, a_R \oplus ks_2$ are correct but the decrypted a_R is not (cf. Protocol 1). This reveals four bits of keystream with a probability of $\frac{1}{256}$. Finally, a secret key of a Mifare Classic smartcard can be extracted within seconds using a combination of card-only attacks as proposed in [16], hence the cards can be considered fully broken.

3.2 Mifare DESFire and DESFire EV1

Mifare DESFire and Mifare DESFire EV1 cards are compliant to Parts 1-4 of ISO 14443A. Their UID is seven bytes long, and they support high baud rates

Reader   **Card**

$$\text{AUTH (02 0A 00)} \longrightarrow$$
$$n_C \in_R \{0,1\}^{64}$$
$$\xleftarrow{\quad b_0 \quad} \qquad b_0 = Enc_{K_C}(n_C)$$

$$n_R \in_R \{0,1\}^{64}$$
$$b_1 = Dec_{K_R}(n_R)$$
$$r_0 = Dec_{K_R}(b_0)$$
$$r_1 = RotLeft_8(r_0)$$

$$b_2 = Dec_{K_R}(r_1 \oplus b_1) \qquad \xrightarrow{\quad b_1, b_2 \quad} \qquad r_2 = Enc_{K_C}(b_2)$$
$$n'_C = RotRight_8(r_2 \oplus b_1)$$

$$\xleftarrow{\text{ERROR (02 AE)}} \qquad \text{if } n'_C \neq n_C$$
$$\text{else if } n'_C = n_C$$
$$r_3 = Enc_{K_C}(b_1)$$
$$r_4 = RotRight_8(r_3)$$

$$r_5 = Dec_{K_R}(b_3) \qquad \xleftarrow{\quad b_3 \quad} \qquad b_3 = Enc_{K_C}(r_4)$$
$$n'_R = RotLeft_8(r_5)$$
$$\text{verify } n'_R \overset{?}{=} n_R$$

Protocol 2. The Mifare DESFire authentication protocol [4]

of up to 848 kBit/s. A communication with the cards can be performed in plain, with an appended Message Authentication Code (MAC), or with full data encryption. Mifare DESFire cards offer 4 kByte of storage and data encryption by hardware DES and 3DES encryption. Mifare DESFire EV1 cards additionally provide AES-128 data encryption and are sold in three variants with 2 kByte, 4 kByte and 8 kByte of non-volatile memory, respectively. Each card holds up to 28 different applications with up to 14 different keys per application. For DESFire, each application may contain up to 16 files, while for DESFire EV1 the maximum number of files is 32. As in Mifare Classic cards, the UID is unchangeably programmed into the card at production time. Depending on the access rights for each application a mutual authentication protocol (see Protocol 2 / Protocol 3), ensuring that the symmetric key of the card K_C and of the reader K_R are identical, has to be completed before reading and manipulation of the data.

Previous to the authentication, an application represented by its Application Identifier (AID) is selected. The reader starts the authentication protocol [4] with an authenticate command together with the key number that is to be used during the authentication. Note that Mifare DESFire cards only perform (3)DES encryptions $Enc_K(\cdot)$ employing the secret key K, hence, DESFire readers always have to use (3)DES decryption $Dec_K(\cdot)$.

As illustrated in Protocol 2, a DESFire card responds to the authentication command with an encrypted 64-bit random nonce n_C. The reader likewise chooses a 64-bit random nonce n_R, decrypts the received n_C, rotates it eight bits to the left and decrypts n_R as well as the rotated n_C. The card verifies if the rotated value equals n_C after reverting the rotation. If so, the card encrypts the first value to obtain n_R, rotates it eight bits to the right and encrypts the result which is then sent to the reader. The rotated and encrypted nonce is verified by the reader and if this final step is successful, both parties are mutually authenticated.

We furthermore reverse-engineered the DESFire EV1 authentication protocol, as presented in Protocol 3, by eavesdropping on genuine protocol runs. We found that the protocol of Mifare DESFire EV1 cards using AES-128 diverges from Protocol 2 as follows. In Protocol 3, en- and decryption are used in the common sense, i.e., data that is to be sent is encrypted and data that was received has to be decrypted. The CBC mode is modified in a way that all en- or decryptions are chained, even though they operate on different cryptograms. The Initialization Vector (IV) is not reset when en- or decrypting a new message, but instead depends on the previous en- or decryption. The nonces are extended to a length of 128 bit to match the block size of AES-128 and the second rotation is executed in the opposite direction on both sides. Again, AES-128 en- and decryption involving the key K are denoted by $Enc_K(\cdot)$ and $Dec_K(\cdot)$, respectively. Apart from that, the protocol equals the authentication protocol of Mifare DESFire cards and thus mutually authenticates both parties on successful execution.

$$\begin{array}{lcr}
\textbf{Reader} & & \textbf{Card} \\
& \underrightarrow{\text{AUTH (02 0A 00)}} & \\
& & n_C \in_R \{0,1\}^{128} \\
r_0 = Dec_{K_R}(b_0) & \underleftarrow{\quad b_0 \quad} & b_0 = Enc_{K_C}(n_C) \\
r_1 = RotLeft_8(r_0) & & \\
n_R \in_R \{0,1\}^{128} & & \\
b_1 = Enc_{K_R}(n_R \oplus b_0) & & \\
b_2 = Enc_{K_R}(r_1 \oplus b_1) & \underrightarrow{\quad b_1, b_2 \quad} & r_2 = Dec_{K_C}(b_1) \\
& & r_3 = Dec_{K_C}(b_2) \\
& & n'_C = RotRight_8(r_3 \oplus b_1) \\
& \underleftarrow{\text{ERROR (02 AE)}} & \text{if } n'_C \neq n_C \\
& & \text{else if } n'_C = n_C \\
& & r_4 = RotLeft_8(r_2 \oplus b_0) \\
r_5 = Dec_{K_R}(b_3) & \underleftarrow{\quad b_3 \quad} & b_3 = Enc_{K_C}(r_4 \oplus b_2) \\
n'_R = RotRight_8(r_5 \oplus b_2) & & \\
\text{verify } n'_R \stackrel{?}{=} n_R & &
\end{array}$$

Protocol 3. The Mifare DESFire EV1 authentication protocol

Security of Mifare DESFire / EV1. The non-invasive side-channel attacks on RFID devices presented in [15] allow to extract secret information from contactless cards by measuring the electromagnetic emanations of a card while it carries out a cryptographic operation. The focus is on devices that make use of DES or 3DES and the first successful key-recovery attack on such devices is accomplished. In a discussion with the authors, we came to know that the attacks have been improved since and are applicable to Mifare DESFire cards. With about $1\,000\,000$ measurements they are able to fully recover the 3DES key stored on a Mifare DESFire card. Note that their side-channel attack is currently *not* applicable to DESFire EV1, which has been certified according to Common Criteria EAL 4+. However, efficient attacks might come up in the future or the secret key could obtained by other means, e.g., by exploiting weaknesses of the backend system.

4   Software Implementations

In this section we detail on our software implementations for emulating several cards with Chameleon.

4.1   Mifare Classic Emulator

The attacks detailed in Sect. 3.1 imply that an adversary can easily read out the secret keys and all content of a Mifare Classic card. To produce a duplicate, the adversary can write all previously read data to a blank Mifare Classic card. This results in an almost perfect clone, differing from the original only in the single block containing the read-only UID of the blank card. If the UID is verified by a contactless system (compare with [16]), this type of card-cloning becomes useless in practice. To allow for perfect clones, we implemented the features of Mifare Classic on Chameleon. Thus, we have complete control of the content of every memory block, including the previously unchangeable manufacturer block.

Optimized Crypto1. A first approach to emulate a Mifare Classic card on an AVR ATmega32 microcontroller [22] revealed difficulties in complying to the timing requirements given in ISO 14443. After a command is issued by the reader, the card has to reply within 4.8 ms, or the reader will reach a timeout and abort the connection. Compiling the open-source Crypto1 C-library [6] for an 8-bit microcontroller results in inefficient code regarding the underlying platform. Hence, in [22] the time limit of 4.8 ms set in the protocol is exceeded with 11.7 ms for an 18-byte encryption, neglecting all other necessary computations, such as encoding the encrypted data. Since an 18-byte encryption is required every time when reading or writing a block with appended CRC checksum, the existing implementation is not suitable.

It became obvious that a significant speedup of Crypto1 is essential for a successful Mifare Classic emulation. Hence, we implemented the cipher from the scratch in AVR assembly. This allows to optimize the code for an 8-bit platform and make use of special commands that may not be considered by the C compiler. Using instructions to access bits of registers directly, the amount of clock cycles required for an encryption was reduced, amongst others by replacing inefficient shifting and masking operations to access single bits with instructions that allow accessing a particular bit in one clock cycle (e.g., SBRC, BST, BLD). We further implemented the non-linear filter functions f_a, f_b and f_c of Crypto1 with lookup tables to avoid time consuming boolean AND, OR and XOR operations. In the first stage, f_a is used two and f_b three times with a 4-bit input of the state LFSR. Their output is used to generate a 5-bit input to f_c, which in turn generates one bit of keystream. For both f_a and f_b, we created a dedicated lookup table that includes the respective shifting of the output. Thereby, the input of f_c can be easily obtained by ORing the five outputs of f_a and f_b. This speed advantage comes at the cost of storing one bit of information in one byte of memory. Finally, the lookup table f_c is a simple 5-bit input, 1-bit output table. The overall size of the lookup tables is 112 byte, formed by two 16-byte tables for f_a, three 16-byte tables for f_b and one 32-byte table for f_c. With respect to the 192 kByte size of the program memory, the tables are negligibly small.

Furthermore, we applied the idea of precomputation. When the nonce n_C is fixed before the authentication protocol is executed, the card is able to precompute the corresponding answers a_R and a_C which saves time during the authentication process. Precomputation of keystream bits is not possible because of two reasons. Firstly, since the sector to be accessed by the reader cannot be predicted, it is not clear which key has to be loaded into the LFSR. Secondly, the random reader nonce n_R that only becomes known during the authentication process is an input to the cipher.

4.2 Mifare DESFire (EV1) Emulator

Similarly to the Mifare Classic implementation, we additionally implemented the authentication protocols of both Mifare DESFire and Mifare DESFire EV1, as given in Sect. 3.2. For encryption, Mifare DESFire cards use DES/3DES in CBC mode, whereas Mifare DESFire EV1 cards can use either DES/3DES or AES-128 in CBC-mode.

4.3 Practical Results

Before carrying out security analyses in the real-world, we thoroughly tested our emulators in our laboratory. The reliability and accurate timing behaviour of our emulator was successfully verified with different RFID readers, including an ACG passport reader and a Touchatag [23] reader. Further tests with real-world systems are described in Sect.5.

Mifare Classic. With the optimized implementation of Crypto1 detailed in Sect. 4.1, we successfully emulated Mifare Classic 1K cards with varying content. Table 1 summarizes the execution times for the relevant operations which are now all well within the limits specified in ISO 14443. All features, e.g., authentication, encrypted read and write of blocks, or specifying an arbitrary UID, are fully functional with the used readers.

Table 1. Execution times of crucial Crypto1 functions

Command	Execution time	Explanation
setup_crypto1()	98 μs	Initializes the cipher
auth_crypto1()	542 μs	Keystream for the authentication
crypto1_1()	8.3 μs	Generates 1 bit of keystream
crypto1_8()	49 μs	Generates 8 bits of keystream
crypto1_32()	186 μs	Generates 32 bits of keystream

Mifare DESFire (EV1) Likewise, we tested our DESFire (EV1) emulations from Sect. 4.2. Table 2 shows the execution times for the needed cryptographic functions using the hardware accelerators of the ATxmega. Note that the first call to an en-/decryption function involves some overhead for the initial setup. After that, subsequent blocks can be processed faster. For reference, we included the runtime both for a single block and for ten data blocks in Table 2.

According to [4], an original Mifare DESFire card answers 690 μs (9356 clock cycles at 13.56 MHz) after b_1, b_2 was received when Protocol 2 is executed. During this time, two 3DES encryptions are performed (one encryption of two blocks

Table 2. Execution times of 3DES and AES-128 en-/decryption functions

Command	Block count	Execution time
TripleDES_CBC_Enc()	1 block	14.1 μs
TripleDES_CBC_Enc()	10 blocks	85.1 μs
AES128_CBC_Enc()	1 block	35.9 μs
AES128_CBC_Enc()	10 blocks	270.2 μs
AES128_CBC_Dec()	1 block	58.4 μs
AES128_CBC_Dec()	10 blocks	304.9 μs

and one encryption of a single block). Our implementation performs about three times faster than a genuine card, with 219 μs (5932 clock cycles at 27.12 MHz) to produce a valid answer b_3 after b_1,b_2 was received.

A genuine DESFire EV1 card replies with b_3 approx. 2.2 ms after having received b_1,b_2. In contrast, our implementation only consumes about 438 μs and is thus faster by a factor of five. As we are able to en-/decrypt faster than both DESFire cards, encrypting or MACing data which is the most critical part for Mifare Classic does not pose a problem in the context of emulating DESFire (EV1) cards. For both Mifare DESFire and Mifare DESFire EV1, our implementation performed successfully with the readers in our laboratory. As with the emulation of Mifare Classic cards, we are able to equip our emulator with a UID that is free of choice.

We conclude that the ATxmega microcontroller on our current hardware revision is powerful enough to handle the amount of computation that is needed for the emulation of the simple Mifare Classic cards and also for more sophisticated contactless smartcards using 3DES or AES.

5   Real-World Attacks

We successfully employed the Chameleon to bypass the security mechanisms of several real-world systems, for example, we utilized the Mifare Classic emulation to fake a card that is accepted by a widespread payment system. In the following, we summarize the characteristics of this system and then detail on the attacks carried out with our hardware.

5.1   A Vulnerable Contactless Payment System

For the identification of a customer of the payment system analyzed in [16], in addition to the UID each card contains a card number chosen by the system integrator. The credit balance is stored in plain in a value block on the card, without any extra security measures. The credit can be increased by means of cash or a credit card at charging terminals, while the cash registers are equipped with RFID readers to decrease the credit according to the balance due. The contactless cards furthermore allow to open doors and grant access to restricted areas.

The system can be easily spoofed, because all cards issued have identical secret keys. Hence, once the secret keys of one card have been recovered, the content of any card in the system can be read out or modified. The authors were able to carry out payments by copying the content of original payment cards to blank Mifare Classic cards. The so obtained cards are not exact clones, since the UIDs of the blank cards are different from that of the genuine ones, as detailed in Sect. 4.1. Consequently, the fraud could be easily detected in the back-end by verifying the correctness of the UID of a card on each payment.

The authors of [16] mention that the existence of a device that can fully clone a card including the UID would allow for devastating attacks, but suppose that these devices, if available, will be very costly so buying and using them for micropayments would not be profitable. With our developed hardware, the presence of an arbitrary valid card, e.g., an exact clone including the UID, can be simulated with minimal effort and cost, as shown in the following.

5.2   Electronically Spoofing a Contactless Payment System

A powerful type of attack that can be conducted with the Chameleon is called state-restoration. Even if the credit value was stored encryptedly on the payment card, e.g., using AES with an individual key per card, the content can be simply reset to the original credit value by dumping the full content of the card before paying and reprogramming the card (respectively our card emulation device) with the previous content after the payment.

As a first step to conduct this attack, we extracted the secret keys using the methods described in Sect. 3.1. Then, we dumped the content of a genuine card, including the UID, and copied it to our emulation device, thereby creating an exact clone. Hiding the device in a wallet, we consequently were able to carry out contactless payments. The credit value was stored in the EEPROM of our emulator and is decreased according to the balance due. As a result, the remaining credit displayed to the cashier appears to be correct and our device was accepted as genuine. The Chameleon allows to recharge the balance to its original value by restoring the initial dump when the attacker presses a push button. Finally, unlimited payments could be carried out with our device. Our practical tests furthermore showed that the Chameleon allows to open doors when cloning a valid card of an employee. However, if the fraud occuring due to the state restoration attack would be detected on the long term, the card number and/or the UID could be blacklisted and blocked for future payments.

For a more powerful attack, we programmed the Chameleon to generate a new random UID and card number for each payment. In our practical tests with the payment system, our emulator now appeared like a new card every time. Again, we were able to carry out payments, but this time, the device cannot be blacklisted and blocked in the backend.

In a similar manner, we were able to spoof a copy-and-print service that relies on contactless smartcards. The printers and copy stations are equipped with RFID readers that decrease the credit stored on the Mifare Classic card according to the amount of copies or printings carried out. By repeatedly using the service and comparing the content of the card between the payments, we found the block in which the amount of remaining credit was stored, again without any encryption. We hence programmed our card emulator to simulate the original card such that the credit appears to be lowered on each payment. However, the previous state of the card, i.e., charged to a high credit value, can again be restored by pressing a button on our hardware. As a consequence, we gain an unlimited amount of copies with our hardware.

Since cards of other customers can be read out from a distance[2], the Chameleon can also be used to clone their cards in a real-world scenario. Reading out the relevant sectors takes less than 100 ms. Several cards of other customers can be stored in the Chameleon and hence payments can be carried out with cloned cards that already exist in the payment system. Note that the original card of the customer remains unmodified and thus still contains the original credit value. Accordingly, a financial damage will only occur for the payment institution, while the customer is not affected. Altogether, taking the above illustrated devastating attacks and its low cost into account, the Chameleon can clearly be profitable for a criminal.

6   Conclusion

We present a microcontroller-based, freely programmable emulator for ISO 14443 compliant RFIDs that allows to simulate various contactless smartcards at a very low cost. The device works autonomously, operated from a battery, and its card-sized antenna fits into slots of most readers for contactless smartcards. Due to its small dimensions, the emulator can be used covertly, e.g., hidden in the purse, and is well-suited for real-world attacks. Our hardware can be connected to a PC by means of a USB interface and the non-volatile memory of the microcontroller allows amongst others to monitor the communication with an RFID reader and store the acquired data in order to reverse-engineer unknown protocols.

We exposed the protocol of Mifare DESFire EV1 cards, implemented the (3)DES and AES block ciphers as required, and present the first successful emulation of Mifare DESFire and DESFire EV1 cards in the literature. The current software further includes the emulation of Mifare Classic cards, based on a highly optimized variant of the Crypto1 stream cipher. The firmware of our device is not limited to Mifare cards but can be adapted to support other contactless smartcards and their respective protocols, e.g., the electronic passport and cards from other manufacturers.

We tested the emulations with different RFID readers and show that our implementations of the ciphers and protocols meet the timing requirements of all protocols and that the performance in most cases is even faster than that of original cards. In all our tests, the emulator could not be distinguished from a genuine card. The device proved to be a valuable tool for the security analysis of contactless technology and can be used to practically identify security weaknesses of real-world RFID systems.

---

[2] Modified RFID readers allow for reading distances up to 30 cm.

Since secret keys of Mifare Classic cards and Mifare DESFire cards can be extracted by means of mathematical cryptanalysis and side-channel analysis, respectively, our emulator poses a severe threat for many commercial applications, if it was used by a criminal. To demonstrate the capabilities of our findings we perform several real-world attacks, amongst others on a contactless payment system. We emulate exact clones (including the UID) of Mifare cards, successfully spoofed an access control system and carried out payments. Furthermore, we implemented a mode of operation in which our emulator appears as a new card with a new UID and new content on every payment, which hinders detection of fraud in the backend.

With contactless payment, ticketing and access control systems being omnipresent today, it is crucial to realize that only strong cryptography, together with sound protocol design and protection against implementation attacks can ensure long-term security. Bug-fixes for broken systems based on false assumptions on certain device characteristics, e.g., UID-based protection schemes for Mifare Classic, are a fatal design choice, as we demonstrate that exact cloning of cards is feasible at a very low cost.

References

1. Atmel. ATmega32 Data Sheet,
   http://www.atmel.com/dyn/resources/prod_documents/doc2503.pdf
2. Atmel. ATxmega192A3 Data Sheet,
   http://www.atmel.com/dyn/resources/prod_documents/doc8068.pdf
3. Atmel. AVR XMEGA A Manual,
   http://www.atmel.com/dyn/resources/prod_documents/doc8077.pdf
4. Carluccio, D.: Electromagnetic Side Channel Analysis for Embedded Crypto Devices, Diplomarbeit, Ruhr-University Bochum (March 2005)
5. Courtois, N.: The Dark Side of Security by Obscurity and Cloning Mifare Classic Rail and Building Passes, Anywhere, Anytime. In: SECRYPT 2009, pp. 331–338. INSTICC Press (2009)
6. Crapto1. Open Implementation of Crypto1 (2008),
   http://code.google.com/p/crapto1
7. de Koning Gans, G., Hoepman, J., Garcia, F.: A Practical Attack on the MIFARE Classic. In: Grimaud, G., Standaert, F.-X. (eds.) CARDIS 2008. LNCS, vol. 5189, pp. 267–282. Springer, Heidelberg (2008)
8. Federal Office for Information Security, Germany. Advanced Security Mechanisms for Machine Readable Travel Documents – Extended Access Control,
   http://www.bsi.de/fachthem/epass/EACTR03110_v110.pdf
9. Future Technology Devices International Ltd. FT245R Datasheet,
   http://www.ftdichip.com/Support/Documents/DataSheets/ICs/DS_FT245R.pdf
10. Garcia, F., de Koning Gans, G., Muijrers, R., Van Rossum, P., Verdult, R., Schreur, R., Jacobs, B.: Dismantling MIFARE Classic. In: Jajodia, S., Lopez, J. (eds.) ESORICS 2008. LNCS, vol. 5283, pp. 97–114. Springer, Heidelberg (2008)
11. Garcia, F., van Rossum, P., Verdult, R., Schreur, R.: Wirelessly Pickpocketing a Mifare Classic Card. In: Symposium on Security and Privacy, pp. 3–15. IEEE, Los Alamitos (2009)

12. IAIK Graz. HF Demo Tag,
    http://www.iaik.tugraz.at/content/research/rfid/tag_emulators
13. ISO/IEC 14443-A. Identification Cards - Contactless Integrated Circuit(s) Cards
    - Proximity Cards - Part 1-4 (2001), http://www.iso.ch
14. Kasper, T., Carluccio, D., Paar, C.: An Embedded System for Practical Security
    Analysis of Contactless Smartcards. In: Sauveron, D., Markantonakis, K., Bilas,
    A., Quisquater, J.-J. (eds.) WISTP 2007. LNCS, vol. 4462, pp. 150–160. Springer,
    Heidelberg (2007)
15. Kasper, T., Oswald, D., Paar, C.: EM Side-Channel Attacks on Commercial Con-
    tactless Smartcards Using Low-Cost Equipment. In: Youm, H.Y., Yung, M. (eds.)
    WISA 2009. LNCS, vol. 5932, pp. 79–93. Springer, Heidelberg (2009)
16. Kasper, T., Silbermann, M., Paar, C.: All You Can Eat or Breaking a Real-World
    Contactless Payment System. In: Sion, R. (ed.) FC 2010. LNCS, vol. 6052, pp.
    343–350. Springer, Heidelberg (2010)
17. Nohl, K., Evans, D.: Reverse-engineering a Cryptographic RFID Tag. In: USENIX
    Security Symposium, pp. 185–193 (2008)
18. NXP. About MIFARE (2001), http://mifare.net/about/
19. NXP. Mifare Classic 1K MF1 IC S50 Functional Specification (2008),
    http://www.nxp.com
20. OpenPICC. Programmable RFID-tag, http://www.openpcd.org/openpicc.0.html
21. Proxmark III. A Radio Frequency IDentification Tool, http://www.proxmark.org/
22. Silbermann, M.: Security Analysis of Contactless Payment Systems in Practice.
    Diplomarbeit, Ruhr-University Bochum (November 2009)
23. Touchatag. Touchatag RFID Reader, http://www.touchatag.com/
24. Verdult, R.: Proof of Concept, Cloning the OV-Chip Card,
    http://www.sos.cs.ru.nl/applications/rfid/2008-concept.pdf

A Authentication Protocols

This appendix provides the commands and the exact binary format for the authentication protocols used in this paper. Note that for DESFire (EV1), the message format according to ISO 14443A part 4 (including the 16-bit CRC) is taken into account in the following.

A.1 Mifare Classic Authentication Protocol

Table 3. Authentication protocol between a reader R and a Mifare Classic card C

#	Direction	Protocol Message	Explanation
1	R → C	60, sector (1 byte), CRC1 CRC2 (2 byte)	Auth ‖ sector ‖ CRC
2	C → R	4 byte	n_C
3	R → C	4 byte, 4 byte	$n_R \oplus \mathrm{ks}_1$ ‖ $a_R \oplus \mathrm{ks}_2$
4	C → R	4 byte	$a_C \oplus \mathrm{ks}_3$

A.2 Mifare DESFire Authentication Protocol

Table 4. Authentication protocol between a reader R and a Mifare DESFire card C

#	Direction	Protocol Message	Explanation
1	R → C	02 0A, key (1 byte), CRC1 CRC2	Auth ‖ key number ‖ CRC
2	C → R	02 AF, 8 byte, CRC1 CRC2	Card nonce ‖ b_0 ‖ CRC
3	R → C	03 AF, 8 byte, 8 byte, CRC1 CRC2	Reader response ‖ b_1 ‖ b_2 ‖ CRC
4	C → R	03 00, 8 byte, CRC1 CRC2	Success ‖ b_3 ‖ CRC

A.3 Mifare DESFire EV1 Authentication Protocol

Table 5. Authentication protocol between a reader R and a Mifare DESFire EV1 card C

#	Direction	Protocol Message	Explanation
1	R → C	02 AA, key (1 byte), CRC1 CRC2	Auth ‖ key number ‖ CRC
2	C → R	02 AF, 16 byte, CRC1 CRC2	Card nonce ‖ b_0 ‖ CRC
3	R → C	03 AF, 16 byte, 16 byte, CRC1 CRC2	Reader response ‖ b_1 ‖ b_2 ‖ CRC
4	C → R	03 00, 16 byte, CRC1 CRC2	Success ‖ b_3 ‖ CRC

B   Schematics

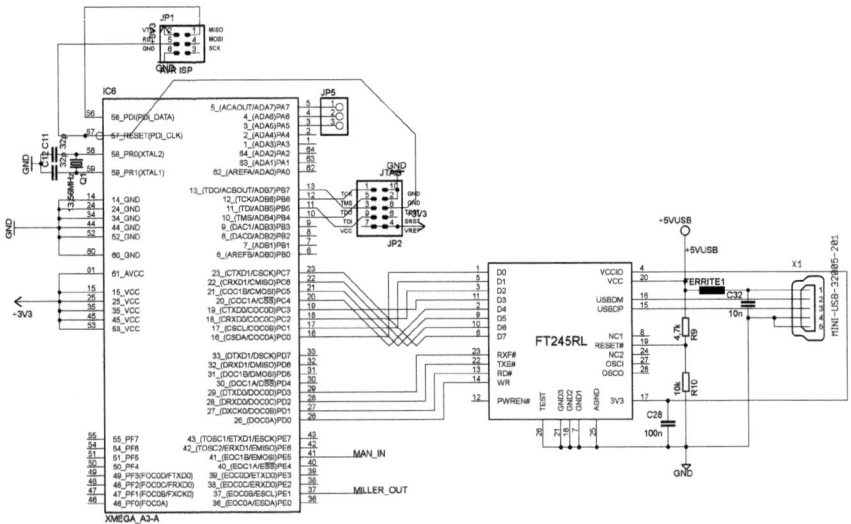

Fig. 3. Schematics of the microcontroller and the USB interface

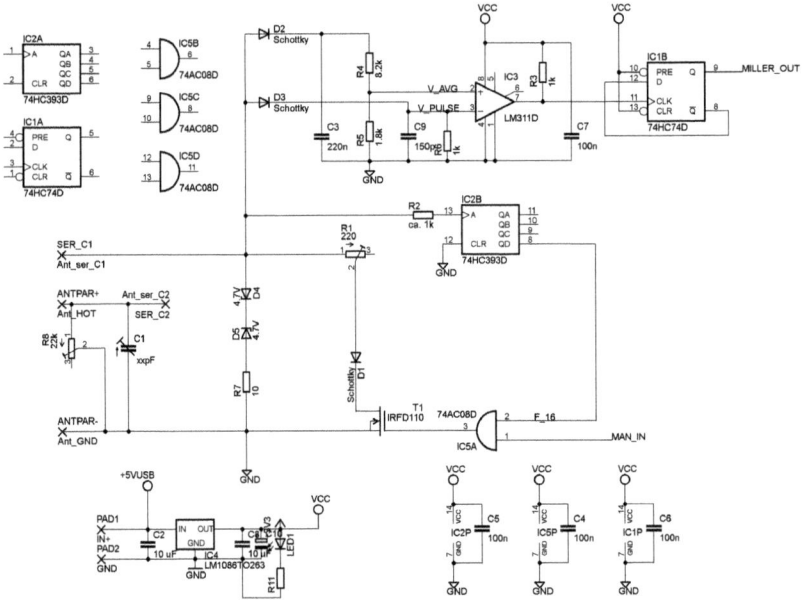

Fig. 4. Schematics of the power supply and the (de)modulation circuitry

Revisiting Address Space Randomization

Zhi Wang[1], Renquan Cheng[2], and Debin Gao[2]

[1] College of Information Technology and Science, Nankai University, China
[2] School of Information Systems, Singapore Management University, Singapore

Abstract. Address space randomization is believed to be a strong defense against memory error exploits. Many code and data objects in a potentially vulnerable program and the system could be randomized, including those on the stack and heap, base address of code, order of functions, PLT, GOT, etc. Randomizing these code and data objects is believed to be effective in obfuscating the addresses in memory to obscure locations of code and data objects. However, attacking techniques have advanced since the introduction of address space randomization. In particular, return-oriented programming has made attacks without injected code much more powerful than what they were before. Keeping this new attacking technique in mind, in this paper, we revisit address space randomization and analyze the effectiveness of randomizing various code and data objects.

We show that randomizing certain code and data objects has become much less effective. Typically, randomizing the base and order of functions in shared libraries and randomizing the location and order of entries in PLT and GOT do not introduce significant difficulty to attacks using return-oriented programming. We propose a more general version of such attacks than what was introduced before, and point out weaknesses of a previously proposed fix. We argue that address space randomization was introduced without considering such attacks and a simple fix probably does not exist.

Keywords: Address space randomization, return-oriented programming, software exploit.

1   Introduction

Address Space Randomization (ASR) has been proposed as a technique to fight against memory error exploits [2,3,4]. Most of these techniques obfuscate addresses in memory to obscure the location of code and data objects, including those on the stack and heap, static data, PLT, GOT, and etc. An attacker would then have a hard time finding out the addresses of code and data objects. This in turn makes the result of invalid memory access unpredictable. For example, randomizing the base of the stack and introducing random sized gaps between successive stack frames could make it difficult for an attack to locate or overwrite the return address; randomizing the locations of the PLT and GOT could make it difficult for an attack to access system functions such as `execve()` after subverting the program's control flow and therefore limit what a successful exploit could perform.

K.-H. Rhee and D. Nyang (Eds.): ICISC 2010, LNCS 6829, pp. 207–221, 2011.
© Springer-Verlag Berlin Heidelberg 2011

However, attacking techniques have advanced a lot since the introduction of address space randomization. In particular, return-oriented programming [14] has made attacks without injected code more powerful, in many cases able to perform arbitrary computation. This raises the question of whether randomizing certain code and data objects is still as effective as what we believed. In this paper, we show that randomizing the base and order of functions in shared libraries and randomizing the location and order of entries in PLT and GOT do not introduce significant difficulty to attacks using return-oriented programming. In particular, we present an attack on a system in which the library base addresses, the order of library functions, and the PLT and GOT are randomized. In the course of presenting the attack, we also detail a few improvements to return-oriented programming to make our attack more effective. We continue to show that a previously proposed fix of encrypting GOT might not work in many cases. We argue that address space randomization was introduced without considering such attacks, and a simple fix probably does not exist.

Note that what we study here is more than returning to randomized lib(c) as shown in a previous work [13]. Besides the attack we propose here being more general, i.e., we consider a system where the order of library functions are also randomized, we strive to study the effectiveness of randomizing various code and data objects rather than proposing a particular attack. We analyze the root cause of attacks using return-oriented programming, point out weaknesses of mitigation techniques in the previous work [13], and argue that randomizing such code and data objects are just ineffective and no simple fix exists. To support our analysis, we evaluate a number of commonly used application programs and show that encrypting GOT is, in fact, not effective in stopping the attack, since there are enough gadgets found in the binary program itself to exercise the attack and returning to libc is not needed.

We caution the readers from drawing from our analysis more than what it deserves. We are not trying to show that address space randomization is not effective in general. On the other hand, since there are many code and data objects that can be randomized, our analysis shows that randomizing some of these does not necessarily improve the system security because of the new attacking technique. Address space randomization is certainly effectively in, e.g., making it difficult for an attack to exploit a vulnerability to subvert the program's control flow. What we show in this paper is that after an attack manages to subvert the program's control flow, the difficulty of causing the program to execute in a manner of his choosing using return-oriented programming is not much affected by randomizing the base and order of functions or location and order of PLT and GOT.

In summary, the paper makes the following contributions.

- Propose and implement a general attack on an address space randomization system where the base and order of library functions and location and order of entries in PLT and GOT are randomized.
- Propose a few improvements to the return-oriented programming to make our attack more effective.

- Analyze limitations of the previously proposed attack mitigation technique of encrypting GOT.
- Discuss on the effectiveness of randomizing the base and order of functions and location and order of entries in PLT and GOT.

The rest of the paper is organized as follows. In Section 2, we outline the background and discuss some related work in this area. Section 3 presents an overview and intuition of our attack. We detail the implementation of our attack in Section 4. Section 5 discusses the limitation of a previously proposed attack mitigation technique and our experimental results on it, and discusses the implications. We conclude in Section 6.

2   Background and Related Work

There are many code and data objects that can be randomized [2,3,4]. Table 1 presents a summary of the important ones and the specific data to be randomized.

Table 1. Code and data objects to be randomized

Code and data objects	What to randomize
Stack-resident variables	Base of stack
	Gaps between stack frames
Heap-resident variables	Base of heap
	Gaps between heap allocations
Static variables	Order of static variables
Program code	Addresses of function call targets
	Position independent code
Functions in library	Base of library
	Order of functions in library
	Gaps between functions in library
Entries in PLT and GOT	Locations of PLT and GOT
	Order of entries in PLT and GOT

Randomizing these code and data objects is effective in stopping some particular types of attacks or steps in some attacks. In this paper, we try to analyze the effectiveness of randomizing some of these data in making attacks difficult. In particular, our analysis shows that randomizing functions in library and entries in PLT and GOT is ineffective. We support this by presenting our general attack on an address space randomization system and analyzing an attack mitigation technique previously proposed.

To understand how these randomization helps in making attacks difficult, we briefly describe the two steps an attack usually needs to perform. First, it needs to find a way to exploit the vulnerability to subvert the program's control flow. Second, it needs to cause the program to execute in a manner of his choosing. Traditionally, the first step could be done by overflowing a buffer on the stack and overwriting a return address, although many other techniques, e.g., heap [8] and integer overflows [18] and format string vulnerabilities [16], could be used.

The second step can be done by executing injected code [12] or performing a return-to-libc attack.

Address space randomization [2,3,4] and a variant of it [17] are proposed to make both steps discussed above difficult. For example, in order to overwrite a return address on the stack to subvert the program's control flow, an attack needs first to locate the return address. If the base of the stack is randomized, the location of the return address is no longer the same on different executions of the same program and therefore the attack will be difficult. A brief summary of the randomizing techniques to make it difficult to subvert the program's control flow follows.

- Introducing shadow stack for buffer-type variables;
- Randomizing the base of the stack and heap;
- Introducing random sized gaps between successive stack frames and heap allocations;
- Avoiding calls using absolute addresses by transforming them into function pointers.

Address space randomization can also make it difficult for an attack to perform arbitrary computation after the attack subverts the program's control flow. For example, making memory spaces non-writable or non-executable could stop injected code execution. Randomizing functions in the binary and shared library could make return-to-libc attacks difficult. Here is a summary of randomizing techniques to make this step difficult.

- Making certain memory spaces non-writable or non-executable;
- Randomizing the order of functions in the binary and shared libraries;
- Introducing random sized gaps and inaccessible pages between functions in the binary and shared libraries;
- Randomizing the order of static variables;
- Randomizing the location of PLT and GOT;
- Randomizing the order of entries in PLT and GOT;
- Uses position independent code in the program.

In this paper, we assume that an attack has successfully subverted the vulnerable program's control flow (first step), and try to evaluate how effective address space randomization is in making the second step difficult, i.e., in making it difficult for the attack to perform arbitrary computation.

Our attack uses the idea of return-oriented programming [14,6]. Return-oriented programming fits the requirement of the attack well because it does not need to execute any injected code. Only a large number of short instruction sequences from either the original program or libc is to be executed in order for the attack to perform arbitrary computation. However, our attack is more challenging than return-oriented programming on a normal (non-randomized) machine in that the addresses of the short instruction sequences are randomized and unknown to the attacker. Although return-oriented programming has been extended to a number of different environments [5,9,10,11,7], it is non-trivial how it can be applied on address space randomization systems.

Perhaps the work to surgically return to randomized libc [13] is the closest to our work in this paper. In this work, Roglia et al. introduced an attack on address space randomization assuming the base of the libc library is randomized. The attack surgically finds the address of a libc function by reading entries in PLT and GOT using return-oriented programming. The attack we present in this paper uses the same strategy, but differs in that it also assumes that the order of library functions are randomized. Roglia et al. also proposed an attack mitigation technique of encrypting the GOT. In this paper, we argue that such a technique might not work on programs where enough gadgets are found in the program binary itself and libc is not needed for the attack. We demonstrate this by analyzing a few commonly used application programs and show that an attack on them indeed does not require the use of libc. In general, this paper is not just about introducing an attack on address space randomization, but to study the effectiveness of randomizing certain code and data objects, and to argue that randomizing them is ineffective to defend against attacks using return-oriented programming, and a simple fix does not exist.

The effectiveness of address space randomization on 32-bit architectures has been analyzed previously [15]. In this work, a brute force attack is proposed to guess the libc text segment offset in order to perform a return-to-libc attack. Experiments show that such an attack is effective on a 32-bit system where the vulnerable service automatically restarts after crashing. Our attack is different from this attack in that we derandomize the addresses in an efficient way without brute forcing. Therefore, our attack has a wider application on systems where counter-measures are in place to fight against brute force attacks.

3   Attack on Address Space Randomization

As shown in Section 2, there are many code and data objects that can be randomized to make different attacks or attack steps difficult. Although return-oriented programming [14] has made attacks without injected code more powerful, in many cases able to perform arbitrary computation, intuitively it does not work well on address space randomization systems because the locations of gadgets are randomized and hard to be found.

In this section, however, we show that randomizing the base of the library, order of library functions, entries in PLT and GOT is ineffective in defending against attacks using return-oriented programming. We show this by presenting an attack on an address space randomization system where we assume that position independent code is not in use in the binary program. This assumption is valid in most existing computing systems because recompilation is needed to generate position independent code. We show that our attack is able to execute arbitrary computation after subverting the control flow of the program. This attack uses the same strategy of the one presented by Roglia et al. [13]. However, here we assume that the order of library functions is randomized whereas Roglia et al. only considers the randomized base address.

In the rest of this section, we first give an intuition of the attack we propose and an overview of the steps involved. In Section 4, we detail the implementation of the attack and a few improvements we introduce to make return-oriented programming more effective in our attack.

3.1   Attack Intuition

As many memory pages are made non-writable or non-executable in an address space randomization system, our attack tries to use existing code in the system to perform arbitrary computation. A typical way of performing such an attack is to use return-to-libc attacks to transfer control to system function `execve()`. Recall that we assume that the first step of the attack to subvert the control flow of the program, see Section 2, has been done. Therefore, the most important next step is to locate the address of a system call in existing code (e.g., in libc) and then transfer control over there.

Randomizing base address of the library and order of library functions. Randomizing the base address of libc and the order of libc functions are definitely effective in making our attack more difficult, since the address of these function has been randomized and cannot be pre-computed in our attack.

Randomizing entries in PLT and GOT. PLT (procedure linkage table) and GOT (global offset table) play crucial roles in resolution of library functions, and therefore is a potential target of our attack. As shown in Figure 1, GOT stores the address of libc functions, while PLT contains entries that jump to the addresses stored in GOT.

The dependency between randomizing PLT/GOT and randomizing library base address and functions was well documented — if an attacker knows the location and offsets of PLT, then the address of libc functions can be found even if the base address of libc and order of libc functions are randomized [4].

We have seen the dependency between randomizing libc and randomizing PLT/GOT because addresses of libc functions are used in PLT/GOT. By the same token, entries of PLT/GOT are used by other parts of the program, in particular, by `call` instructions in the code segment. If an attack can locate

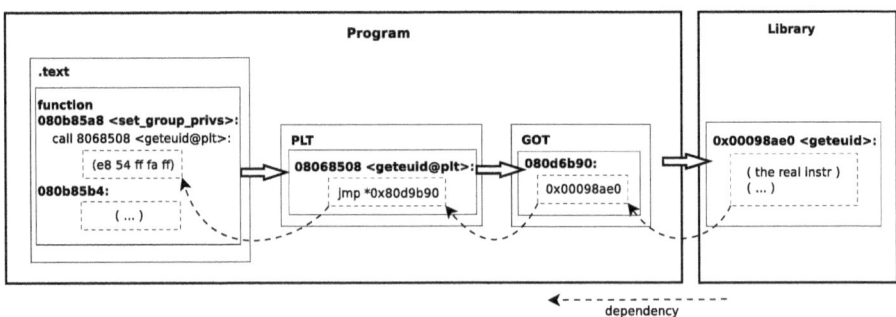

Fig. 1. PLT and GOT in a dynamically linked ELF executable

such `call` instructions in the program, theoretically the target of the `call` would reveal the location and offset in PLT/GOT, too. This analogy can also be seen from Figure 1.

Another way to look at such an attack is that no matter how well code and data objects are randomized, the randomized object would need to be accessible by the original program anyway to enable execution of the program. Addresses of libc functions are randomized, but the randomized addresses are used in PLT/GOT to allow libc functions to be called; by the same token, PLT and GOT can be randomized, but the randomized addresses are used in `call` instructions to allow functions to be called, too. If our attack is able to locate the `call` instructions and find out the target of the call, we can find the address of libc functions indirectly.

3.2   Attack Overview

To demonstrate the chain of dependencies, we propose our attack to perform arbitrary computation when the binary program does not make use of position independent code, i.e., when the attacker has access to the vulnerable program for static analysis. In such a scenario, the attacker can easily locate the `call` instructions by disassembling the code segment. However, finding out the (randomized) target of the call still remains nontrivial since it requires a memory read operation to be executed. Recall that 1) we assume that memory pages are non-writable or non-executable, and therefore executing injected code is not an option; 2) libc function addresses have not been found, and therefore return-to-libc is not an option either.

However, with the advances of return-oriented programming [14], such an attack becomes possible. Return-oriented programming fits the requirement of the attack well because it does not need to execute any injected code. Instead, it can make use of short instruction sequences from the original program (not the libc since the randomized libc addresses have not been found yet) to perform

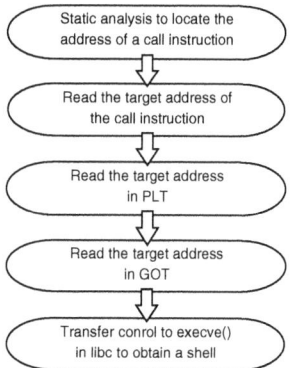

Fig. 2. Overview of our attack

the read operation (and some others; see Section 4). Figure 2 shows the steps involved in our attack.

After the control flow of the program is subverted (our assumption), our return-oriented programming code will first read the target of a `call` instruction whose address is known by static analysis of the vulnerable program. After that, we locate the address and offset through PLT and GOT. Once the entry in PLT and GOT is located, we read the entry to find out the corresponding libc function, and eventually we can use the short code sequences inside libc. In the end, the address of the libc function can be used to obtain a shell for arbitrary computation by making a system call. Note that our attack works well when the order of library functions is randomized, which a previously proposed attack does not consider [13].

4   Attack Implementation

As discussed in Section 3.2, there are a few steps involved in a successful attack, and each step requires some instructions to be executed. In this section, we first explain in more details what instructions are needed in each step, and then present a realization of executing these instructions using a few improvements to the return-oriented programming. We demonstrate our attack with an example on `apache-2.2.15`.

4.1   Instructions Needed to be Executed in Our Attack

The first step in our attack is to find the static address of a useful `call` instruction in the code segment of the vulnerable program. There are typically many `call` instructions in the code segment, and what we need is 1) one that calls a libc function; and 2) the corresponding libc function makes a system call. We need the second requirement in order to make sure that we can later make use of the system call to execute `execve()` for arbitrary computation. The one that we choose is `call geteuid` at `0x80b85af` in apache-2.2.15 (see Figure 3). Note that many other `call` instructions could be used.

Finding target address of the call instruction. As shown in Figure 3, the target address of the `call` instruction is represented as an offset (`0xfffaff54`) of

```
080b85a8 <set_group_privs>:
  80b85a8:    55                  push    %ebp
  80b85a9:    89 e5               mov     %esp,%ebp
  80b85ab:    53                  push    %ebx
  80b85ac:    83 ec 34            sub     $0x34,%esp
  80b85af:    e8 54 ff fa ff      call    8068508 <geteuid@plt>
  80b85b4:    ...
```

Fig. 3. call instruction in the code segment

the address of the next instruction (0x080b85b4). Therefore, in order to obtain the target address of the `call` instruction (0x08068508), our attack needs two instructions, i.e., a memory read instruction (at an address of our choosing) to read the offset, and an add instruction to add the offset to the address of the next instruction (static).

Finding jump target address in PLT. Every entry in PLT has 3 instructions that correspond to 16 bytes; see Figure 4. What we are interested in the jump target in is the first instruction, assuming that the program has been executing for a while and lazy linking has already initialized the address of the GOT entry in the first instruction. To find the jump target (0x08d06b90), we need another add instruction to find the address of the jump target (offset of 2 bytes at 0806850a) and another memory read instruction to read the jump target address.

```
08068508 <geteuid@plt>:
 8068508:      ff 25 90 6b 0d 08    jmp    *0x80d6b90
 806850e:      68 20 17 00 00       push   $0x1720
 8068513:      e9 a0 d1 ff ff       jmp    80656e0 <_init+0x30>
```

Fig. 4. Entry in PLT

Finding the address of the libc function in GOT This step is simple, as the jump target found in PLT contains exactly the address of the libc function; see Figure 5. Therefore, we need only a memory read instruction here.

```
080d6b90 <_GLOBAL_OFFSET_TABLE_+2972>: e0 8a 09 00
```

Fig. 5. Entry in GOT

Making a system call Once the address of the libc function (`geteuid`) is found, we can make a system call by transferring control to an instruction inside the libc function. Figure 6 shows the instructions inside `geteuid`, in which the fourth instruction `call %gs:0x10` is the new system call instruction in Linux. We first initialize four register values (`eax`, `ebx`, `ecx`, `edx`) and then transfer control to this instruction. So our attack in this step simply needs register initiation instructions.

4.2   Finding Gadgets to Realize the Instructions Needed

In this subsection, we outline how the instructions needed in our attack are realized by return-oriented programming [14]. The idea of return-oriented programming is to use gadgets (short code sequences ended by `ret`, or by `jmp <reg>` [6]).

```
00098ae0 <geteuid>:
 98ae0:      55                        push    %ebp
 98ae1:      89 e5                     mov     %esp,%ebp
 98ae3:      b8 c9 00 00 00            mov     $0xc9,%eax
 98ae8:      65 ff 15 10 00 00 00      call    *%gs:0x10
 98aef:      5d                        pop     %ebp
 98af0:      c3                        ret
```

Fig. 6. System call in libc

Note that in our attack, these gadgets have to be found in the vulnerable program except in the last step after the libc function address has been found. This makes our attack more challenging than return-oriented programming in general where useful gadgets can be easily found in the large libc library.

Since the vulnerable program is usually relatively small when compared to the libc library, we might not be able to locate the gadgets we want. We propose and use a few techniques to expand the set of useful candidate gadgets. We do not further discuss how the last step of our attack can be implemented by finding useful gadgets in libc since it has been well discussed in the return-oriented programming paper [14].

Alternative instructions. There could be multiple different instructions that serve what we need in the operations. Table 2 shows some candidate gadgets of different instructions for the same purpose needed in our attack. Note that they are just some examples, and each of them could have different variations, e.g., by using different registers.

Table 2. Useful gadgets with alternative instructions

Operations	Useful gadgets
Memory reading	<mov (%eax), %eax; ret;>
Addition	<add %ebp, %ebx; ret;>
	<lea (%eax, %ecx, 1), %eax; ret;>
Register writing	<pop %eax; ret;>
	<xchg %eax, %edx; ret;>

Combination of instructions. Besides using gadgets of different instructions, we can also combine different instructions (their corresponding gadgets) together to realize the intended operation. For example, <or (%eax), %ebx; ret;> or's the value at a memory address (specified by eax) with another register (ebx). It serves the purpose of memory reading if ebx happens to be zero. Even if ebx is not zero, this gadget can be combined with a register writing to set ebx to be zero first. Table 3 gives some examples of such combinations.

Instructions with side-effects. Some instructions in a gadget might have no effect in the execution context or might have side effects that can be reversed by other gadgets. Although these instructions (and the corresponding gadgets)

Table 3. Useful gadgets by combining instructions

Operations	Useful gadgets
Memory reading	`<register writing>` `<or (%eax), %ebx; ret;>`
Addition	`loop: <inc %eax; ret;>`

Register writing	`<mov $const, %eax; ret;>`
	`<lea ($const), %eax; ret;>`
	`<addition>`

make our analysis more complicated, taking them into consideration helps us find more useful gadgets. For example, in searching for gadgets to pop data from the stack to a register, we only managed to find `<pop eax; ret;>` and `<pop ecx; ret;>` directly from `apache-2.2.15`. After analyzing instructions with some side-effects, we managed to find `<pop ebx; pop ebp; ret;>` and `<pop edx; push eax; std; dec ecx; ret;>` with one and three instructions with side-effects in the middle, respectively.

4.3   Attacks on Apache and Other Programs

With the techniques discussed in Section 4.2, we search the binary code of `apache-2.2.15` and other programs to see if gadgets needed could be found using the Galileo algorithm [14]. The number of gadgets found for different operations are presented in Table 4.

Table 4. Number of gadgets found

Programs	Memory reading	Addition	Register writing
apache-2.2.15 (695 KB)	2	7	34
vsftpd-2.2.2 (116 KB)	1	3	47
bind-9.7.0 (486 KB)	3	1	17
sendmail-8.14.3 (806 KB)	1	4	14
mplayer-1.0~rc3 (4 MB)	5	19	117
firefox-3.6.3 (50 KB)	0	1	13

Table 4 shows that we manage find the needed gadgets from `apache`, `vsftpd`, `bind`, `sendmail`, and `mplayer`, while relatively small programs, e.g., `firefox`[1], may not provide enough useful gadgets.

To try out our attack on `apache-2.2.15` on a real system, we downloaded the address space randomization proposed by Bhatkar et al. and migrated the code to a PAX-enabled Ubuntu 10.04 desktop computer. We configure the system such that base address of the library, order of library functions, PLT and GOT are randomized. We then use `gdb` to overflow a buffer of `apache-2.2.15` on the stack with our attack code. The attack successfully creates a shell for arbitrary computation. Appendix A shows the shell code that we use in this attack. Since

---

[1] Firefox is a large program, but its binary file, /usr/lib/firefox-3.6.3/firefox-bin (under Ubuntu-10.04), is only of 50 KB as most functionality is provided in libraries.

it is possible to find the needed gadgets from various programs as shown in Table 4, we believe that our attack can be generalized to be applied on other vulnerable programs. We leave this as our future work.

4.4 Discussions of Our Attack

What we propose is a more general attack which works even when the order of library functions is randomized, which is different from a previously proposed attack [13].

Other considerations of our attack. In the discussions above, we have not considered a level of indirection address space randomization might have introduced, namely converting direct function calls to indirect ones with function pointers. Our attack works in the same way when function pointers are used; in fact, the attack could even be simplified in some cases because offsets might not be used in indirect calls.

Limitations of our attack. There are a few limitations of our attack. First, we assume that the control flow of the vulnerable program can be subverted. This might not be true as address space randomization could make such subverting very difficult. However, this assumption does not hinder our analysis less important because a security system should not rely on the single point of protection and should try to make attacks difficult even when the first line of defense fails. Second, we assume that the attacker has access to the vulnerable program to do static analysis and position independent code is not in use. Our attack relies on this assumption because we wouldn't be able to locate the `call` instruction should this assumption be invalid. Third, we might not be able to find enough useful gadgets from the vulnerable program. Although we have shown programs meeting our attack requirement, it remains future work to study other ways of finding useful gadgets to generalize our attack.

Extension of our attack. The idea of our attack could be extended to make stack randomization ineffective, if instructions like `mov eax, esp` could be found by using return-oriented programming. We tried using the Galileo algorithm [14] to search for it, but could not find one in our experiments. Theoretically, this is possible especially when searching on various sections that are marked executable, e.g., `.plt`, `.text`, `.fini`, `.rodata`, `.eh_frame_hdr`, and `.eh_frame`. We leave this as future work.

5 Possible Mitigation Techniques and Discussions

Roglia et al. proposed a few mitigation techniques to defend against attacks that dereference and overwrite GOT [13], which include using position independent code, self-randomization of the program, and encrypting GOT. Although such techniques could defend against our attack presented as well, we try to ask a deeper question: is address space randomization weak in randomizing GOT only and therefore becomes effective once the mitigation techniques are in place, or is it true that randomizing some of the code and data objects (e.g., base and order of library functions) is simply ineffective when return-oriented programming is used in an attack?

Before we try to answer this question, we first revisit our attack presented in Section 3 and Section 4 and see if exploiting GOT is the only way for the attack to succeed. The answer is definitely not. We try to derandomize the address of libc functions simply because the library has a larger code base which could be analyzed offline and usually contains more useful gadgets for return-oriented programming. However, in many cases, all an attack wants is simply to be able to make a system call (with values of the attacker's choice on a few registers), which might be possible with only gadgets from the vulnerable program itself without making use of the library. We perform an analysis on some commonly used application programs by using the Galileo algorithm [14] and our improvements on it (see Section 4.2) to search for gadgets that allow an attack to make a system call. Results (see Table 5) show that some programs, such as the vulnerable version of Ghostscript [1], could be attacked by only gadgets from the program.

In an attack using return-oriented programming with gadgets in the program binary only, even fewer gadgets could be required. For example, to execute the **execve()** system call, we only need to write four registers (**eax**, **ebx**, **ecx**, **edx**) and then execute the system call instruction. Only these two categories of instructions (and the corresponding gadgets) are needed. Table 5 shows the number of gadgets found for a few application programs.

Table 5. Number of gadgets found in some large programs

Programs	Register writing	syscall (int80 or call *%gfs:0x10)
gs-8.61 (11 MB)	34	130
mencoder-4.3.2 (8.7 MB)	47	5
emacs-23 (11 MB)	143	15
qemu-0.11.1 (2.1 MB)	23	10
qmake 2.01a (3.8 MB)	27	4

This shows that GOT is actually not the most important weaknesses in address space randomization in view of attacks using return-oriented programming. Rather, because address space randomization was proposed well before return-oriented programming was introduced, it was not designed to defend against return-oriented programming and therefore it is not surprising that the randomization of some of the code and data objects is simply not effective to defend against return-oriented programming. We argue that the randomization of base and order of library functions and the location and order of entries in PLT and GOT are typical examples. Mitigation techniques like encrypting GOT does not actually make address space randomization secure against return-oriented programming.

6   Conclusion

In this paper, we demonstrate our attack on randomizing the base address of library, order of library functions, and entries in PLT and GOT with return-oriented programming under the assumption that the attacker has a copy of the vulnerable program for static analysis. Besides introducing this more general attack and proposing improvements to return-oriented programming to make the attack more effective, we also evaluate an attack mitigation technique previously proposed. Results show that dereferencing GOT is actually not a necessary step in the attack, and therefore encrypting GOT does not make address space randomization secure against return-oriented programming.

References

1. CVE-2008-0411, Ghostscript (8.61 and earlier) zseticcspace() Stack-based Buffer Overflow Vulnerability
2. PaX (2001), http://pax.grsecurity.net
3. Bhatkar, S., DuVarney, D.C., Sekar, R.: Address obfuscation: an efficient approach to combat a broad range of memory error exploits. In: Proceedings of the 12th USENIX Security Symposium (USENIX Security 2003) (2003)
4. Bhatkar, S., Sekar, R., DuVarney, D.C.: Efficient techniques for comprehensive protection from memory error exploits. In: Proceedings of the 14th USENIX Security Symposium (USENIX Security 2005) (2005)
5. Buchanan, E., Roemer, R., Shacham, H., Savage, S.: When good instructions go bad: generalizing return-oriented programming to risc. In: Proceedings of the 15th ACM Conference on Computer and Communications Security, CCS 2008 (2008)
6. Checkoway, S., Davi, L., Dmitrienko, A., Sadeghi, A.-R., Shacham, H., Winandy, M.: Return-oriented programming without returns. In: Proceedings of the 17th ACM Conference on Computer and Communications Security, CCS (2010)
7. Checkoway, S., Feldman, A.J., Kantor, B., Halderman, J.A., Felten, E.W., Shacham, H.: Can dres provide long-lasting security? the case of return-oriented programming and the avc advantage. In: Proceedings of the 2009 Electronic Voting Technology Workshop/Workshop on Trustworthy Elections, EVT/WOTE 2009 (2009)
8. Solar Designer. JPEG COM marker processing vulnerability (2000), http://www.openwall.com/articles/JPEG-COM-Marker-Vulnerability
9. Francillon, A., Castelluccia, C.: Code injection attacks on harvard-architecture devices. In: Proceedings of the 15th ACM Conference on Computer and Communications Security, CCS 2008 (2008)
10. Hund, R., Holz, T., Freiling, F.C.: Returnoriented rootkits: Bypassing kernel code integrity protection mechanisms. In: Proceedings of the 18th USENIX Security Symposium (USENIX Security 2009) (2009)
11. Kornau, T.: Return oriented programming for the arm architecture. Master's thesis, Ruhr-University Bochum, Germany (2009)
12. Aleph One. Smashing the stack for fun and profit. Phrack magazine (1996), http://www.phrack.com/issues.html?issue=49&id=14
13. Roglia, G.F., Martignoni, L., Paleari, R., Bruschi, D.: Surgically returning to randomized lib(c). In: Proceedings of the 25th Annual Computer Security Applications Conference, ACSAC 2009 (2009)

14. Shacham, H.: The geometry of innocent flesh on the bone: return-into-libc without function calls (on the x86). In: Proceedings of the 14th ACM conference on Computer and Communications Security, CCS 2007 (2007)
15. Shacham, H., Page, M., Pfaff, B., Goh, E., Modadugu, N., Boneh, D.: On the effectiveness of address-space randomization. In: Proceedings of the 11th ACM Conference on Computer and Communications Security, CCS 2004 (2004)
16. Scut/team teso. Exploiting format string vulnerabilities (2001),
    http://team-teso.net
17. Xu, J., Kalbarczyk, Z., Iyer, R.K.: Transparent runtime randomization for security. In: Symposium on Reliable and Distributed Systems, SRDS (2003)
18. Zalewski, M.: Remote vulnerability in ssh daemon crc32 compensation attack detector (2001) (Bindview)

A   Shell Code of Our Attack

```
00000000   e8 01 05 08 d0 85 0b 08   3d fe 06 08 bf bf bf bf
00000010   3d fe 06 08 bf bf bf bf   ac c6 0c 08 ac c6 0c 08
00000020   3d fe 06 08 bf bf bf bf   3d fe 06 08 bf bf bf bf
00000030   ac c6 0c 08 ac c6 0c 08   ac c6 0c 08 ac c6 0c 08
00000040   ac c6 0c 08 ac c6 0c 08   ac c6 0c 08 ac c6 0c 08
00000050   51 63 0a 08 e8 01 05 08   58 f4 ff bf 51 63 0a 08
00000060   66 b4 08 08 bf bf bf bf   e8 01 05 08 08 80 04 08
00000070   3d fe 06 08 bf bf bf bf   51 63 0a 08 e8 01 05 08
00000080   60 f4 ff bf 51 63 0a 08   66 b4 08 08 bf bf bf bf
00000090   e8 01 05 08 60 f4 ff bf   51 63 0a 08 a8 02 05 08
000000a0   5c f4 ff bf c2 85 06 08   64 f4 ff bf bf bf bf bf
000000b0   e8 01 05 08 5c 82 04 08   3d fe 06 08 bf bf bf bf
000000c0   bf bf bf bf 64 f4 ff bf   bf bf bf bf 2f 62 69 6e
000000d0   2f 73 68 00
```

Evaluation of a Spyware Detection System Using Thin Client Computing

Vasilis Pappas, Brian M. Bowen, and Angelos D. Keromytis

Department of Computer Science, Columbia University
{vpappas,bmbowen,angelos}@cs.columbia.edu

Abstract. Spyware – malicious software that passively collects users' information without their knowledge – is a prevalent threat. After a spyware program has collected and possibly analyzed enough data, it usually transmits such information back to its author. In this paper, we build a system to detect such malicious behaving software, based on our prior work on detecting crimeware. Our system is specifically designed to fit with thin-client computing, which is popular in some corporate environments. We provide implementation details, as well as experimental results that demonstrate the scalability and effectiveness of our system.

Keywords: Spyware, Thin Client Computing.

1 Introduction

Spyware has traditionally targeted individual consumers for purposes of conducting fraud and identity theft. Much of the defense has typically been left to anti-virus software operating on individual consumers' PCs and the financial institutions themselves who monitor for suspicious activity in an attempt to mitigate financial loss. More recently, the enterprise as has become the target [16] for spyware where the attackers' goal is to pilfer corporate information including webmail accounts, VPN accounts, and other enterprise credentials. One study conducted by RSA's FraudAction Anti-Trojan division found that almost all Fortune 500 companies have shown activity from the Zeus Trojan [12], one of the largest botnets. Given that many existing trojans and malware samples evade detection by traditional anti-virus software most of the time [12], there is demand for new approaches that can be applied at scalable levels within an enterprise.

In prior work [4], we developed a system that was designed to detect spyware proactively through the use of tamper resistant decoys. The system is intended to complement traditional signature and anomaly based defense systems rather than replace them. The system works by injecting decoys made up of monitored information that triggers alerts during exploitation. The system makes the malware's task significantly harder by requiring it to distinguish real actions from simulated actions to in order to avoid decoys. We demonstrated the system's ability to detect spyware using various types decoy credentials including those

K.-H. Rhee and D. Nyang (Eds.): ICISC 2010, LNCS 6829, pp. 222–232, 2011.
© Springer-Verlag Berlin Heidelberg 2011

for PayPal, a large bank, and Gmail. The implementation relied upon an out-of-host software agent to drive user-like interactions in a virtual machine, seeking to convince malware residing within the guest OS that it has captured legitimate credentials. The system successfully demonstrated that decoys can be used for detecting spyware on a single host.

In this work, we explore and demonstrate the scalability of the approach across many hosts, making this work applicable to enterprise environments. Specifically, we address threats within a thin-client based environment and propose a novel architecture for bait injection on thin clients. The maturity of thin-clients has increased their usage in corporate computing environments, making this approach especially applicable [9,7]. In this system, we rely on virtualized mouse and keyboard devices to inject decoy actions and credentials to an innumerable number of hosts with very low network and CPU overhead.

In summary, the contributions for this work include:

- An extension of an already proven system that aims to proactively detect malware on a single host to one that scales to service any number of hosts.
- A thin-client based architecture that supports the injection of bait information to and from a scalable number of servers and clients.
- A demonstration of the thin-client based architecture showing that it provides reasonable performance.
- The results of experiments that examine how these new systems induce malware to exfiltrate information.

Organization: Section 2 presents previous work, related to ours. In Section 3 we describe our original system and we detail our new scalable architecture based on thin client computing. We then present our evaluation results in Section 4 and conclude in Section 5.

2   Related

The use of manually injected human input for generating network requests has been shown to be useful by Borders *et al.* [3] for detecting malware. The aim of their system is to is to thwart malware that attempts to blend in with normal user activity to avoid anomaly detection systems. Chandrasekaran *et al.* [5] expanded upon this system and demonstrated an approach to randomizing generated human input to foil potential analysis techniques that may be employed by malware. Work by Holz *et al.* [8] investigated keyloggers and dropzones, relied on executing maleware in CWSandbox [13] and automating user input with AutoIt[1] for the purpose of detecting harvesting channels. Since AutoIt resides within the host, attackers are provided with a simple means of detecting and avoiding it. In prior work, we demonstrated a platform for the automatic generation and injection of bait information designed to convince malware it has captured legitimate credentials [4]. In addition, we adapted our original system

---

[1] http://www.autoitscript.com

to personal workstation environments where the convenience of virtualization is usually absent [10]. In contrast to all prior work, this effort is focused on designing a system for the large-scale injection of decoys to detect malware that may otherwise go undetected.

Taint analysis is another technique that has been used to detect credential stealing malware [6,15]. This approach works well, but does so with a cost of a 10-20 times slowdown. Taint analysis systems also contain components that reside on the guest, which is undesirable because they can be used by malware to detect and elude the injected decoys. Our system aims to be undetectable by malware residing within so that it is not easily avoided.

The authors of [14] evaluated a number of different remote screen protocols. Although this is not directly related to the goals of our system, it is closely related to the evaluation of our system's application to thin clients.

3   Architecture

In this section, we begin by briefly presenting the goal and architecture of our original system. We then detail an architecture that demonstrates how the same approach can be scaled to handle a large number of hosts in a thin client environment, which is achieved by exploiting its centralized computation nature.

3.1   Original System

The ultimate goal of our technique is to detect crimeware using tamper resistant injection of believable decoys. In summary, we can detect the existence of credential stealing malware by (i) impersonating a user login to a sensitive site (using decoy credentials) and (ii) detecting whether this specific account was accessed by anyone else except for our system. That would be a clear evidence that the credentials were stolen and somebody tried to check the validity and/or the value of that account. Our technique depends on the following properties:

- **Out-of-host Detection.** Our system must live outside of the host to be protected. This prerequisite is for the tamper resistance feature of our system.
- **Believable Actions.** The replayed actions must be indistinguishable by any malware in the host we protect so as to not be easily eluded.
- **Injection Medium.** There must be a medium, able to transmit user like actions (mouse, keyboard, etc.) to the protected host.
- **Verification Medium.** Optionally, but highly preferable, there should be a medium that can be used to verify the injected actions. This can actually be the same medium as above, if possible.

Our original system's implementation was on a personal VM-based environment. More precisely, in order to fulfill the *Out-of-host Detection* requirement, our

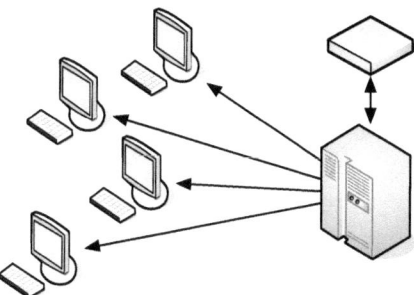

Fig. 1. Thin client environment – our system is on the top left corner

system resided on the host operating system and operated on the guest operating system(s). To verify the *Believability* of the replayed actions, we conducted a user study which concluded that the actions generated by our system were indeed indistinguishable. Moreover, as an *Injection Medium*, we utilized the X server of the host operating system to replay the actions. Finally, by slightly modifying the component of the virtual machine manager that was responsible for drawing the screen, we were able to verify the actions by checking the color value of selected pixels.

The original system relied on a language for the creation of believable actions. It is worth noting here that the approach is generic enough to be used as-is in the application bellow. This stands because the injection medium is flexible enough to support replaying of believable actions, although there could be cases where the believability of the actions can be degraded due to artifacts of the injection medium itself.

3.2 Thin Clients

The environment we chose to apply our technique to is thin clients, which, although they have been around for a long time, they are recently becoming more and more prominent in corporate networks. The main benefits of choosing such a setup are low cost, easy maintenance and energy efficiency.

A typical thin client setup consists of two main components: (i) a central virtual machine host (can be one physical server or more) and (ii) a collection of "dummy" computers connected to that host over a local and fast network. All the computation is offloaded to the central server, leaving the user terminals responsible only for transmitting user actions (keyboard, mouse, etc.) and remotely displaying the screen output of the virtual machine. Each user is then able to access and use virtual machines hosted on the central server, using these terminals (thin clients).

The application of our technique in this case was straightforward. In summary, we deployed our system like an ordinary thin client that periodically connects to each hosted virtual machine and injects decoy credentials. It is trivial to show that this type of application satisfies all the properties, previously introduced.

First, the *out-of-host* property is covered by deploying our system as a thin client and not inside the VMs under protection. Second, all the remote access protocols used in thin client environments provide a medium both for *injection* and *verification*. Figure 1 depicts what we previously described. On the lower right corner is the central server, on the left side, the thin clients and on the top right corner, our system. As our system only needs to communicate with the central server, we can safely adjust its proximity to it, reducing network overhead imposed on intermediate links.

In our prototype implementation, we assumed that there is a Linux version of the client part of the remote access protocol. For instance, in our evaluation (Section 4.1) we used VNC [11], which is a standard remote access protocol. Although this is not a requirement, it greatly improves scalability, because it allows us to easily initiate many remote access sessions, concurrently. Overall, the implementation was similar to our original system with the primary exception being that we leveraged out-of-the-box tools, as opposed to customizing. The main motivation behind that was to make our system as generic as possible and thus easily portable to other remote access protocols. More precisely, we used a vanilla version of GNU Xnee[2] for the injection of the previously recorded believable user actions, both mouse and keyboard. These actions were injected in a full screen view of the client side remote access software, Xvnc here. For the verification, we used the ImageMagick software suite[3]. More specifically, we made use of the `import` utility in order to grab arbitrary portions of the screen and the `compare` utility, to count the absolute number of different pixels. Finally, in order to enable the capability of concurrently injecting to multiple virtual machines, and thus the scalability of the system, we leveraged the Virtual Frame buffer (part of the X server). By doing this, we could simultaneously execute many full screen remote access sessions, each in a distinct X server (using the `xvfb-run` utility).

4   Evaluation

Our evaluation is divided in three parts, Subsection 4.1 examines the performance and scalability factors of our technique, when applied to a thin client environment. Next, we present the results on an exfiltration study we did using a relatively large number of malware samples. Finally, we discuss some real "hits" we had during the evaluation of our system.

4.1   Performance

In order to evaluate the performance and scalability of our system in a thin client setup, we set up such an environment in our lab. Using that as a testbed, we measured both the overhead and the limits of our system.

More precisely, we used three Dell PowerEdge R410 servers, each having 8 CPU cores, 24Gb of memory and 1 TB of storage. For the virtualization layer,

---

[2] Website: http://www.gnu.org/software/xnee/
[3] Website: http://www.imagemagick.org/

we chose to use Xen[2,1] because it has built-in remote guest access through VNC. We installed the Xen hypervisor 4.0.0 on top of Ubuntu 10.04 server edition. On each server we hosted 32 virtual machines, running Windows XP SP2 as their guest operating system. In total, our setup was comprised of 96 virtual machines. Our prototype was also running on a virtual machine (on top of a different host), with just one CPU and 1 GB of memory.

Memory: The amount of memory required by our system is proportional to the number of concurrent sessions. Each virtual frame buffer consumes its number of pixels times the number of bytes to encode the color for each of them. For example, during our evaluation, the screen settings on the Windows guests were set to 800x600 pixels using 32-bit colors. This equals to $800 * 600 * 4 = 1,920,000$ bytes, or $\sim$2 MB. The total memory consumption for the whole 96 VM set is $\sim$176 MB.

Scalability: In the first part of our evaluation we examine the scalability of our system. In order to do that, we monitored both the network and CPU utilization, under various workloads – in terms of simultaneous injections. More precisely, the different workloads we used were 24, 48, 72 and 96 concurrent injections using our bait credentials. As for the VNC settings, we used the default values (full color and *hextile* encoding).

Figure 2(a) shows the CPU utilization under each workload. In this figure, we observe two expected things. First, the CPU load is proportional to the number of concurrently replayed sessions. Second, we notice an increase in the total duration. This increase is the result of failed verification attempts, which leads to more wait periods. These verification failures are caused both because of the virtual machine host's high load and network level congestion which causes poor refresh rates in VNC.

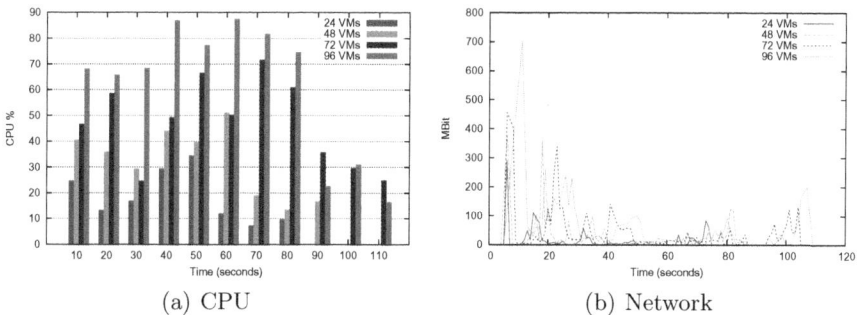

(a) CPU   (b) Network

Fig. 2. CPU and network utilization when simultaneously replaying to 24, 48, 72 and 96 VMs (using full color and HEX encoding). As expected, both metrics are proportional to the number of the VMs.

The other resource we measured, in order to analyze the scalability of our system, is network utilization. Figure 2(b) shows the total network usage, under different workloads. In general, we see that network usage is high in the beginning of the injection sessions (first 30 seconds) and decreases afterwards. This is caused because VNC transmits only the portions of the screen that have been changed. In the very beginning, the whole screen has to be transmitted (first peak) and right after, Internet Explorer is started in maximized mode (high usage around the 20th second). Although the network utilization may seem forbiddingly high at times, we have to keep in mind that (i) we try to measure the scalability in the worst case scenario – that is all the injection sessions are initiated simultaneously – and (ii) this is a prototype unoptimized implementation, using of-the-self tools. The most important thing to keep from this measurement is that our system, even under these conditions, was robust enough to sustain and adapt to the workload increases.

Optimizations: After we demonstrated the scalability and adaptability to resource variations, we experimented with application level optimizations. Although we could achieve much better overall performance by developing custom injection and verification tools, we wanted to examine the benefits of tweaking parameters of the remote access protocol – VNC in this case. There are two such parameters that are related to the quality of the transmitted screen view. These are: (i) color depth and the encoding algorithm used. The different options for color depth are: 8, 256 or full colors. Each time something has changed on the screen, VNC transmits the surrounding rectangle of that portion, encoded in one of the following ways:

- RAW. This is the simplest out of all the encoding schemes. As its name implies, rectangles are transmitted in *width* x *height* pixel values.
- HEXTILE. In this case, the rectangles to be transmitted are firstly partitioned in 16x16 tiles. Then, each of them is either sent raw (as above) or using a variant of Rise-and-Run-length-Encoding, where a sequence of identical pixels are compacted to a single color value and repeat count.
- ZRLE. Finally, this encoding scheme combines a form of the previous one with Zlib compression.

In order to measure the benefits and tradeoffs of the different encodings and color depths, we evaluated four typical combinations. These were full color-RAW, full color-HEXTILE, 8 colors-HEXTILE and 8 color-ZRLE. For each combination, we concurrently injected bait credentials to the whole VM set – the 96 of them. As before, we collected CPU and network utilization statistics. Figure 3(a) shows the CPU usage under the different encoding-color depth pairs. Using full color yields slightly higher CPU utilization, but, overall the benefit seems negligble. On the other hand, network utilization (shown in Figure 3(b)) is indeed affected by the different encoding-color depth combinations. As expected, using full color and RAW encoding is the most network demanding scheme. Switching to HEXTILE encoding clearly results to a first improvement. Finally, lowering the color depth reduces network utilization even more. It is interesting to see that the encoding scheme does not play such a big role when using just a few colors. Hence, it would be sufficient to use even HEXTILE instead of ZRLE, in order to save a few CPU cycles.

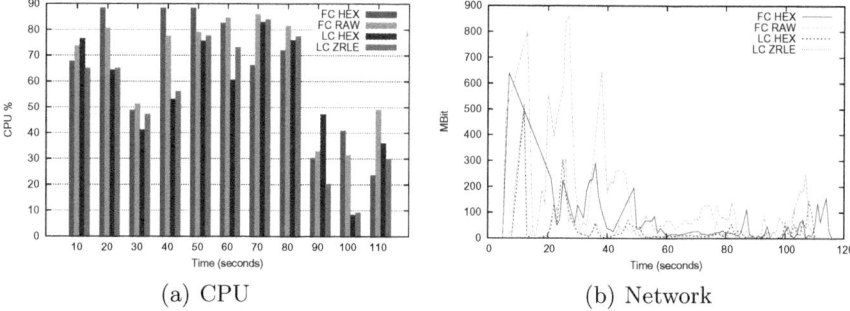

(a) CPU   (b) Network

Fig. 3. CPU and network utilization when replaying to all 96 VMs, using different combinations of color depth and encoding schemes. RAW encoding is clearly the most demanding. As for the low color depth ones, there is no big difference between HEX and ZRLE.

4.2 Exfiltration Study

In order to demonstrate the threat posed by credential stealing spyware, we conducted a study using a relatively large number of distinct samples. For the purposes of our study, we used variations of the Zeus (also known as ZBot) malware which is notorious for its credential stealing capabilities. All the samples were downloaded from Zeus Tracker[4].

In previous work, we also did provide a similar study, but somehow more limited, as each malware sample was only active on a VM for a small amount of time – order of a few tens of minutes. In our current study, we installed each malware sample on a separate VM, running on the virtualized infrastructure we built in order to simulate a thin client environment. By keeping each malware active for a relatively long period (weeks or even months) we want to explore two probable phenomena, not covered by our previous study. Firstly, we want to examine whether there are malware instances that wait for a period of time before exfiltrating the stolen credentials, and secondly, it would be interesting to see whether instances not intended to exfiltrate, get updated in a later time to do so. Both of these cases, if existent, would require a larger time window than our previous study, to happen.

We bootstrapped the study by installing all the malware samples available at the Zeus Tracker, and also we automated the procedure of installing new samples as they are made available. In total, during the study there were 108 Zeus malware instances installed on distinct VMs running on our Dell servers for a period of 3 to 4 weeks. During that time, we periodically injected both Paypal and anonymous bank's bait credentials. The component that monitors for external login attempts to the bait accounts was running for the next few months, as login attempts can occur even months after the credentials are stolen – based on our previous study. Along with the injections, we also monitored the

---

[4] Site: https://zeustracker.abuse.ch

Table 1. Top10 domain names / IP addresses that malware communicate with (left). Top10 script names that exfiltrated data are "dropped" to (right).

#	Domain / IP address	Count	Dropzone Script	Count
1	varxx.com	29808	/xt/gate.php	29808
2	nevereversite.ru	18890	/gate321.php	18890
3	95.224.124.151:555	17101	/temp/stuk.php	17820
4	65.60.36.114	13218	/~ataactc1/z/gate.php	13218
5	podgorz.org	9599	/zuo/zsweb_cleaned/gate.php	9599
6	iesahnaepi.ru	8042	/y93/_gate.php	6238
7	wifahquaht.ru	4763	/cp11/zengate.php	4243
8	community.infinitie.net	3436	/cp01/zengate.php	2945
9	esvr3.ru	2945	/k1o/_gate.php	2892
10	phaizeipeu.ru	2702	/cache/lang_cache/web/s.php	2888

network traffic in order to see which of the malware samples have already started exfiltrating data.

Even in a such a short time period, we already encountered thousands of suspicious data transmissions. More precisely, we saw that from 74 out of the 108 VMs, outbound HTTP POST messages were transmitted to websites other than the ones we are navigating to while injecting, or even to raw IP addresses. These are most probably drop zones for the credentials stolen by the malware samples and/or configuration or command updates. In total, we recorded 134,302 such requests. The body of each POST message is in binary format, most probably encrypted in some way. Table 1 contains both the top 10 host names / IP addresses that exfiltrated data were sent to and the top 10 script names in the POST messages that handle the data, along with the number of times they appeared in our logs. By examining the counter values on both lists, we see that there are cases where there is an one to one match between host names and script names. After looking into these cases, we saw that these script names were only accessed on these host names. On the other hand, in the rest of the cases, where host name counters do not match script name counters, some scripts with the same name were installed on different hosts and some host names had more than one scripts installed.

4.3   Feasibility Study

In total, we encountered two hits on the bait accounts from the 108 installed malware samples (described in the previous subsection). The first one was on an account from the anonymous bank, after 26 days. The second hit was a Paypal account access almost two months after (57 days). These results show that our technique is indeed effective, which does validate that our new architecture is working.

As far as the number of hits is concerned, it does raise some interesting questions. On one hand, it could be normal for only a ~2% of the accounts to be accessed. Some of the dropzones could be inactive or offline. Or, some malware samples may be unable to steal the accounts from the financial services we used, or their owners were not interested to these type of accounts, etc. On the other hand, the low hits percentage could be due to the nature of our study. One thing that we have to keep in mind is the fact that all the malware samples we used were downloaded from Zeus Tracker. As the attackers get more and more sophisticated and cautious, it would be no surprise to us that they could discard any credentials reported by malware samples that have been published in sites like Zeus Tracker. Similarly, as our main goal was the performance and scalability evaluation, the injection of the bait credentials was periodical and simultaneous to all the accounts and all the VMs were connected to the Internet through a single public IP address (NAT). It would be trivial for an attacker with several malware instances to filter out our credentials as suspicious, because they are all reported from the same IP address, periodically and simultaneously.

5   Conclusion

We presented the application of our spyware detection technique for a common setup in multiuser enterprise environments. We demonstrated it for thin client environments where we utilized out-of-the-box tools to implement our tamper resistant bait injection and action verification. The system was designed to be generic and portable to different remote access protocol stacks to make it generally applicable.

We experimentally demonstrated the scalability of our system when applied to a thin client environment. Our results showed that our system can successfully operate concurrently on a scalable number of VMs. Finally, the study we conducted using more than a hundred of malware samples revealed a number of different relationships between the malware samples and the dropzones. In addition, the relatively small number of bait account accesses from the attackers raises some interesting questions about their sophistication.

Acknowledgements. This work was supported by the NSF through Grants CNS-09-14312 and CNS-04-26623, and ONR through MURI Contract N00014-07-1-0907. Any opinions, findings, conclusions or recommendations expressed herein are those of the authors, and do not necessarily reflect those of the US Government, ONR or the NSF.

References

1. Xen website, http://www.xen.org/
2. Barham, P., Dragovic, B., Fraser, K., Hand, S., Harris, T., Ho, A., Neugebauer, R., Pratt, I., Warfield, A.: Xen and the art of virtualization. In: SOSP 2003: Proceedings of the Nineteenth ACM Symposium on Operating Systems Principles, pp. 164–177. ACM, New York (2003)
3. Borders, K., Zhao, X., Prakash, A.: Siren: Catching evasive malware. In: Proc. of the IEEE Symposium on Security and Privacy, Oakland, CA, USA, pp. 78–85 (May 2006)

4. Bowen, B.M., Prabhu, P., Kemerlis, V.P., Sidiroglou, S., Keromytis, A.D., Stolfo, S.J.: BotSwindler: Tamper resistant injection of believable decoys in VM-based hosts for crimeware detection. In: Jha, S., Sommer, R., Kreibich, C. (eds.) RAID 2010. LNCS, vol. 6307, pp. 118–137. Springer, Heidelberg (2010)
5. Chandrasekaran, M., Vidyaraman, S., Upadhyaya, S.: SpyCon: Emulating User Activities to Detect Evasive Spyware. In: Proc. of the Performance, Computing, and Communications Conference (IPCCC), pp. 502–509 (May 2007)
6. Egele, M., Kruegel, C., Kirda, E., Yin, H., Song, D.: Dynamic spyware analysis. In: Proc. of the USENIX Annual Technical Conference, Santa Clara, CA, USA, pp. 233–246 (June 2007)
7. Fest, G.: Why thin is back in (March 2010),
   http://www.americanbanker.com/usb_issues/120_3/
   why-thin-is-back-in-1014707-1.html
8. Holz, T., Engelberth, M., Freiling, F.: Learning More About the Underground Economy: A Case-Study of Keyloggers and Dropzones. In: Backes, M., Ning, P. (eds.) ESORICS 2009. LNCS, vol. 5789, pp. 1–18. Springer, Heidelberg (2009)
9. Lohr, S.: Thin-client boom, finally? (July 2007),
   http://bits.blogs.nytimes.com/2007/07/26/thin-client-boom-finally/
10. Pappas, V., Bowen, B.M., Keromytis, A.D.: Crimeware swindling without virtual machines. In: Burmester, M., Tsudik, G., Magliveras, S., Ilić, I. (eds.) ISC 2010. LNCS, vol. 6531, pp. 196–202. Springer, Heidelberg (2011)
11. Richardson, T.: The rfb protocol, version 3.8,
    http://realvnc.com/docs/rfbproto.pdf
12. The Security Division of EMC RSA. Malware and enterprise. White paper (April 2010)
13. Willems, C., Holz, T., Freiling, F.: Toward Automated Dynamic Malware Analysis Using CWSandbox. In: Proc. of the IEEE Symposium on Security and Privacy (S&P), pp. 32–39 (March 2007)
14. Jae Yang, S., Nieh, J., Selsky, M., Tiwari, N.: The performance of remote display mechanisms for thin-client computing. In: ATEC 2002: Proceedings of the General Track of the Annual Conference on USENIX Annual Technical Conference, pp. 131–146. USENIX Association, Berkeley (2002)
15. Yin, H., Song, D., Egele, M., Kruegel, C., Kirda, E.: Panorama: Capturing System-wide Information Flow for Malware Detection and Analysis. In: Proc. of the 14th ACM Conference on Computer and Communications Security, pp. 116–127 (2007)
16. Zetter, K.: Google hack attack was ultra sophisticated, new details show (January 2010), http://www.wired.com/threatlevel/2010/01/operation-aurora/

A Comparative Usability Evaluation of Traditional Password Managers

Ambarish Karole[1], Nitesh Saxena[1], and Nicolas Christin[2]

[1] Polytechnic Institute of New York University
[2] Carnegie Mellon University

Abstract. Proposed in response to the growing number of passwords users have to memorize, password managers allow to store one's credentials, either on a third-party server (online password manager), or on a portable device (portable password manager) such as a mobile phone or a USB key. In this paper, we present a comparative usability study of three popular password managers: an online manager (LastPass), a phone manager (KeePassMobile) and a USB manager (Roboform2Go). Our study provides valuable insights on average users' perception of security and usability of the three password management approaches. We find, contrary to our intuition, that users overall prefer the two portable managers over the online manager, despite the better usability of the latter. Also, surprisingly, our non-technical pool of users shows a strong inclination towards the phone manager. These findings can generally be credited to the fact that the users were not comfortable giving control of their passwords to an online entity and preferred to manage their passwords themselves on their own portable devices. Our results prompt the need for research on developing user-friendly and secure phone managers, owing to the ubiquity of mobile phones.

1 Introduction

Typical credentials employed for user authentication fall into following categories of authentication "factors": (1) *Something You Know*," such as passwords or PINs, (2) "*Something You Have*," such as a token or a card, and (3) *Something You Are*," such as biometrics; or combinations thereof. Of these, passwords or PINs are the most widely deployed, for authentication to remote servers, ATMs and mobile phones.

For over more than a decade, users have been asked to memorize an increasing number of passwords [1] to authenticate to various online services. While users can usually easily memorize a couple of passwords, the current explosion of the number of passwords each user has to maintain is severely testing the limits of their cognitive abilities [2]. This leads to "weak" choices in practice. For example, users often tend to choose short and "low-entropy" passwords [3,4], enabling offline dictionary attacks and brute-forcing attempts, or they write passwords down or use the same password at multiple sites [5].

Password Managers (PMs) attempt to solve this conundrum by having a computing device, rather than the user herself, store (and optionally, generate) passwords, and then later deliver or recall them to the user whenever access is needed. To this end, a number of password management schemes have been proposed and are used currently.

K.-H. Rhee and D. Nyang (Eds.): ICISC 2010, LNCS 6829, pp. 233–251, 2011.
© Springer-Verlag Berlin Heidelberg 2011

We can broadly distinguish between three categories of password managers: desktop manager, online manager and portable manager. A desktop manager (e.g., Mozilla Firefox, Apple MacOS Keychain, RoboForm [6]) stores strong passwords on the user's desktop (i.e., on the terminal used for authentication) while an online manager (e.g., LastPass [7] and Mozilla Weave Sync [8]) stores them on remote third-party server(s).[1] A portable manager, on the other hand, stores strong passwords on user's portable device. Among portable managers, we can further identify two different types: phone-based password managers (e.g., KeePassMobile for J2ME enabled devices [10] and OpenIntents Safe for Android [11]) and USB-based password managers (e.g., Roboform2Go for USB devices [6]).

In each of these approaches, the strong passwords are typically protected using a master password; at the time of recalling a specific password, the user simply types in her master password. If a user is mobile and uses multiple terminals for authentication (e.g., her desktop at home and her laptop in the office), a desktop manager would not offer any portability to the user. We, therefore, do not consider desktop managers to be of much benefit on their own.

The online and portable managers have their own pros and cons. An online manager, although portable, requires the user to trust the third-party service provider(s). Since user's passwords would typically be encrypted using her master password and then stored on remote server(s), they might be vulnerable to offline dictionary attacks. Imagine if all users were to use a remote manager, the passwords corresponding to all of them might be susceptible to an adversarial break-in at the end of the server(s). Moreover, often proprietary, a remote manager might not offer the users any transparency in outsourcing their sensitive information and how this information has been protected.

A portable manager can possibly be more trusted since it can be locally managed by the user on her own trusted portable device. However, all existing phone managers typically involve displaying a (long and possibly random) password on the portable device, which the user is simply asked to copy onto the terminal. Typing in a such a password might have poor usability. USB managers do not have this drawback, but they may not offer a desired level of portability and accessibility to a modern user.

The goal of this paper is to formalize an evaluation of existing password managers, by comparing them in terms of security, ease of use, necessity and level of acceptance, as perceived by an average web user. To that effect, we present a comparative usability study of three popular password managers: an online manager (LastPass), a phone manager (KeePassMobile) and a USB manager (Roboform2Go).

Our study was performed with a sample of users controlled with respect to technical background (i.e, computer science students vs. non-technical "average" users). We find, contrary to our intuition, that users overall preferred the two portable managers over the online manager, despite the better usability of the latter. Surprisingly, the online manager was the last choice for non-technical people, who mostly preferred the

---

[1] Rather than storing passwords, another password management approach (e.g., PwdHash [9]) derives passwords on-the-fly, based on a master password and a specific variable, e.g. the URL of the website to authenticate to. From the usability perspective, this approach and desktop/online managers are equivalent, in that they only require a master password to be memorized/recalled.

phone manager. Also, technical people were more inclined towards the USB manager in comparison to the online manager. These findings can generally be credited to the fact that the users were not comfortable giving control of their passwords to an online entity and preferred to manage their passwords themselves on their own portable devices.

We note that the only prior work that directly relates to our study, to the best of our knowledge, is by Chiasson et. al [12]. The study [12] evaluates two desktop managers – PwdHash [9] and Password Multiplier [2], and points out underlying usability problems with these two managers. Our study, on the other hand, aims at evaluating and comparing three different types of traditional password management approaches, with a particular focus on mobile users.

2   Background and Research Questions

In this section, we discuss the three password managers in more details and compare them based on their usability and security characteristics. This background information will serve as a foundation to frame the research questions that we aim at answering via our study, and to come up with the usability and security measures across which the password managers will be compared. We provide a side-by-side comparison of an online manager, a phone manager and a USB manager in Figure 1.

	Strong authentication	Trusted terminal	Third-party trust	Server-side modifications	Client-side modifications	Observation resistance	Automated or manual?	Master password	Portability	Fall-back
Online Manager	Optionally (if random)	Yes	Yes	No	Yes	No	Automated	Yes	Yes	If master password is lost
Phone Manager	Optionally (if random)	Yes	No	No	Optionally (for backup)	No	Manual	Yes	Yes	If phone is lost
USB Manager	Optionally (if random)	Yes	No	No	Optionally (for backup)	No	Automated	Yes	Yes	If USB drive is lost

Fig. 1. Comparison of Password Management Methods

As discussed in Section 1, online password managers incorporate remote third-party servers for password storage. Portable managers, on the other hand, consists of a credential listing on users' personal portable devices, e.g., a mobile phone and USB drive.

One example of software that falls into the category of online manager is LastPass [7]. LastPass is a proprietary extension for the Mozilla Firefox web browser which locally encrypts user credentials using 256 bit AES prior to transmitting them to Last-Pass's data centers via SSL. Though their key generation algorithm is not described, LastPass's encryption and decryption is protected using a master password which is not transmitted beyond the local terminal. A similar online password management extension for Firefox is Mozilla Weave Sync [8]. Weave is an open source solution which operates by encrypting browser data with asymmetric cryptography; this allows users to share selected browser data with others if desired. Though each user's private key is stored locally as well as on remote Weave servers, in both cases this key is encrypted with a user specified passphrase. As is the case with LastPass's master password, this passphrase is used locally and not transmitted to or stored on the remote server.

These online password managers introduce some drawbacks, however. Foremost among these is the issue of trust. This class of managers asks users to trust a remote server or group of servers with their sensitive data. When a remote server is employed, the password encrypted with a master password is sent across the internet, making it much more likely for a malicious entity to capture and store it for later offline dictionary attacks (master password is still user-chosen). Furthermore, should an adversary manage to break in to one of these servers they would be able to gain access to all the encrypted passwords for every user stored on that server. Again, the fact that these credentials are stored as ciphertexts alleviates this issue somewhat, but the threat of a later offline attack on this data remains. In contrast, an offline attack on a portable password manager of a user only exposes that particular user's passwords.

An additional consideration pertaining to remote credential storage is the flexibility of authentication. Because these remote servers manage passwords for many users, authentication with a user name and password prior to credential retrieval is a necessity. Portable managers, on the other hand, never requires a user name due to the personal nature of a user's mobile device.

Also, as noted in [13], there are several flaws and challenges associated with with managing credentials through remote servers. Although users desire the additional security benefits online servers can provide, users are unwilling to compromise on usability to improve security. Thus remote servers must be careful not to add security at the cost of detracting from the overall user experience. Client side software must be easy to download and install, and should be tightly integrated with the browser or operating system to prevent users from cutting corners that could potentially lead to social engineering attacks.

Several portable managers exist for various mobile phone platforms, such as KeePassMobile for J2ME enabled devices [10] and OpenIntents Safe for Android [11]. While uncomplicated, users of these alternatives must *manually* transfer their password by reading it off their mobile device and typing it on their terminal's keyboard. This may be clumsy in terms of usability, but also restricts the security of the password management solution by limiting the length of passwords that can be used to that which a user is capable of correctly reading and typing during each authentication.

USB managers (e.g., RoboForm2Go [6]), being personal, offer a similar level of trust as provided by phone managers. One potential advantage of a USB manager over phone manager is that the password recalling process is automated. However, mobile phones appear, at first glance, potentially more appealing to users. USB devices indeed do not serve any additional purpose other than providing data storage, while mobile phones are increasingly playing the role of a "digital swiss army knife."

Strong authentication in existing passwords managers is achieved through the use of randomly generated password strings. Most existing solutions provide users with the option of either storing their pre-existing, non-random credentials or generating new random passwords at registration time. If existing passwords are stored then the solution does not provide any measure of additional security, only the convenience of password recall.

All password management approaches trust the intermediary terminal with the user's plaintext credentials, i.e., passwords. This is due to the inherent difficulty of authenticating without introducing server-side modifications.

Our discussion above raises several questions that we intend to answer through our study. These include:

- How do the three PMs compare in terms of usability? The usability can be measured with respect to perceived toughness, satisfaction and ease of use.
- How do the three PMs compare in terms of security and protection of passwords? This covers giving control of passwords to a program and perceived security.
- How do the three PMs compare in terms of their perceived necessity and acceptance? In other words, would the users be willing to adopt them in practice?
- How do the three PMs compare in terms of all security and usability measures taken together?
- How do the three PMs compare across a diverse set of users categorized based on background (technical or non-technical)? Also, what is the effect of different users' background on each PM?

3 Study Preliminaries

Password Manager Implementations: Our goal is to compare the three PMs – USB manager (denoted as USB henceforth), phone manager (Phone) and online manager (Online) – in terms of their usability and security, as perceived by average users. We also intend to evaluate each PM according to several underlying tasks, including registration, login from a personal computer, login from a remote computer, change password, and login with a changed password (these tasks will be explained in Section 4.2). This implies that each user would need to execute all these tasks to evaluate a PM, which might lead to a lengthy overall experimentation period per user. This in turn might cause user fatigue and influence the results of the study. To avoid this, it was paramount that no more than one PM of each type (USB, Phone and Online) is selected for the study. This necessitated that only those PM implementations are selected that are representative of their respective PM category.

As discussed previously, a number of commercial and popular options exist that can be used in our study. These include (to name a few) LastPass [7] and Mozilla Weave Sync [8] as Online managers; KeePassMobile for J2ME enabled devices [10] and Open-Intents Safe for Android [11] as Phone managers; Roboform2Go [6] and HandyPassword [14] as USB Managers. Numerous other implementations exist, as listed in an online survey of PMs [15]. Fortunately, the user actions involved in all PM implementations of a given category are roughly very similar to one other. In other words, for example, to login using any of these USB Managers, the user simply needs to connect her USB drive to the USB port of her computer terminal, and type in her master password to unlock the password to be recalled. To login using any of the Phone managers, the user needs to first unlock her phone with a master password and then copy the password – to be recalled – displayed on the phone's screen onto the keypad of the terminal. Similarly, in order to login using an Online Manager, the user only needs to type in her master password on the terminal, the rest of the process being farily oblivious to the user. The only distinction among these PMs are the underlying software interfaces.

According to the reviews available online [15][16], we chose Roboform2Go as our USB manager, KeePassMobile as the Phone manager and LastPass as the Online manager. Based on their popularity, we believe these three PMs are quite suitable for our usability study which aims at comparing the three PM categories (USB, Phone and Online). We also believe that our selection, and our use of existing and deployed implementations was a better approach than trying to pursue our study with our own (likely unpolished) research prototypes of the PMs.

Devices: We used common devices that most users are quite familiar with. We used Imation 2GB USB 2.0 thumb drive [17] – as our USB manager – with RoboForm2Go software. We chose Nokia 5310 mobile phone [18] as our Phone manager installed with KeePassMobile. We used a Dell Desktop as our primary authentication terminal and a Sony Laptop for the purpose of login from another terminal (see Section 4.2).

Browser: Based on its popularity [19], Mozilla Firefox was used as the Internet browser throughout our study. Participants were instructed to authenticate, using the three password managers, to a popular web email service – Gmail.

4   Usability Testing Details

Having made a selection of a password manager for each category (as discussed in Section 3), we are now ready to start the usability study. The most obvious method to record responses from a user is through the use of a 5-point Likert scale, in addition to open-ended and multiple choice personal preference questionnaires. The questionnaires were handed over to a user depending on which stage of testing he/she was at. The *During Test* questionnaire was posed after the respondent finished performing each one of the five tasks common to all the three password managers (these tasks will be discussed in Section 4.2). The *Post Test* questionnaires, on the other hand, were asked after all the three password managers had been tested by each user. Based on our discussion in Section 2, we decided to evaluate and compare the password managers with respect to the following usability and security measures. (A similar set of measures have previously been used in the study of [12]).

• During Test –

1. **Toughness:** how tough it was to execute each task? (1 question was posed)
2. **Satisfaction:** how satisfied the users felt with each task? (1 question was posed)

• Post Test –

1. **Giving Control:** how users felt while giving control into the hands of a software/tool to manage their passwords? (4 questions were posed)
2. **Perceived Ease:** did users find the password manager easy to use? (5 questions were posed)

3. **Perceived Necessity:** did users deem the password manager necessary and acceptable? (2 questions were posed for all PMs. For Phone and USB, 1 additional question was posed regarding the accessibility to mobile devices.)
4. **Perceived Security:** did users find the password manager secure? (4 questions were posed)

The users were also posed a few open-ended questions, in each of the above questionnaires, in order to poll for their opinions about any perceived problems with the password managers and suggestions for possible improvement.

Finally, a *Final Test* questionnaire was also presented to each user polling which password manger he/she preferred the most and asking about their order of preference based on the level of (1) security, (2) convenience and (3) overall experience.

The main challenge we faced was the sheer number of questions which each user needed to answer, potentially leading to lazy respondent behavior and user fatigue. Care was taken so as to minimise both the number of questions and to discard any questions which showed a tendency of not being answered genuinely (or honestly).

4.1 Study Participants

We recruited a total of 20 participants: 10 technical and 10 non-technical users. In the rest of this paper, we will refer to our technical users as Students, and non-technical users as Non-Students, because all technical people were students while all non-technical people were non-students. The participants were recruited on a first-come first-serve basis from respondents to emails. Prior to recruitment, each participant was briefed on the estimated amount of time required to complete the tests and on the importance of completing the tests in its entirety.

The student participants were all university students, studying towards undergraduate, Master's and Ph.D. degrees in Computer Science or closely related fields. This group of our users represented a fairly young, well educated and technology-savvy sample of user population.

The other group, consisting of the non-students, had an average age difference of nearly two decades from that of the students. This group was tested to gain insights into whether such a group – differing in terms of full time occupation – had any impact on the choices made with respect to the password managers.

There is an obvious concern that, if a technology-savvy group (students) does not react well to a password management approach, the approach will perform a lot worse with average users; or on the contrary, if a password manager that fares well with students, it might not perform equally well with average users. This concern was our prime motivation to categorize the respondents into students and non-students.

Our non-students ranged from help-desk personnel, technicians, real estate agents, restaurant workers to housewives. In addition to the students vs. non-students distinction, our sample was also controlled, as much as possible, in terms of other important user-centric characteristics, i.e., gender and age. This was done in order to evaluate the password managers among a diverse user population pool. The gender split was: 60% male and 40% female for both students and non-students. Our test users were divided into three age groups: 40% Young (less than 24 years old), 40% Middle-Aged (25-44 years old) and 20% Old (45-54 years old). In addition to the students and non-students category, we have also pursued gender-centric and age-centric analysis. However, due to space constraints, we only restrict ourselves to the former in the rest of this paper, which we believe is most important to our study.

Gender, age and other information was collected through a *Pre Test* questionnaire completed by our participants prior to starting the test process. None of the study participants reported any physical impairment that could interfere with their ability to complete a given task.

4.2 Testing Process

Our study was conducted at two testing locations, one on-campus (at our university) for the students and the other off-campus for non-students. These two venues were chosen solely for the purpose of convenience to the targeted participant groups. Same devices (USB drive and phone) and computer terminal (see Section 3) were used at both locations giving rise to consistent test set-up across all users. Our study lasted for a duration of about two months.

An overview of the testing process was given to each respondent prior to the study and due care was taken to minimise any scope of *explicit* "priming" of respondents considering a security-focused nature of our study.[2] Such a priming in terms of security can possibly result in skewed (over-alert) participant behavior and in biased results, as demonstrated by prior research [20].

As mentioned previously, after administering the Pre Test questionnaire, the respondents were asked to perform five tasks corresponding to each password manager. Any possible user errors in performing the above tasks were taken note of by the test administrator (no such errors were observed throughout our study, however).

1. **Register** involves registering with a password manager the password, username and other information for a particular web site.
2. **Login** involves login to a web site, whose password has already been saved with a password manager.
3. **Second Login** is similar to the Login task, only difference being the computer is not the same as the one used in the previous task. This task is aimed at judging the portability of the password manager from terminal to terminal.
4. **Change Password** involves changing the password, both with the website and password manager.
5. **Login with New Password** involves repeating the login task but with the new password.

As mentioned in Section 3, the test set-up comprised of a desktop computer which acted as the primary computer for Login, Change Password and Login with New Password, and a laptop for Second Login. This set-up, consisting of two computers, was used in order to closely mimic the tasks akin to a realistic password manager setting.

A randomly chosen 8 character master password was provided to each test user, which he/she was asked to memorize and use throughout the experiments.

---

[2] Since the study was about password managers, it was neither possible nor meaningful to avoid implicit priming.

The respondents were instructed, in advance, to fill-in During Test questionnaires after each of the above task was completed. As mentioned earlier, the questionnaire consisted of two simple Likert Scale type questions and two open-ended questions. The order in which the password managers were presented to the users was randomized so as to avoid any possible learning effects.

Following the During Test, the respondents were required to fill out the Post Test questionnaire for each password manager. This too comprised of Likert Scale questions followed by a few open ended questions. The purpose of this questionnaire was to judge the changes in attitudes (opinions) of the respondents towards the password managers after having worked with all three of them.

Finally, the last part of the questionnaire (Final Test) was administered to the respondents. Here the participants ranked the best of the three password managers which they felt were most appealing with respect to their overall experience with them, the level of convenience and security provided. This part of the questionnaire also consisted of a few open-ended questions. These were aimed to better understand the motivating reasons for a respondent to choose a particular password manager, which could not be captured by the Likert scales or the multiple choice questions.

5   Test Results and Analysis

In this section, we present and analyze our During Test and Post Test Likert scale logged observations. We also discuss the final preferences provided by our test subjects for the three PMs evaluated. We present two types of comparison in our analysis throughout:

- **Within-Subjects Comparisons:** This analysis would tell us how the three password managers (USB, Phone and Online) fare with one another, corresponding to the entire user sample as well as corresponding to students and non-students.
- **Between-Subjects Comparisons:** Using this analysis, we intend to understand the effect of occupation (student vs. non-student) on the usability and security of three PMs.

Recall that the During Test data is aimed at evaluating the usability of each PM in terms of two measures: Toughness and Satisfaction. The Post-Test data, on the other hand, is for investigating the PMs with respect to usability and security measures: Giving Control, Perceived Ease, Perceived Necessity and Perceived Security. We analyze the PMs based on these individual measures first, followed by a principle component and cluster analysis that evaluates the combined effect of different measures.

In the remainder of this section, we discuss our results and their interpretations. Unless stated otherwise, statistical significance is reported at the 5% level.

5.1   During Test Analysis

Figure 2 shows the *average* Likert ratings regarding Toughness and Satisfaction of the three PMs, for Students and Non-students (also shown are the collective average ratings for All Users taken together as well as those corresponding to All PMs).

Within-Subjects Comparisons: Observing the bars of the graph along the X-axis, we find that Phone is deemed the toughest, followed by USB and then Online, for all

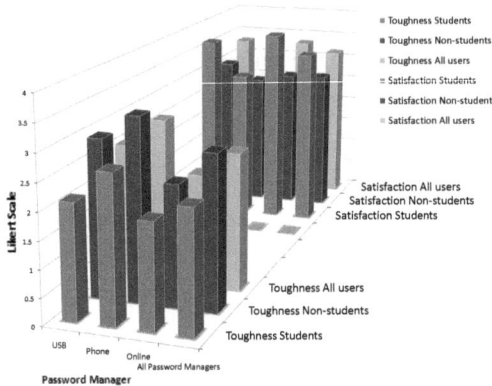

Fig. 2. During Test Toughness and Satisfaction

our users (Students, Non-Students, All Users). In terms of satisfaction, on the other hand, among both Students and All Users, Online was preferred over USB, followed by Phone. For Non-Students, however, USB was the first choice, preferred slightly more than Online and Phone. These results are intuitive because both Online and USB require a minimal amount of effort from the users and are supposed to be quite fast in comparison to Phone due to manual transfer of password.

Based on paired t-tests, we found the following statistical differences. Students found Phone tougher than USB ($p = 0.0103$) and tougher than Online ($p = 0.0028$). Students also found USB more satisfying than Phone ($p = 0.0006$), Online more satisfying than USB ($p = 0.0238$) and Online more satisfying than Phone ($p < 0.0001$). Non-Students also found Phone tougher than USB ($p = 0.009$) and tougher than Online ($p = 0.003$), and USB tougher than Online ($p = 0.020$). In terms of satisfaction levels for Non-Students, we did not find any statistical difference; the ratings were quite close for different PMs. For All Users, Phone was deemed tougher than USB ($p = 0.043$) and Online ($p = 0.00013$), and USB was found to be tougher than Online ($p = 0.0444$). Also, for All Users, USB was more satisfactory than Phone ($p = 0.0498$), and Phone was more satisfactory than Online ($p = 0.009$).

Between-Subjects Comparisons: Observing the graph of Figure 2 along the Y-axis, we notice that Non-Students consistently found the three PMs tougher to use when compared to Students. Likewise, Students found the three PMs more satisfying than Non-Students. The reason for this is simple: Students are expected to be much more technologically savvy compared to Non-Students.

Based on unpaired t-tests, following significant differences were noticed: Non-Students found USB tougher ($p = 0.0008$) and less satisfactory ($p < 0.0001$) compared to Students. Non-Students also found Phone tougher compared to Non-Students ($p = 0.003$). Students were highly more satisfied with Online ($p < 0.0001$).

Usability Measures Taken Together: In the previous subsection, we considered the usability of PMs in terms of two measures: Toughness and Satisfaction. Although the two measures were usually negatively correlated with each other (Pearson correlation coefficient was found to be -0.771, when considering data from all users), in certain cases the correlation was not entirely clear. In order to understand an overall impact of Toughness and Satisfaction on the usability of PMs, we pursued principle component (PCA) and cluster analysis. Due to ease of readability, we do not include the details regarding this analysis (a similar analysis, however, is later discussed for Post Test measures in Section 5.2). We only depict the results (using Agglomerative Hierarchical Clustering) of this analysis in Figure 3.

For All users, we obtain that Online $\succ$ USB $\succ$ Phone, and USB and Phone are clustered together (we use '$\succ$' to denote preference). A similar and independent PCA and cluster analysis for Students and Non-Students indicate the following. For students, Online $\succ$ USB $\succ$ Phone, and Online and USB form a cluster of their own. On the other hand, for Non-Students, Online $\succ$ USB $\succ$ Phone, and USB and Phone are clustered with each other. These results are intuitive and very much inline with our observations shown in Figure 2, which we discussed in the previous subsection.

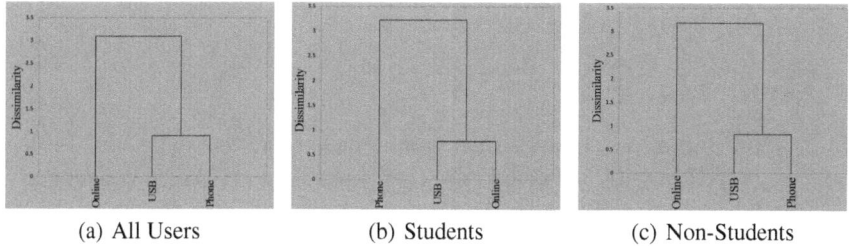

  (a) All Users   (b) Students   (c) Non-Students

Fig. 3. During Test Cluster Analysis Based On Principal Components (Dissimilarity vs. PM)

Usability of Individual Tasks: As we explained in Section 4.2, in our experiments, each PM was tested for several different processes, namely, Register, Login, Second Login, Change Password, Login with New Password. Since usability of a PM depends on all these processes, we compare the three PMs based across these processes.

Figure 4 depicts the average Likert scale Toughness ratings for different processes corresponding to Students, Non-Students and All Users. In this plot, first three bars for each process correspond to Students (USB, Phone, Online, resp.), next three bars correspond to Non-Students (USB, Phone, Online, resp.) and last three bars correspond to All Users (USB, Phone, Online, resp.). The Satisfaction ratings were generally inversely related to the Toughness ratings and are not shown in this paper.

Let us first compare the three PMs across different processes. We note that for each process, in general, Phone is tougher than the other two PMs. Between USB and Online, the former is deemed tougher, for all processes. This analysis conforms well with our overall analysis of During Test data presented in previous subsections.

There are a few exceptions to the above claim, however. Login and Change Password have the same average ratings for both USB and Online for Students Students deemed Login with New Password as equally tough for USB and Phone. Register was also rated at a equal level of toughness by Non-Students. For Second Login, Students found USB

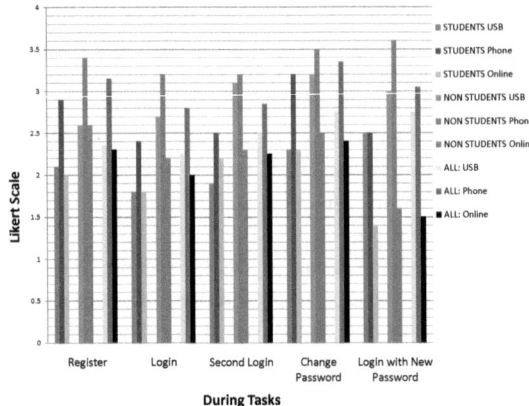

Fig. 4. During Test Toughness Per Task

less tougher than Online. For Second Login and Change Password, Non-Students rated Phone as only slightly tougher than USB.

5.2 Post Test Analysis

Figure 5 shows the average Likert Post Test ratings regarding Giving Control, Perceived Ease, Perceived Necessity, and Perceived Security, for Students and Non-students (also shown are the collective ratings for All users taken together). We discuss the observations made from these ratings as follows.

Within-Subjects Comparisons:

• **Giving Control:** Looking at the Giving Control ratings (Figure 5(a)), we find that Students order of preference is USB, followed by Phone and Online. Non-Students, on the other hand, like the Phone the best, followed by USB and Online. Collectively looking at All Users, USB is an overall winner, which seems slightly better than Phone, which in turn is much better than Online. In general, users felt that USB and Phone provide a better sense of control compared to Online. This is a surprising finding and is perhaps due to the fact that managing the passwords locally on their own devices gave users a sense of control and authority.

Based on paired t-tests, we found the following statistical differences. Students found USB better than Phone ($p = 0.0049$) and USB better than Online ($p = 0.016$). Non-Students preferred Phone over Online ($p = 0.0001$), and USB over Online ($p = 0.0009$). Non-Students prefer USB over Online ($p = 0.001$) and Phone over Online ($p = 0.030$). All Users prefer USB over Online ($p < 0.0001$), and Phone over Online ($p = 0.00024$).

• **Perceived Ease:** Looking at the Perceived Ease ratings (Figure 5(b)), we find that Students order of preference is USB, followed by Online and then Phone. Non-Students, on the other hand, like the Phone the best, followed by Online and then USB. Collectively looking at All Users, Phone is an overall winner, which is slightly better than USB, which in turn is slightly better than Online. Here we can see a clear split across

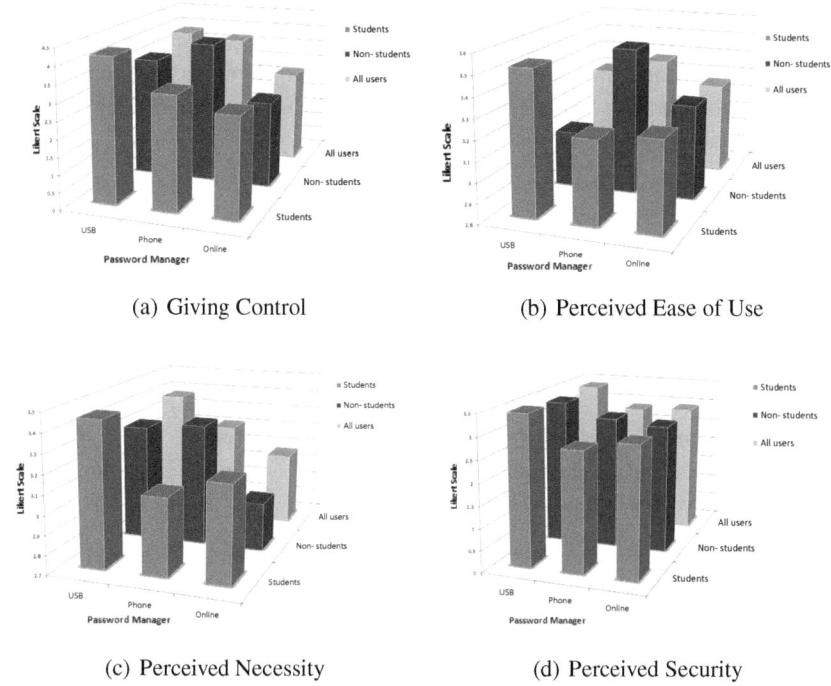

(a) Giving Control   (b) Perceived Ease of Use

(c) Perceived Necessity   (d) Perceived Security

Fig. 5. Post Test – Students vs. Non-Students

Students and Non-Students: the former still preferred Online or Phone, whereas the latter found the Phone as the easiest. Paired t-tests, however, did not lead to any significant statistical differences with respect to Perceive Ease.

• **Perceived Necessity:** Looking at the Perceived Necessity (Figure 5(c)), we find that Students order of preference is USB, followed by Online and then Phone. Non-Students, on the other hand, like the Phone the best, followed by USB and then Online. Collectively looking at All Users, USB is an overall winner, which is somewhat better than Phone, which in turn is quite better than Online. These findings are somewhat similar to those in case of Perceived Ease, which means that necessity of a PM was based on its ease. Based on paired t-tests, Non-Students found USB better than Online ($p = 0.0448$). No other significant statistical differences were found.

• **Perceived Security:** From the average ratings corresponding to Perceived Security (Figure 5(d)), we can see that USB is generally preferred by all users, and there is not much to choose between Phone and Online (although Students prefer Online slightly more so than Non-Students, who slightly prefer Phone).

According to paired t-tests, Students found USB better than Phone ($p = 0.0132$), and USB better than Online ($p = 0.03$). All Users found USB more secure than Phone ($p = 0.0089$) and USB more secure than Online ($p = 0.012$). No other significant statistical differences were found.

Between-Subjects Comparisons

• **Giving Control:** Looking at the Giving Control ratings (Figure 5(a)), Students prefer USB and Online more than Non-Students, however, Non-Students are more inclined to use Phone compared to Students. Based on unpaired t-tests, Non-Students preferred Phone more than Students ($p = 0.011$).

• **Perceived Ease:** From Perceived Ease ratings (Figure 5(b)), we observe that Students prefer USB much more than Non-Students, however, Non-Students are much more inclined to use Phone compared to Students. Both somewhat equally prefer Online.

• **Perceived Necessity:** Looking at the Perceived Necessity (Figure 5(c)), Students prefer USB and Online more than Non-Students, however, Non-Students are more inclined to use Phone compared to Students. This is very similar to users' perception of ease as discussed above.

• **Perceived Security:** Looking at the Perceived Necessity (Figure 5(d)), there is not much distinction between the rating of Students and Non-Students. For Phone, however, Non-Students provided higher ratings.

Accessibility to Portable Devices: In response to whether the users would have their phone and USB drive handy while accessing a web site, the average ratings (with standard deviations) were as shown in Table 1. The ratings imply that there is not much to choose between USB drive and phone when looking at All Users. Students, on the other hand, rated USB drive as more accessible compared to phone, whereas Non-Students gave higher ratings to phone. This is perhaps one of the reasons why Students had a stronger inclination towards using USB PM and why Non-Students preferred Phone.

Table 1. Average ratings (with standard deviation) for accessibility of USB and Phone

	All Users	Students	Non-Students
USB	3.4 (1.095)	3.8 (0.422)	3 (1.414)
Phone	3.4 (1.046)	3.1 (0.876)	3.7 (1.160)

All Usability and Security Measures Taken Together: A usable PM should perform well with respect to all (not just one of the) usability measures we discussed so far, i.e., Giving Control, Perceived Ease, Perceived Necessity and Perceived Security. To this end, we performed linear cross-correlations among the PMs across these usability measures. We first present a complete analysis over data acquired from all test subjects. Table 2 shows the correlation coefficients and their respective statistical significance.

The coefficients from less than -0.5 and more than 0.5 are generally regarded as large [21] and in line with the findings of [22], we cannot regard any of our usability measures as sufficiently correlated with others that they could be justifiably omitted. On the other hand, since the measures are mildly correlated, it motivates us to also look at them as a whole as we show next.

Table 2. Cross-Correlation of Usability Measures

	Control	Ease	Necessity	Security
Control	1	0.287	0.130	0.337
Ease	0.287	1	0.248	0.368
Necessity	0.130	0.248	1	0.374
Security	0.337	0.368	0.374	0.287

Principal Component and Cluster Analysis: Table 3 lists the four principal components, denoted PC1, PC2, PC3 and PC4, that explain 100% of the variance in the logged data. As the first two components, i.e., PC1 and PC2, together explain nearly 70% of the variance, and PC3 and PC4 have eigenvalues that are less than 1, i.e., explaining less variance than one original variable [23], we disregard PC3 and PC4 in the following analysis. Table 4 shows the factor loadings of PC1 and PC2. As shown, PC1 factors in all usability measures positively and more in comparison to PC2, while PC2 has a negative rating for Giving Control and Perceived Ease. This means that high PC1 score for a PM would indicate its good usability and security, whereas low score for PC2 may indicate better control and ease.

Table 3. Principle Components of Usability Measures

	PC1	PC2	PC3	PC4
Eigenvalue	1.885	0.877	0.686	0.553
Proportion of Variance	47.119	21.913	17.155	13.814
Cumulative Proportion	47.119	69.031	86.186	100.000

Table 4. Factor Loadings of PC1 and PC2

	PC1	PC2
Giving Control	0.619	-0.634
Perceived Ease	0.702	-0.130
Perceived Necessity	0.621	0.671
Perceived Security	0.789	0.085

Table 5 depicts how each method scores with respect to PC1 and PC2. We find that a high PC1 score for USB indicates its superiority as a PM. Online is considered to have poorest overall usability due to low PC1 score and Phone has a mediocre level of usability. Figure 6(a) shows how methods form two clusters (using Agglomerative Hierarchical Clustering), one consisting of USB and Phone together and another consisting of Online. The figure indicates that our methods can be partitioned into two classes, with good and poor usability overall. Methods with good usability are USB and Phone. Online exhibits poor overall usability and security.

248   A. Karole, N. Saxena, and N. Christin

Table 5. Scores for each PM with respect to PC1 and PC2

	PC1	PC2
USB	0.612	-0.075
Phone	-0.013	-0.244
Online	-0.599	0.319

A similar and independent analysis for Students and Non-Students indicate the following (results shown in Figures 6(b) and 6(c), resp.). For students, USB is better compared to both Phone and Online, which form a cluster together, whereby Online is better than Phone. On the other hand, for Non-Students, USB and Phone form a cluster with each other (Phone is better than USB) which compares favorably with Online.

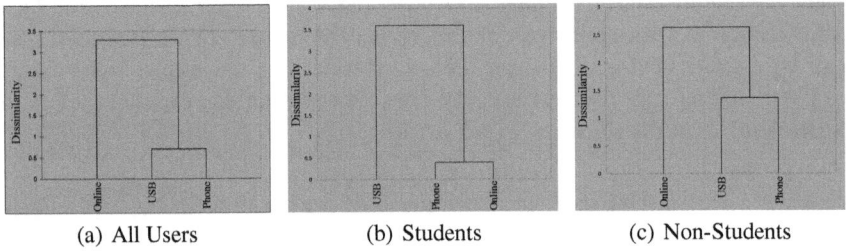

(a) All Users   (b) Students   (c) Non-Students

Fig. 6. Post Test Cluster Analysis Based On Principal Components (Dissimilarity vs. PM)

In summary, our Post-Test analysis shows that, for all of our users, Online was surprisingly the last choice (despite its better usability as indicated via our During Test analysis in Section 5.1). Users either preferred USB or Phone. This can be credited to the fact that users felt that managing their passwords locally on their own devices gives them a sense of control and authority, as shown by their ratings for Giving Control.

5.3 Final User Preferences

After having performed the usability experiments with the PMs, we also polled for users' final preference based on their experience. We posed the subjects three questions and asked them to rank the PMs based of their order of preference:

1. Which password manager do you prefer the most?
2. Which password manager according to you offers better security and protection of your passwords?
3. Which password manager according to you is most convenient to use?

The responses we received are depicted in Table 6. While a large fraction of Non-Students shows a strong liking for Phone, most Students' preference was either USB or Online (although most of them selected Phone to be most secure). In short, our overall analysis, presented in Section 5.2, conforms well with the final preferences provided by our users (i.e, All Users, Students and Non-Students).

Table 6. % of Users Who Preferred a Particular PM (All, S, NS denote All Users, Students, Non-Students)

PM	Prefer the Most			Secure			Convenient		
	All	S	NS	All	S	NS	All	S	NS
USB	40%	40%	40%	25%	30%	20%	35%	30%	40%
Phone	35%	20%	50%	55%	40%	70%	30%	20%	40%
Online	25%	40%	10%	20%	30%	10%	35%	50%	20%

5.4   Answers to Open-Ended Questions

Few of our test users responded to the open-ended questions that we posed. We quote below some of the interesting feedback that we received.

– **Q**: From your understanding what does Roboform2Go [USB] do?
  A: Manages, stores passwords and makes them portable. We need to know and remember only one master password, rest is taken care of by the software.
– **Q**: From your understanding, what does Lastpass [Online] do?
  A: Stores passwords on a central server, so makes it "real time" portable but more vulnerable towards the attacks from cyber criminals.
– **Q**: Do you have any suggestions for Lastpass? What would make it more useful or easier to use?
  A1: Lastpass server should force users to change the passwords more frequently to make it cyber-attacks proof
  A2: It is good but what if [it] does not respect our privacy and [does] not follow the code of conduct?
– **Q**: Why do you prefer this particular type of password manager [USB]?
  A: [It is] Easy to handle, portable, comparatively safe.
– **Q**: Do you think that using a password manager would make it easier to manage your passwords?
  A: Yes, but [it] is not an absolute necessity.
– **Q**: Why do you prefer this particular type of password manager [Phone]?
  A1: My Phone is the most secure of the devices and I always have it present with me wherever I go.
  A2: [It provides] Better sense of security.
– **Q**: Why do you prefer this particular type of password manager [Online]?
  A1: [It is] Easy and no need to carry anything
  A2: It is more efficient because no [additional] hardware is required.

To summarize, we find that users are aware of the importance of security of their passwords and would be inclined to use password managers. They expressed concerns regarding off-shoring their passwords to a remote entity due to security and privacy reasons and may prefer to use their own devices for managing their passwords. They, however, may not deem password managers as an absolute necessity.

5.5   Discussion and Summary

We now provide a summary of our most notable findings. Our during test analysis shows that across all users and across non-students, Online ≻ USB ≻ Phone, and USB and

Phone are clustered together. For students also, Online $\succ$ USB $\succ$ Phone, but Online and USB form a cluster of their own. This finding can be termed intuitive since on-line PM was expected to possess better usability than the two portable PMs. Moreover, non-students generally found the three PMs tougher and less satisfactory compared to Students. The reason for this is simple: students are much more technologically savvy compared to non-students.

Post test analysis, on the other hand, reveals surprising facts. For all of our users, the order of preference turned out to be USB $\succ$ Phone $\succ$ Online, with USB and Phone clustered together. For students, the order was USB $\succ$ Online $\succ$ Phone, whereby USB and Online form a cluster together. On the other hand, for non-students, the order of preference is Phone $\succ$ USB $\succ$ Online, and USB and Phone form a cluster. In general, we found that the portable managers are preferred over the online manager.

The above turn-around from the during test to post test can be credited to the fact that the users were not comfortable giving Control of their passwords to an online entity and preferred to manage their passwords themselves on their own portable devices. This preference reversal from during test to post test results was also confirmed by users' final preferences about the three PMs.

We also observe that the non-students had a much stronger liking for Phone compared to students while looking at overall post test data, and in terms of giving control of their passwords. Being less tech-savvy, non-students perhaps felt much more comfortable and safe while copying in their passwords (from the phone to authentication terminal) manu-ally as opposed to letting a device (USB or remote server) doing it for them automatically.

6   Conclusions and Future Direction

We presented a comparative usability study of three notable traditional password man-agers. Contrary to our intuition, overall the two portable managers were preferred over the online manager, despite the better usability of the latter. Surprisingly, the online manager was the last choice for non-technical people, who mostly preferred the phone manager. Also, technical people were more inclined towards the USB manager in com-parison to the online manager. These findings can generally be credited to the fact that the users were not comfortable giving control of their passwords to an online entity and preferred to manage their passwords themselves on their portable devices.

Based on our results, we can conclude that portable managers represent a more promising password management approach than online managers. The latter provide a higher degree of confidence to users in managing their passwords. However, current portable managers (especially phone managers) do not offer the usability as expected by average users, thus motivating the need for usable portable managers in the future. Owing to an ever increasing "always on, always with me" mobile phone usage trend, we believe that developing user-friendly and secure phone managers is an interesting and important research direction.

References

1. Gabber, E., Gibbons, P.B., Matias, Y., Mayer, A.J.: How to make personalized web browsing simple, secure, and anonymous. In: Proceedings of Financial Cryptography 1997, Anguilla, West Indies, pp. 17–32 (February 1997)

2. Halderman, A., Waters, B., Felten, E.: A convenient method for securely managing passwords. In: Proceedings of the 2005 World Wide Web Conference, Chiba, Japan, pp. 471–479 (May 2005)
3. Morris, R., Thompson, K.: Password security: a case history. Commun. ACM 22(11), 594–597 (1979)
4. Yan, J., Blackwell, A., Anderson, R., Grant, A.: Password memorability and security: Empirical results. IEEE Security and Privacy 2(5), 25–31 (2004)
5. Adams, A., Sasse, M.A.: Users are not the enemy. Commun. ACM 42(12), 40–46 (1999)
6. Siber Systems. Roboform password manager (2009), http://www.roboform.com
7. LastPass. Lastpass password manager (2009), https://lastpass.com
8. Mozilla Labs. Weave sync (2009), http://labs.mozilla.com/projects/weave
9. Ross, B., Jackson, C., Miyake, N., Boneh, D., Mitchell, J.C.: Stronger password authentication using browser extensions. In: USENIX Security Symposium (2005)
10. Reichl, D.: Keepassmobile (2009), http://www.keepassmobile.com
11. Openintents safe (2009), http://www.openintents.org/en/node/205
12. Sonia Chiasson, P., van Oorschot, C., Biddle, R.: A usability study and critique of two password managers. In: USENIX Security Symposium (2006)
13. Dhamija, R., Dusseault, L.: The seven flaws of identity management: Usability and security challenges. IEEE Security and Privacy (2008)
14. Handypassword,
    http://www.handypassword.com/login_password_manager_terms/
    usb_password_manager.shtml
15. Pc magazine: Password managers & form fillers,
    http://www.pcmag.com/article2/0,2817,1791459,00.asp
16. Password management software review (2009),
    http://password-management-software-review.toptenreviews.com/
17. Imation 2gb usb thumb drive: Specifications,
    http://www.pcmall.com/p/Imation-Removable-Hard-Drives/
    productãpno῀17643p̃dp.cggiicj
18. Nokia 5310 mobile phone: Specifications,
    http://europe.nokia.com/find-products/
    devices/nokia-5310-xpressmusic
19. Browser statistics,
    http://www.w3schools.com/browsers/browsers_stats.asp
20. Schechter, S.E., Dhamija, R., Ozment, A., Fischer, I.: The emperor's new security indicators. In: IEEE Symposium on Security and Privacy (2007)
21. Cohen, J., Cohen, P., West, S.G., Aiken, L.S.: Applied multiple regression/correlation analysis for the behavioral sciences (1983)
22. Frokjaer, E., Hertzum, M., Hornbaek, K.: Measuring usability: are effectiveness, efficiency, and satisfaction really correlated. In: SIGCHI Conference on Human Factors in Computing Systems (2000)
23. Kaiser, H.F.: The application of electronic computers to factor analysis. Educational and Psychological Measurement 20(1), 141–151 (1960)

An Adversarial Evaluation of Network Signaling and Control Mechanisms

Kangkook Jee[1], Stelios Sidiroglou-Douskos[2],
Angelos Stavrou[3], and Angelos Keromytis[1]

[1] Department of Computer Science, Columbia University
{jikk,angelos}@cs.columbia.edu
[2] Computer Science and Artificial Intelligence Laboratory, MIT
stelios@csail.mit.edu
[3] Department of Computer Science, George Mason University
astavrou@gmu.edu

Abstract. Network signaling and control mechanisms are critical to coordinate such diverse defense capabilities as honeypots and honeynets, host-based defenses, and online patching systems, any one of which might issue an actionable alert and provide security-critical data. Despite considerable work in exploring the trust requirements of such defenses and in addressing the distribution speed of alerts, little work has gone into identifying how the underlying transport systems behave under adversarial scenarios.

In this paper, we evaluate the reliability and performance trade-offs for a variety of control channel mechanisms that are suitable for coordinating large-scale collaborative defenses when under attack. Our results show that the performance and reliability characteristics change drastically when one evaluates the systems under attack by a sophisticated and targeted adversary. Based on our evaluation, we explore available design choices to reinforce the reliability of the control channel mechanisms. To that end, we propose ways to construct a control scheme to improve network coverage without imposing additional overhead.

1 Introduction

The prevalence and effectiveness of large-scale malware phenomena (worms, botnets, web-based malware) has led to the development of several automated defenses that detect new threats and generate various kinds of fixes such as patches, filters. The security literature is rife with distributed security systems [7,5] which assume that reliable, scalable and robust Content Distribution Network (CDN) functionality is universally available. To date, the primary metrics of effectiveness have been propagation time (latency and throughput) and node coverage in the presence of "natural" phenomena such as churn. However, the conspicuous absence of an adversarial analysis, both in terms of performance impact and security guarantees (*e.g.*, susceptibility to man-in-the-middle attacks), is of particular concern as the control channel for security data is a very attractive target for adversaries. This is especially true for systems that make design decisions that favor performance over robustness (*e.g.*, using a centralized tracker in BitTorrent).

We argue that such a narrow view of system performance is inadequate and even dangerous in the presence of malicious adversaries. In other areas of security (spam,

K.-H. Rhee and D. Nyang (Eds.): ICISC 2010, LNCS 6829, pp. 252–265, 2011.

honeypots and honeynets, anti-virus), we have seen active targeting of protection mechanisms and, occasionally, their hijacking and use for malicious purposes. Instead, we need to consider system behavior in the presence of intelligent, targeted interference by botnets and other malware. At a minimum, these systems must be able to withstand attacks that seek to disrupt their primary function: the timely and reliable delivery of security-critical data to all benign participating nodes and users.

To this end, we conduct an evaluation of control channel mechanisms that have been proposed for use in distributing security-critical data at massive scale. Specifically, we evaluate different approaches of centralized, distributed, and hybrid designs in presence of global adversary. We recognize that this is only part of the security-oriented evaluation criteria that such systems should be subjected to; however, we strive for an in-depth analysis of a particular aspect of system behavior rather than a shallower examination of more features. A key contribution of our work is a detailed analysis of existing control channel mechanisms in a number of realistic adversarial scenarios. Rather than limiting our measurements to simple latency and throughput characterizations, metrics of *coverage*, *latency*, and *control efficiency* are considered. We use these to investigate the trade-offs between system performance and resilience to certain type of attacks. Thus, our work explores the spectrum of possible design choices when creating and deploying a distribution mechanism for security-critical data.

As a result of evaluation, we find that centralized designs introduce fragile failure points, centralized entities, or hierarchical indirection, that can cripple performance and reliability when attacked. Distributed mechanisms also cease to function upon failure of nodes more than a certain threshold. Furthermore, the attacker can escalate his impact on distributed mechanisms by taking advantage of heterogeneity of network knowledge among participants. Extending reliability to some extent, the hybrid mechanism still inherits the shortcomings of both centralized and distributed systems. To maximize the reliability benefit of the hybrid mechanism, we explore the design choices available on integrating two contrasting schemes without sacrificing control efficiency.

The road-map of the paper is as follows. After discussing background work (Section 2), the adversarial scenarios are provided in Section 3. In Section 4, we show how we implemented control mechanisms for evaluation. Section 5 delivers evaluation results. Our analysis on these results are presented in Section 6 and the paper is concluded in Section 7.

2 Background

2.1 Control and Signaling Approaches

In contrast to the data transfer channel, the control channel performs its task by signaling small sized management packets to the participating peers. The signaling channel is responsible for: *i)* peer join and leave, *ii)* locating objects *iii)* resource scheduling and allocation *vi)* authentication, integrity, and authenticity and *v)* application specific tasks – for instance, a system for alert distribution raises an alarm of urgent security events using this channel. Traditionally, there were two fundamental but contrasting schools of thought regarding the design and implementation of the signaling mechanisms – centralized and distributed. Recently, there are attempts to leverage the strengths of both distributed and centralized schemes while avoiding some of their weaknesses by using a hybrid approach.

Centralized Schemes. These simple and efficient mechanisms require one or a small set of centralized entities to coordinate the operations of the entire system. However, the scalability of the system is limited by the network and processing capacity of the control nodes. As a workaround, a hierarchical control network [27] consisting of super-nodes (SNs) was proposed. The control plane is implemented by adding layer of super-nodes which act as the leaders and are in charge of their own sub-networks. Unfortunately, selecting the right super nodes and the size of the clustering for each sub-network is still an open problem. This is further exacerbated in dynamic environments with many joins and leaves. Moreover, akin to the pure centralized solution, each super-node is single point of failure to its own sub-network.

Distributed Schemes. This class of mechanisms is designed to mitigate the scalability problems of the centralized design. Their design can be accomplished using either structured or unstructured overlay networks. Distributed Hash Tables (DHTs) [25,13,22] is a structured overlay solution that are leveraging the power of consistent hashing [9]. On the other hand, the gossip-based information sharing protocols [18], also known as epidemic or flooding protocols, process requests from clients in unstructured way. The core implementation relies in flooding search requests to peering neighbors. Nevertheless, despite their numerous benefits, distributed solutions also come with their own limitations. DHT-based approaches do not work well in practice [20] as their performance is severely influenced by even a small fraction of slow performing nodes. Moreover, the gossip protocol becomes very costly as the size of network grows and has difficulty in locating information with low availability. In following evaluations, we use DHTs to implement distributed control channel.

Hybrid Schemes. These signaling mechanisms attempt to combine the advantages of the centralized and the distributed (DHTs) design principles. During normal operations, a hybrid system uses a fast and efficient centralized channel. It can, however, switch to a slower but also more robust distributed channel to resolve capacity overload or even node failures. There is a wealth of recent research on hybrid designs [29,8,10] all of which share the same basic design principles with minor modifications. Moreover, there are systems that attempt to combine two distributed mechanisms of structured (DHTs) and unstructured (gossip protocol) to achieve better search efficiency [12,28].

2.2 Reliability Analysis of Signaling Channels

There exist some previous works that focus on analyzing the reliability of network systems and the security protection of control channels. For reliability of centralized systems, there is work relate to the stability of super-node networks. Yang et al. [27] suggest general guidelines in designing super-node networks and about principles for reliable design. Mitra et al. [15] propose an analytic framework that correlates super-nodes' fraction and their network connectivity with reliability. This work also considers a global adversary of different knowledge and power. It does not, however, address hybrid schemes or provide any comparison between systems. Distributed control systems are designed to be more reliable but they introduce new threats and vulnerabilities which exploit the specifics of each architecture. To counter the shortcomings, researchers proposed a number of implementations [21] and theoretical studies [11,2] with the aim to improve the reliability of distributed systems. Specific example of control channel which supervises the entire system's operation and becomes a viable target to the adversary is the BitTorrent tracker network. Although originally designed as a centralized control, extensions have been proposed to enhance the reliability of the tracker by having distributed tracker or multiple trackers. Unlike previous studies on BitTorrent [4,19] where the primary interests were performance related factors such as latency and fairness of resource utilization, recent studies [16,17,14] focus more on the system's reliability.

2.3   Secure Message Propagation Systems

The goal of alert distribution systems is to deliver small size messages to many participants under a strict time constraint. Ever since fast, self-replicating worms (for instance Slammer and Nimda viruses) crippled the Internet, there have been many theoretical studies [30,1,26,24] to build an alert distribution system which can compete against such worms. The outcome of this line of research was guidelines regarding how fast the patch propagation should be. However, none of these works consider scenarios of active adversary who also wants to take over the alert propagation processes.

In addition, RapidUpdate [23] is a research performed by research groups of commercial security vendors. It offers a specific solution to their own alert propagation model. The goal of the system is to propagate small sized alert messages (less than 200K) and meet distribution deadlines. Having assistance from peers, the RapidUpdate tries to alleviate the workload of servers/vendors. Another work [7] by a major software vendor quantifies the performance of the world's biggest patch distribution system – Microsoft's Windows update. Based on trace analysis, this work delivers interesting observations on traffic characteristics of patch distribution and end-user's behavioral patterns. Nonetheless, no previous study considers the presence of a sophisticated adversary that attempts to disrupt the operation of the alert distribution network.

3   Application Environment and Adversarial Scenarios

The key element of our work is the evaluation of different mechanisms for implementing a rapid and reliable alert distribution system in the adversarial context. Previous analyses of such systems were largely done without taking into account sophisticated (or, in many cases, even simple) adversaries who might seek to disrupt the operation of the system. Such disruption may, for example, be attempted in parallel with an attack, so as to maximize its impact and minimize the effectiveness of any defenses.

The goal of the adversaries would be to delay distribution and delivery of such alerts, or to prevent their delivery altogether to as large a fraction of the nodes as possible. We consider different adversaries, at varying levels of sophistication and resources. For generality, our evaluation considers the *impact* such adversaries would have on the system, in terms of inhibiting communications to/from some fraction of nodes.

The sophistication of the adversaries in our threat model is determined in terms of their ability to collect reconnaissance on the internal structure of the alert distribution mechanism and focus their attack. Thus, at a high level, we distinguish between two types of adversaries:

- Adversary with **random attack**: Unsophisticated adversaries who can inhibit communication to/from randomly selected nodes. The fraction of nodes they can bring down depends on the level of resources available to them.
- Sophisticated adversaries, who exploit knowledge of the system structure to target nodes such that they maximize the impact of their disruption. We further consider two sub-types of such adversaries:

  1. Adversary with **targeted attack:** Attackers that know and exploit the high-level structure of the network topology. Such attackers, for example, know the identity of and target the super-nodes or other, relatively "fixed" important nodes in the system.
  2. Adversary with **degree dependent attack:** More powerful adversaries that somehow have detailed topology information about a large part or all of the distribution mechanism. Such knowledge includes, for example, the complete connectivity graph of the participating nodes (or a large fraction thereof).

For all type of the above schemes, selected victim nodes are taken out from the system as a consequence of the attacks.

4   Implementation

For our evaluation, three different alert distribution systems were implemented on Over-Sim [3] network simulation framework. Here, we describe how we implemented the simulation modules. We first talk about the design choices for the signaling channels and the various reliability parameters that we explored. Then, we cover communication models considered for alert distribution systems.

4.1   Control Channel

Centralized System. In the case of centralized control, we employed a super node (SN) network. Among many configuration parameters [27] for the SN network, we carefully identified the ones that affect the robustness of the overall network: the size of sub-network (*cluster size*) and number of super-node replicas (*k-redundancy*). The *cluster size* was tested using a range of different values. The same holds for *k-redundancy*. However, in our graphs, we present only the case where *k-redundancy* is two. We did so because other values of *k-redundancy* do not notably change the system's behavior beyond the one captured by the graphs. We configured the rest of the parameters unchanged as these parameters have an effect only on the network performance.

Distributed System. For distributed control, we chose Chord [25] to implement a decentralized alert notification system. Chord was selected for two reasons. First, Chord's ring-based routing structure and ID space has been well-studied allowing us to compare our performance results with others when the network is not under attack. This validates our approach beyond the results of a mere simulation. Second, the structural differences among variants of DHT implementations are not discernible in terms of robustness. Indeed, most of the hash-based systems use a common architecture that employs key-based routing [6]. Among many configurable parameters for Chord, we considered *successor list size* to be the most important one to the reliability and stability of the system. This was varied with different values to see its impact on the system's maintenance cost and reliability.

Hybrid System. This model aims to achieve better network performance similar to the centralized systems while maintaining the reliability of the purely decentralized approaches. In hybrid systems, all nodes initially join both a decentralized and a centralized signaling channel. For instance, a super-node in the hybrid network is the centralized entity for its sub-network as well as a regular participant in the DHT channel. Therefore, the hybrid designs inherit all their configuration parameters. Moreover, peers in the hybrid network can utilize the primary (centralized) and secondary (decentralized) signaling channels either in *serial* or *parallel*. In our implementation of hybrid systems, frequent operations such as querying were done first using the centralized and then the decentralized signaling path. This increases performance under normal operations while maintaining robustness in case of attacks. However, for less frequent but more critical functions, such as publishing new information, we used both channels at the same time to increase resilience without severely impacting the performance of the network.

4.2   Models for Alert Distribution

Publish-Subscribe Model. In this model, peers have the option to subscribe to certain classes of security events. *Polling* and *pushing* are available choices to implement this model. For our experiments, we used the *polling* model with 30 seconds of polling interval. This is a cost-effective and easy-to-implement solution, widely adopted by most vendors for their online patching system.

Distributed Sensors Model. In this model, participants with proper permission can be sensors who can detect security incidents and initiate the alert propagation process. This is typical model used to deploy large scale defense posture but it also comes with issues of trust – the security information's integrity and node authentication. For our experiment, only nodes with proper permission can publish new message to subscribers. Their integrity is examined by super-nodes, in the case of centralized and hybrid mechanisms, or peering nodes in charge of the ID segment, for distributed schemes.

5   Evaluation

In the section, we describe evaluation results for the alert distribution systems implemented with three different control mechanisms. First, we explain the evaluation metrics and then we talk about the reason behind the choice of the Oversim simulation framework. Lastly, we discuss our evaluation results with and without global adversaries. For each evaluation instance, all results are averaged over at least 10 iterations.

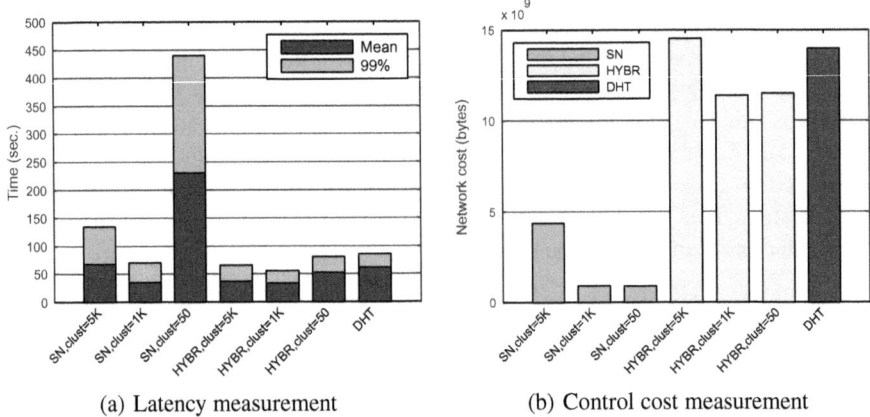

(a) Latency measurement   (b) Control cost measurement

Fig. 1. The evaluation results under normal operations. Figure (a) depicts the mean notification time and the completion time for 99% of the nodes and for different control plane mechanisms. Figure (b) presents the total network cost in terms of bytes for the test duration of 600 seconds.

5.1   Evaluation Metrics

To evaluate the reliability and efficiency of the different control plane mechanisms, we introduce three metrics: *coverage*, *latency*, and *control efficiency*. *Coverage* is measured by enumerating the number of nodes that receive the alert message when the system is under attack. Alternatively, all alert messages that are not delivered within the duration of the experiment instance are regarded as failures. *Latency* is defined to be the period of time that it takes for alerts to reach each participant from the moment that an alert message is dispatched. *Control efficiency* is the cost to utilize the control mechanisms. This is calculated by summing the total number of bytes required for network operations during the experiment.

5.2   Evaluation Design

To validate the network behavior under adversarial conditions, we implemented three control mechanisms using the OverSim [3] simulation framework. The use of simulation was mandated for the following reasons:

Scalability. We were interested in observing the behavior of large scale networks implementing signaling systems in the presence of network-wide malicious attacks. Having tens of thousands of number of participants, Oversim framework enabled us to quantify the design parameters that really influence the behavior of the system.

Global Adversary. Emulating global adversary in a real-world large-scale testbed is a costly and time-consuming task and it does not allow repetition of experiments. The simulation framework not only helped us to instantiate this size of network but also provided the interface to implement a more precise behavior for the global adversary.

5.3 Evaluation under Normal Operation

To establish a baseline for our experimental results, we first measured the latency and control efficiency of the mechanisms without considering a global adversary. For test topologies of 20,000 nodes, each test instance was measured for 600 seconds of simulation time. Each test instance contained an alert notification event and the same size (40KB) of control messages were propagated to all participants. The size is derived from the average size of Microsoft patches [7]. SN network, implemented for centralized mechanism, was configured with different *cluster sizes* and *k-redundancy* was fixed to 2 for all test cases. DHT network was used for the distributed control mechanism and its *successor list size* was set to five. Hybrid network inherited parameter from both systems.

Latency Measurement. The latency results are shown in Figure 1(a). On the X-axis, from the left to right, we have results for the SN network, Hybrid network, and DHT network. The SN network and hybrid network are configured with different cluster sizes. For each bar, the dark portion represents the average time for notification and the gray part represents the time until 99% of the nodes are notified. Large variance was observed for the latency results of SN network. With different cluster sizes, mean latency ranged from 35 to 230 seconds. Populated sub-networks (lower-layer) accounted for delays in the case of large cluster size (5,000). For smaller cluster size (50), having more super-nodes made the upper-layer network the bottleneck. In contrast, for hybrid network, we observed small variance in latency and less delays. This is because the secondary, distributed channels masked the errors or failures of the primary channel. Mean latencies ranged only from 33 to 51 seconds. Not having a secondary channel, the DHT network took longer than the worst case of hybrid network. However, the latency remained relatively low (61 seconds).

Control Cost Measurement. Figure 1(b) represents the control cost of different mechanisms to propagate alert messages of the same size (40KB). SN network, thanks to its simple implementation, required the least amount of packets to maintain its control channel and signaling operations. However, in the case of larger cluster size (5,000), many number of network errors and retries introduced rapid increase in cost. DHT network required larger amount of control traffic to maintain its distributed data structures. Hybrid network with large cluster size (5,000) required even more and was the most expensive control channel due to excessive numbers of network errors from its primary channel. However, with the proper choice of cluster size, hybrid network could spare its control cost to become a more efficient solution than the DHT network.

5.4 Evaluation of Adversarial Scenarios

In adversarial scenarios, we again used the topology of 20,000 nodes with longer simulation duration of 1,200 seconds to carefully observe the system's reaction to malicious activities. Nodes that could not be notified within this time duration were regarded as a delivery failure. Two different cluster sizes were plotted for SN network and hybrid network – 50 to represent a small cluster size and 5,000 for large cluster size. During the experiment, the alert propagation event was triggered at 100 seconds of the simulation time and the attack from the adversary was launched five seconds prior to the event.

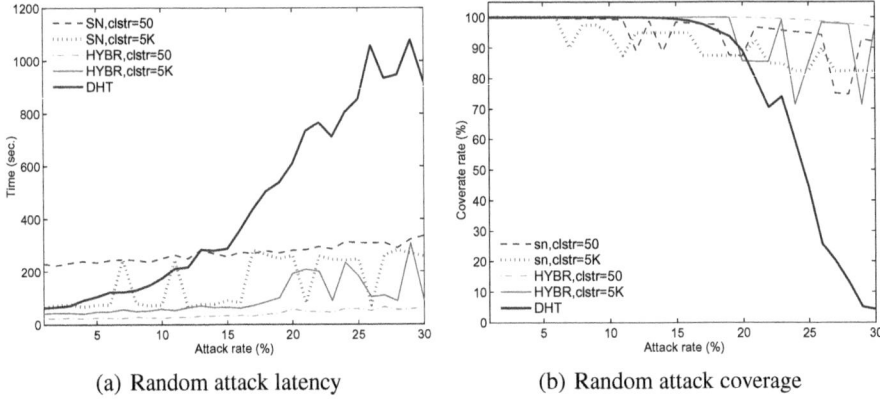

(a) Random attack latency   (b) Random attack coverage

Fig. 2. Figures (a) and (b) illustrate the latency and coverage for the random attack scenario respectively and for different attack intensities. In (a), Y-axis shows average notification time and in (b), Y-axis shows the percentage of nodes which successfully received the alert message.

Random Attack Evaluation. In this attack scenario, the adversary randomly selects its victims varying its attack ranges (0% ~ 30%). The latency and coverage results against this attack are shown in Figure 2(a) and Figure 2(b) respectively. For DHT network, both latency and coverage were most severely impacted by this attack. Having acceptable latency from its initial stage, DHT network's latency steadily increased. Network failures that impacted coverage started approximately around 17 ~ 18%. The coverage results dropped rapidly from that point onwards. SN network showed better results than DHT network in terms of both metrics. It is interesting to note that the centralized mechanism with a little redundancy configuration (*k-redundancy*=2) showed better coverage results than the distributed system. In the DHT network, by distributing certain amount of connections to all participants, each node's failure had some influences on the system's connectivity. This resulted in network disintegration and gradual deterioration of latency beyond a certain threshold. This result is consistent with the observation that DHT network's performance is severely influenced by even a small fraction of slow performing nodes [20]. In the SN network, failures of all SN replicas for a sub-network significantly deteriorate system's latency and coverage. But, in the case of random attack, probability to hit all replicas in the same group is exponentially low in regards to *k-redundancy* parameter. Irregular spikes in its latency and coverage results indicates this type of failures where *k-redundancy* is two. Hybrid-network, by having dual channels, showed improved coverage and latency results. While the hybrid network showed smoother results than the SN network overall, systems with smaller cluster size had better latencies and reduced traffic irregularities.

Targeted Attack Evaluation. In this attack, the adversary takes one step further by targeting nodes of *explicit* importance – super-nodes for the SN network and the hybrid network. After selecting all available target nodes, the attacker randomly select the rest of her victims. DHT does not expose any *explicit* targets. Thus, all the victims are selected randomly. The attack in this case becomes identical to the random attack. Coverage result against this attack are presented in Figure 3(a). Unlike the DHT network, whose results didn't changed much from the random attack result, the SN network is seriously impacted by this attack. Having all super-nodes eliminated, the system stopped being operational from the very initial stage of attack, less than or around 4%. Similarly, not having the benefits of its primary SN channel, latency and coverage results of the hybrid network soon converged to that of the DHT network.

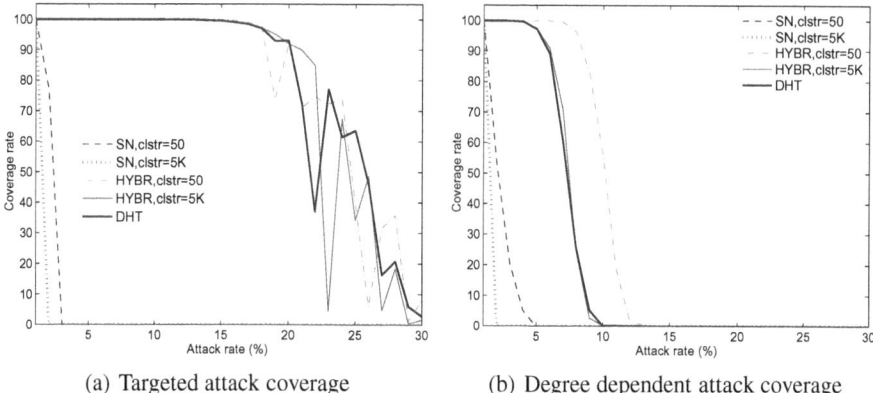

(a) Targeted attack coverage   (b) Degree dependent attack coverage

Fig. 3. Figure (a) and (b) present coverage results for the targeted attack and degree dependent attack respectively and for different attack intensities. Y-axis shows the percentage of nodes who successfully received the alert message.

Degree Dependent Attack Evaluation. In this attack, the attacker can identify nodes not only of *explicit*, but also of *implicit* importance. For this, she considers each node's topological significance. Super nodes maintain more state acting as defaults routes for the their clusters and thus are higher priority targets. In Figure 4(a), we depict the connection distribution for the hybrid network (of cluster size 50). We present the number of connections for the super-nodes and regular nodes using different colors. This Figure illustrates how the attacker chooses its victims for degree dependent attack with different attack rates of 1% and 4%. With respective dotted and dashed lines, the nodes with number of connections above the lines will be the victims.

The coverage result against the attack is presented in Figure 3(b). Similar to the targeted attack, SN network's coverage deteriorated from the initial stage of the attack. By choosing nodes with higher connectivity, this attack was highly effective in crippling the DHT network. DHT's coverage starts to drop around 7%~8%. In the case of hybrid network, the coverage was also impacted by the attack. The outcome for a large cluster size (5,000) with few super-nodes, does not show much difference from the DHT network's result. The small cluster size (50) performed better and extended coverage about 4%, because it was able to distribute the SN connections more evenly across the network curtailing the reachability failures due to the attacks.

Quantifying the behavior of the different signaling mechanisms when under different attack scenarios allowed us to make this observation: hybrid network is the efficient solution for both adversarial and normal situations with the following benefits. *i)* latency-wise, it was an efficient solution with less configuration sensitivity. *ii)* with the proper choice of cluster size, the system consumes reasonable amount of control cost which is higher than SN but less than DHT system. *iii)* Under all type of attacks, it showed the best resilience in terms of coverage and latency. Another interesting observation is that SN network, even with less network connections, could show better results than DHT network against the random attack. However its reliability benefit is immediately cancelled by sophisticated and targeted attackers.

6 Analysis

Our evaluation shows that the hybrid network gained the number of reliability benefits by adding a constant number (two) of SN connections to the DHT network. This result indicates that the number of connections and the way it connects participants can seriously impact the reliability of the system. Unfortunately, the number of network connections is constrained by both system and network resources. Therefore, we want to explore the design space that can enhance reliability by only improving the way it connects participants. To that end, we investigate different ways of implementing control systems by using the same number of connections. More concretely, we extend DHTs with the fixed number (two) of connections in different ways to observe how these influence the coverage result. The number of connections is the same one used from the previous evaluations. Of course, this parameter can have significant impact on coverage results. However, for all proposals, we want to demonstrate how we can add connections under the same constraints and maximize the coverage benefits.

To further enhance the system's behavior when attacked, we leverage the benefits of DHT's internal structure with a modified routing table. This technique, which exploits finger-table, is implemented for Chord and is also applicable to other DHT systems.

6.1 Chord Connection Types

The Chord maintains two types of logical network connections. One for the *successor lists* and the other for the *finger table entries*.

- **Successor list** maintains the list of neighboring nodes. It is an important parameter that influences DHT's reliability and its default size is set to five. Having a $O(\log n)$ size of this connection provably guarantees the stability of the system which indicates success rate of lookup request.
- **Finger table** is a core data structure that implements $O(\log n)$ routing of Chord. The upper limit of its size is logarithmic to the size of hash space (in the case of Chord, this is set to 2^{160}).

Unlike previous proposals which naïvely added SN connection to DHT connections, we implement a hybrid network utilizing existing slots of finger tables. SN connections can be replaced with immediately preceding entries in the finger table. This does not increase the required state per node or the total number of connections, but this costs additional hops for lookup activities due to some sub-optimal entries.

Table 1. Control mechanisms and their labels

Label	Control mechanisms
DHT	Chord with successor list size of 5 (default).
DHT S-list	Extend *DHT* by adding 2 connections to successor list. Successor list size is set to 7.
HYBR	Hybrid mechanism that naïvely integrates 2 additional SN connections to *DHT*. This is the hybrid network used from previous evaluations.
HYBR F-table	This extends *HYBR* by integrating 2 additional SN-connections with the finger table.

6.2 Evaluation of Network Coverage

We measured the performance of our proposed modifications in terms of coverage. To that end, we present our experimental results from the degree dependent attack by varying its attack rate (0% ~ 30%) for a network of 20,000 participating nodes. The cluster size for hybrid network was set to 50 to make the effect of SN connections more pronounced. With larger cluster sizes, thus smaller SN connections, we expect coverage results similar to that of a DHT network. Table 1 details specific configurations and their labels used for evaluations. The evaluation results are presented in Figure 4(b).

The last three configurations (*DHT S-list, HYBR, HYBR F-table*) from Table 1 are implemented with the same number (two) of additional connections to the original Chord DHTs (*DHT*). The result for *DHT S-list* shows the limited effects of the two additional success list entries. From Figure 4(b), this improves coverage only by 2 ~ 3%. The result for *HYBR* shows better coverage (5 ~ 6%) than *DHT S-list*. Although SN connections replace connections assigned to the successor list, the structural benefits offered by the SN network are far greater. This is apparent in the *HYBR F-table*, by harnessing the reliability benefits of both successor list entries and SN connections, the coverage

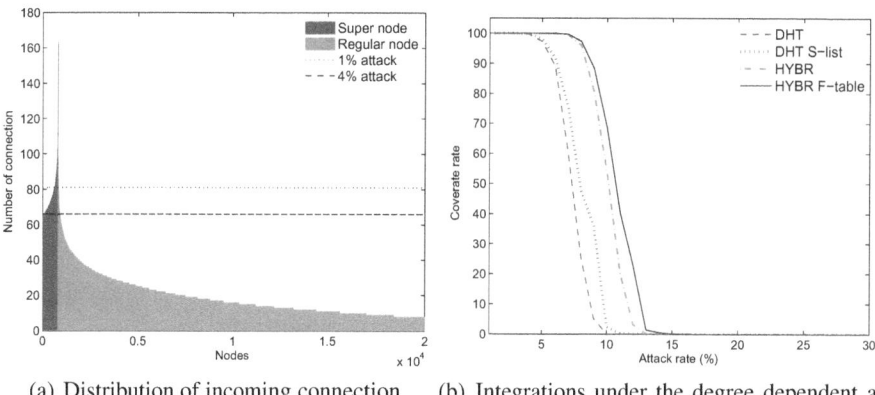

(a) Distribution of incoming connection   (b) Integrations under the degree dependent attack

Fig. 4. (a) enumerates connections for nodes in a hybrid network with cluster size of 50. The dotted and dashed lines show the impact of the degree dependent attack. The attacker choose victims with number of connection above the lines. (b) presents the coverage for different modifications and for the degree dependent attack. In X-axis we vary the attack intensity while Y-axis shows the alert success rate.

is boosted by $7 \sim 9\%$. Furthermore, penalty for having two sub-optimal entries in its finger table is negligible and requires only a small amount of additional lookup calls (3.4%).

The experimental results present interesting insight about the trade-offs between network structure and their impact on reliability. We can deduct that additional entries in the list have limited effect. Thus, it is better to consider other avenues of adding connections in order to enhance system's reliability. Modifying the finger table can be an option to consider because it increases coverage without deteriorating its original functionality.

7   Conclusions

We evaluated alert distribution systems implemented using three control channel mechanisms under different adversarial scenarios. Our evaluation enabled us to draw a number of interesting insights regarding the reliability of the signaling channel. The pure distributed system (DHTs), designed to be robust under attacks, suffers in terms of network performance including latency and coverage. In the case of random attack, DHTs reliability turned out to be worse than that of a super-node based centralized design. To alleviate this, we proposed the integration of centralized and the distributed designs. Our approach consists of structural changes that enable us to seamlessly integrate a SN network and a DHT network. We evaluated a hybrid network design that offered the best coverage and reliability under all type of attack scenarios. We believe that with proper engineering choices, we can further enhance the system's reliability.

Acknowledgements. This work was supported by the NSF through Grant CNS-06-27473, 09-37060 to the Computing Research Association for the CIFellows Project,by ONR through MURI Contract N00014-07-1-0907, and by AFOSR through MURI Contract FA9550-07-1-0527. Any opinions, findings, conclusions or recommendations expressed herein are those of the authors, and do not necessarily reflect those of the US Government, ONR, AFOSR, or the NSF.

References

1. Aspnes, J., Rustagi, N., Saia, J.: Worm versus alert: Who wins in a battle for control of a large-scale network? In: Tovar, E., Tsigas, P., Fouchal, H. (eds.) OPODIS 2007. LNCS, vol. 4878, pp. 443–456. Springer, Heidelberg (2007)
2. Awerbuch, B., Scheideler, C.: Towards a scalable and robust dht. Theory of Computing Systems (2009)
3. Baumgart, I., Heep, B., Krause, S.: Oversim: A flexible overlay network simulation framework. In: Proc. of IEEE GI (2007)
4. Bharambe, A., Herley, C., Padmanabhan, V.: Analyzing and improving a bittorrent network's performance mechanisms. In: Proc. IEEE INFOCOM (2006)
5. Costa, M., Crowcroft, J., Castro, M., Rowstron, A., Zhou, L., Zhang, L., Barham, P.: Vigilante: end-to-end containment of internet worms. In: Proc. of SOSP (2005)
6. Dabek, F., Zhao, B., Druschel, P., Kubiatowicz, J., Stoica, I.: Towards a common api for structured peer-to-peer overlays. In: Kaashoek, M.F., Stoica, I. (eds.) IPTPS 2003. LNCS, vol. 2735, Springer, Heidelberg (2003)

7. Gkantsidis, C., Karagiannis, T., VojnoviC, M.: Planet scale software updates. In: Proc. of SIGCOMM (2006)
8. Hui-shan, L., Ke, X., Ming-wei, X., Yong, C.: S-chord: Hybrid topology makes chord efficient. In: Lorenz, P., Dini, P. (eds.) ICN 2005. LNCS, vol. 3421, pp. 480–487. Springer, Heidelberg (2005)
9. Karger, D., Lehman, E., Leighton, T., Panigrahy, R., Levine, M., Lewin, D.: Consistent hashing and random trees: distributed caching protocols for relieving hot spots on the world wide web. In: Proc. of STOC (1997)
10. Ktari, S., Hecker, A., Labiod, H.: Exploiting power-law node degree distribution in chord overlays. In: Proc. of NGI (2009)
11. Li, J., Stribling, J., Morris, R., Kaashoek, M., Gil, T.: A performance vs. cost framework for evaluating dht design tradeoffs under churn. In: Proc. IEEE INFOCOM (2005)
12. Loo, B., Huebsch, R., Stoica, I., Hellerstein, J.: The case for a hybrid P2P search infrastructure. In: Voelker, G.M., Shenker, S. (eds.) IPTPS 2004. LNCS, vol. 3279, pp. 141–150. Springer, Heidelberg (2005)
13. Maymounkov, P., Mazieres, D.: Kademlia: A peer-to-peer information system based on the XOR metric. In: Druschel, P., Kaashoek, M.F., Rowstron, A. (eds.) IPTPS 2002. LNCS, vol. 2429, p. 53. Springer, Heidelberg (2002)
14. Menasche, D., Rocha, A., Li, B., Towsley, D., Venkataramani, A.: Modeling content availability in peer-to-peer swarming systems. SIGMETRICS Perform. Eval. Rev. (2009)
15. Mitra, B., Peruani, F., Ghose, S., Ganguly, N.: Analyzing the vulnerability of superpeer networks against attack. In: Proc. of CCS (2007)
16. Neglia, G., Reina, G., Zhang, H., Towsley, D., Venkataramani, A., Danaher, J.: Availability in bittorrent systems. In: Proc. IEEE INFOCOM (2007)
17. Piatek, M., Isdal, T., Anderson, T., Krishnamurthy, A., Venkataramani, A.: Do incentives build robustness in bittorrent. In: Proc. of NSDI (2007)
18. Pittel, B.: On spreading a rumor. SIAM Journal on Applied Mathematics (1987)
19. Qiu, D., Srikant, R.: Modeling and performance analysis of bittorrent-like peer-to-peer networks. In: Proc. of SIGCOMM (2004)
20. Rhea, S., Chun, B., Kubiatowicz, J., Shenker, S.: Fixing the embarrassing slowness of opendht on planetlab. In: Proc. of WORLDS (2005)
21. Rhea, S., Geels, D., Roscoe, T., Kubiatowicz, J.: Handling churn in a dht. In: Proc. of the USENIX Annual Technical Conference (2004)
22. Rowstron, A., Druschel, P.: Pastry: Scalable, distributed object location and routing for large-scale peer-to-peer systems. In: IFIP/ACM International Conference on Distributed Systems Platforms, Middleware (2001)
23. Serenyi, D., Witten, B.: Rapidupdate: Peer-assisted distribution of security content. In: Proc. IPTPS (2008)
24. Shakkottai, S., Srikant, R.: Peer to peer networks for defense against internet worms. In: Proc. of Inter-Perf (2006)
25. Stoica, I., Morris, R., Karger, D., Kaashoek, M., Balakrishnan, H.: Chord: A scalable peer-to-peer lookup service for internet applications. SIGCOMM Comput. Commun. Rev. (2001)
26. VojnoviC, M., Ganesh, A.: On the race of worms, alerts, and patches. IEEE/ACM Transactions on Networking (2008)
27. Yang, B., Garcia-Molina, H.: Designing a super-peer network. In: Proc. of ICDE (2003)
28. Zaharia, M., Keshav, S.: Gossip-based search selection in hybrid peer-to-peer networks. In: Proc. of IPTPS (2006)
29. Zhu, Y., Wang, H., Hu, Y.: A super-peer based lookup in structured peer-to-peer systems. In: Proc. of PDCS (2003)
30. Zou, C., Gong, W., Towsley, D.: Worm propagation modeling and analysis under dynamic quarantine defense. In: Proc. of WORM (2003)

Secure Personalized Recommendation System for Mobile User

Soe Yu Maw

Computer University, Myeiktilar, Myanmar
soeyumaw@gmail.com

Abstract. Nowadays, due to the rapid growth of the mobile users, personalization and recommender systems have gained popularity. The recommender systems serve the personalized information to the users according to user preferences or interests and their profiles. Tourism is an industry which had adopted the use of new technologies. Recently, mobile tourism has come into spotlight. Due to the rapid growing of user needs in mobile tourism domain, we concentrated on to gives the personalized recommendation based on multi-agent technology in tourism domain to serve the mobile users [7]. The objective of this paper is to build a secure personalized recommendation system. Attackers can affect the prediction of the recommender system by injecting a number of biased profiles. In this paper, we consider detecting or preventing the profile injection (also called shilling attacks) by using significant weighting and trust weighting that complements to our proposed RPCF Algorithm.

Keywords: Security, Personalization, Recommender System, Collaborative Filtering, Profile Injection Attacks, RPCF Algorithm, Significant Weighting, Trust Weighting.

1 Introduction

The ever-changing trends of our lifestyle require mobility supports [15] which open up new accessibility opportunities for tourism industry. Today, tourism systems are one of the most important application areas for recommender system. To cope with the demand for quality of access, tourism information system should be made ubiquitous, time-aware, location-aware and personalized.

In Modern world, personalization and recommendation systems have gained widespread acceptance and attracted increased public interest in commercial services [6].

Collaborative filtering (CF) provides personalized recommendations, based on suggestions of users with similar preferences. The development of CF algorithms has focused mainly on how to provide accurate recommendations.

Recommender systems based on CF have the issues for the process of finding similar users. An attacker can attempt to influence the behavior of the recommender system for other users by using biased (fake) rating profiles to artificially either promote or demote a target item. Such attacks have been referred to as shilling attacks or profile injection attacks, and attackers as shillers [5]. Since user profiles of shillers

K.-H. Rhee and D. Nyang (Eds.): ICISC 2010, LNCS 6829, pp. 266–277, 2011.

looks very similar to an authentic user, it is a difficult task to correctly identify shilling attacks. These attacks can cause the degradation of user trust in the objectivity and accuracy of the recommender system.

Our system aims at providing personalized recommendation for mobile users based on user profiles for tourism domain. Capturing user profiles naturally involves the processing of personal data such as location data, personal preferences (interests), travel information and so on. However, the processing of user data requires security measures to ensure the user's fundamental rights to privacy. Current recommender systems have the privacy problem [1]. For the privacy concerns, the user's personal data must be protected and proceed in a safe manner. Trust concept can take advantage over recommender system. Local trust, reputation, demographic trust and location aware reputation [13] can be used to construct a trust model.

In [8], we proposed RPCF (Rule-based Personalization with Collaborative Filtering) algorithm which can address the scalability, sparsity and cold-start problem of pure collaborative filtering method and can give the accurate and good quality recommendation to the user. In [9], we proposed the architecture of multi-agent tourism system (MATS) which provides the most relevant and updated information according to the user's interest by using RPCF Algorithm. In [7], we extended MATS for mobile user.

This paper is an extension of the encyclopedia article [7] that gives the secure personalized recommendation based on multi-agent technology in tourism industry to serve the mobile users. This paper pays great attention to security issue. The privacy and trust management are not considered in this paper. The primary contribution of this paper is to detect or prevent the shilling attacks by adopting significant weighting and trust weighting that complements to the RPCF algorithm for giving the accurate recommendation to the user.

The rest of the paper is organized as follows. Section 2 describes the theoretical background of the system and Section 3 presents the security architecture of the personalized recommendation system in detailed. Section 4 points out the experimental results of the system. Section 5 concludes with a summary and suggests directions for future works.

2 System Background

This section describes the system background related with recommender system, personalization system and the attack types.

2.1 Recommender System

Recommender system can be defined as a specific type of information filtering (IF) technique that attempts to present information items (movies, music, books, news, images, web pages, etc.) that are likely of interest to the user. The goal of a recommender system is to generate meaningful recommendations to a collection of users for items or products that might interest them.

Suggestions for books on *Amazon*, or movies on *Netflix*, hotel recommendation on *Tripadvisor* are real-world examples of the operation of industry-strength recommender systems [14]. In order to give recommendation, these systems take different information like product ratings, history of purchase or the customer's interests into account. The term *collaborative filtering* was introduced in the context of the first commercial recommender system, called *Tapestry* [4], which was designed to recommend documents drawn from newsgroups to a collection of users.

Because of the recommender systems are dependent on external sources of information, they are vulnerable to attacks. Recommender systems have proven to be an important response to the information overload problem, by providing users with more proactive and personalized information services. And collaborative filtering techniques have proven to be a vital component of many such recommender systems as they facilitate the generation of high-quality recommendations by leveraging the preferences of communities of similar users.

2.2 Personalization System

Personalization means knowing who the user is and can recognize a specific user based on a user profile [16]. Personalization involves as a process of gathering and storing information about users, analyzing the information and based on the analysis, delivering the information to each user at the right time. User satisfaction is the ultimate aim of personalization.

Personalization can be divided into content-based filtering (customization), rule-based filtering and collaborative filtering.

Content-based filtering is an information seeking process in which contents are selected to satisfy a relatively stable and specific information need. Rule-based personalization use "If-then" process and based on a customer's demographics, past purchases, or product attributes. Collaborative filtering (CF) is one of the most successful recommender techniques. It is the method of making automatic predictions (filtering) about the interests of a user by collecting taste information from many users (collaborating). The underlying assumption of CF approach is that those who agreed in the past tend to agree again in the future.

Challenges in collaborative filtering include scalability, sparsity, cold-start, accuracy and security. We proposed the RPCF algorithm [8] which is the combination of rule-based and collaborative filtering approach to give the recommendation results. Our prior work [9] have addressed the scalability, sparsity and cold-start problem by using RPCF algorithm and the experimental results showed the improvement of accuracy and the quality of recommendation compared with the pure collaborative filtering approach.

The open nature of collaborative recommender systems allows attackers who inject biased profile data to have a significant impact on the recommendations produced. A collaborative recommender database consists of many user profiles, each with assigned ratings to a number of products that represent the user's preferences. A malicious user may insert multiple profiles under false identities designed to bias the recommendation of a particular item for some economic advantage. This may be in the form of an increased number of recommendations for the attacker's product, or fewer recommendations for a competitor's product.

In this work, we are primarily concerned with security issues of collaborative recommender system. Besides addressing the above challenges, our system can be able to make accurate predictions in the presence of shilling attacks, and be effectively applied in fast-growing mobile applications as well.

2.3 Attack Types: Profile Injection Attacks (Shilling Attacks)

An *attack type* is an approach to constructing attack profiles, based on knowledge about the recommender system, its rating database, its products, and/or its users [11]. The set of *filler items* represents a group of selected items in the database that are assigned ratings within the attack profile. Attack types can be characterized according to the manner in which they choose filler items, and the way that specific ratings are assigned.

The profile injection attacks can be classified in two basic categories called *push* attacks and *nuke* attacks [10]. Since shilling profiles looks very similar to an authentic user, it is a difficult task to correctly identify such profiles.

For each of the attack types, it is assumed the objective of the attack is to *push* or *nuke* the recommendations that are made for one particular target item. This item is always included in attack profiles and is assigned the maximum rating (r_{max}) to promote the item or minimum rating (r_{min}) to demote the item as push or nuke attacks, respectively.

Table 1. Push Attack and Nuke Attack Profiles

Push Attack Profile				Nuke Attack Profile			
item$_1$	item$_2$	.....	item$_{m-1}$	target	item$_1$	item$_2$	.....	item$_{m-1}$	target
r_1	r_2	.....	r_{m-1}	r_{max}	r_1	r_2	.....	r_{m-1}	r_{min}

The form of push attack and nuke attack profiles is shown in table 1. An attack profile consists of a *m*-dimensional vector of ratings, where *m* is the total number of items in the system. The rating given to the pushed item, *target*, is r_{max} and is the maximum allowable rating value and the nuke item, target, is r_{min}, the minimum allowable rating value. The ratings r_1 through r_{m-1} are assigned to the corresponding items according to the specific attack model. In our system, the average attack model is used to fill the rating of other items.

The remaining items for attacks profiles are selected for the different attack model as random attack, average attack, bandwagon attack and favorite item attack (consistency or segmented attack). Due to the space limitation, only the average attack model is described.

Average Attack: Filler items are selected uniformly at random from the system item set. Ratings for filler items are assigned based on a more specific knowledge of the domain. In this case, filler items are rated randomly on a normal distribution with mean equal to the average rating of the item being rated and with the standard deviation [2].

3 Architecture of Secure Personalized Recommendation System

As shown in Fig. 1 we provide the deployment of multi-agents technology. In this system, the client device is a handheld device or PDA as a terminal for receiving information from a web server and received signal from the GPS satellite. The detailed explanation of each module can be found in [9], [7].

In this paper, we will focus on to build the secure personalized recommendation system by applying modified RPCF algorithm for detection profile injection attacks (push attack or nuke attack) which is highlighted in interface module of Fig. 1.

The users are required to define their preferences (interests) as user profiles. The user's information can be changed whenever the users want to revise their interests. On receiving the user request, the personalization agent performs the task of gathering user information, interest or preferences explicitly and stores the context of information as the user profiles.

The next step is to analyze the information and apply modified RPCF algorithm. We picked the significant weighting and trust weighting complements to the RPCF Algorithm, which can give the secure personalized recommendations to the users.

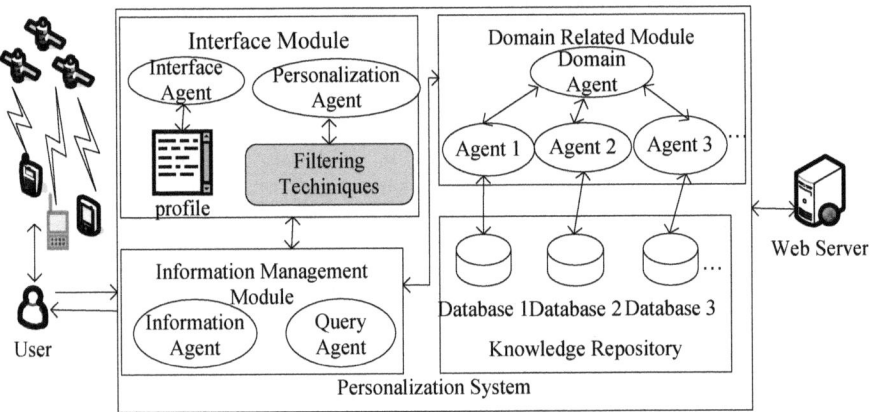

Fig. 1. Architecture of Personalized Recommendation System for Mobile User

3.1 Modified RPCF Algorithm

In this sub-section, we will promote RPCF algorithm with significant weighting and trust weighting to give the accurate recommendation for building secure personalized recommendation system. The main focus of this paper is to detect the profile injection attacks by applying modified RPCF algorithm as shown in Fig. 2.

```
Begin
//Modified RPCF Algorithm
1  get request
     //get the request from the user
```

```
  2  get user location
       //retrieve user location from GPS satellite
  3  get user profile
       //retrieve user information from the user profile or create user profile for new
  user
  4  if (condition) then (action)
       //take the action by the condition of Rule-Based Personalization
  5  search query information from the database
       //search the corresponding information from the database of knowledge
  repository
  6  compute similarity (identify the two items that are most similar)
       //compute similarity using Pearson Correlation Coefficient
  7  compute weight
       //compute weight using Significant Weighting (eq.(1))
  8  compute trust-weight
       //compute weight using Trust Weighting (eq.(3))
  9  compute prediction
       //compute prediction using (eq.(2))
  10compute MAE
       //compute Mean Absolute Error (MAE)
  11display result (recommendation)
       //display the result to the user which is closet to their interest
  End
```

Fig. 2. Modified RPCF Algorithm

3.2   Process of Modified RPCF Algorithm

Fig. 3 shows the recommendation process of modified RPCF algorithm. The modification process made to RPCF algorithm are computing significant weighting after computing similarity and trust weighting before making prediction.

As an example, rule-based filtering process takes the user's request and used the hotel rating to make recommendation to the target user. In this example, hotel rating profile contains the attacks inserted by the attacker and the ratings of the neighborhood users. The collaborative filtering method firstly computes the similarity values of the hotel rating for the targeted user among neighborhood users.

The item-based filtering method is used to filter the item according to the user's request and user-based filtering is used to filter the most appropriate item among the filtered items of the rule-based process based on the similar neighborhood users.

For instance, the user's request is "Hotels in Yangon", the system accept the user's request and retrieve the user's current location from the GPS and search the hotels according to the user's current location by using rule-based filtering. After searching the hotels, the system will give the most appropriate recommendation to the user according to the user's interest by applying collaborative filtering which is computed from neighborhood users.

During this computation process, the significant weighting is computed by using the similarity values of the neighborhood users. According to the significant weighting results the fewer commonly rated items are pushed out the neighborhood although there is a higher degree of the similarity to the target user. It follows that users who have rated a large number of items will belong to more neighborhoods than those users who have rated few items. This is a potential security risk in the context of profile injection attacks.

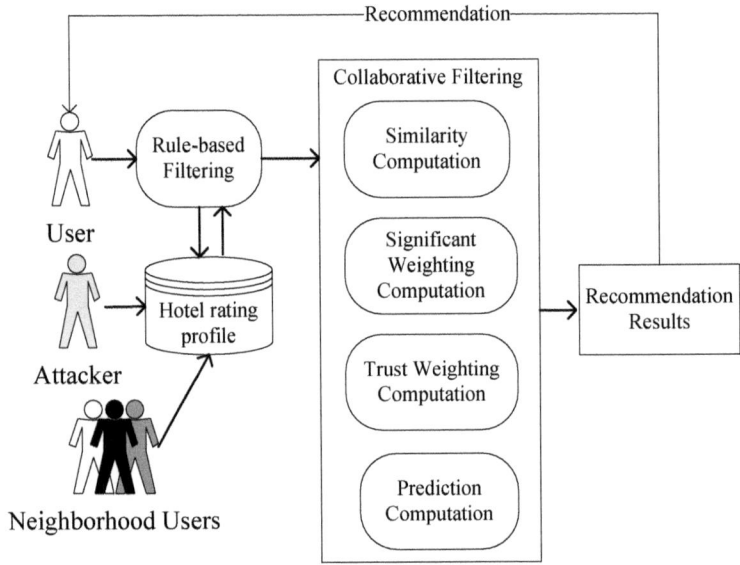

Fig. 3. Recommendation Process by Modified RPCF Algorithm for Hotel Enquiry

The significance weight of a target user u for a neighbor v is computed as:

$$W_{u,v} = \begin{cases} sim_{u,v} * \dfrac{n}{N} & n < N \\ sim_{u,v} & otherwise \end{cases} \tag{1}$$

where n is the number of co-rated items, N is a global constant, and $sim_{u,v}$ is Pearson's correlation coefficient. A prediction for the target user is computed by using equation (2), replacing $sim_{u,v}$ with $w_{u,v}$.

An attack profile with a very large number of filler items will necessarily be included in more neighborhoods, regardless of the rating value. The risk can minimize because a large filler size threshold is required to make the attack successful. In most cases, genuine users rate only a small portion of all recommendable items; therefore, an attack profile with a very large filler size is easier to detect [17].

$$P_{u,i} = \bar{r}_u + \frac{\displaystyle\sum_{i=1}^{m} sim_{u,v}(r_{v,i} - \bar{r}_v)}{\displaystyle\sum_{i=1}^{m} |sim_{u,v_i}|} \tag{2}$$

The profile injection attacks are detected by applying modified RPCF algorithm. Trust model can improve in collaborative filtering [12]. By explicitly calculating a trust value, the reputation of a user can be used as insight into the user's relevance to recommendation. Trust weighting calculates a trust value for every user by cross-validation using equation (3).

$$trust_{u,i} = \frac{\sum_{i=1}^{m} correct_{u,v,i}}{\sum_{i=1}^{m} recommend_{u,v,i}} \tag{3}$$

The system then computes the prediction. To incorporate values from the trust model into recommendation, the system filters the trust value with some threshold value [3]. The robustness of relevance weighting is evaluated as item-trust and similarity using equation (4).

$$w_{u,v,i} = \frac{2 * sim_{u,i} * trust_{v,i}}{sim_{u,i} + trust_{v,i}} \tag{4}$$

where $sim_{u,v}$ is Pearson's correlation coefficient. A prediction for the target user is computed using equation (2), replacing $sim_{u,v}$ with $w_{u,v,i}$.

4   Experimental Results of the Modified RPCF Algorithm

In the experiment, the hotel ratings datasets from Travelocity is used to evaluate RPCF algorithm. The dataset contains 2721 ratings from 40995 users' reviews for 740 hotels. Each user can rate a hotel to express his/her willingness to stay at this hotel and a rating is a number ranging from 1 to 5. A higher score indicates a higher preference.

Table 2. Hotel Rating Profiles showing Push Attack and Nuke Attack

Users	A Push Attack Favoring H10							A Nuke Attack Favoring H12						
	H1	H2	H3	...	H10	...	H15	H1	H2	H3	...	H12	...	H15
Alice	5	2	5	...	?	...	4	5	2	5	...	?	...	4
U1	5	3	3	...	2	...	5	5	3	3	...	4	...	5
U2	4	3	2	...	3	...		4	3	2	...	5	...	
U3	5	4		...	2	...	4	5	4		...	4	...	4
U4		4	3	...	1	...	4		4	3	...	5	...	4
⋮	⋮	⋮	⋮	⋮	⋮	⋮	⋮	⋮	⋮	⋮	⋮	⋮	⋮	⋮
Att 1		1		...	5	...			1		...	1	...	
Att 2	2	1		...	5	...	3	2	1		...	1	...	3
Att 3		1	2	...	5	...			1	2	...	1	...	
Att 4	1		1	...	5	...	2	1		1	...	1	...	2
Att 5		1	3	...	5	...			1	3	...	1	...	

Table 2 shows the hotel ratings of (genuine) users and attackers. In the sample case of push attack, the active user is Alice and the system will predict the rating on H10 using the ratings of the neighborhood profiles. The attacker, Eve, has injected the five attack profiles which give the high rating on H10. The push attack gives the high rating to the target item. In the case of nuke attack, the system will predict the rating on H12 for the active user Alice using the ratings of the neighborhood profiles. The attacker, Eve, has inserted the low rating on H12 to demote the target hotel rating.

Collaborative Filtering produce the personal recommendation by computing the similarity between the ratings of the neighborhood with target user for target item. On account of the shilling attacks, the active user can not get the actual rating of the target item. By applying modified RPCF algorithm, the push attack and nuke attack can be detected.

The evaluation results are depicted in the following figures with comparison of before and after attack detection algorithm is applied for push attack and nuke attack.

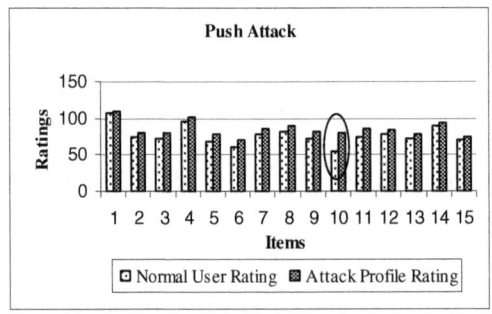

Fig. 4. Comparison of Hotel Ratings Values for Push Attack

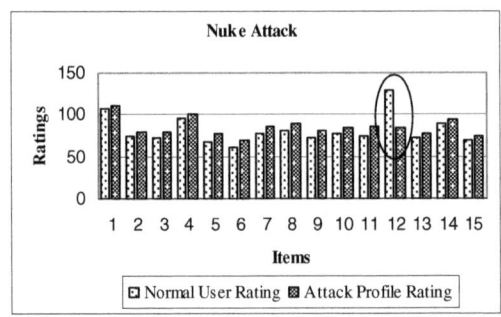

Fig. 5. Comparison of Hotel Ratings Values for Nuke Attack

Fig. 4 and 5 present the comparison of the total ratings for hotels of genuine users and attackers. Attackers inserted the push attack to H10 and nuke attack to H12 which shows the significant difference between the attackers' rating and genuine users rating. The total ratings of the other items are not very different or nearly the same.

In the experiment, the similarity measure between users is computed by means of Pearson Correlation Coefficient. Similarly, the similarity value is weighted and selected the neighboring users that have the highest similarity rating with the active user by using Significant Weighting and then computing the Trust Weighting. Finally, a prediction from the rating of neighbor is computed.

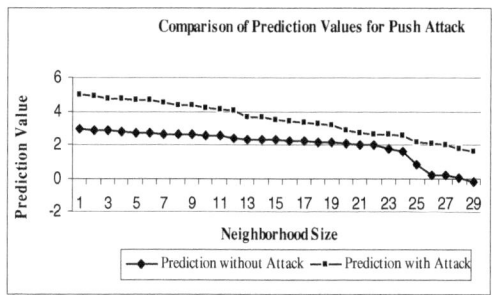

Fig. 6. Comparison of Prediction Values for Push Attack

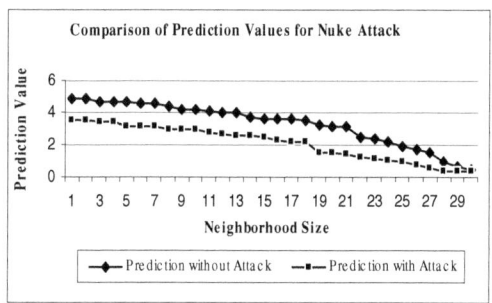

Fig. 7. Comparison of Prediction Values for Nuke Attack

Fig. 6 and 7 show the comparison of prediction values of the active user with and without shilling attack. Before applying attack detection algorithm, the prediction values are very high for the target item which is roughly 5 on account of the push attack. After detecting push attack, the prediction values are actual rating values for the active user. As shown in the comparison of prediction values for nuke attack in Fig. 7, the prediction value is low for the target item and after detecting the nuke attack gives the actual rating value. The neighborhood of the attackers is very sparse with genuine users. The highly neighborhood is removed from the neighbor and then the prediction result is computed. The modified RPCF algorithm can make the actual prediction to the active user by detecting the push or nuke attack.

Finally, the accuracy of a prediction is evaluated by using Mean Absolute Error (MAE). The lower MAE presents the more accurate prediction value. As shown in Fig.8 and 9, while the fluctuation of the MAE value with the attack is high, the MAE value without attack is quite stable. After detecting the push attacks and nuke attacks show the value of MAE is much closed to 0 (zero). This means that modified RPCF algorithm made predictions with fewer (even without) errors. Therefore we can say that the prediction result of modified RPCF algorithm is accurate and the system is robust under the profile injection attacks.

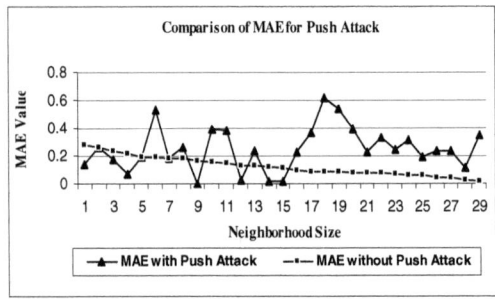

Fig. 8. Comparison of MAE Values for Push Attack

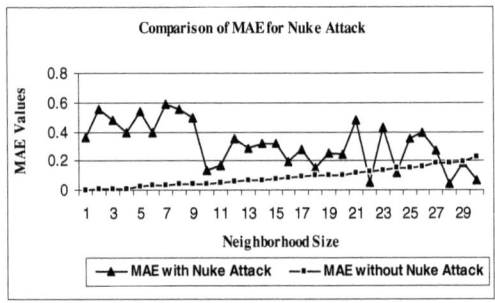

Fig. 9. Comparison of MAE Values for Nuke Attack

5. Conclusion and Future Work

The main purpose of this paper is to build the secure personalized recommendation system by adopting Significant Weighting and Trust Weighting which complements to RPCF algorithm. This algorithm can detect the profile injection attacks and can give the secure personalized recommendation to the user.

The experimental results showed that the increased in recommendation accuracy and improved robustness under profile injection attacks. In this paper, only the modified RPCF algorithm is experimented and implemented for the profile shilling attack, and the issue of false detection is not analyzed and lack of comparison with other algorithm.

For our future work, we will analyze the false detection issue, examine some other detection algorithms and other attack models to build more secure, robust and accurate recommendation system. The comparison with other algorithm will also be considered. To be a good personalized recommender system, we will focus on to the security issues of localization system, wireless network security and mobile security.

Security and privacy are the critical issues for developing a recommender system. In this work, we only consider the security issue. We plan to publish the privacy-preserving and trust management recommender system in our next paper.

References

1. Aïmeur, E., Brassard, G., Fernandez, J.M., Onana, F.S.M.: Alambic: A Privacy-Preserving Recommender System for Electronic Commerce. International Journal of Information Security 7(5), 307–334 (2008)
2. Burke, R., Mobasher, B., Zabicki, R., Bhaumik, R.: Identifying Attack Models for Secure Recommendation. In: Beyond Personalization: A Workshop on the Next Generation of Recommender Systems, San Diego, California (2005)
3. Cheng, Z., Hurley, N.: Analysis of Robustness in Trust-based Recommender Systems. In: RIAO 2010, Paris, France, Copyright CID (2010)
4. Goldberg, D., Nichols, D., Oki, B., Terry, D.: Using Collaborative Filtering to Weave an Information Tapestry. ACM 35(12), 61–70 (1992)
5. Lam, S.K., Riedl, J.: Shilling Recommender Systems for Fun and Profit. In: 13th International World Wide Web Conference (WWW 2004), New York, NY, USA, pp. 393–402 (2004)
6. Loh, S., Lorenzi, F., Saldana, R., Licthnow, D.: A Tourism Recommender System based on Collaboration and Text Analysis. Information Technology & Tourism 6, 157–165 (2004)
7. Maw, S.Y., Thein, N.L.: Multi-Agent Mobile Tourism System. Encyclopedia of Information Science and Technology. In: Information Science Reference, 2nd edn., vol. VI (Mu-Q), pp. 2722–2727. IGI Global Publishing, Hershey (2008)
8. Maw, S.Y., Naing, M.-M., Thein, N.L.: RPCF Algorithm for Multi-Agent Tourism System. In: IEEE International Symposium on Micro-NanoMachatronics and Human Science (MHS 2006), Nagoya, Japan, pp. 533–538 (2006)
9. Maw, S.Y., Naing, M.-M.: Multi-Agent Tourism System (MATS). In: Social Information Retrieval System: Emerging Technologies and Applications for Searching the Web Effectively. Ch. XV, pp. 289–310. IGI Global Publishing, Hershey (2007)
10. Mobasher, B., Burke, R., Bhaumik, R., Williams, C.: Effective Attack Models for Shilling Item-based Collaborative Filtering Systems. In: WebKDD Workshop (2005)
11. Mobasher, B., Burke, R., Bhaumik, R., Sandvig, J.: Attacks and Remedies in Collaborative Recommendation. IEEE Intelligent Systems 22(3), 56–63 (2007)
12. O'Donovan, J., Smyth, B.: Trust in Recommender Systems. In: 10th International Conference on Intelligent User Interfaces (IUI 2005), pp. 167–174. ACM Press, San Diego (2005)
13. Quan, Q., Hinze, A.: Trust-based Recommendations for Mobile Tourists in TIP. Working Paper: 13/2008, Hamilton, New Zealand (2008)
14. Sammut, C., Webb, G.: Encyclopedia of Machine Learning. Ch. 00338, pp. 1–9. Springer, Heidelberg (2010)
15. Shon, T., Choi, W.: An Analysis of Mobile WiMAX Security: Vulnerabilities and Solutions. In: Enokido, T., Barolli, L., Takizawa, M. (eds.) NBiS 2007. LNCS, vol. 4658, pp. 88–97. Springer, Heidelberg (2007)
16. Web site personalization, http://www.128.ibm.com/developerworks/websphere/library/techarticles/hipods/personalize.html
17. Williams, C., Bhaumik, R., Burke, R., Mobasher, B.: The Impact of Attack Profile Classification on the Robustness of Collaborative Recommendation. In: WebKDD Workshop, ACM SIGKDD Conference on Data Mining and Knowledge Discovery (KDD 2006), Philadelphia (2006)

Protecting White-Box AES with Dual Ciphers

Mohamed Karroumi

Technicolor, Security & Content Protection Labs
1 avenue de Belle Fontaine, 35576 Cesson-Sévigné Cedex, France
mohamed.karroumi@technicolor.com

Abstract. In order to protect AES software running on untrusted platforms, Chow et al. (2002) designed a white-box implementation. However, Billet et al. (2004) showed that the secret key can be extracted with a time complexity of 2^{30}. In this paper, we present an improved white-box implementation of AES. We use dual ciphers to modify the state and key representations in each round as well as two of the four classical AES operations, SubBytes and MixColumns. We show that, with 61200 possible dual ciphers the complexity of Billet et al. attack is raised to 2^{91}. Interestingly, our white-box implementation does not require more memory space than that of Chow et al. implementation.

Keywords: White-box cryptography, dual cipher, AES, block ciphers, implementation.

1 Introduction

A cryptographic algorithm intended to run on a malicious host is, by definition, prone to the reverse engineering attacks. The adversary is actually able to monitor its execution as well as any intermediate results generated during the computation. The white-box attack context was introduced by Chow et al. in [8] as a setting where the adversary is allowed to make observations about the software and to examine or alter the software intermediate results. In order to protect AES in such context, they implement the AES encryption or decryption algorithm in a white-box fashion [9]. However, Billet et al. showed in [4] that the secret key can be extracted from Chow et al. implementation with a time complexity of about 2^{30}. Michiels et al. presented in [14] a generic attack against white-box ciphers. They prove that a family of Substitution Linear-Transformation (SLT) ciphers with special properties of the diffusion matrix cannot be secured by the method of Chow et al. The attack they present is partially based on the one of Billet et al.

There is therefore a need for an improved implementation that makes such attacks more difficult in a white-box context. In [3], Billet and Gilbert proposed a traceable block cipher from which they can derive from a meta-key many equivalent keys for the same instance of a cipher. Their construction presents the advantage of making computationally difficult the calculation of the meta-key when knowing the derived keys. This has been cryptanalyzed by Faugere

K.-H. Rhee and D. Nyang (Eds.): ICISC 2010, LNCS 6829, pp. 278–291, 2011.
© Springer-Verlag Berlin Heidelberg 2011

et al. [11]. In [6], Bringer et al. showed how to improve the security of the traceable block cipher by adding some perturbations in its description. The perturbation idea was exploited by the same authors to improve the white-box AES implementation [7]. Indeed, Bringer et al. added perturbations to the global rounds of AES in order to make its algebraic structure inaccessible. In this instance, the constants of SubBytes operation are made non-standard and unknown to the adversary.

In this paper, we propose to build upon Chow et al. ideas to create a version of white-box AES that better resists to the Billet et al. attack. Our approach changes the algebraic structure in each round of AES in addition to the mixing bijections. The use of different algebraic structure for the same instance of an iterative block cipher was already proposed in [3]. The intrinsic structure of the block cipher used in [3] is however based on Matsumoto-Imai multivariate scheme [13] that is different from that of AES. We propose a different method that works with the AES building block structure and that improves the security of the white-box implementation. Our modification concerns all the operations that involve constants within one round (i.e. InvSubBytes, InvMixColumns and key schedule operations). Moreover, all the elements of a state as well as the round subkeys are transformed in order to fit the modified structure in each round.

Our solution relies on using the dual ciphers [2,15,5] and raises the complexity of Billet et al. attack from 2^{30} to 2^{91}. The structure changes make the original cipher more intricate for the adversary such that he has to repeat the attack of Billet et al. for all possible combinations of dual ciphers. Although raising the attack complexity to 2^{91} operations does not provide theoretical security for a 128-bit AES decryption key, it is useful from a practical perspective. In addition to providing the white-box AES with a protection against practical attacks, our design implementation is comparable in time and space requirements to that of Chow et al.

The rest of this paper is organized as follows. Section 2 describes the white-box AES implementation proposed by Chow et al. In Section 3 we review the Billet et al. attack as well as the generic attack proposed by Michiels et al. In Section 4 our improvements to the white-box implementation are detailed, and in Section 5 we propose an enhanced implementation that is better resistant against the attacks. Finally, we conclude in Section 6.

2   AES White-Box Implementation

For most cryptographic applications, a program is supplied with an AES decryption algorithm together with a decryption key. As the decryption key must be kept secret and inaccessible to the user, the AES decryption algorithm (which is different from the encryption algorithm) has, in some cases, to be white-box implemented. This is for example the case when the application is expected to run on an open platform. In other cases, one may want the white-box implementation to be backward compatible with a legacy implementation (non white-box) used previously. For these reasons, we give in this paper a description of a white-box implementation for the decryption case (but note that our technique can be applied to the encryption case as well). The AES-128 will serve for illustration purposes; but this can be adapted for AES-192 or AES-256.

The first step of the white-box AES implementation (WB-AES), is to convert AES into a series of look-up tables and to hide the secret keys into these tables. Compared with a standard AES implementation [10], the operations of the WB-AES rounds are slightly modified without impacting the input or the output of the round.

Algorithm 1. Regular Implementation of AES Decryption

$S \leftarrow$ AddRoundKey(S,ExpandedKey[10])
$S \leftarrow$ InvShiftRows(S)
$S \leftarrow$ InvSubBytes(S)
$S \leftarrow$ AddRoundKey(S,ExpandedKey[9])
for $i = 9$ **downto** 1 **do**
   $S \leftarrow$ InvMixColumns(S)
   $S \leftarrow$ InvShiftRows(S)
   $S \leftarrow$ InvSubBytes(S)
   $S \leftarrow$ AddRoundKey(S,ExpandedKey[$i - 1$])
end for

InvSubBytes (IS) and AddRoundKey are combined in a single step, and the subkey (calculated using the AES key expansion) is integrated into InvSubBytes by creating the byte input/output look-up tables T^r of round r. InvShiftRows is implemented by providing shifted input data to the generated tables. The first round table is slightly different from the other rounds as there is an additional AddRoundKey. The T-boxes T^r are defined as follows:

$$T_{i,j}^1(x) := IS(x \oplus K_{i,j}^0) \oplus K_{i,j}^1 , \qquad i \in [0..3], j \in [0..3]$$
$$T_{i,j}^r(x) := IS(x) \oplus K_{i,j}^r , r \in [2..10], i \in [0..3], j \in [0..3].$$

$K_{i,j}^r$ represents the subkey byte number $4 \cdot i + j$ of the round r and $K_{i,j}^{10}$ is the decryption key[1] K. In total we have $10 \times 16 = 160$ T-boxes.

InvMixColumns operates on the AES state one column at a time. This can be implemented by multiplying a 32×32 matrix IMC and a vector in $GF(2)$. Multiplication of a 32-bit vector by the IMC matrix is performed by four separate multiplications and three 32-bit XORs. To avoid large tables, the matrix IMC is divided into four 32×8 matrices ($IMC_0, \ldots, IMC_3$) and the multiplication is performed separately with each matrix.

---

[1] To avoid confusion, we change the subkey numbering to make it correspond to round numbering in the decryption process. ExpandedKey[10], ..., ExpandedKey[0] are renamed to $K^0, \ldots, K^{10}$.

The rounds' boundaries are also modified. Let a round begin with the T-boxes computations followed by `InvMixColumns` and finally `InvShiftRows`. Under this condition, the last round does not contain any `InvMixColumns` operation.

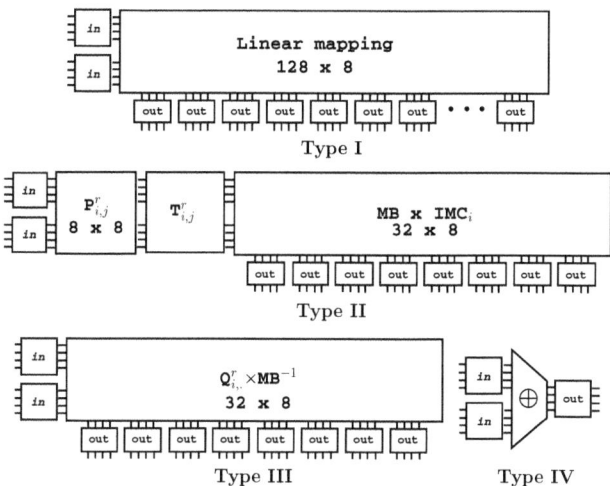

Fig. 1. Mixing Bijection Tables

The next step is to compose each table with random bijections. A mixing bijection $P_{i,j}^r$ (8×8 matrix in $GF(2)$) is inserted before the T-boxes $T_{i,j}^r$ and an affine bijection MB (32×32 matrix in $GF(2)$) is inserted after `InvMixColumns` step (type II table). MB is a non-singular matrix with 4×4 sub-matrices of full rank. The mixing bijections of the next round $P_{i,j}^{r+1}$ and the affine bijection of the current round MB are inverted (resp. $Q_{i,j}^r$ and MB^{-1}) and combined in an additional set of look-up tables (type III). In order to avoid large look-up tables, the mapping MB^{-1} is split into four blocks, just like `InvMixColumns` matrix. The XORs are computed by type IV tables. Finally, the external input and output encodings are implemented. It consists in selecting two 128×128 mixing bijection matrices F and G, defined over $GF(2)$, whose all aligned 4×4 sub-matrices are of full rank. Multiply F by the inverted input mixing bijections of the first round (i.e. $Q_{i,j}^0, i \in [0..3], j \in [0..3]$), then insert the resulting bijection F' prior to the first round. G is inserted after the last `AddRoundKey` step. The external encodings are implemented in type I tables.

3   Known Attacks

3.1   Billet et al. Attack

In the following, we list the different steps of the attack and their complexity. We refer the reader to [4] for more details. Instead of locally inspecting the tables, Billet et al. looks at the input and the output of the composition of the tables for a round. A round consists of type II tables, type III tables and supporting type IV tables, i.e. four 4-byte input and 4-byte output mappings. $P_{i,j}^r$ (resp. $Q_{i,j}^r$) is the composition of two concatenated 4-bits to 4-bits input (resp. output) encodings and one 8-bits to 8-bits mixing bijection. $P_{i,j}^r$ and $Q_{i,j}^r$ cancel each other between two consecutive rounds (i.e. $Q_{i,j}^r = \text{Inv}(P_{i,j}^{r+1})$, $r \in [0..9]$).

It was shown that by analyzing a round input/output the non-linear 4-byte mapping $P_{i,j}^r$ and $Q_{i,j}^r$ can be reduced to one where they are affine. Removing non-linear part requires 2^{24} computation steps for each mapping.

To recover the affine mapping $Q_0^r = A_0 \oplus q_0$ where A_0 is linear and q_0 is a constant, it was first shown in [4] that there exists a unique linear mapping L and a constant c, such that for all $x_0 \in GF(2^8)$

$$y_i(x_0, 0, 0, 0) = L(y_j(x_0, 0, 0, 0)) \oplus c$$

where y_i is a function of (x_0, x_1, x_2, x_3) defined as follows:

$$y_i(x_0, x_1, x_2, x_3) = Q_i^r(\alpha_{i,0} T_0^r(P_0^r(x_0)) \oplus \alpha_{i,1} T_1^r(P_1^r(x_1)) \oplus \\ \alpha_{i,2} T_2^r(P_2^r(x_2)) \oplus \alpha_{i,3} T_3^r(P_3^r(x_3))) \tag{1}$$

($\alpha_{i,j}$ are the coefficients of InvMixColumns) and that (L, c) can be determined with a complexity lower than 2^{16} by solving an over-defined linear system of equations, involving 2048 equations and 72 unknowns.

Second, by determining the characteristic polynomial of L and knowing the coefficients of InvSubBytes operation, it is possible to determine A_0 with a time complexity of about 2^{24}. From the knowledge of $\alpha_{i,j}$ values, constant q_0 is recovered at the same time by setting the four variables in Equation (1) to a zero value. Billet et al. show that from the knowledge of the linear part of Q_0^r, the linear parts of Q_1^r, Q_2^r and Q_3^r can be computed with a time complexity of 2^{16} for each part. All Q_i^r of a round can be recovered similarly. As $Q_i^r = \text{Inv}(P_i^{r+1})$, P_i^{r+1} is recovered at the same time. Once the mappings are recovered, the bytes of a subkey round (embedded in the T-boxes) can be retrieved. The bytes are however in a shuffled order. Nevertheless, computing the Q_i^r for two consecutive rounds makes it possible to get another shuffled subkey. Constraints in the AES key schedule algorithm enable retrieving both subkeys in the correct order as well as all other round subkeys.

Finally, the complexity for recovering the affine part of the Q_i^r is $2^{16} + 2^{24} + 3 \cdot 2^{16} \approx 2^{24}$. The non-affine part of the Q_i^r can be recovered with the same time complexity of 2^{24}. Hence, P_i^r and Q_i^r can be determined in 2^{25} steps. The attack is performed for two consecutive rounds. The total complexity of the attack is bounded by $2 \cdot 4 \cdot 4 \cdot 2^{25} = 2^{30}$.

3.2   Michiels et al. Attack

Michiels et al. presented in [14] another type of attack against white-box ciphers. They interestingly remarked that the diffusion operator of a block cipher makes the white-box implementation relatively vulnerable to attacks. The diffusion operator in the case of AES decryption algorithm is represented by the InvMixColumns operation. The attack they propose is composed of three steps.

- The first consists in removing the non-linear part of the mixing bijection encodings. This step is partially based on the Billet et al. method.
- The second consists in guessing the linear part of the encodings. This step uses a method for solving linear equivalence problem for matrices (LEPM) [5,12].
- The third is the extraction of the secret key information by algebraic analysis.

In the second step cryptanalysis, a round function is described as a Substitution Affine-Transformation (SAT) cipher round function. A round function in a SAT cipher is a cascade of T-boxes T_i^r, followed by an affine transformation b^r. These components T_i^r and b^r can be computed by an adversary provided that InvMixColomns operation is known. The last step performs the round key extraction. This is achieved by obtaining the equivalence between the computed T_i^r and the inverse S-boxes IS_i^r of the AES decryption algorithm. The algebraic equations that need to be solved are:

$$T_i^r = c_i^r \circ IS_i \circ d_i^r ,$$

where c_i^r, d_i^r are the affine functions that describe the affine relation between T_i^r and IS_i^r. Function c_i^r depends on b^r and d_i^r contains the key addition operation. Solving these equations leads to the secret decryption key.

From these two attacks, we learn that the input and output mixing bijection encodings do not sufficiently hide the rounds' operations. This is especially the case if the parameters of the round operations are publicly known. Indeed, both attacks are based on the fact that coefficients of InvSubBytes and InvMixColumns are known. Also, they both have a similar complexity when applied against a white-box AES implementation. Consequently, raising the complexity in the context of Billet et al. attack makes the system to be more difficult to break with Michiels et al. attack too.

4   Our White-Box Implementation

4.1   General Idea

AES is based on simple algebraic operations over the finite field $GF(2^8)$. If we change all the constants in AES, including the irreducible polynomial, matrix coefficients, affine transformations, we could create new dual ciphers. It is mentioned in [2] that 240 new dual ciphers of AES can be so created. The list of these 240 dual ciphers can be found in [1]. There are even more AES dual ciphers according to [15,5]. In [5], authors expand the set of 240 ciphers to a set of $61,200$ representations that are dual to the AES.

Outputs of AES and dual AES are correlated. There exists a linear transformation Δ that maps a byte state of AES into a byte state of a dual AES, i.e. $X_{dual} = \Delta(X)$. The same transformation maps also the AES input or output like the plaintext P, the ciphertext C and the decryption key K into the dual AES input or output (i.e. $P_{dual} = \Delta(P)$, $C_{dual} = \Delta(C)$ and $K_{dual} = \Delta(K)$). Other transformations can be built to map a state of any dual AES into a state of another different dual AES. An algorithm described in [5] permits to compute an affine equivalence for two S-boxes S_1 and S_2.

We present here a method that uses multiple different dual AES within the same WB-AES implementation.

- We choose a random dual representation for every AES round (10 in total).
- InvSubBytes constants and InvMixColumns matrix of a given round are replaced by the one of the corresponding dual AES.
- To construct the new T-boxes ($T'_{i,j}$), a key is expanded through all dual AES key expansions and for each round, we select the corresponding dual subkey.

With these modifications, each round takes a byte state of the corresponding dual AES and outputs a byte state for the same dual AES. In order to keep, for a given input, the output of the overall implementation unchanged, the round input and output have to be encoded with the linear transformation Δ. The encoding Δ is used such that a byte state at the input matches the modifications made in the round internal operations. Considering a round building block B as a combination of four lookup operations using the new T-boxes and a multiplication by the new matrix IMC', the encoding will correspond to a composition $\Delta \circ B \circ \Delta^{-1}$.

As the white-box mixing bijections are built using the same principle, our idea is to incorporate the Δ-encodings within these mixing bijections. There are two possible strategies. The first uses a single encoding $\Delta_r \times \Delta_{r-1}^{-1}$ to perform at the same time the output Δ-decoding of the previous round and the input Δ-encoding of the current round and to combine it with the mixing bijection $P_{i,j}^r$. The second uses one encoding $\Delta_{r+1} \times \Delta_r^{-1}$ to perform at the same time the output Δ-decoding of the current round and the input Δ-encoding of the next round and to combine it with the inverse mixing bijection $Q_{i,j}^r$. The two strategies are illustrated in Figure 2. Both strategies are similar from a security perspective. Nevertheless, the first strategy requires changing only type I and II tables, whereas the second requires modifications in type I, type II and also in type III tables. Thus, we choose the first for the description.[2]

- We multiply the linear transformation of the first round (i.e. Δ_1) by the mixing bijections of the first round (i.e. $P_{i,j}^1$) resulting in mixing bijections inserted in type II tables of the first round.
- The linear transformation of the previous round Δ_{r-1} is inverted and left-multiplied with the linear transformation of the current round (i.e. $\Delta_r \times \Delta_{r-1}^{-1}$). The result of the multiplication is combined with the mixing bijections $P_{i,j}^r$ for r in $[2..10]$.
- The linear transformation Δ_{10} is inverted and combined with the external decoding G after the last round.

---

[2] Another strategy consists in combining Δ_r with the mixing bijection $P_{i,j}^r$ and Δ_r^{-1} with the inverse mixing bijection $Q_{i,j}^r$ but it brings nothing as the input before type II tables and the output after type III tables do not change.

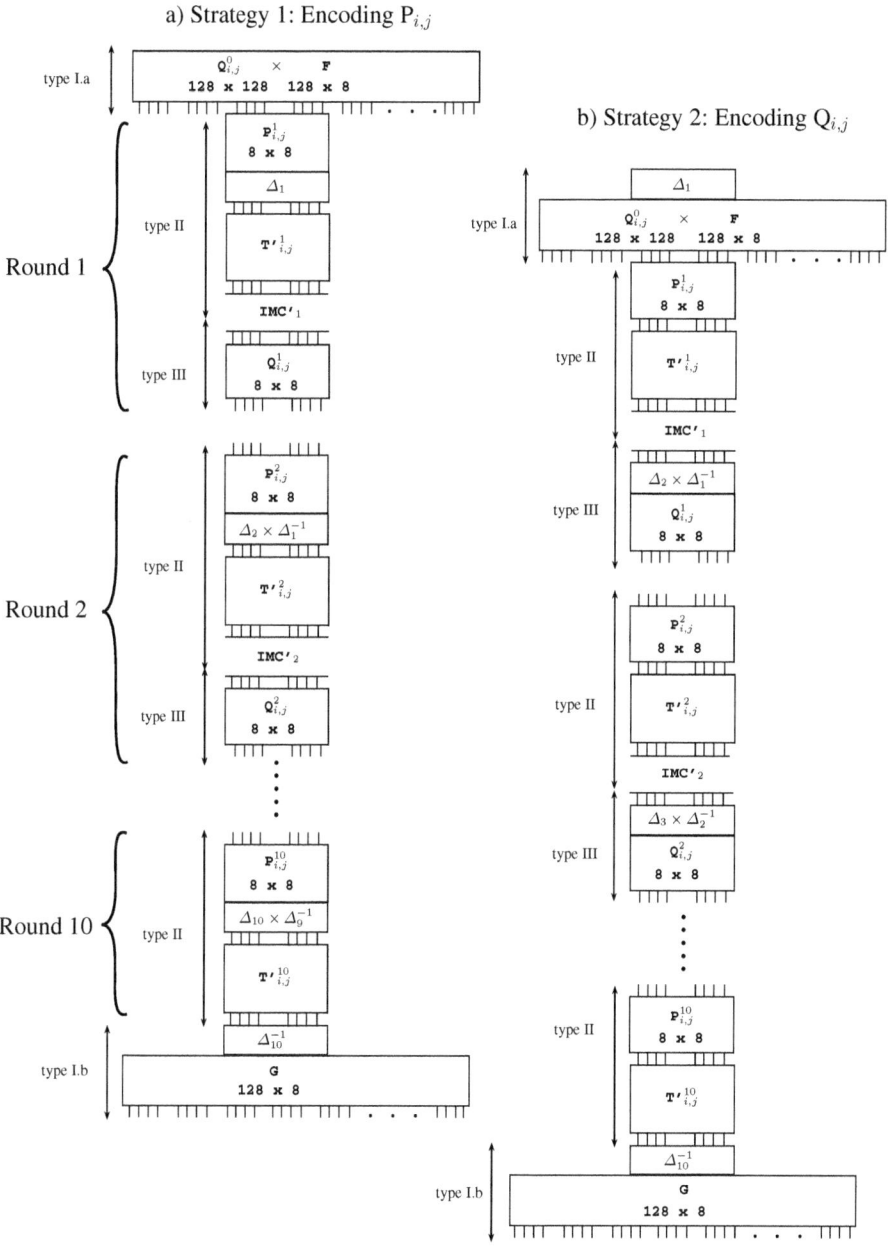

Fig. 2. Encoding The Mixing Bijections

Doing so, all rounds intermediate values will be different from those of classical WB-AES. Value of `InvSubBytes` and `InvMixColumns` constants are not fixed but vary from round to round depending on the dual parameters beeing used. For a given input, these modifications do not change the output of our WB-AES, when compared to Chow et al. implementation's output.

4.2   Construction of the New Tables

Each dual AES (D-AES) representation is allocated an index from 1 to 61200. Choose randomly 10 values $\sigma_r \in_R \{1, \ldots, 61200\}$ for $r = 1, \ldots, 10$ without repetition. This random value permits to select, for a round number r of AES, associated D-AES in which operations are performed. Let $\Delta_{\sigma_r} : GF(2^8) \rightarrow GF(2^8)$ the linear transformation that maps a byte state of AES into a byte state of D-AES number σ_r. Δ_{σ_r} can be represented as an invertible matrix M_r of size 8×8 in $GF(2)$ which maps a representation of a byte of the state array of AES into a byte of the state array of D-AES(σ_r). The inverse mapping $\Delta_{\sigma_r}^{-1}$ is obtained by inverting the matrix M_r in $GF(2)$.

The New T-Boxes. `InvSubBytes` operation can be represented in an algebraic way:

$$IS : GF(2^8) \rightarrow GF(2^8), x \mapsto IS(x) = A \cdot x + b$$

where A is a matrix transformation and b is a constant vector. The non-linear transformation is replaced by $IS^{\sigma_r}(x) = (M_r \cdot A \cdot M_r^{-1}) \cdot x + M_r \cdot b$.

A dual subkey byte is obtained by $K_{i,j}^{\sigma_r} = M_r \cdot K_{i,j}$ from AES subkey byte. The new look-up tables T^{σ_r} of round r are built as follows:

$$T_{i,j}^{\sigma_1}(x) := IS^{\sigma_1}(x \oplus K'^0_{i,j}) \oplus K_{i,j}^{\sigma_1} , \quad (i,j) \in [0..3]^2$$
$$T_{i,j}^{\sigma_r}(x) := IS^{\sigma_r}(x) \oplus K_{i,j}^{\sigma_r} , r \in [2..10], (i,j) \in [0..3]^2$$

where $K'^0_{i,j} = \Delta_{\sigma_1}(K^0_{i,j})$, and $K_{i,j}^{\sigma_r} = \Delta_{\sigma_r}(K^r_{i,j})$ for $r \in [1..10]$. The transformations IS^{σ_r} are modified from original `InvSubBytes` according to the matrix representing Δ_{σ_r}.

The New IMC Matrix. Any constant c in `InvMixColumns` is replaced by $M_r \cdot c$. This means that polynomial constants of `InvMixColumns` are replaced by $M_r \cdot 0b, M_r \cdot 0d, M_r \cdot 0e$ and $M_r \cdot 09$ leading to a new matrix $IMC' = IMC^{\sigma_r}$. IMC' is then combined with a 32×32 random matrix MB just like in the Chow et al. implementation.

Encoding the Mixing Bijections. In our new design of WB-AES, only the mixing bijections in type I and II tables are modified.

For type II tables, we multiply the dual transformation of the current round Δ_{σ_r} with the inverse dual transformation of the previous round, i.e. $\Delta_{\sigma(r)} \times \Delta_{\sigma_{r-1}}^{-1}$ with r in [2..10]. Next, we multiply the the result of $\Delta_{\sigma_r} \times \Delta_{\sigma_{r-1}}^{-1}$ with the input mixing bijections of the current round $P_{i,j}^r$, which gives us the new mixing

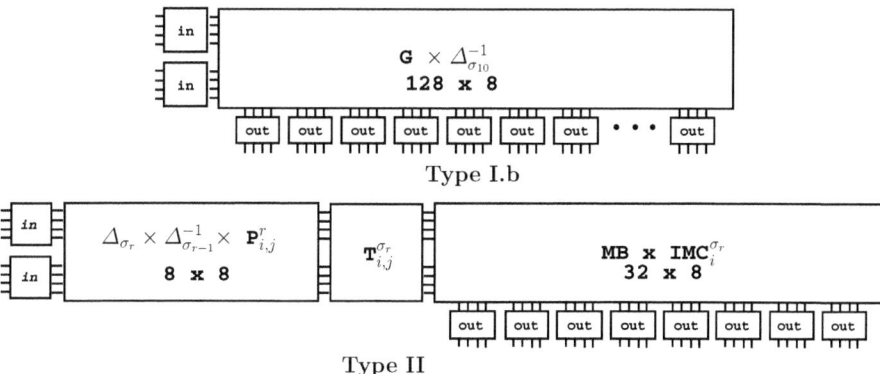

Fig. 3. New Type I.b and II Tables

bijections $P'^r_{i,j} = \Delta_{\sigma_r} \times \Delta^{-1}_{\sigma_{r-1}} \times P^r_{i,j}$. For the first round, we multiply the first dual transformation with $P^1_{i,j}$; the mixing bijections are then $P'^1_{i,j} = \Delta_{\sigma_1} \times P^1_{i,j}$.

Regarding type I tables, two encodings F and G are put around the initial white-box implementation. F and G are both randomly chosen as 128×128 matrices in $GF(2)$ in which all aligned 4×4 sub-matrices are of full rank. Prior to be inserted, F is left-multiplied by $Q^0_{i,j}, i \in [0..3], j \in [0..3]$ (just like in the original implementation) and G is multiplied by $\Delta^{-1}_{\sigma_{10}}$. This last operation is new and has first the effect to undo the Δ-encoding of the last round. Finally, the resulting 128×128 matrices F' and G' are split into 128×8 tables and inserted respectively before the first and the after last `AddRoundKey` operations. These tables are followed by 4-bit to 4-bit non-linear input decodings and output encodings implemented with type IV tables (we omit to describe these encodings here).

Proposition 1 (correctness). *Any $T^{\sigma_r}_i, r \in [1..10]$ as constructed above for D-AES(σ_r), receives as input an AES vector state transformed by Δ_{σ_r}.*

Proof. Let $(x_0, \ldots, x_{31})$ be an input word for the round r, for which dual cipher is D-AES(σ_r). Let $(y_0, \ldots, y_{31})^{\sigma_r}$ be the output after the table of type III, which serves as input to type II table of the next round. We have then $(y_0, \ldots, y_{31})^{\sigma_r} = Q^r((z_0, \ldots, z_{31})^{\sigma_r})$, where $(z_0, \ldots, z_{31})^{\sigma_r}$ is the output after the `InvMixColumns` operation.

When the data is to be processed with type II tables of the round $r + 1$, the mixing bijection P^{r+1} is first applied to the vector $(y_0, \ldots, y_{31})$. We have then as input for the round $r + 1$ the following

$$= \Delta_{\sigma_{r+1}} \circ \Delta^{-1}_{\sigma_r} \circ P^{r+1}((y_0, \ldots, y_{31})^{\sigma_r})$$
$$= \Delta_{\sigma_{r+1}} \circ \Delta^{-1}_{\sigma_r} \circ P^{r+1}(Q^r((z_0, \ldots, z_{31})^{\sigma_r}))$$
$$= \Delta_{\sigma_{r+1}} \circ \Delta^{-1}_{\sigma_r}((z_0, \ldots, z_{31})^{\sigma_r})$$
$$= \Delta_{\sigma_{r+1}}((z_0, \ldots, z_{31}))$$
$$= (z_0, \ldots, z_{31})^{\sigma_{r+1}} .$$

This shows that the vector is in the correct dual state before being interpreted by the T-box in the round $r + 1$ and holds for r in $[1..9]$. We can show similarly that the vector being interpreted by the T-boxes of the first round is in the correct dual state.

We ignored the input encodings and the output decodings implemented by type IV tables because the input decodings before type II tables cancel the output encodings after type III tables.   □

5   Security Analysis

5.1   Attacking Our Implementation

Billet et al. attack supposes that classical AES constants in `InvSubBytes` or `InvMixColumns` coefficients are known. Knowing `InvSubBytes` parameters is helpful for computing A_0 whereas `InvMixColumns` coefficients, which are based on the four numbers 0x0b,0x0d,0x0e,0x09, are helpful for determining (L, c) and the constants q_i. Furthermore, an attacker that is able to guess mappings Q_i^r for a round r gets only a shuffled round subkey. To recover the decryption key, the attacker has to guess the mappings for two consecutive rounds. In our implementation, the subkeys for two consecutive rounds are not related anymore and were derived from algorithms that use different algebraic structures. Indeed, `InvSubBytes` constants and `InvMixColumns` coefficients as well as constants in the key schedule algorithm differ for any two rounds depending on the dual cipher used amongst the 61200 possible ones.

An attacker who observes the inputs of all tables in this implementation would have access to the encoded version $y_i = \Delta_{\sigma_r}(x_i)$ of each byte state value x_i, $i = 0, \ldots, 15$. Here Δ_{σ_r} is a secret linear mapping used as input encoding for the T-boxes $T_{i,j}^{\sigma_r}$. To reconstruct the byte in the standard AES state, all the combinations have to be checked by calculating $z_i^k = \Delta_k^{-1}(y_i)$, $i = 0, \ldots, 15$ and $k = 1, \ldots, 61200$. Then the attacker repeats the attack of Billet et al. twice for all 61200 possible vector states $(z_0^k, \ldots, z_{15}^k)$. This raises the attack complexity to at least 2^{16} more computation steps, which makes the complexity of the attack to be 2^{46}.

In the context of a Michiels et al. attack, our implementation makes the diffusion operator to be a variable and thus prevents its vulnerability. Indeed the diffusion operator depends on the varying dual ciphers, which make steps 2 and 3 of the attack more difficult, i.e. it is more difficult to find out what are the cascaded T_i^r and b^r as well as the affine relation between T_i^r and IS_i^r by using a linear equivalence solver for matrices. This way the attacker needs to discover more information to realize a successful attack.

5.2   Improving the Resistance

We give in the following a generalization of our construction that provides a better resistance against the attacks. We have shown in Section 4 how to implement 10 different dual ciphers in the same white-box implementation. Indeed, we changed the dual cipher at the round level (to ease the description). It is possible to use even more dual ciphers. Since each 4 bytes output of a round depends only on 4 bytes of input to that round, a different dual AES cipher may be used for each of the four mappings in a round, which means that up to $4 \cdot 10 = 40$ different dual ciphers can be used in a given white-box AES implementation. If we let y_i is the i-th output byte of type III tables of the round r then we have:

$$y_0, y_4, y_8, y_{12} \text{ depend on } x_0, x_1, x_2, x_3 \ ;$$
$$y_1, y_5, y_9, y_{13} \text{ depend on } x_4, x_5, x_6, x_7 \ ;$$
$$y_2, y_6, y_{10}, y_{14} \text{ depend on } x_8, x_9, x_{10}, x_{11} \ ;$$
$$y_3, y_7, y_{11}, y_{15} \text{ depend on } x_{12}, x_{13}, x_{14}, x_{15} \ .$$

Without loss of generality, let $\Delta_{\sigma_0}^{(r)}, \ldots, \Delta_{\sigma_3}^{(r)}$ be the four different transformation matrix associated to the dual ciphers used in round r. Let the bytes $(x_0, \ldots, x_3)^{\sigma_0} = \Delta_{\sigma_0} \cdot (x_0, \ldots, x_3)^t$. Using $\Delta_{\sigma_1}^{(r)}, \Delta_{\sigma_2}^{(r)}$ and $\Delta_{\sigma_3}^{(r)}$ we get $(x_4, \ldots, x_7)^{\sigma_1}, (x_8, \ldots, x_{11})^{\sigma_2}$ and $(x_{12}, \ldots, x_{15})^{\sigma_3}$. The resulting bytes are taken as input of type II tables for which the T-boxes were built as follows:

$$T_{i,j}^1(x_{4 \cdot i+j}) := IS_i^{\sigma_i}(x_{4 \cdot i+j} \oplus K'^{0}_{i,j}) \oplus K_{i,j}^{\sigma_i}, \qquad (i,j) \in [0..3]^2$$
$$T_{i,j}^r(x_{4 \cdot i+j}) := IS_i^{\sigma_i}(x_{4 \cdot i+j}) \oplus K_{i,j}^{\sigma_i}, \ r \in [2..10], (i,j) \in [0..3]^2,$$

where $K'^{0}_{i,j} = \Delta_{\sigma_i}^{(1)}(K_{i,j}^0)$ and $IS_i^{\sigma_i}$ for $i \in [0..3]$ are modified from original InvSubBytes according to the matrix representing $\Delta_{\sigma_i}^{(r)}$ for the round r. Now, as data are shifted (to implement InvShiftRows) as input to type III tables, care should be taken as to which product $\Delta \times \Delta^{-1}$ is to be combined with which of type II tables of the next round to have the correct input state. As illustrated in Figure 4, we then encode the mixing bijections of type II tables of round $r + 1$ as follows:

$$z_{4 \cdot i+j}^{(1)} = \Delta_{\sigma_i}^{(1)} \times P_{i,j}^1(y_{4 \cdot i+j}), \ (i,j) \in [0..3]^2$$
$$z_{4 \cdot i+j}^{(r+1)} = \Delta_{\sigma_i}^{(r+1)} \times (\Delta_{\sigma_j}^{(r)})^{-1} \times P_{i,j}^{r+1}(y_{4 \cdot i+j}), \ r \in [1..9], (i,j) \in [0..3]^2 \ .$$

It can be noted that product $\Delta_{\sigma_i}^{(r+1)} \times (\Delta_{\sigma_j}^{(r)})^{-1}$ changes for each of the 16 tables in a round $r + 1$, for r in $[1..9]$. Similarly, we modify the mixing bijections in type I.b table as:

$$z_{4 \cdot i+j} = G \times (\Delta_{\sigma_j}^{(10)})^{-1}(y_{4 \cdot i+j}), \ i \in [0..3], j \in [0..3] \ .$$

In the case of Billet et al. attack, an attacker would need to put each 4 bytes output in the standard AES state. To do so, he has to check $61200^4 \approx 2^{63}$ combinations. The complexity for recovering mixing bijections for a round would be then $4 \cdot 2^{25} \cdot 2^{63} = 2^{90}$. For two rounds, the complexity is bounded by 2^{91} computation steps.

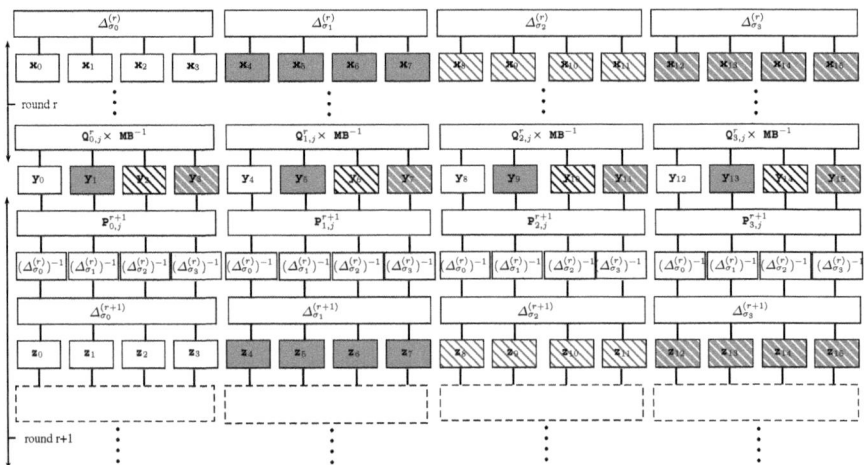

Fig. 4. 4-byte Encoding Method

6   Conclusion

This paper proposed a new white-box implementation for AES. The implementation shares many features with that of Chow et al. when considering the hiding of the key using random bijections. However, and contrary to Chow et al. our construction makes `InvSubBytes` and `InvMixColumns` operations variable by using further sets of coefficients. These coefficients are taken from dual representations of AES.

We illustrated two different ways for modifying the mixing bijections; the one which involves the minimal changes to the Chow et al. implementation was fully detailed. The modifications apply to type I and II tables. The way these tables are constructed better protects the white-box implementations against known attacks. Remarkably, the proposed implementation does not impact the code size. Further, the overall performance is unchanged compared to previously proposed implementations. Yet it raises the expected security level from 2^{30} to 2^{91}, offering a good security margin for practical applications.

Acknowledgment. I am very grateful to Amaël Grevin for his implementation of an earlier solution and to Marc Joye for helpful discussions. I am also grateful to Eric Diehl and the anonymous referees for their insightful comments on previous versions of this article. Finally, I wish to thank Alain Durand and Davide Alessio for some useful suggestions.

References

1. Barkan, E., Biham, E.: The book of Rijndaels. Cryptology ePrint Archive, Report 2002/158 (2002), http://eprint.iacr.org/2002/158
2. Barkan, E., Biham, E.: In how many ways can you write Rijndael? In: Zheng, Y. (ed.) ASIACRYPT 2002. LNCS, vol. 2501, pp. 160–175. Springer, Heidelberg (2002)

3. Billet, O., Gilbert, H.: A traceable block cipher. In: Laih, C.-S. (ed.) ASIACRYPT 2003. LNCS, vol. 2894, pp. 331–346. Springer, Heidelberg (2003)
4. Billet, O., Gilbert, H., Ech-Chatbi, C.: Cryptanalysis of a white box AES implementation. In: Handschuh, H., Hasan, M.A. (eds.) SAC 2004. LNCS, vol. 3357, pp. 227–240. Springer, Heidelberg (2004)
5. Biryukov, A., De Cannière, C., Braeken, A., Preneel, B.: A toolbox for cryptanalysis: Linear and affine equivalence algorithms. In: Biham, E. (ed.) EUROCRYPT 2003. LNCS, vol. 2656, pp. 33–50. Springer, Heidelberg (2003)
6. Bringer, J., Chabanne, H., Dottax, E.: Perturbing and protecting a traceable block cipher. In: Leitold, H., Markatos, E.P. (eds.) CMS 2006. LNCS, vol. 4237, pp. 109–119. Springer, Heidelberg (2006)
7. Bringer, J., Chabanne, H., Dottax, E.: White box cryptography: Another attempt. Cryptology ePrint Archive, Report 2006/468 (2006), http://eprint.iacr.org/2006/468
8. Chow, S., Eisen, P.A., Johnson, H., van Oorschot, P.C.: White-box cryptography and an AES implementation. In: Nyberg, K., Heys, H.M. (eds.) SAC 2002. LNCS, vol. 2595, pp. 250–270. Springer, Heidelberg (2003)
9. Chow, S., Eisen, P.A., Johnson, H., van Oorschot, P.C.: A white-box des implementation for DRM applications. In: Feigenbaum, J. (ed.) DRM 2002. LNCS, vol. 2696, pp. 1–15. Springer, Heidelberg (2003)
10. Daemen, J., Rijmen, V.: The Design of Rijndael: AES - The Advanced Encryption Standard. Springer, Heidelberg (2002)
11. Faugère, J.-C., Perret, L.: Polynomial equivalence problems: Algorithmic and theoretical aspects. In: Vaudenay, S. (ed.) EUROCRYPT 2006. LNCS, vol. 4004, pp. 30–47. Springer, Heidelberg (2006)
12. Fuller, J., Millan, W.: Linear redundancy in s-boxes. In: Johansson, T. (ed.) FSE 2003. LNCS, vol. 2887, pp. 74–86. Springer, Heidelberg (2003)
13. Matsumoto, T., Imai, H.: Public quadratic polynomial-tuples for efficient signature-verification and message-encryption. In: Günther, C.G. (ed.) EUROCRYPT 1988. LNCS, vol. 330, pp. 419–453. Springer, Heidelberg (1988)
14. Michiels, W., Gorissen, P., Hollmann, H.D.L.: Cryptanalysis of a generic class of white-box implementations. In: Avanzi, R.M., Keliher, L., Sica, F. (eds.) SAC 2008. LNCS, vol. 5381, pp. 414–428. Springer, Heidelberg (2009)
15. Raddum, H.: More dual Rijndaels. In: Dobbertin, H., Rijmen, V., Sowa, A. (eds.) AES 2005. LNCS, vol. 3373, pp. 142–147. Springer, Heidelberg (2005)

$\mathcal{E}$-MACs: Towards More Secure and More Efficient Constructions of Secure Channels

Basel Alomair and Radha Poovendran

Network Security Lab (NSL)
University of Washington-Seattle
{alomair,rp3}@uw.edu

Abstract. In cryptography, secure channels enable the confidential and authenticated message exchange between authorized users. A generic approach of constructing such channels is by combining an encryption primitive with an authentication primitive (MAC). In this work, we introduce the design of a new cryptographic primitive to be used in the construction of secure channels. Instead of using general purpose MACs, we propose the employment of special purpose MACs, named "$\mathcal{E}$-MACs". The main motive behind this work is the observation that, since the message must be both encrypted and authenticated, there can be a redundancy in the computations performed by the two primitives. If this turned out to be the case, removing such redundancy will improve the efficiency of the overall construction. In addition, computations performed by the encryption algorithm can be further utilized to improve the security of the authentication algorithm. In this work, we show how $\mathcal{E}$-MACs can be designed to reduce the amount of computations required by standard MACs based on universal hash functions, and show how $\mathcal{E}$-MACs can be secured against key-recovery attacks.

Key words: Confidentiality, authenticity, message authentication code (MAC), authenticated encryption, encrypt-and-authenticate, universal hash families

1 Introduction

There are two main approaches for the construction of secure cryptographic channels: a dedicated approach and a generic approach. In the dedicated approach, a cryptographic primitive is designed to achieve authenticated encryption as a standalone system (see, e.g., [6, 18, 23, 32, 35, 45]). In the generic approach, an authentication primitive is combined with an encryption primitive to provide message integrity and confidentiality (see, e.g., [14, 21, 51]).

Generic compositions can be constructed in three different ways: encrypt-and-authenticate ($E\&A$), encrypt-then-authenticate (EtA), and authenticate-then-encrypt (AtE). In the $E\&A$ composition, the plaintext is passed to the encryption algorithm to get the corresponding ciphertext, the plaintext is passed to the MAC algorithm to get the corresponding tag, and the resulting ciphertext-tag

K.-H. Rhee and D. Nyang (Eds.): ICISC 2010, LNCS 6829, pp. 292–310, 2011.

pair $(\mathcal{E}(M), \mathrm{MAC}(M))$ is transmitted to the intended receiver. In the EtA composition, the plaintext is passed to the encryption algorithm to get a ciphertext, the resulting ciphertext is passed to the MAC algorithm to get a tag, and the resulting $(\mathcal{E}(M), \mathrm{MAC}(\mathcal{E}(M)))$ is transmitted to the intended receiver. In the AtE composition, the plaintext is passed to the MAC algorithm to get a tag, the resulting tag is appended to the plaintext message and the result is passed to the encryption algorithm, and the resulting $(\mathcal{E}(M, \mathrm{MAC}(M)))$ is transmitted to the intended receiver. The transport layer of SSH uses a variant of $E\&A$ [51], IPsec uses a variant of EtA [14], while SSL uses a variant of AtE [21].

Over dedicated primitives, generic compositions possess several design and analysis advantages due to their modularity and the fact that encryption and authentication schemes can be designed, analyzed, and replaced independently from each other [38]. Further, and most important, generic compositions can allow for faster implementations of authenticated encryption when fast encryption algorithms, such as stream ciphers, are combined with fast MACs, such as universal hash functions based MACs [38].

The $E\&A$ composition has a parallelizable advantage over the EtA and the AtE constructions. The fact that the encryption and authentication operations can be performed simultaneously can further increase the efficiency of the generic composition. On the other hand, the $E\&A$ composition imposes an extra requirement on the MAC algorithm. As opposed to the EtA and AtE compositions, the tag in the $E\&A$ composition is a function of the plaintext message (not the ciphertext as in EtA) and is sent in the clear (not encrypted as in AtE). Therefore, the tag must be at least as confidential as the ciphertext since, otherwise, the secrecy of the plaintext can be compromised by an adversary observing its corresponding tag. This implies that generic compositions are more involved than just combining an encryption algorithm and a MAC algorithm. Indeed, in [38] and [5], the security of different generic compositions of authenticated encryption systems is analyzed. Using a secure encryption algorithm (secure in the sense that it provides privacy against chosen-plaintext attacks) and a secure MAC (secure in the sense that it provides unforgeability against chosen-message attacks), it was shown that only the EtA will guarantee the construction of secure channels. Therefore, special attention must be paid to the design of secure channels if the $E\&A$ or the AtE compositions are used.

Although significant efforts have been devoted to the design of dedicated authenticated encryption primitives, and the analysis of the generic compositions, no effort has been made to design new primitives that utilize the special characteristics of the generic compositions. In this paper, we provide the first such work. Specifically, we introduce the design of special purpose MACs to be used in the construction of $E\&A$ compositions. The driving motive behind this work was the intuition that MACs used in the generic composition of authenticated encryption systems, unlike standard MACs, can utilize the fact that messages to be authenticated must also be encrypted. That is, since both the encryption and authentication algorithms are applied to the same message, there might be a redundancy in the computations performed by the two primitives. If this turned out to be the case, removing such redundancy can improve the efficiency of the overall composition.

One class of MACs that is of a particular interest, due its fast implementation, is the class of MACs based on universal hash-function families. In universal hash-function families based MACs, the message to be authenticated is first compressed using a universal hash function in the Wegman-Carter style [13, 49] and, then, the compressed image is processed with a cryptographic function. Indeed, processing messages using universal hash functions is faster than processing them block by block using block ciphers. Combined with the fact that processing short strings is faster than processing longer ones, it becomes evident why universal hash functions based MACs are the fastest for message authentication [48].

Recently, however, Handschuh and Preneel [27] discovered a vulnerability in universal hashing based MACs. They demonstrated that once a collision in the hashing phase occurs, secret key information can be exposed, allowing subsequent forgeries to succeed with high probabilities. Their attack is not directed to a specific universal hash family and can be applied to all such MACs. The recommendations of the work in [27] are not to reuse the universal hash function key, thus going back to the impractical use of universal hash families for unconditionally secure authentication, or proceeding with the less efficient, yet more secure, block cipher based MACs.

CONTRIBUTIONS. In this paper, we propose the deployment of a new cryptographic primitive for the construction of secure channels using the $E\&A$ composition. We introduce the design of $\mathcal{E}$-MACs, Message Authentication Codes for $\mathcal{E}$ncrypted messages. By proposing the first instance of $\mathcal{E}$-MACs, we show how the structure of the $E\&A$ system can be utilized to increase the efficiency and security of the authentication process. In particular, we show how a universal hash function based $\mathcal{E}$-MAC can be computed with fewer operations than what standard universal hash functions based MACs require. That is, we will demonstrate that universal hash functions based $\mathcal{E}$-MACs can be implemented without the need to apply any cryptographic operation to the compressed image. Moreover, we will also show how $\mathcal{E}$-MACs can further utilize the special structure of the $E\&A$ system to improve the security of the authentication process. More specifically, we will show how universal hash functions based $\mathcal{E}$-MACs can be secured against the key-recovery attack, to which standard universal hash functions based MACs are vulnerable. Finally, we will show that the extra confidentiality requirement on $\mathcal{E}$-MACs can be achieved rather easily, again, by taking advantage of the $E\&A$ structure.

2   Related Work

Many standard MACs that can be used in the construction of authenticated encryption schemes have appeared in the literature. Standard MACs can be block ciphers based, cryptographic hash functions based, or universal hash functions based. CBC-MAC is one of the most known block cipher based MACs specified in FIPS publication 113 [19] and the International Organization for Standardization

ISO/IEC 9797-1 [29]. CMAC, a modified version of CBC-MAC, is presented in the NIST special publication 800-38B [15], which was based on OMAC of Iwata and Kurosawa[31]. Other block cipher based MACs include, but are not limited to, XOR-MAC [2] and PMAC [46]. The security of different MACs has been exhaustively studied (see, e.g., [3, 43]).

HMAC is a popular example of the use of iterated cryptographic hash functions to design MACs [1], which was adopted as a standard [20]. Another cryptographic hash function based MAC is the MDx-MAC of Preneel and Oorschot [42]. HMAC and two variants of MDx-MAC are specified in the International Organization for Standardization ISO/IEC 9797-2 [30]. Bosselaers *et al.* described how cryptographic hash functions can be carefully coded to take advantage of the structure of the Pentium processor to speed up the authentication process [11].

The use of universal hash families was pioneered by Wegman and Carter [13, 49] in the context of designing unconditionally secure authentication. The use of universal hash functions for the design of computationally secure MACs appeared in [7, 8, 9, 17, 26, 33, 40]. The basic concept behind the design of computationally secure universal hash functions based MACs is to compress the message using universal hash functions and then process the compressed output using a cryptographic function. The key idea is that processing messages using universal hash functions is faster than processing them block by block using block ciphers. Then, since the hashed image is typically much shorter than the message itself, processing the hashed image with a cryptographic function is faster then processing the entire message.

Since in many practical applications both message confidentiality and authenticity are sought, the design of authenticated encryption schemes has attracted a lot of attention historically. Variety of earlier schemes based on adding some redundancy to messages before cipher block chaining (CBC) encryption were found vulnerable to attacks [5]. Establishing secure channels by means of generic constructions of authenticated encryption schemes was of particular interest. The security relations among different notions of security in authenticated encryption schemes was studied in detail in [5]. In [12], it was shown that EtA schemes build secure channels and, in [38], the security of the three generic construction methods is analyzed.

In a different direction, block ciphers that combine encryption and message authentication have been proposed in the literature. Proposals that use simple checksum or manipulation detection code (MDC) have appeared in [22, 34, 41]. Such simple schemes, however, are known to be vulnerable to attacks [32]. Other dedicated schemes that combine encryption and message authenticity include [6, 18, 23, 32, 35, 45]. In [32], Jutla proposed the integrity aware parallelizable mode (IAPM), an encryption scheme with authentication. Gligor and Donescu proposed the XECB-MAC [23]. Rogaway *et al.* [45] proposed OCB: a block-cipher mode of operation for efficient authenticated encryption. Kohno *et al.* [35] proposed a high-performance conventional authenticated encryption mode (CWC), which the NIST standard Galois/Counter Mode (GCM) was based on [16].

3   Preliminaries

A message authentication scheme consists of a signing algorithm $\mathcal{S}$ and a verifying algorithm $\mathcal{V}$. The signing algorithm might be probabilistic, while the verifying one is usually not. Associated with the scheme are parameters ℓ and N describing the length of the shared key and the resulting authentication tag, respectively. On input an ℓ-bit key K and a message M, algorithm $\mathcal{S}$ outputs an N-bit string τ called the authentication tag, or the MAC of M. On input an ℓ-bit key K, a message M, and an N-bit tag τ, algorithm $\mathcal{V}$ outputs a bit, with 1 standing for accept and 0 for reject. We ask for a basic validity condition, namely that authentic tags are accepted with probability one. That is, if $\tau = \mathcal{S}(K, M)$, it must be the case that $\mathcal{V}(K, M, \tau) = 1$ for any K, M, and τ.

In general, an adversary in a message authentication scheme is a probabilistic algorithm $\mathcal{A}$, which is given oracle access to the signing and verifying algorithms $\mathcal{S}(K, \cdot)$ and $\mathcal{V}(K, \cdot, \cdot)$ for a random but hidden choice of K. $\mathcal{A}$ can query $\mathcal{S}$ to generate a tag for a plaintext of its choice and ask the verifier $\mathcal{V}$ to verify that τ is a valid tag for the plaintext. Formally, $\mathcal{A}$'s attack on the scheme is described by the following experiment:

1. A random string of length ℓ is selected as the shared secret.
2. Suppose $\mathcal{A}$ makes a signing query on a message M. Then the oracle computes an authentication tag $\tau = \mathcal{S}(K, M)$ and returns it to $\mathcal{A}$. (Since $\mathcal{S}$ may be probabilistic, this step requires making the necessary underlying choice of a random string for $\mathcal{S}$, anew for each signing query.)
3. Suppose $\mathcal{A}$ makes a verify query (M, τ). The oracle returns the decision $d = \mathcal{V}(K, M, \tau)$ to $\mathcal{A}$.

The adversary's attack is a (q_s, q_v)-attack if during the course of the attack $\mathcal{A}$ makes no more than q_s signing queries and no more than q_v verify queries. The outcome of running the experiment in the presence of an adversary is used to define security. As in [5], we say that the MAC algorithm is weakly unforgeable against chosen-message attacks (WUF-CMA) if $\mathcal{A}$ cannot make a verify query (M, τ) which is accepted for an M that has not been queried to the signing oracle $\mathcal{S}$. We say that the MAC algorithm is strongly unforgeable against chosen-message attacks (SUF-CMA) if $\mathcal{A}$ cannot make a verify query (M, τ) which is accepted regardless of whether or not M is *new*, as long as the tag has not been attached to the message by the signing oracle.

As in fast MACs, the proposed $\mathcal{E}$-MAC is based on universal hash-function families. A family of hash functions $\mathcal{H}$ is specified by a finite set of keys $\mathcal{K}$. Each key $k \in \mathcal{K}$ defines a member of the family $\mathcal{H}_k \in \mathcal{H}$. As opposed to thinking of $\mathcal{H}$ as a set of functions from A to B, it can be viewed as a single function $\mathcal{H} : \mathcal{K} \times A \to B$, whose first argument is usually written as a subscript. A random element $h \in \mathcal{H}$ is determined by selecting a $k \in \mathcal{K}$ uniformly at random and setting $h = \mathcal{H}_k$.

There has been a number of different definitions of universal hash families (see, e.g., [13, 26, 36, 37, 44, 47, 49]). We give below a formal definition of one class of universal hash families called ϵ-almost universal [9].

Definition 1. *Let* $\mathcal{H} = \{h : A \to B\}$ *be a family of hash functions and let* $\epsilon \geq 0$ *be a real number.* $\mathcal{H}$ *is said to be* ϵ-*almost universal, denoted* ϵ-AU, *if for all distinct* $M, M' \in A$, *we have that* $\Pr_{h \leftarrow \mathcal{H}}[h(M) = h(M')] \leq \epsilon$. $\mathcal{H}$ *is said to be* ϵ-*almost universal on equal-length strings if for all distinct, equal-length strings* $M, M' \in A$, *we have that* $\Pr_{h \leftarrow \mathcal{H}}[h(M) = h(M')] \leq \epsilon$.

4   The Proposed $\mathcal{E}$-MAC

4.1   Overview

Semantic security (or equivalently indistinguishability under chosen plaintext attacks (IND-CPA) [24]) is the only assumption we make on the underlying encryption algorithm. In fact, secure deterministic encryption algorithms suffices for our construction. However, since semantic security is a basic requirement in most applications, we will assume the use of a semantically secure encryption.

As in fast MACs in the literature, the proposed $\mathcal{E}$-MAC utilizes universal hash-function families in the Wegman-Carter style [13, 49]. However, as opposed to universal hash functions based MACs, we will show that $\mathcal{E}$-MACs can be secure without any post computation on the compressed image. (Recall that universal hash functions based MACs have two rounds of computations: 1. message compression using universal hash functions and, 2. output transformation, which in most practical applications a pseudorandom function applied to the compressed image [9, 27].) That is, as will be shown in the remaining of this section, the structure of the authenticated encryption system can be utilized to eliminate the need to employ pseudorandom function families. Thus, improving the speed of the MAC and reducing the required amount of shared key information (the key needed to identify the pseudorandom function).

Before we proceed with the detailed description of the proposed $\mathcal{E}$-MAC, we emphasize that the proposed universal hash family used for the implementation of the proposed $\mathcal{E}$-MAC is not the only possible solution. In fact, any ϵ-almost-Δ-universal (ϵ-AΔU) hash family, such as the MMH family of Halevi and Krawczyk [26] and the NH family of Black *et al.* [9], will satisfy the security requirements detailed in Section 5. (The ϵ-AΔU is a stronger notion than ϵ-AU given in Definition 1; interested readers may refer to [26] for a formal definition of ϵ-AΔU hash families.)

Furthermore, different assumptions about the underlying encryption algorithm may lead to different constructions of $\mathcal{E}$-MACs. That is, whether the encryption is a stream cipher, cipher block chaining (CBC) mode block cipher, electronic code book (ECB) mode block cipher, etc., can have an impact on the design and performance of the composition. We only show here how the semantic security of the underlying encryption algorithm can be utilized to improve the efficiency and security of message authentication. Further improvements in $\mathcal{E}$-MACs performance using specific modes of operations is left for a continuing research in this direction.

4.2   Description

Instantiation. Fix an encryption primitive $\mathcal{E}$ that is semantically secure. Based on a security parameter N, legitimate users agree on an N-bit long prime integer p. Let $K = (k_1, k_2, \ldots, k_B)$, for k_i's drawn uniformly and independently from $\mathbb{Z}_p^*$, be the shared secret key that will be used for message authentication. As in typical universal hash functions, depending on the values of N and B, the key can be long. One way to generate such a key is via a pseudorandom generator, e.g., [10, 28]. In such a case, only the seed of the pseudorandom generator is required to be distributed to the legitimate parties. As in symmetric-key cryptographic systems, the shared secret is distributed to the legitimate users via a secure channel. With the knowledge of the shared secret, legitimate users can exchange subsequent messages, over insecure channels, in an authenticated and confidential way. (Observe that the encryption key $K_{\mathcal{E}}$ in our setup is independent of the authentication key K.) Only the shared keys are assumed to be secret; all other parameters such as N, B, and p are publicly known.

Authentication. Without loss of generality, we assume the message can be divided into B-1 blocks of length N-bits, that is $M = m_1 || m_2 || \ldots || m_{B-1}$. (We overload m_i to denote both the binary string in the ith block and the integer representation of the ith block as an element of $\mathbb{Z}_p$; the distinction between the two representations will be omitted when it is clear from the context.) For every message M to be encrypted and authenticated, the sender draws an integer r uniformly at random from $\mathbb{Z}_p$ anew for each message (this r represents the coin tosses of $\mathcal{S}$). We emphasize that r must be independent of all r's generated to authenticate other messages. The sender encrypts $M||r$ and transmits the resulting ciphertext $c = \mathcal{E}(M||r)$ to the receiver (the symbol "$||$" denotes the concatenation operation), along with the the N-bit long tag of message M computed as:

$$\tau = \sum_{i=1}^{B-1} k_i m_i + k_B r \mod p, \tag{1}$$

where m_i denotes the ith block of message M.

Remark 1. A misconception about universal hash-function families is that the authentication key needs to be as long as the longest message to be authenticated. Obviously, if this was true, universal hashing will be impractical for most applications. In the literature, there exist standard techniques to hash arbitrary-length messages using a fixed-length key. The first such technique was proposed by Wegman and Carter in [50], and later refined by Halevi and Krawczyk in [26]. The work of Black *et al.* [9] provides a different generic algorithm to transform any hash function that is ϵ-AU on *equal-length* messages, h, to a hash function that is ϵ-AU on *arbitrary-length* messages, h^*. However, for a lack of space and for a better continuity of the main ideas of the paper, we omit going into the details of such techniques. (Interested readers may refer to [9, 26, 50] for more information.) Therefore, we emphasize that the key $K = (k_1, k_2, \ldots, k_B)$ can be used to authenticate arbitrary-length messages.

Remark 2. Clearly, as will be formally proven in Section 5, the bound on the probability of successful forgery is dependent on the security parameter N. Depending on application, one might require lower bounds on probability of successful forgery. A straightforward way is to increase the security parameter to give lower probability of successful forgery. Another method is to hash the same message multiple times with independent keys. This, however, will require a much longer key. A well-studied and more efficient method is to use the Toeplitz-extension on the hash function [36, 39]. (See, e.g., [9] for a detailed use of Toeplitz-extension to increase the security of MACs based on universal hash functions.) Again, we omit describing this topic since it is out of the scope of this work and refer interested readers to [9, 26, 36, 39] for more details.

Verification. Upon receiving a ciphertext-tag pair, (c, τ), the receiver calls the corresponding decryption algorithm $\mathcal{D}$ to extract the plaintext $M||r$. To verify the integrity of $M||r$, the receiver computes $\sum_{i=1}^{B-1} k_i m_i + k_B r$ and authenticates the message only if the computed value is congruent to the received τ modulo p. Formally, the following integrity check must be satisfied for the message to be authenticated:

$$\tau \stackrel{?}{\equiv} \sum_{i=1}^{B-1} k_i m_i + k_B r \mod p. \tag{2}$$

Remark 3. We emphasize that the random nonce, r, requires no key management. It is generated by the sender as the coin tosses of the signing algorithm and delivered to the receiver via the ciphertext. In other words, it is not a shared secret and it needs no synchronization.

5 Security Analysis

5.1 Security of the Proposed $\mathcal{E}$-MAC

Assume that message M has been encrypted with any semantically secure encryption scheme. For the rest of the paper, we will refer to M and τ as the message and the tag generated at the transmitter's end, respectively; while M' and τ' represent the message and the tag at the receiver's end, respectively. For ease of notation, we will refer to the plaintext message as the concatenation of M and r. The following lemmas are the main ingredient for the security of the proposed $\mathcal{E}$-MAC.

Lemma 1. *Let m_i and k_i be the ith message block and i^{th} key, respectively. For a modified message block $m_i' \not\equiv m_i \mod p$, the probability that $k_i m_i' \equiv k_i m_i \mod p$ is zero.*

Lemma 1 is a direct consequence of the fact that, for a prime integer p, $\mathbb{Z}_p$ is a field.

Lemma 2. *Let k_1 and k_2 be two secret keys in the proposed $\mathcal{E}$-MAC. The probability to choose two nonzero integers δ_1 and δ_2 in $\mathbb{Z}_p$ such that $k_1\delta_1 \equiv k_2\delta_2$ mod p is at most $1/(p-1)$.*

Proof. Fix a $\delta_1 \in \mathbb{Z}_p^*$. Since every nonzero element in $\mathbb{Z}_p$ is invertible, the resulting $(k_1\delta_1 \mod p)$ will be uniformly distributed over $\mathbb{Z}_p^*$. Similarly, the resulting $(k_2\delta_2 \mod p)$ is uniformly distributed over $\mathbb{Z}_p^*$. Since k_1 and k_2 are assumed to be secret, the probability that $k_1\delta_1 \equiv k_2\delta_2 \mod p$ is $1/(p-1)$, and the lemma follows. □

Lemma 3. *Authentication tags are statistically independent of their corresponding messages, and different authentication tags are mutually independent.*

The proof of Lemma 3 is provided in Appendix A. Now we can proceed with the proofs of the main claims of the proposed $\mathcal{E}$-MAC. Recall that applying a cryptographic function to the compressed image is an essential operation for the security of standard universal hash functions based MACs. Without such a cryptographic operation, the key of the universal hash function can be exposed (by chosen message attacks, for instance). We now formally prove two important claims about $\mathcal{E}$-MACs. Namely, the semantic security of the used encryption algorithm suffices to protect the key of the proposed $\mathcal{E}$-MAC, and tags do not reveal any information about the plaintext that is not revealed by the corresponding ciphertext.

Theorem 1. *An adversary able to extract any information about the proposed $\mathcal{E}$-MAC's secret key, or extract any information about the plaintext from authentication tags is able to break the semantic security of the underlying encryption algorithm.*

Proof. By Lemma 3, each tag is independent of its corresponding message and the secret key. Therefore, by only observing a single tag, the adversary cannot reveal any information about the authenticated message or the secret key. Furthermore, also by Lemma 3, different authentication tags are mutually independent. Therefore, the observation of multiple tags gives the adversary no extra information than what a single tag gives individually. This holds as long as the coin tosses, the r's, remain secret. The transmitted r's, however, are generated internally and encrypted with the semantically secure algorithm $\mathcal{E}$. Therefore, no information about the proposed $\mathcal{E}$-MAC's key nor the authenticated messages can be exposed, unless the adversary can extract secret information about the r's, which can be done only by breaking the semantic security of the encryption algorithm. □

Remark 4. In addition to its important statement regarding the secrecy of transmitted messages, Theorem 1 presents an important statement regarding the secrecy of the nonce r. Clearly, if some r's are revealed, partial key information can be exposed. Other than attacking the system in a non-cryptographic way, Theorem 1 states that the only way to expose secrete information about the r's is by breaking the semantic security of the encryption algorithm. That is, from a cryptographic point of view, it is safe to assume that no information about the r's will be revealed.

This shows how $\mathcal{E}$-MACs can take advantage of the $E\&A$ structure to improve authentication efficiency and satisfy their secrecy requirement. All that is needed is to generate a random string, append it to the encrypted plaintext message, and use it to encrypt the authentication tag. Therefore, as claimed earlier, no post-processing of the compressed image is required and the secrecy requirement of $\mathcal{E}$-MACs can be achieved, without expensive computational effort. This result is not surprising. In fact, it supports the main motive behind this work, namely the intuition that post-processing the compressed image by a cryptographic function can be replaced by computations performed by the encryption algorithm (i.e., post-processing is redundant in such compositions).

We will now state the main theorem regarding the probability of successful forgery against the proposed $\mathcal{E}$-MAC.

Theorem 2. *Let Σ denotes the proposed $\mathcal{E}$-MAC and let $\mathcal{A}$ be an adversary making a (q_s, q_v)-attack on Σ. Given the semantic security of the underlying encryption scheme, $\mathcal{A}$'s advantage of successful forgery is at most*

$$\mathsf{Adv}_{\mathcal{A}}^{\Sigma} = \begin{cases} \dfrac{q_v}{p} & \text{if } q_s = 0 \\[3mm] \dfrac{q_v}{p-1} & \text{if } q_s > 0. \end{cases} \tag{3}$$

Proof. From the proof of Lemma 3 (equation (16)), the tag is uniformly distributed over $\mathbb{Z}_p$. Hence, if the adversary makes no signing queries, the probability of forging a valid tag is $1/p$.

Assume that the adversary has queried the signing oracle $\mathcal{S}(K, \cdot)$ for q_s times and recorded $(M_1, \tau_1), \cdots, (M_{q_s}, \tau_{q_s})$. By Theorem 1, given the semantic security of the encryption algorithm, no information about the $\mathcal{E}$-MAC's secret key is revealed by the observed τ_i's.

Now, consider calling the query $\mathcal{V}(K, M', \tau')$, where M' and τ' are any message-tag pair of the adversary's choice. We aim to bound the probability of successful forgery for an M' that has not been queried to the signing oracle; that is, $M' \neq M_i$ for any $i = 1, \cdots, q_s$. We break the proof into two cases: queried tag and unqueried tag. For ease of notations, r_i will be denoted as the Bth block of the ith message, that is, $r_i = m_{i_B}$.

QUERIED TAG $(M', \tau' = \tau_q)$: Assume that $\tau' = \tau_q$ for a $q \in \{1, \cdots, q_s\}$. This case represents the event that a collision in the hashing operation occurs. Then, $\mathcal{V}(k, M', \tau') = 1$ if and only if the following holds:

$$\sum_{\ell=1}^{B} k_\ell\, m'_\ell \overset{?}{\equiv} \tau' \equiv \tau_q \equiv \sum_{\ell=1}^{B} k_\ell\, m_\ell \quad \bmod p, \tag{4}$$

where m'_ℓ denotes the ℓth block of M' and m_ℓ denotes the ℓth block of M_q (note that we write m_ℓ instead of m_{q_ℓ} for ease of notations since no distinction between different messages is necessary). We will analyze equation (4) by considering the following three cases: M' and M_q differ by a single block, M' and M_q differ by two blocks, or M' and M_q differ by more than two blocks.

1. Assume that only a single message block is different. Since addition is commutative, assume without loss of generality that the first message block is different; that is, $m'_1 \not\equiv m_1 \mod p$. Since only the first message block is different, equation (4) is equivalent to

$$k_1 m'_1 \equiv k_1 m_1 \mod p. \tag{5}$$

Therefore, by Lemma 1, the probability of successful forgery given a single block difference is *zero*.

2. Assume, without loss of generality, that the first two message blocks are different; i.e., $m'_1 \equiv m_1 + \delta_1 \not\equiv m_1 \mod p$ and $m'_2 \equiv m_2 + \delta_2 \not\equiv m_2 \mod p$. Then, equation (4) is equivalent to

$$k_1 \delta_1 + k_2 \delta_2 \equiv 0 \mod p. \tag{6}$$

Therefore, by Lemma 2, the probability of successful forgery given that exactly two message blocks are different is at most $1/(p-1)$.

3. Assume that more than two message blocks are different, i.e., $m'_i \equiv m_i + \delta_i \not\equiv m_i \mod p$; $\forall i \in I \subseteq \{1, 2, \cdots, B\}; |I| \geq 3$. Then, equation (4) is equivalent to

$$k_i \delta_i + \sum_{\substack{j \in I \\ j \neq i}} k_j \delta_j \equiv 0 \mod p, \tag{7}$$

for some $i \in I$. Therefore, using Lemma 2 and the fact that $\sum_{j \in I, j \neq i} k_j \delta_j$ can be congruent to *zero* modulo p, the probability of success is at most $1/p$. (The difference between this case and the case of exactly two blocks is that, even if the δ's are chosen to be nonzero integers, $\sum_{j \in I, j \neq i} k_j \delta_j$ can still be congruent to zero modulo p.)

From the above three cases, the probability of successful forgery when the forged tag has been outputted by the signing oracle is at most $1/(p-1)$.

UNQUERIED TAG (M', τ'): Assume now that the tag τ' is different than all the recorded tags; that is, $\tau' \neq \tau_q$ for all $q = 1, \cdots, q_s$. If τ' is independent of the recorded tags, then the probability of successful forgery is $1/p$ (using the fact that the tag is uniformly distributed over $\mathbb{Z}_p$). Assume, however, that τ' is a function of τ_q, for a $q \in \{1, \cdots, q_s\}$. Let $\tau' \equiv \tau_q + \gamma \mod p$ for some $\gamma \in \mathbb{Z}_p \backslash \{0\}$ of the adversary's choice. (Note that, γ can be a function of any value recorded by the adversary.) Then, $\mathcal{V}(K, M', \tau') = 1$ if and only if the following congruence holds:

$$\sum_{\ell=1}^{B} k_\ell m'_\ell \overset{?}{\equiv} \tau' \equiv \tau_q + \gamma \equiv \sum_{\ell=1}^{B} k_\ell m_\ell + \gamma \mod p, \tag{8}$$

where m'_ℓ denotes the ℓth block of M' and m_ℓ denotes the ℓth block of M_q. Bellow we analyze equation (8) by considering two cases: M' and M_q differ by a single block, or M' and M_q differ by more than one block.

1. Without loss of generality, assume that M' and M_q differ in the first block only. That is $m'_1 \equiv m_1 + \delta \not\equiv m_1 \mod p$ and $m'_i \equiv m_i \mod p$ for all $i = 2, \cdots, B$. Then, equation (8) is equivalent to

$$k_1\delta \equiv \gamma \quad \mod p. \tag{9}$$

   Therefore, by Lemma 2, the probability of success is at most $1/(p-1)$.

2. Assume now that M' and M_q differ by more than one block. That is, $m'_i \equiv m_i + \delta_i \not\equiv m_i \mod p$; $\forall i \in I \subseteq \{1, 2, \cdots, B\}$; $|I| \geq 2$. Then, equation (8) is equivalent to

$$\sum_{i \in I} k_i\delta_i \equiv \gamma \quad \mod p. \tag{10}$$

   By Lemma 2 and the fact that $\sum_{i \in I} k_i\delta_i$ can be congruent to *zero* modulo p, the probability of success is at most $1/p$.

From the above two cases, the probability of successful forgery when the forged tag has not been outputted by the signing oracle is at most $1/(p-1)$.

Therefore, given that $\mathcal{A}$ has made at least one signing query, $\mathcal{A}$'s probability of successful forgery for each verify query is at most $1/(p-1)$.   □

Remark 5. Observe that the case of queried tag implies that the used hash family is $(\frac{1}{p-1})$-AU. Similarly, the case of unqueried tag implies that the used hash family is $(\frac{1}{p-1})$-AΔU.

Observe further that the proposed $\mathcal{E}$-MAC is strongly unforgeable under chosen message attacks (SUF-CMA). Recall that SUF-CMA requires that it be computationally infeasible for the adversary to find a new message-tag pair after chosen-message attacks even if the message is not new, as long as the tag has not been attached to the message by a legitimate user [5]. To see this, let (M, τ) be a valid message tag pair. Assume that the adversary is attempting to authenticate the same message with a different tag τ'. For the (M, τ') pair to be authenticated, $\sum_i k_i m_i + k_B r' \mod p$ must be equal to τ'. That is, given τ', r' must be set to $k_B^{-1}(\tau' - \sum_i k_i m_i) \mod p$ for the tag to be authenticated. By Theorem 1, however, the adversary cannot expose the $\mathcal{E}$-MAC's key. Therefore, Theorem 2 holds whether or not the message is new, as long as the tag has not been attached to the message by the signing oracle.

5.2   Security of the *E&A* Composition

In [5], Bellare and Namprempre defined two notions of integrity in authenticated encryption schemes, integrity of plaintexts (INT-PTXT) and integrity of ciphertexts (INT-CTXT). INT-PTXT implies that it is computationally infeasible for an adversary to produce a ciphertext decrypting to a message which the sender had never encrypted, while INT-CTXT implies that it is computationally infeasible for an adversary to produce a ciphertext not previously produced by the sender, regardless of whether or not the corresponding plaintext is *new*.

Although the work of [5] shows that the $E\&A$ composition is generally insecure, the results do not apply to all variants of $E\&A$ constructions. For instance, the $E\&A$ composition does not provide indistinguishability under chosen plaintext attacks (IND-CPA) because there exist secure MACs that reveal information about the plaintext ([5] provides a detailed example). Obviously, if such a MAC is used in the construction of an $E\&A$ system, the resulting composition will not provide IND-CPA. Unlike standard MACs, however, it is a basic requirement of $\mathcal{E}$-MACs to be as secret as the used encryption algorithm. Indeed, Theorem 1 guarantees that the proposed $\mathcal{E}$-MAC does not reveal any information about the plaintext that is not revealed by the ciphertext.

Another result of [5] is that the generic $E\&A$ does not provide INT-CTXT. (Although the notion of INT-PTXT is the more natural security requirement [5] while the interest of the stronger INT-CTXT notion is more in the security implications shown in [5].) The reason why $E\&A$ compositions generally do not provide INT-CTXT is that one can come up with a secure encryption algorithm with the property that a ciphertext can be modified without changing its decryption [5]. Obviously, when such an encryption algorithm is combined with the proposed $\mathcal{E}$-MAC to construct an $E\&A$ system, since the tag is computed as a function of the plaintext, only INT-PTXT is reached.

In practice, however, it is possible to construct an $E\&A$ system that does provide INT-CTXT. For instance, a sufficient condition for the proposed $\mathcal{E}$-MAC to provide INT-CTXT for the composed system is to be used with a secure one-to-one encryption algorithm. To see this observe that any modification of the ciphertext will correspond to modifying the plaintext (since the encryption is one-to-one). Therefore, by Theorem 2, modified ciphertexts can only be accepted with negligible probabilities. Indeed, secure $E\&A$ systems have been constructed in practice. A popular example of such constructions is SSH [51], which uses a variant of $E\&A$ the has been proven to be secure in [4].

So far, we have shown that $\mathcal{E}$-MACs can be used to replace standard MACs in the construction of $E\&A$ systems with two additional properties: they can have provable confidentiality and they can be more efficient (observe that the tag of the proposed $\mathcal{E}$-MAC is the output of the universal hash function; no post-processing was performed). What we will show next is that $\mathcal{E}$-MACs can have another security advantage. More specifically, we will show that $\mathcal{E}$-MACs can utilize the structure of the $E\&A$ system to achieve better resilience to a new attack on universal hash functions based MACs; namely, the key-recovery attack [27].

6 $\mathcal{E}$-MACs and Key Recovery

Recently, Handschuh and Preneel [27] showed that, compared to block cipher based, MACs based on universal hash functions have a key-recovery vulnerability. In principle, a small probability of successful forgery on authentication codes is always possible. However, the work in [27] demonstrates that, for universal hash functions based MACs, once a successful forgery is achieved, subsequent forgeries can succeed with high probabilities. The main idea in their attacks is to look for a collision in the message compression phase. Once a message that causes a collision is found, partial information about the hashing keys can be exposed. Using this key information an attacker can forge valid tags for fake messages. We give a detailed example below.

Example 1. Consider the universal hash family presented in this paper. Assume an adversary calling the signing oracle on $M = m_1||m_2$, thus obtaining its authentication tag τ. The adversary now can call the verification oracle with $M = m_2||m_1$ and the same tag τ. Obviously, the verification will pass if and only if $k_1 \equiv k_2 \mod p$ (in which case $k_1 m_1 + k_2 m_2 \equiv k_2 m_1 + k_1 m_2 \mod p$).

Although the verification will pass with a small probability, the adversary can continuously call the verification oracle with $M = m_2||\alpha_i m_1$, for different α_i's until the message is authenticated. Let $M = m_2||\alpha m_1$ be the message that passes the verification test, for some $\alpha \in \mathbb{Z}_p^*$. Then, the relation

$$k_1 \equiv \beta k_2 \mod p, \tag{11}$$

where $\beta = (\alpha m_1 - m_2)(m_1 - m_2)^{-1}$ is exposed. With this knowledge, a man in the middle can always replace the first two blocks, $m_1||m_2$, of any future message M with $\beta^{-1} m_2||\beta m_1$ without violating its tag. This is because $k_1(\beta^{-1} m_2) + k_2(\beta m_1) = k_2 m_2 + k_1 m_1$ regardless of values of m_1 and m_2.

Handschuh and Preneel [27] defined three classes of weak keys in universal hash functions. Each class can be exploited in a way similar to the one discussed in the above example to substantially increase the probability of successful forgery after a single collision. This attack is shared by all universal hash based MACs [27]. As per [27], the recommended mitigations to this attack are to use the less efficient block cipher based MACs, or not to reuse the same hashing key for multiple authentication.

Compared to standard MACs, however, $\mathcal{E}$-MACs can utilize the structure of the $E\&A$ system to overcome the key-recovery problem discovered in [27]. Consider the $\mathcal{E}$-MAC proposed in Section 4, and recall that a random number $r \in_R \mathbb{Z}_p$ is generated internally in the $E\&A$ process. In the basic construction of Section 4, the goal of r is to encrypt the authentication tag. However, the random r can play a pivotal role in key-recovery security.

In the basic construction in Section 4, the universal hashing key is $K = k_1||k_2||\cdots||k_B$ and the authentication tag is computed as:

$$\tau = \sum_{i=1}^{B-1} k_i m_i + k_B r \mod p. \tag{12}$$

Now, with the same shared key, consider another use of r. More specifically, let the authentication tag be computed as follows:

$$\tau = \sum_{i=1}^{B-1} (k_i \oplus r) m_i + k_B r \mod p. \tag{13}$$

In other words, r can be used to randomize the key in every authentication call.

Assume the same attack described in Example 1 and let $M = m_2||\alpha m_1$ passes the verification test, for some $\alpha \in \mathbb{Z}_p^*$. This time, however,

$$k_1' \equiv \beta k_2' \mod p, \tag{14}$$

where $k_1' = k_1 \oplus r$, $k_2' = k_2 \oplus r$, and $\beta = (\alpha m_1 - m_2)(m_1 - m_2)^{-1}$ is the relation revealed to the adversary. For any future authentication, the sender will generate a new random number r' that is independent of r. Thus, the keys that will be used for authentication will be k_1'' and k_2'', where $k_i'' = k_i \oplus r'$ for $i = 1, 2$. That is, from the standpoint of key-recovery attacks, by using equation (13) instead of equation (12), different authentication tags are computed with different keys. Therefore, finding a collision in the message compression phase does not lead to information leakage about the keys, as long as the same nonce does not authenticate different messages. (Note that there is no need to randomize k_B since it is independent of the message to be authenticated.)

Remark 6. This shows how the system can be designed to utilize the authenticated encryption application to increase the robustness of universal hash functions based $\mathcal{E}$-MACs. This could not have been achieved without the use of the fresh random number r that was secretly delivered to the verifier as part of the ciphertext.

7   Conclusion and Future Work

In this work, we studied the encrypt-and-authenticate generic composition of secure channels. We introduced $\mathcal{E}$-MACs, a new symmetric-key cryptographic primitive that can be used in the construction of $E\&A$ compositions. By taking advantage of the $E\&A$ structure, the use of $\mathcal{E}$-MACs is shown to improve the efficiency and security of the authentication operation. More precisely, since the message to be authenticated is encrypted, universal hash functions based $\mathcal{E}$-MACs can designed without the need to apply cryptographic operations on the compressed image, since this can be replaced by operations performed by the encryption algorithm. Further, by appending a random string at the end of the plaintext message, two security objectives have been achieved. First, the random string is used to encrypt the authentication tag so that the secrecy of the plaintext is not compromised by its tag. Second, the random string can be used to randomize the secret key of the used $\mathcal{E}$-MAC so that it will be secure against key-recovery attacks.

Since this is only the first work in this direction, bringing more research can only contribute positively towards the design of more efficient and more secure authentication. One specific direction that is yet to be investigated is the use of encryption algorithms that provide more than just semantic security. In particular, since most secure block ciphers are pseudorandom permutations, using block ciphers operated in different modes is a promising direction for more improvements in the design of $\mathcal{E}$-MACs.

References

1. Bellare, M., Canetti, R., Krawczyk, H.: Keying hash functions for message authentication. In: Koblitz, N. (ed.) CRYPTO 1996. LNCS, vol. 1109, pp. 1–15. Springer, Heidelberg (1996)
2. Bellare, M., Guerin, R., Rogaway, P.: XOR MACs: New methods for message authentication using finite pseudorandom functions. In: Coppersmith, D. (ed.) CRYPTO 1995. LNCS, vol. 963, pp. 15–28. Springer, Heidelberg (1995)
3. Bellare, M., Kilian, J., Rogaway, P.: The Security of the Cipher Block Chaining Message Authentication Code. Journal of Computer and System Sciences 61(3), 362–399 (2000)
4. Bellare, M., Kohno, T., Namprempre, C.: Breaking and provably repairing the SSH authenticated encryption scheme: A case study of the Encode-then-Encrypt-and-MAC paradigm. ACM Transactions on Information and System Security 7(2), 241 (2004)
5. Bellare, M., Namprempre, C.: Authenticated encryption: Relations among notions and analysis of the generic composition paradigm. Journal of Cryptology 21(4), 469–491 (2008)
6. Bellare, M., Rogaway, P., Wagner, D.: The EAX mode of operation. In: Roy, B., Meier, W. (eds.) FSE 2004. LNCS, vol. 3017, pp. 389–407. Springer, Heidelberg (2004)
7. Bernstein, D.: Floating-point arithmetic and message authentication. Unpublished manuscript (2004), http://cr.yp.to/hash127.html
8. Bernstein, D.: The Poly1305-AES message-authentication code. In: Gilbert, H., Handschuh, H. (eds.) FSE 2005. LNCS, vol. 3557, pp. 32–49. Springer, Heidelberg (2005)
9. Black, J., Halevi, S., Krawczyk, H., Krovetz, T., Rogaway, P.: UMAC: Fast and Secure Message Authentication. In: Wiener, M. (ed.) CRYPTO 1999. LNCS, vol. 1666, pp. 216–233. Springer, Heidelberg (1999)
10. Blum, L., Blum, M., Shub, M.: A Simple Unpredictable Pseudo-random Number Generator. SIAM Journal on Computing 15, 364 (1986)
11. Bosselaers, A., Govaerts, R., Vandewalle, J.: Fast hashing on the Pentium. In: Koblitz, N. (ed.) CRYPTO 1996. LNCS, vol. 1109, pp. 298–312. Springer, Heidelberg (1996)
12. Canetti, H., Krawczyk, H.: Analysis of Key-Exchange Protocols and Their Use for Building Secure Channels. In: Pfitzmann, B. (ed.) EUROCRYPT 2001. LNCS, vol. 2045, pp. 453–472. Springer, Heidelberg (2001)
13. Carter, J., Wegman, M.: Universal classes of hash functions. In: Proceedings of the Ninth Annual ACM Symposium on Theory of Computing STOC 1977, pp. 106–112. ACM, New York (1977)
14. Doraswamy, N., Harkins, D.: IPSec: the new security standard for the Internet, intranets, and virtual private networks. Prentice Hall, Englewood Cliffs (2003)
15. Dworkin, M.: Recommendation for block cipher modes of operation: The CMAC mode for authentication (2005)
16. Dworkin, M.: NIST Special Publication SP800-38D defining GCM and GMAC (2007)
17. Etzel, M., Patel, S., Ramzan, Z.: Square hash: Fast message authentication via optimized universal hash functions. In: Wiener, M. (ed.) CRYPTO 1999. LNCS, vol. 1666, pp. 234–251. Springer, Heidelberg (1999)

18. Ferguson, N., Whiting, D., Schneier, B., Kelsey, J., Kohno, T.: Helix: Fast encryption and authentication in a single cryptographic primitive. In: Johansson, T. (ed.) FSE 2003. LNCS, vol. 2887, pp. 330–346. Springer, Heidelberg (2003)
19. FIPS 113. Computer Data Authentication. Federal Information Processing Standards Publication, 113 (1985)
20. FIPS 198. The Keyed-Hash Message Authentication Code (HMAC). Federal Information Processing Standards Publication, 198 (2002)
21. Freier, A., Karlton, P., Kocher, P.: The SSL Protocol Version 3.0 (1996)
22. Gligor, V., Donescu, P.: Integrity-Aware PCBC Encryption Schemes. In: Proceedings of Security Protocols: 7th International Workshop, Cambridge, Uk, April 19-21, 1999 (2000)
23. Gligor, V., Donescu, P.: Fast Encryption and Authentication: XCBC Encryption and XECB Authentication Modes. In: Matsui, M. (ed.) FSE 2001. LNCS, vol. 2355, pp. 20–92. Springer, Heidelberg (2002)
24. Goldwasser, S., Micali, S.: Probabilistic encryption. Journal of Computer and System Sciences 28(2), 270–299 (1984)
25. Gubner, J.: Probability and random processes for electrical and computer engineers. Cambridge University Press, Cambridge (2006)
26. Halevi, S., Krawczyk, H.: MMH: Software message authentication in the Gbit/second rates. In: Biham, E. (ed.) FSE 1997. LNCS, vol. 1267, pp. 172–189. Springer, Heidelberg (1997)
27. Handschuh, H., Preneel, B.: Key-Recovery Attacks on Universal Hash Function Based MAC Algorithms. In: Wagner, D. (ed.) CRYPTO 2008. LNCS, vol. 5157, pp. 144–161. Springer, Heidelberg (2008)
28. Hastad, J., Impagliazzo, R., Levin, L., Luby, M.: A Pseudorandom Generator from Any One-Way Function. SIAM Journal on Computing 28(4), 1364–1396 (1999)
29. ISO/IEC 9797-1. Information technology – Security techniques – Message Authentication Codes (MACs) – Part 1: Mechanisms using a block cipher (1999)
30. ISO/IEC 9797-2. Information technology – Security techniques – Message Authentication Codes (MACs) – Part 2: Mechanisms using a dedicated hash-function (2002)
31. Iwata, T., Kurosawa, K.: omac: One-key cbc mac. In: Johansson, T. (ed.) FSE 2003. LNCS, vol. 2887, pp. 129–153. Springer, Heidelberg (2003)
32. Jutla, C.: Encryption modes with almost free message integrity. Journal of Cryptology 21(4), 547–578 (2008)
33. Kaps, J., Yuksel, K., Sunar, B.: Energy scalable universal hashing. IEEE Transactions on Computers 54(12), 1484–1495 (2005)
34. Kohl, J., Neuman, C.: The Kerberos Network Authentication Service (V5). Technical report, RFC 1510, (September 1993)
35. Kohno, T., Viega, J., Whiting, D.: CWC: A high-performance conventional authenticated encryption mode. In: Roy, B., Meier, W. (eds.) FSE 2004. LNCS, vol. 3017, pp. 408–426. Springer, Heidelberg (2004)
36. Krawczyk, H.: LFSR-based hashing and authentication. In: Desmedt, Y.G. (ed.) CRYPTO 1994. LNCS, vol. 839, pp. 129–139. Springer, Heidelberg (1994)
37. Krawczyk, H.: New hash functions for message authentication. In: Guillou, L.C., Quisquater, J.-J. (eds.) EUROCRYPT 1995. LNCS, vol. 921, pp. 301–310. Springer, Heidelberg (1995)
38. Krawczyk, H.: The order of encryption and authentication for protecting communications (or: How secure is SSL?). In: Kilian, J. (ed.) CRYPTO 2001. LNCS, vol. 2139, pp. 310–331. Springer, Heidelberg (2001)

39. Mansour, Y., Nisan, N., Tiwari, P.: The computational complexity of universal hashing. In: Proceedings of the Twenty-Second Annual ACM Symposium on Theory of Computing–STOC 1990, pp. 235–243. ACM, New York (1990)
40. McGrew, D., Viega, J.: The security and performance of the Galois/Counter Mode (GCM) of operation. In: Canteaut, A., Viswanathan, K. (eds.) INDOCRYPT 2004. LNCS, vol. 3348, pp. 343–355. Springer, Heidelberg (2004)
41. Meyer, C., Matyas, S.: Cryptography: A New Dimension in Computer Data Security. John Wiley & Sons, Chichester (1982)
42. Preneel, B., Van Oorschot, P.: MDx-MAC and building fast MACs from hash functions. In: Coppersmith, D. (ed.) CRYPTO 1995. LNCS, vol. 963, pp. 1–14. Springer, Heidelberg (1995)
43. Preneel, B., Van Oorschot, P.: On the security of iterated message authentication codes. IEEE Transactions on Information theory 45(1), 188–199 (1999)
44. Rogaway, P.: Bucket hashing and its application to fast message authentication. Journal of Cryptology 12(2), 91–115 (1999)
45. Rogaway, P., Bellare, M., Black, J.: OCB: A Block-Cipher Mode of Operation for Efficient Authenticated Encryption. ACM Transactions on Information and System Security 6(3), 365–403 (2003)
46. Rogaway, P., Black, J.: PMAC: Proposal to NIST for a parallelizable message authentication code (2001)
47. Stinson, D.: Universal hashing and authentication codes. Designs, Codes and Cryptography 4(3), 369–380 (1994)
48. van Tilborg, H.: Encyclopedia of cryptography and security. Springer, Heidelberg (2005)
49. Wegman, M., Carter, J.: New classes and applications of hash functions. In: 20th Annual Symposium on Foundations of Computer Science–FOCS 1979, pp. 175–182. IEEE, Los Alamitos (1979)
50. Wegman, M., Carter, L.: New hash functions and their use in authentication and set equality. Journal of Computer and System Sciences 22(3), 265–279 (1981)
51. Ylonen, T., Lonvick, C.: The Secure Shell (SSH) Transport Layer Protocol. Technical report, RFC 4253 (2006)

A   Proof of Lemma 3

Proof. Throughout this proof, random variables will be represented by bold font symbols, whereas the corresponding non-bold font symbols represent specific values that can be taken by these random variables. Let the secret key $K = k_1 || k_2 || \cdots || k_B$ be fixed. Then, for any tag $\tau \in \mathbb{Z}_p$ computed according to equation (1), and any plaintext message M, the following holds:

$$\Pr(\boldsymbol{\tau} = \tau | \boldsymbol{M} = M) = \Pr \left(\boldsymbol{r} = (\tau - \sum_{i=1}^{B-1} k_i m_i) \, k_B^{-1} \right) = \frac{1}{p}, \tag{15}$$

where m_i denotes the ith block of the message M. Equation (15) holds by the assumption that r is drawn uniformly from $\mathbb{Z}_p$. The existence of k_B^{-1}, the multiplicative inverse of k_B in the integer field $\mathbb{Z}_p$, is a guaranteed since k_B is not the zero element. Furthermore, as a direct consequence of the fact that $\mathbb{Z}_p$ is a field, for an r drawn uniformly at random from $\mathbb{Z}_p$, the resulting $(k_B r \mod p)$ is uniformly distributed over $\mathbb{Z}_p$. Consequently, for any plaintext message M, since the tag is a result of adding $(k_B r \mod p)$ to $(\sum_i k_i m_i \mod p)$, and since $(k_B r \mod p)$ is uniformly distributed over $\mathbb{Z}_p$, the resulting tag is uniformly distributed over $\mathbb{Z}_p$. That is, for any fixed value $\tau \in \mathbb{Z}_p$, the probability that the tag will take this specific value is given by:

$$\Pr(\boldsymbol{\tau} = \tau) = \frac{1}{p}. \tag{16}$$

Combining Bayes' theorem [25] with equations (15) and (16) yields:

$$\Pr(\boldsymbol{M} = M | \boldsymbol{\tau} = \tau) = \frac{\Pr(\boldsymbol{\tau} = \tau | \boldsymbol{M} = M) \Pr(\boldsymbol{M} = M)}{\Pr(\boldsymbol{\tau} = \tau)} = \Pr(\boldsymbol{M} = M). \tag{17}$$

Equation (17) implies that the tag τ gives no information about the plaintext M since τ is statistically independent of M. Similarly, one can show that the tag is independent of the secret key.

Now, let τ_1 through τ_ℓ represent the tags for messages M_1 through M_ℓ, respectively. Further, let r_1 through r_ℓ be the coin tosses of the signing algorithm $\mathcal{S}$ for the authentication of messages M_1 through M_ℓ, respectively. Recall that r_i's are *mutually independent* and *uniformly* distributed over $\mathbb{Z}_p$. Then, for any possible values of the messages M_1 through M_ℓ with arbitrary joint probability mass function, and all possible values of τ_1 through τ_ℓ, we get:

$$\Pr(\boldsymbol{\tau}_1 = \tau_1, \cdots, \boldsymbol{\tau}_\ell = \tau_\ell) = \sum_{M_1, \cdots, M_\ell} \Pr(\boldsymbol{\tau}_1 = \tau_1, \cdots, \boldsymbol{\tau}_\ell = \tau_\ell | \boldsymbol{M}_1 = M_1, \cdots, \boldsymbol{M}_\ell = M_\ell)$$

$$\Pr(\boldsymbol{M}_1 = M_1, \cdots, \boldsymbol{M}_\ell = M_\ell)$$

$$= \sum_{M_1, \cdots, M_\ell} \Pr\left(\boldsymbol{r}_1 = (\tau_1 - \sum_{i=1}^{B-1} k_i m_{1_i}) \, k_B^{-1}, \cdots, \boldsymbol{r}_\ell = (\tau_\ell - \sum_{i=1}^{B-1} k_i m_{\ell_i}) \, k_B^{-1}\right)$$

$$\Pr(\boldsymbol{M}_1 = M_1, \cdots, \boldsymbol{M}_\ell = M_\ell) \tag{18}$$

$$= \sum_{M_1, \cdots, M_\ell} \Pr\left(\boldsymbol{r}_1 = (\tau_1 - \sum_{i=1}^{B-1} k_i m_{1_i}) \, k_B^{-1}\right) \cdots \Pr\left(\boldsymbol{r}_\ell = (\tau_\ell - \sum_{i=1}^{B-1} k_i m_{\ell_i}) \, k_B^{-1}\right)$$

$$\Pr(\boldsymbol{M}_1 = M_1, \cdots, \boldsymbol{M}_\ell = M_\ell) \tag{19}$$

$$= \sum_{M_1, \cdots, M_\ell} \frac{1}{p} \cdots \frac{1}{p} \Pr(\boldsymbol{M}_1 = M_1, \cdots, \boldsymbol{M}_\ell = M_\ell) \tag{20}$$

$$= \Pr(\boldsymbol{\tau}_1 = \tau_1) \cdots \Pr(\boldsymbol{\tau}_\ell = \tau_\ell), \tag{21}$$

where m_{j_i} denotes the ith block of the jth message M_j. Equation (19) holds due to the independence of the r_i's; equation (20) holds due to the uniform distribution of the r_i's; and equation (21) holds due to the uniform distribution of the τ_i's. Therefore, authentication tags are mutually independent, and the lemma follows.   □

On Equivalence Classes of Boolean Functions

Qichun Wang[1,2,*] and Thomas Johansson[2]

[1] The Shanghai Key Lab of Intelligent Information Processing, School of Computer Science, Fudan University, Shanghai 200433, P.R. China
032018023@fudan.edu.cn
[2] Dept. of Electrical and Information Technology, Lund University,
P.O. Box 118, 221 00 Lund, Sweden
Thomas.Johansson@eit.lth.se

Abstract. In FSE 2010, Rønjom and Cid put forward a nonlinear equivalence for Boolean functions and demonstrated that many cryptographic properties are not invariant among functions within the same equivalence class by providing some special examples. Their paper presented the idea and many problems were left open.

In this paper, we investigate equivalence of Boolean functions more deeply using a new method and discuss the number of Boolean functions in each equivalence class. We investigate further the cryptographic properties including algebraic immunity, algebraic degree and nonlinearity of equivalence classes, and deduce tight bounds on them. We find that there are many equivalence classes of Boolean functions with optimum algebraic immunity, optimum algebraic degree and a good nonlinearity. Moreover, we discuss how to construct equivalence classes with desired properties and show that it is possible to construct practical Boolean functions such that their equivalence classes have guaranteed cryptographic properties.

Keywords: Stream ciphers, Boolean functions, equivalence, algebraic immunity, nonlinearity.

1 Introduction

In recent years, algebraic attacks and fast algebraic attacks have received a lot of attention in the cryptographic community. They might be efficient against LFSR-based stream ciphers as well as on block ciphers [10,11,12,13,19]. To resist algebraic attacks, the Boolean functions used in stream ciphers should have high algebraic immunity. Up to now, many classes of Boolean functions achieving optimum algebraic immunity have been introduced [1,2,3,4,5,6,7,8,9]. To resist other attacks, the Boolean functions should also be balanced, have high algebraic degree, high nonlinearity and good immunity to fast algebraic attacks [10,14,16,33]. In [7], Carlet and Feng proposed an infinite class of balanced functions with

* Research supported by 973 Program with No. 2010CB327906, NSFC with No. 60873178, 60772131 and 60832001, and Grant No.09DZ2271800 of Shanghai Committee of Science and Technology.

K.-H. Rhee and D. Nyang (Eds.): ICISC 2010, LNCS 6829, pp. 311–324, 2011.

optimum algebraic degree, optimum algebraic immunity and a much better non-linearity than all the previously obtained infinite classes of functions. However, the lower bound they deduced is not enough to resist fast correlation attacks [14,15]. In [8], Tu and Deng introduced another class of balanced functions with optimum algebraic degree, optimum algebraic immunity and a provable good nonlinearity. However, also these functions are weak against fast algebraic attacks [16,21]. It seems very hard to construct Boolean functions achieving all the necessary criteria.

Moreover, even if we can find many Boolean functions satisfying all the necessary criteria, they may still be vulnerable to those attacks. In fact, Rønjom and Cid put forward a nonlinear equivalence of Boolean functions used in a filter generator and demonstrated that many cryptographic properties may be not invariant among functions in the same equivalence class by providing some special examples [17]. Therefore, Boolean functions with good cryptographic properties may be equivalent to some functions that are not good and we should assess the resistance of a cipher against some types of attacks by investigating the whole equivalence class. They gave the basic idea and many problems were left open.

In this paper, we investigate equivalence of Boolean functions more deeply using another method and discuss the number of Boolean functions in each equivalence class. We investigate further the cryptographic properties including algebraic degree, algebraic immunity and nonlinearity of the equivalence classes, and deduce tight bounds on them. We find that there are many equivalence classes of Boolean functions with optimum algebraic immunity, optimum algebraic degree and a good nonlinearity. Moreover, we discuss how to construct equivalence classes with desired properties and show that it is possible to construct practical Boolean functions such that their equivalence classes have guaranteed cryptographic properties.

The paper is organized as follows. In Section 2, the necessary background is established. We represent sequences and Boolean functions by generator matrices in Section 3. In Section 4, we then introduce equivalence of Boolean functions and investigate the number of Boolean functions in each equivalence class. We investigate further the cryptographic properties including algebraic degree, algebraic immunity and nonlinearity of the equivalence classes in Section 5, and deduce tight bounds on them. In Section 6, we give some equivalence classes of Boolean functions with optimum algebraic immunity, optimum algebraic degree and a good nonlinearity. In Section 7, we discuss how to construct equivalence classes with desired properties and show that it is possible to construct practical Boolean functions such that their equivalence classes have guaranteed cryptographic properties. We end in Section 8 with a few conclusions.

2   Preliminaries

Let $\mathbb{F}_2^n$ be the n-dimensional vector space over the finite field $\mathbb{F}_2$. A Boolean function of n variables is a function from $\mathbb{F}_2^n$ into $\mathbb{F}_2$. We denote by B_n the set of all n-variable Boolean functions.

Any $f \in B_n$ can be uniquely represented as a multivariate polynomial in $\mathbb{F}_2[x_1, \cdots, x_n]$,

$$f(x_1, ..., x_n) = \sum_{K \subseteq \{1,2,...,n\}} a_K \prod_{k \in K} x_k,$$

which is called its algebraic normal form (ANF). The algebraic degree of f, denoted by $\deg(f)$, is the number of variables in the highest order term with nonzero coefficient.

A Boolean function is affine if there exists no term of degree strictly greater than 1 in the ANF and the set of all affine functions is denoted by A_n.

Let

$$1_f = \{x \in \mathbb{F}_2^n | f(x) = 1\}, \ 0_f = \{x \in \mathbb{F}_2^n | f(x) = 0\}.$$

The cardinality of 1_f, denoted by $wt(f)$, is called the Hamming weight of f. The Hamming distance between two functions f and g, denoted by $d(f, g)$, is the Hamming weight of $f + g$. We say that an n-variable Boolean function f is balanced if $wt(f) = 2^{n-1}$.

Let $f \in B_n$. The nonlinearity of f, denoted by $nl(f)$, is its distance from the set of all n-variable affine functions, i.e.,

$$nl(f) = \min_{g \in A_n} d(f, g).$$

The r-order nonlinearity, denoted by $nl_r(f)$, is its Hamming distance to the set of all n-variable functions of degree at most r.

The nonlinearity of an n-variable Boolean function is upper bounded by $2^{n-1} - 2^{n/2-1}$, and a function is said to be bent if it can achieve this bound. Clearly, bent functions exist only for n even and it is known that the algebraic degree of a bent function is upper bounded by $\frac{n}{2}$ [18].

For any $f \in B_n$, a nonzero function $g \in B_n$ is called an annihilator of f if $fg = 0$, and the algebraic immunity of f, denoted by $\mathcal{AI}(f)$, is the minimum value of d such that f or $f + 1$ admits an annihilator of degree d. It is known that the algebraic immunity of an n-variable Boolean function is upper bounded by $\lceil \frac{n}{2} \rceil$ [19].

To resist algebraic attacks, a Boolean function f used in stream ciphers should have a high algebraic immunity, which implies that the nonlinearity of f is also not very low since [26]

$$nl(f) \geq 2 \sum_{i=0}^{AI(f)-2} \binom{n-1}{i}.$$

Many bounds on higher order nonlinearities have also been deduced [27,28,29,30,31,32].

Let $f \in B_n$. If there are functions g of low degree and h of reasonable degree such that $fg = h$, then f is considered to be weak against fast algebraic attacks. The fast algebraic immunity of an n-variable Boolean function f, denoted by $\mathcal{FAI}(f)$, is defined as

$$\mathcal{FAI}(f) = \min_{g \in B_n} \{2\mathcal{AI}(f), \deg g + \deg(fg)\},$$

where $1 \leq \deg g < \mathcal{AI}(f)$ [24]. To resist fast algebraic attacks, the Boolean functions used in stream ciphers should have high fast algebraic immunity and it is known that $\mathcal{FAI}(f) \leq n$ [10,24].

Two Boolean functions f and g are said to be affine equivalent if there exist $A \in GL_n(\mathbb{F}_2)$ and $b \in \mathbb{F}_2^n$ such that $g(x) = f(Ax + b)$. Clearly, algebraic degree, algebraic immunity and nonlinearity are all invariant under affine transformation.

In what follows, $q = 2^n$ and we denote an element of $\mathbb{F}_2^n$ by a column vector. For simplicity, we consider a stream cipher being a filter generator that consists of an LFSR of length n that generates an m-sequence and a filter function $f \in B_n$.

3   Representations of Sequences and Boolean Functions

Let the register generate an m-sequence of period $q - 1$, and the sequence $\{s_t\}$ obey the recursion

$$\sum_{j=0}^{n} m_j s_{t+j} = 0, \quad m_j \in \mathbb{F}_2,$$

where $m_0 = m_n = 1$. That is, $m(x) = m_0 + m_1 x + ... + m_{n-1}x^{n-1} + x^n$ is its generator polynomial, and is primitive. The (transpose) companion matrix M (we call it the generator matrix of the sequence) is

$$M = \begin{pmatrix} 0 & 1 & \cdots & 0 & 0 \\ 0 & 0 & \cdots & 0 & 0 \\ \cdots & \cdots & \cdots & \cdots & \cdots \\ 0 & 0 & \cdots & 0 & 1 \\ m_0 & m_1 & m_2 & \cdots & m_{n-1} \end{pmatrix}.$$

Let $(s_t, s_{t+1}, \ldots, s_{t+n-1})^T$ denote the state of the register at time t. Then the next state is determined by $(s_{t+1}, s_{t+2}, \ldots, s_{t+n})^T = M(s_t, s_{t+1}, \ldots, s_{t+n-1})^T = M^{t+1}(s_0, s_1, \ldots, s_{n-1})^T$. If the initial state of the register is b, then the sequence can be represented by $S = (b, Mb, ..., M^{q-2}b)$. Here b can be any nonzero n-dimensional column vector and hence there are exactly $q - 1$ such S which correspond to $q - 1$ different m-sequences. Let $b_0 = (1, 0, \cdots, 0)^T$. Then these sequences can be represented by

$$S_k = (M^k b_0, M^{k+1}b_0, ..., M^{k+q-2}b_0),$$

where $0 \leq k \leq q - 2$.

Since the number of primitive polynomials of degree n is $\phi(q-1)/n$, there are $\phi(q - 1)/n$ LFSRs generating m-sequences, and different LFSRs correspond to different sequences. Therefore, there are exactly $(q - 1)\phi(q - 1)/n$ m-sequences, and each sequence can be represented by

$$S_{jk} = (M_j^k b_0, M_j^{k+1}b_0, ..., M_j^{k+q-2}b_0),$$

where $1 \leq j \leq \phi(q-1)/n$, M_j is the generator matrix of the sequence and $0 \leq k \leq q-2$. Clearly, $M_j^k b_0$ is the initial state of the LFSR.

Let $T = \mathbb{F}_2^n - \{0\}$. Clearly, there are exactly two $f \in B_n$ such that $1_f \cap T = T_0$, where T_0 is a subset of T. We denote these two functions by f_1 and f_2. Then f_1 differs from f_2 only when $x = 0$ and $f_1 = f_2 + (x_1+1)(x_2+1)\cdots(x_n+1)$. Given any LSFR, the keystream generated by using f_1 or f_2 as the filter function is the same. Therefore, f_1 and f_2 can be viewed as the same function and we consider the set

$$B_n^* = B_n / \{0, (x_1+1)(x_2+1)\cdots(x_n+1)\}.$$

Then any $f \in B_n^*$ can be represented by its support set in the form

$$1_f = \{M_j^{i_1} b_0, M_j^{i_2} b_0, \ldots, M_j^{i_w} b_0\},$$

where M_j is the generator matrix of the register and $0 \leq i_1 < i_2 < \ldots < i_w \leq q-2$.

4   Equivalence of Boolean Functions

Rønjom and Cid put forward nonlinear equivalence of Boolean functions in [17], which can be defined as follows.

Definition 1. *Let z be a keystream generated by a filter generator where the LFSR has primitive feedback polynomial $g_1(x)$ and the filter function is f_1. $f_2 \in B_n^*$ is said to be equivalent to f_1 ($f_1 \sim f_2$) if there exists an LFSR which filtered by f_2 will generate the same keystream. Particularly, if the two LFSRs have the same generator polynomial, we say that f_1 and f_2 are linear equivalent and denote it by $f_1 \sim_L f_2$. Otherwise, f_1 and f_2 are said to be nonlinear equivalent which is denoted by $f_1 \sim_N f_2$.*

Example 1. Let the generator polynomial be $m_1(x) = x^5 + x^2 + 1$, the initial state be $(1,0,0,0,0)^T$, and the filter function be $f_1(x) = x_1 x_3 + x_1 x_4 + x_2 x_4 + x_3 x_4 + x_4 x_5 + x_5$. Then a full period of keystream is

$$(0100111110111000101011010000110).$$

Let the generator polynomial of another LFSR be $m_2(x) = x^5 + x^4 + x^3 + x^2 + 1$, the initial state be $(1,0,0,0,0)^T$, and the filter function be $f_2(x) = x_3 + x_4 + x_5$. Then the keystream generated will be the same as the above one. Therefore, $f_1 \sim_N f_2$. Clearly, $\deg(f_1) = 2$, $\mathcal{AI}(f_1) = 2$ and $nl(f_1) = 12$, while f_2 is a linear function.

Theorem 1. *Given an LFSR and the filter function $f \in B_n^*$, let $|1_f| = w$, where $1 \leq w < q-1$. Then the number of $g \in B_n^*$ such that $g \sim_L f$ is $q - \frac{w}{m}$, where m is a positive divisor of w and $\frac{w}{(w,q-1)}|m$.*

Proof. Let the LFSR be represented by S_{jk_1}, and write

$$1_f = \{M_j^{k_1+i_1}b_0, M_j^{k_1+i_2}b_0, \ldots, M_j^{k_1+i_w}b_0\},$$

where M_j is the generator matrix of the register and $0 \leq i_1 < i_2 < \ldots < i_w \leq q-2$. Clearly, $g \sim_L f$ if and only if there exists a $0 \leq k_2 \leq q-2$ such that

$$1_g = \{M_j^{k_2+i_1}b_0, M_j^{k_2+i_2}b_0, \ldots, M_j^{k_2+i_w}b_0\}.$$

If $k_1 \neq k_2$, then $f = g$ if and only if there exists an $1 \leq m \leq w-1$ such that

$$k_2 + i_s = k_1 + i_{s\oplus m} \pmod{q-1},$$

for $1 \leq s \leq w$, where

$$s \oplus m = \begin{cases} s+m, & \text{if } s+m \leq w, \\ s+m-w, & \text{otherwise.} \end{cases}$$

That is, for $1 \leq s \leq w$,

$$k_2 - k_1 = i_{s\oplus m} - i_s \pmod{q-1}. \tag{1}$$

Let k_0 be the number satisfying the equations (1) such that

$$k_0 - k_1 = \min_k\{k - k_1 \pmod{q-1}\},$$

where k satisfy the equations (1). Then a number k_2 satisfies the equations (1) if and only if $k_0 - k_1 | (k_2 - k_1 \pmod{q-1})$. Therefore, there are exactly $\frac{q-1}{k_0-k_1}$ such k_2, and each k_2 corresponds to a function which is equal to f. Since

$$k_0 - k_1 = i_{m+1} - i_1 = i_{m+2} - i_2 = \ldots = i_w - i_{w-m}$$
$$= i_1 - i_{w-m+1} + q - 1 = \ldots = i_m - i_w + q - 1,$$

we have $w(k_0 - k_1) = m(q-1)$. Therefore, the number of g satisfying $g \sim_L f$ is $q-1$ or $q - \frac{w}{m}$, where $1 \leq m \leq w-1$ and $\frac{w}{(w,q-1)}|m$, and the result follows.

Particularly, if $(q-1,w) = 1$, then $\frac{w}{(w,q-1)}|m$ implies $m = w$. We have the following two corollaries:

Corollary 1. *Let $f \in B_n^*$ and $|1_f| = w$, where $1 \leq w < q-1$ and $(q-1,w) = 1$. Then the number of $g \in B_n^*$ such that $g \sim_L f$ is $q-1$.*

Corollary 2. *Let $f \in B_n^*$ and $|1_f| = w$, where $1 \leq w < q-1$. Then there are $\phi(q-1)/n - 1$ classes of $g \in B_n^*$ such that $g \sim_N f$, and every class contains $q - \frac{w}{m}$ functions which are linear equivalent to each other, where m is a positive divisor of w and $\frac{w}{(w,q-1)}|m$.*

5   Cryptographic Properties of Equivalence Classes

Let $f \in B_n$ and

$$1_f = \{M_1^{k_1+i_1} b_0, M_1^{k_1+i_2} b_0, ..., M_1^{k_1+i_w} b_0\},$$

where M_1 is the generator matrix of the sequence and $0 \le i_1 < i_2 < ... < i_w \le q - 2$. Clearly, any $g \sim f$ can be represented by

$$1_g = \{M_j^{k_2+i_1} b_0, M_j^{k_2+i_2} b_0, ..., M_j^{k_2+i_w} b_0\},$$

where $1 \le j \le \phi(q - 1)/n$, M_j is a generator matrix and $0 \le k_2 \le q - 2$. Let $\overline{f} = \{g \in B_n | g \sim f\}$ denote the equivalence class to which f belongs. Since the same keystream can be generated by using Boolean functions of the same equivalence class as filter functions, to assess the resistance to a certain type of attack, we should define the corresponding cryptographic criteria by using the weakest equivalent function. Define

$$\deg(\overline{f}) = \min_{g \in \overline{f}} \deg(g), \quad \mathcal{AI}(\overline{f}) = \min_{g \in \overline{f}} \mathcal{AI}(g),$$

and

$$nl(\overline{f}) = \min_{g \in \overline{f}} nl(g).$$

To resist many kinds of attacks, $\deg(\overline{f})$, $\mathcal{AI}(\overline{f})$ and $nl(\overline{f})$ should all be high. Moreover, every function equivalent to f should have a good immunity against fast algebraic attacks. Clearly, for the function f_1 in Example 1, we have $\deg(\overline{f_1}) = 1$, $\mathcal{AI}(\overline{f_1}) = 1$ and $nl(\overline{f_1}) = 0$, which are very bad.

The number of variables of f, denoted by $var(f)$, is the number of variables appearing in its ANF. Define

$$var(\overline{f}) = \min_{g \in \overline{f}} var(g)$$

To be implemented efficiently, $var(\overline{f})$ should be small.

Lemma 1. *Let $f \in B_n$. Then we have $\deg(\overline{f}) \le n - 1$, which is a tight bound.*

Proof. Since $(x_1 + 1)(x_2 + 1) \cdots (x_n + 1)$ is of degree n, one of f and $f + (x_1 + 1)(x_2 + 1) \cdots (x_n + 1)$ is of degree n, and the other is not. Therefore, $\deg(\overline{f}) \le n - 1$. In the next section, we can find that there exist many f such that $\deg(\overline{f}) = n - 1$, and the result follows.

Lemma 2. *Let $f \in B_n$. If there is a balanced function g such that $g \sim f$, then $\deg(\overline{f}) \le var(\overline{f}) - 1$.*

Proof. Let $var(\overline{f}) = m$. If $m = n$, then by Lemma 1 the result follows. If $m < n$, then there is an $h \in B_n$ such that the number of variables appearing in its ANF is m. Therefore, $|1_h| = k2^{n-m}$, where k is an integer. Since h is equivalent to a balanced function, we have $|1_h| = 2^{n-1} + c$, where $c = 0$ or ± 1. Therefore, $c = 0$ and h is balanced. We regard h as an m-variable function, that is, $h \in B_m$. Then h is balanced and $\deg(h) \le m - 1$.

Lemma 3. *Let $f \in B_n$. Then we have $\mathcal{AI}(\bar{f}) \leq \lfloor \frac{n}{2} \rfloor$, which is a tight bound.*

Proof. For n odd, f and $f + (x_1+1)(x_2+1)\cdots(x_n+1)$ cannot be both balanced. Say, f is not balanced. Therefore, we have $|1_f| < 2^{n-1}$ or $|1_{f+1}| < 2^{n-1}$. Since the number of coefficients of a function with degree at most $\lfloor \frac{n}{2} \rfloor$ is 2^{n-1}, there exists a function g of degree at most $\lfloor \frac{n}{2} \rfloor$ such that $fg = 0$ or $(f+1)g = 0$. For n even, $\lfloor \frac{n}{2} \rfloor = \lceil \frac{n}{2} \rceil$. Hence, $\mathcal{AI}(\bar{f}) \leq \lfloor \frac{n}{2} \rfloor$. In the next section, we can find that there exist many f such that $\mathcal{AI}(\bar{f}) = \lfloor \frac{n}{2} \rfloor$, and the result follows.

Lemma 4. *Let $f \in B_n$. Then we have*

$$nl(\bar{f}) \leq 2^{n-1} - 2^{n/2-1} - 1.$$

Proof. Clearly, the theorem is true for n odd. For n even, bent functions have the maximum nonlinearity and the highest possible degree for a bent function is $\frac{n}{2}$. Since one of f and $f + (x_1+1)(x_2+1)\cdots(x_n+1)$ is of degree n, they can not be both bent. Therefore, $nl(\bar{f}) \leq 2^{n-1} - 2^{n/2-1} - 1$.

Remark 1: For small even n, it is easy to find equivalence classes whose nonlinearity can achieve the bound. However, for general n, we can not find an equivalence class such that the bound can be achieved. We do not know whether this bound can be tight for an infinite class and we leave this as an open problem.

Fast algebraic immunity and higher order nonlinearities of an equivalence class can be defined in a similar way:

$$\mathcal{FAI}(\bar{f}) = \min_{g \in \bar{f}} \mathcal{FAI}(g) \text{ and } nl_r(\bar{f}) = \min_{g \in \bar{f}} nl_r(g).$$

We do not want to discuss the bounds on $\mathcal{FAI}(\bar{f})$ and $nl_r(\bar{f})$, since it may be hard to deduce tight bounds on them. Anyway, to resist fast algebraic attacks, $\mathcal{FAI}(\bar{f})$ should be high.

In practice, we should use equivalence classes with good cryptographic properties and choose the functions of them which can be implied efficiently as filter functions.

6   Equivalence Classes with Optimum Algebraic Degree, Optimum Algebraic Immunity and a Good Nonlinearity

Given an LFSR and its generator matrix M_1, let $f_1 \in B_n^*$ and

$$1_{f_1} = \{M_1^i b_0 | i = 0, 1, ..., 2^{n-1} - 1\}.$$

This function has been investigated by many papers and has good cryptographic properties (see [7,20,22,23]). We now investigate the equivalence class to which f_1 belongs. Clearly, any $g \sim f_1$ can be represented by

$$1_g = \{M_j^{k+i} b_0 | i = 0, 1, ..., 2^{n-1} - 1\},$$

where $1 \leq j \leq \phi(2^n - 1)/n$, M_j is a generator matrix and $0 \leq k \leq 2^n - 2$.

Theorem 2. *Let f_1 be defined as above. Then* $\deg(\overline{f_1}) = n - 1$, $\mathcal{AI}(\overline{f_1}) = \lfloor \frac{n}{2} \rfloor$
and

$$nl(\overline{f_1}) > 2^{n-1} - (\frac{\ln 2}{3}(n-1) + \frac{3}{2})2^{\frac{n}{2}} - 1.$$

In other words, the equivalence class to which f_1 belongs has optimum algebraic degree, optimum algebraic immunity and a good nonlinearity.

Proof. Let $g \sim f_1$ and

$$1_g = \{M_j^{k+i}b_0 | i = 0, 1, ..., 2^{n-1} - 1\},$$

where M_j is a generator matrix and $0 \leq k \leq 2^n - 2$. Then we have

$$1_{g(M_j x)} = \{M_j^{k+i-1}b_0 | i = 0, 1, ..., 2^{n-1} - 1\},$$

and

$$1_{g(x)+g(M_j x)} = \{M_j^{k-1}b_0, M_j^{k+2^{n-1}-1}b_0\}.$$

Therefore, $\deg(g(x) + g(M_j x)) = n - 1$. Since $g(M_j x)$ is affine equivalent to $g(x)$ and algebraic degree is affine invariant, we have $\deg(g(x)) = \deg(g(M_j x)) = n - 1$. Hence, $\deg(g(x) + (x_1 + 1)(x_2 + 1) \cdots (x_n + 1)) = n$ and $\deg(\overline{f_1}) = n - 1$. Similar to the proof of Theorem 3 in [20], we have

$$\min\{\mathcal{AI}(g(x)), \mathcal{AI}(g(x) + (x_1 + 1)(x_2 + 1) \cdots (x_n + 1))\} = \lfloor \frac{n}{2} \rfloor.$$

Therefore, $\mathcal{AI}(\overline{f_1}) = \lfloor \frac{n}{2} \rfloor$. From Proposition 6 of [20], we have

$$nl(g) > 2^{n-1} - (\frac{\ln 2}{3}(n-1) + \frac{3}{2})2^{\frac{n}{2}}.$$

Therefore,

$$nl(g + (x_1 + 1) \cdots (x_n + 1)) > 2^{n-1} - (\frac{\ln 2}{3}(n-1) + \frac{3}{2})2^{\frac{n}{2}} - 1,$$

and the result follows.

Let $f_2 \in B_n^*$ and

$$1_{f_2} = \{b_0, M_1 b_0, ..., M_1^{r-1}b_0, M_1^{r+1}b_0, ..., M_1^{2^{n-1}}b_0\},$$

where $2^{n-1} - \frac{n-1}{2} < r < 2^{n-1}$. Any $g \sim f_2$ can be represented by

$$1_g = \{M_j^{k+i}b_0 | i = 0, 1, ..., r - 1, r + 1, ..., 2^{n-1}\},$$

where $1 \leq j \leq \phi(2^n - 1)/n$, M_j is a generator matrix and $0 \leq k \leq 2^n - 2$.
Similar to the analysis of f_1, the equivalence class to which f_2 belongs has optimum algebraic degree, optimum algebraic immunity and a good nonlinearity.

7   Construction of Equivalence Classes with Desired Properties

We now discuss how to construct equivalence classes with desired properties. Let $f \in B_n^*$ and $1_f = \{M_j^i b_0 | i \in E\}$, where E is suitable chosen so that the equivalence class that f belongs to has desired properties. Clearly, any $g \sim f$ can be represented by

$$1_g = \{M_j^{k+i} b_0 | i \in E\},$$

where $1 \leq j \leq \phi(2^n - 1)/n$, M_j is a generator matrix and $0 \leq k \leq 2^n - 2$. If we want $\mathcal{AI}(\overline{f}) > r$, then we should investigate the following matrices

$$\begin{pmatrix} M_j^{i_1 l_1} b_1 \; M_j^{i_1 l_2} b_1 \cdots \; M_j^{i_1 l_t} b_1 \\ M_j^{i_2 l_1} b_1 \; M_j^{i_2 l_2} b_1 \cdots \; M_j^{i_2 l_t} b_1 \\ \cdots \qquad \cdots \qquad \cdots \qquad \cdots \\ M_j^{i_s l_1} b_1 \; M_j^{i_s l_2} b_1 \cdots \; M_j^{i_s l_t} b_1 \end{pmatrix}, \tag{2}$$

where $0 \leq l_1 < l_2 < ... < l_t < 2^n - 1$, $wt(l_j) \leq r$ and $t = \sum_{i=0}^r \binom{n}{i}$, $i_1 < i_2 < ... < i_s$ are all in E or not and

$$s = \begin{cases} 2^{n-1} & \text{if } i_1, ..., i_s \in E \\ 2^{n-1} - 1 & \text{if } i_1, ..., i_s \notin E. \end{cases}$$

It is easily found that $\mathcal{AI}(\overline{f}) > r$ if and only if all these matrices are of rank t. Let $M_j^i b_0 = (b_{ij1}, b_{ij2}, ..., b_{ijn})^T$. If we want $\deg(\overline{f}) \geq d$, then we should investigate the following functions

$$\sum_{i \in E} \prod_{k=1}^n (x_k + b_{ijk} + 1) + c_1 \cdot x_1 x_2 \cdots x_n,$$

where $c_1 = 0$ or 1. Clearly, $\deg(\overline{f}) \geq d$ if and only if all these functions are of degrees at least $n - 1$. Let $|E| = 2^{n-1}$ or $2^{n-1} - 1$. Then

$$nl(f) = 2^n - \max_{h \in A_n} (|1_f \cap 0_h| + |0_f \cap 1_h|)$$

$$= 2^n - 2 \max_{h \in A_n} |1_f \cap 0_h| - c_2 = 2^n - 2 \max_{h \in A_n} |0_f \cap 1_h| - c_2.$$

where $c_2 = 0$ or ± 1. If we want $nl(\overline{f})$ to be high, then $\max_{h \in A_n} (|1_g \cap 0_h|)$ should be low, where $g \sim f$. Therefore, we should investigate the sets

$$\{M_j^i b_0 | i \in E\} \cap 0_h$$

and the number of elements of these sets should be small, where $1 \leq j \leq \phi(2^n - 1)/n$, M_j is a generator matrix and $h \in A_n$. The equivalence classes to which f_1 or f_2 belongs are examples with optimum algebraic degree, optimum algebraic immunity and a good nonlinearity. In a similar way we can construct other equivalence classes with desired properties. In fact, it may be not easy

Table 1. Cryptographic Properties of the Equivalence Class $\overline{f}$

GP	VAR	$\mathcal{AI}$	DEG	NL
10110001	8	4	7	106
11000110	5	3	4	80
10001110	8	4	6	104
10111000	8	4	7	102
10110100	8	4	7	106
10010110	8	4	7	106
11110011	8	4	7	102
11100111	8	4	6	100
10010101	8	4	7	106
11010100	8	4	6	108
10110010	8	4	7	100
10100110	8	4	7	104
10101111	8	4	6	104
11111010	8	4	7	108
11100001	8	4	7	104
11000011	8	4	6	104

to construct equivalence classes satisfying all the necessary criteria. Anyway, the approach described above may be promising, which can be seen from the following example.

Example 2. Let $f \in B_8$. We want to construct an equivalence class $\overline{f}$ such that $\deg(\overline{f}) \geq 4$, $\mathcal{AI}(\overline{f}) \geq 3$ and $nl(\overline{f}) \geq 58$ (this is the lowest possible nonlinearity of a 8-variable function with the optimum algebraic immunity). Moreover, we want $var(\overline{f}) \leq 5$ (that is, there is a function of the equivalence class such that its ANF contains at most 5 variables). Let $1_f = \{M^i b_0 | i \in E\}$, where M is the generator matrix and $b_0 = (10000000)^T$. There are many sets E such that the matrices in (2) are of rank 2. We choose

$$E = [\ 1\ 5\ 9\ 14\ 15\ 17\ 19\ 20\ 21\ 29\ 30\ 31\ 32\ 36\ 38\ 39\ 40\ 41\ 42\ 43\ 45\ 46\ 47\ 51$$
$$52\ 58\ 59\ 60\ 61\ 62\ 63\ 64\ 66\ 67\ 70\ 71\ 72\ 73\ 76\ 77\ 82\ 83\ 84\ 85\ 88\ 90\ 95$$
$$96\ 97\ 98\ 99\ 103\ 105\ 109\ 111\ 112\ 113\ 114\ 115\ 116\ 126\ 127\ 133\ 136$$
$$137\ 138\ 140\ 142\ 144\ 145\ 146\ 158\ 159\ 160\ 164\ 165\ 166\ 167\ 168\ 169$$
$$170\ 171\ 173\ 176\ 177\ 178\ 179\ 180\ 181\ 182\ 183\ 184\ 185\ 188\ 189\ 190$$
$$191\ 193\ 194\ 196\ 198\ 201\ 202\ 203\ 206\ 207\ 208\ 209\ 210\ 211\ 212\ 213$$
$$214\ 215\ 219\ 220\ 221\ 222\ 224\ 237\ 244\ 245\ 246\ 249\ 250\ 251\ 252\ 253].$$

There are exactly $\phi(255)/8 = 16$ classes of $g \in B_8$ such that $g \sim f$, and every class is characteristiced by a generator polynomial. We can find some cryptographic properties of these classes from Table 1. In the table, GP denotes generator polynomial (we represent $f(x) = x^8 + a_8 x^7 + a_7 x^6 + a_6 x^5 + a_5 x^4 + a_4 x^3 +$

$a_3x^2 + a_2x + a_1$ by $a_1a_2a_3a_4a_5a_6a_7a_8$), VAR denotes number of variables appearing in ANF, $\mathcal{AI}$ denotes algebraic immunity, DEG denotes algebraic degree and NL denotes nonlinearity. Clearly, $var(\overline{f}) = 5$, $\mathcal{AI}(\overline{f}) = 3$, $\deg(\overline{f}) = 4$ and $nl(\overline{f}) = 80$. It is an equivalence class with desired properties. We now pick out a function g such that it can be implemented efficiently. Let M_1 be the generator matrix corresponding to the primitive polynomial $x^8 + x^6 + x^5 + x + 1$. Take $1_g = \{M_1^i b_0 | i \in E\}$. From Table 1, we have $var(g) = 5$. In fact,

$$g = x_1x_2x_3x_5 + x_1x_2x_4x_5 + x_1x_3x_4x_5 + x_1x_2x_3 + x_1x_3x_5 + x_1x_4x_5$$
$$+ x_2x_3x_4 + x_2x_4x_5 + x_3x_4x_5 + x_1x_2 + x_1 + x_3 + x_5.$$

It is simple to implement and in effect acting as a 5-variable function.

Remark 2: All functions of the equivalence class $\overline{g}$ have nonlinearity at least 80, which is greater than the lowest possible nonlinearity 58 of an 8-variable function with optimum algebraic immunity. If we regard g as a 5-variable Boolean function, then it is balanced and has optimum algebraic degree 4, optimum algebraic immunity 3 and nonlinearity 10. Clearly, $var(\overline{g}) = var(g)$, $\mathcal{AI}(\overline{g}) = \mathcal{AI}(g)$, $\deg(\overline{g}) = \deg(g)$ and $nl(\overline{g}) = 8nl(g)$.

Remark 3: It may be very hard to construct an infinite class of equivalence classes satisfying all the necessary criteria. However, for a given n, the example shows that it is possible to construct practical Boolean functions (in few variables $m < n$) such that their equivalence classes with respect to $\mathbb{F}_2^n$ have guaranteed cryptographic properties.

8   Conclusion

In this paper, we investigated equivalence of Boolean functions and discussed the number of Boolean functions in each equivalence class. We investigated further the cryptographic properties including algebraic immunity, algebraic degree and nonlinearity of the equivalence classes, and deduced tight bounds on them. We found that there are many equivalence classes of Boolean functions with optimum algebraic degree, optimum algebraic immunity and a good nonlinearity. Moreover, we discussed how to construct equivalence classes with desired properties and showed that it is possible to construct practical Boolean functions such that their equivalence classes have guaranteed cryptographic properties.

   Previous constructed Boolean functions with good cryptographic properties may be equivalent to some functions which are not good and we must assess the resistance of a Boolean function against some types of attacks by investigating the whole equivalence class. In practice, we should use equivalence classes with good cryptographic properties and choose the functions of them which can be implemented efficiently as filter functions.

References

1. Carlet, C., Dalai, D.K., Gupta, K.C., Maitra, S.: Algebraic immunity for crypto-graphically significant Boolean functions: analysis and construction. IEEE Trans. Inf. Theory 52(7), 3105–3121 (2006)
2. Braeken, A., Preneel, B.: On the algebraic immunity of symmetric Boolean functions. In: Maitra, S., Veni Madhavan, C.E., Venkatesan, R. (eds.) INDOCRYPT 2005. LNCS, vol. 3797, pp. 35–48. Springer, Heidelberg (2005)
3. Dalai, D.K., Maitra, K.C., Maitra, S.: Cryptographically significant Boolean functions: Construction and analysis in terms of algebraic immunity. In: Gilbert, H., Handschuh, H. (eds.) FSE 2005. LNCS, vol. 3557, pp. 98–111. Springer, Heidelberg (2005)
4. Dalai, D.K., Maitra, S., Sarkar, S.: Baisc theory in construction of Boolean functions with maximum possible annihilator immunity. Des. Codes Cryptogr 40(1), 41–58 (2006)
5. Li, N., Qi, W.-F.: Construction and Analysis of Boolean Functions of $2t+1$ Variables with Maximum Algebraic Immunity. In: Lai, X., Chen, K. (eds.) ASIACRYPT 2006. LNCS, vol. 4284, pp. 84–98. Springer, Heidelberg (2006)
6. Pasalic, E.: Almost Fully Optimized Infinite Classes of Boolean Functions Resistant to (Fast) Algebraic Cryptanalysis. In: Lee, P.J., Cheon, J.H. (eds.) ICISC 2008. LNCS, vol. 5461, pp. 399–414. Springer, Heidelberg (2009)
7. Carlet, C., Feng, K.: An infinite class of balanced functions with optimal algebraic immunity, good immunity to fast algebraic attacks and good nonlinearity. In: Pieprzyk, J. (ed.) ASIACRYPT 2008. LNCS, vol. 5350, pp. 425–440. Springer, Heidelberg (2008)
8. Tu, Z., Deng, Y.: A Conjecture on Binary String and its Application on constructing Boolean Functions of Optimal Algebraic Immunity. Des. Codes Cryptogr (2010), Online First Articles. doi:10.1007/s10623-010-9413-9
9. Carlet, C., Feng, K.: An Infinite Class of Balanced Vectorial Boolean Functions with Optimum Algebraic Immunity and Good Nonlinearity. In: Chee, Y.M., Li, C., Ling, S., Wang, H., Xing, C. (eds.) IWCC 2009. LNCS, vol. 5557, pp. 1–11. Springer, Heidelberg (2009)
10. Courtois, N.: Fast Algebraic attacks on stream ciphers with linear feedback. In: Boneh, D. (ed.) CRYPTO 2003. LNCS, vol. 2729, pp. 176–194. Springer, Heidelberg (2003)
11. Courtois, N., Bard, G.: Algebraic Cryptanalysis of the Data Encryption Standard. In: Galbraith, S.D. (ed.) Cryptography and Coding 2007. LNCS, vol. 4887, pp. 152–169. Springer, Heidelberg (2007)
12. Courtois, N., Pieprzyk, J.: Cryptanalysis of block ciphers with overdefined systems of equations. In: Zheng, Y. (ed.) ASIACRYPT 2002. LNCS, vol. 2501, pp. 267–287. Springer, Heidelberg (2002)
13. Armknecht, F., Krause, M.: Algebraic Attacks on Combiners with Memory. In: Boneh, D. (ed.) CRYPTO 2003. LNCS, vol. 2729, pp. 162–175. Springer, Heidelberg (2003)
14. Meier, W., Staffelbach, O.: Fast correlation attacks on stream ciphers. In: Günther, C.G. (ed.) EUROCRYPT 1988. LNCS, vol. 330, pp. 301–314. Springer, Heidelberg (1988)
15. Johansson, T., Jönsson, F.: Fast Correlation Attacks through Reconstruction of Linear Polynomials. In: Bellare, M. (ed.) CRYPTO 2000. LNCS, vol. 1880, pp. 300–315. Springer, Heidelberg (2000)

16. Carlet, C.: On a weakness of the Tu-Deng function and its repair. Cryptology ePrint Archive 2009/606 (2009), http://eprint.iacr.org/

17. Rønjom, S., Cid, C.: Nonlinear Equivalence of Stream Ciphers. In: Hong, S., Iwata, T. (eds.) FSE 2010. LNCS, vol. 6147, pp. 40–54. Springer, Heidelberg (2010), http://www.isg.rhul.ac.uk/~ccid/publications/NL-equivalence.pdf

18. Rothaus, O.S.: On bent functions. J. Comb. Theory A20(3), 300–305 (1976)

19. Courtois, N., Meier, W.: Algebraic attacks on stream ciphers with linear feedback. In: Biham, E. (ed.) EUROCRYPT 2003. LNCS, vol. 2656, pp. 345–359. Springer, Heidelberg (2003)

20. Wang, Q., Peng, J., Kan, H., Xue, X.: Constructions of Cryptographically Significant Boolean Functions Using Primitive Polynomials. IEEE Trans. Inf. Theory 56(6), 3048–3053 (2010)

21. Wang, Q., Johansson, T.: A Note on Fast Algebraic Attacks and Higher Order Nonlinearities. In: Lai, X., Yung, M., Lin, D. (eds.) INSCRYPT 2010. LNCS, vol. 6584, pp. 404–414. Springer, Heidelberg (2011)

22. Rizomiliotis, P.: On the security of the Feng-Liao-Yang Boolean functions with optimal algebraic immunity against fast algebraic attacks. Des. Codes Cryptogr, http://www.springerlink.com/content/yj27532v5481857v/

23. Feng, K., Liao, Q., Yang, J.: Maximal values of generalized algebraic immunity. Des. Codes Cryptogr 50(2), 243–252 (2009)

24. Liu, M., Lin, D.: Fast Algebraic Attacks and Decomposition of Symmetric Boolean Functions. ArXiv: 0910.4632v1 [cs.CR]

25. Kavut, S., Yucel, M.: Generalized Rotation Symmetric and Dihedral Symmetric Boolean Functions - 9 variable Boolean Functions with Nonlinearity 242. In: Boztaş, S., Lu, H.-F. (eds.) AAECC 2007. LNCS, vol. 4851, pp. 321–329. Springer, Heidelberg (2007)

26. Lobanov, M.S.: Tight bounds between algebraic immunity and high-order nonlinearities. Diskretn. Anal. Issled. Oper 15(6), 34–47 (2008)

27. Carlet, C.: On the higher order nonlinearities of algebraic immune functions. In: Dwork, C. (ed.) CRYPTO 2006. LNCS, vol. 4117, pp. 584–601. Springer, Heidelberg (2006)

28. Carlet, C., Mesnager, S.: Improving the Upper Bounds on the Covering Radii of Binary Reed-Muller Codes. IEEE Trans. Inf. Theory 53(1), 162–173 (2007)

29. Carlet, C.: Recursive Lower Bounds on the Nonlinearity Profile of Boolean Functions and Their Applications. IEEE Trans. Inf. Theory 54(3), 1262–1272 (2008)

30. Lobanov, M.S.: Tight bounds between algebraic immunity and nonlinearities of high orders. Cryptology ePrint Archive 2007/444 (2007), http://eprint.iacr.org/

31. Carlet, C.: On the Higher Order Nonlinearities of Boolean Functions and S-Boxes, and Their Generalizations. In: Golomb, S.W., Parker, M.G., Pott, A., Winterhof, A. (eds.) SETA 2008. LNCS, vol. 5203, pp. 345–367. Springer, Heidelberg (2008)

32. Mesnager, S.: Improving the Lower Bound on the Higher Order Nonlinearity of Boolean Functions With Prescribed Algebraic Immunity. IEEE Trans. Inf. Theory 54(8), 3656–3662 (2008)

33. Rønjom, S., Helleseth, T.: A New Attack on the Filter Generator. IEEE Trans. Inf. Theory 53(5), 1752–1758 (2007)

Public Discussion Must Be Back and Forth in Secure Message Transmission

Takeshi Koshiba and Shinya Sawada

Division of Mathematics, Electronics and Informatics,
Graduate School of Science and Engineering, Saitama University
255 Shimo-Okubo, Sakura, Saitama 338-8570, Japan
{koshiba,s09mm320}@mail.saitama-u.ac.jp

Abstract. Secure message transmission (SMT) is a two-party protocol between a sender and a receiver over a network in which the sender and the receiver are connected by n disjoint channels and t out of n channels can be controlled by an adaptive adversary with unlimited computational resources. If a public discussion channel is available to the sender and the receiver to communicate with each other then a secure and reliable communication is possible even when $n \geq t + 1$. The round complexity is one of the important measures for the efficiency for SMT. In this paper, we revisit the optimality and the impossibility for SMT with public discussion and discuss the limitation of SMT with the "unidirectional" public channel, where either the sender or the receiver can invoke the public channel, and show that the "bidirectional" public channel is necessary for SMT.

Keywords: secure message transmission, public discussion, round complexity.

1 Introduction

Dolev, Dwork, Waarts and Yung [7] introduced *Secure Message Transmission* protocols to address the problem of delivering a message from a sender S to a receiver R in a network guaranteeing *privacy* and *reliability*. In the network, S is connected to R by n node-disjoint paths, referred to as *wires*, up to $t < n$ of which may be maliciously controlled by the adversary with unlimited computational resources.

A *perfectly* secure message transmission (PSMT for short) guarantees that R always receives message sent by S and the adversary A learns nothing about it. It was shown that PSMT is possible if and only if $n \geq 2t + 1$. For the detail and progress of PSMT, you may see [7,17,19,2,10,14].

Franklin and Wright [11] relaxed the security requirement for PSMT protocols and considered *probabilistic* security in which a privacy parameter ε and a reliability parameter δ upper-bound the advantage of the adversary in violating the privacy and the probability that R fails to recover message sent by S, respectively. We refer to secure message transmission protocols in the relaxed setting as (ε, δ)-SMT protocols and consider that PSMT is a special case of SMT where $\varepsilon = \delta = 0$, that is, $(0,0)$-SMT. For further investigation of SMT protocols, you may see [20,9,1,15].

K.-H. Rhee and D. Nyang (Eds.): ICISC 2010, LNCS 6829, pp. 325–337, 2011.

Garay and Ostrovsky [12] studies a model called Secure Message Transmission *by Public Discussion* (SMT-PD) as an important building block for achieving secure multi-party computation [3,6] on sparse networks. (A similar setting was studied earlier by Franklin and Wright [11]). In this model, in addition to the wires in the standard SMT, S and R can access to a *public* channel which the adversary cannot alter but eavesdrop.

Generally speaking, the public channel is more expensive to implement than wires. Since there are several ways to implement the public channel even on the sparse networks [8,21,4,5], the costs are still expensive. Thus, it is desirable to minimize the use of this expensive resource. One of the typical efficiency measures of SMT-PD protocols is the number of *rounds* where each round is one message flow between S and R. Shi, Jiang, Safavi-Naini and Tuhin [18] (and Garay, Givens and Ostrovsky [13] also) gave an SMT-PD protocol with 3 rounds, 2 of which invoke the public channel. We say that an SMT-PD protocol is of (r, r')-round if the SMT-PD protocol satisfies that r is the total number of rounds and r' ($\leq r$) is the number of rounds in which the public channel is invoked. Through this paper, we assume that $t + 1 \leq n \leq 2t$, since an efficient (i.e., polynomial-time) "2-round" PSMT protocol is provided by Kurosawa and Suzuki [14] when $n \geq 2t + 1$. Shi *et al.* also showed that $(3, 2)$-round is optimal for SMT-PD protocols (in case of $n \geq t + 1$) by showing the impossibility of useful $(3, 1)$-round SMT-PD protocols.

In this paper, we revisit the lower-bounds on the round complexity for SMT-PD protocols. As mentioned, known round-optimal SMT-PD protocols invoke the public channel twice: once is for the public transmission from S to R and the other time is for that from R to S. This means that two types of implementation for the public channel are necessary. If the message flows over public channel are one-way in SMT-PD protocols, it is sufficient that we implement only one type of the public channel, which is more desirable. We first discuss the limitation of SMT with the "unidirectional" public channel, where either the sender or the receiver can invoke the public channel, and show that the "bidirectional" public channel is necessary for SMT.

To discuss the directions of the message flows over the public channel, we consider separately the number of rounds where the message flows are from S to R over the public channel and the number of rounds where the message flows are from R to S. We say that an SMT-PD protocol is of (r, r_1, r_2)-round if the SMT-PD protocol is of r rounds and satisfies that r_1 (resp., r_2) is the number of rounds in which the message flows are from S to R (resp., from R to S). We review some previous results by using our terminology. Known round-optimal SMT-PD protocols [18,13] are of $(3, 1, 1)$-round. Shi *et al.* [18] showed that $(r, 1, 0)$-round protocol must satisfy that $\varepsilon + \delta \geq 1 - 1/|M|$; and $(r, 0, 1)$-round protocol must satisfy that $\delta \geq (1 - 1/|M|)/2$, where $r \geq 3$ and M denotes the message space. We generalize their results and prove that (*i*) $(r, r', 0)$-round protocol must satisfy that $\varepsilon + \delta \geq 1 - 1/|M|$; and (*ii*) $(r, 0, r')$-round protocol must satisfy that $\delta \geq (1 - 1/|M|)/2$, where $r \geq 3$ and $r' \leq r$. The statement (*i*) says that any $(r, r', 0)$-round protocol is not useful at all. The statement (*ii*) does not say anything about the privacy. Actually, we provide an upper-bounding $(0, \delta)$-SMT-PD protocol ($\delta \approx 1/2$ when $n = 2t$), which says that the bound in (*ii*) is almost tight.

2 Preliminaries

2.1 Model and Notations

Network and Adversary Models. We assume a *synchronous*, connected point-to-point *incomplete* network. Two parties $\mathcal{S}$ and $\mathcal{R}$ are connected by n node-disjoint paths, called *wires*. In addition to the wires, we assume that there is an authentic and reliable *public channel* between $\mathcal{S}$ and $\mathcal{R}$. Messages over this channel are publicly accessible and correctly delivered to the recipient. All wires are *bidirectional* but we consider that public channels are *unidirectional*. SMT-PD protocols proceed in rounds. In each round, one party may send a message on each wire and the public channel, while the other party will only receive the sent messages. The sent messages will be delivered before the next round starts.

The adversary $\mathcal{A}$ is computationally *unbounded*. $\mathcal{A}$ can corrupt nodes on paths between $\mathcal{S}$ and $\mathcal{R}$. A wire is said to be *corrupted* if at least one node on the path is corrupted. We assume that up to $t \leq n - 1$ wires can be corrupted by the adversary. $\mathcal{A}$ can *eavesdrop*, *alter* or *block* messages sent over the corrupted wires. $\mathcal{A}$ is assumed to be *adaptive*, that is, she can corrupt wires during the protocol execution based on the communication exchanged so far.

Notations. Let M be the message space in which $\mathcal{S}$ selects a message. Let M_S denote the secret message of $\mathcal{S}$, and M_R the message output by $\mathcal{R}$. We denote the null string by $\perp$. We write $u \leftarrow U$ to denote that a value u is uniformly chosen from a set U.

2.2 Definitions

The *statistical distance* of two random variables X and Y over a set U is defined as

$$\Delta(X, Y) = \frac{1}{2} \sum_{u \in U} \left| \Pr[X = u] - \Pr[Y = u] \right|.$$

Lemma 1. *Let X and Y be two random variables over a set U. The advantage of any computationally unbounded algorithm $\mathcal{D} : U \rightarrow \{0, 1\}$ to distinguish X and Y is*

$$\left| \Pr[\mathcal{D}(X) = 1] - \Pr[\mathcal{D}(Y) = 1] \right| \leq \Delta(X, Y).$$

In an *execution* of an SMT protocol Π, $\mathcal{S}$ wants to send $M_S \in M$ to $\mathcal{R}$ privately and reliably. We assume that $\mathcal{R}$ *always* outputs a message $M_R \in M$ at the end of the protocol.

An execution is completely determined by the random coins of all the parties including the adversary, and the message distribution of M_S. For $P \in \{\mathcal{S}, \mathcal{R}, \mathcal{A}\}$, the *view* of P includes the random coins of P and the message that P receives. We denote by $V_A(m, c_A)$ the view of $\mathcal{A}$ when the protocol is run with $M_S = m$ and $\mathcal{A}$'s randomness $C_A = c_A$.

Definition 1. A protocol Π between $\mathcal{S}$ and $\mathcal{R}$ is an (ε, δ)-*secure message transmission by public discussion* (SMT-PD) if the following two conditions are satisfied:

– **Privacy:** For every two messages $m_0, m_1 \in M$ and $c_A \in \{0,1\}^*$, Π satisfies that

$$\Delta(V_A(m_0, c_A), V_A(m_1, c_A)) \le \varepsilon,$$

where the probability is taken over the randomness of $\mathcal{S}$ and $\mathcal{R}$.
– **Reliability:** $\mathcal{R}$ recovers the message M_S with probability larger than $1 - \delta$. In other words, it holds that

$$\Pr[M_R \ne M_S] \le \delta,$$

where the probability is over the randomness of all the parties $\mathcal{S}$, $\mathcal{R}$ and $\mathcal{A}$, and the distribution of M_S.

Observe that the above definition is oblivious of the message distribution. In other words, any SMT-PD protocol must be secure with the same privacy and reliability parameters regardless of the concrete distribution over M.

3   Known Results on the Round Complexity

In this section, we review some of known results on the round complexity for SMT-PD protocols (when $n \le 2t$).

Theorem 1. ([11,18]) *Suppose that $n \le 2t$. Then the following statements hold.*

1. *For any values $r \ge r' \ge 1$, it is impossible to construct (r, r')-round $(0,0)$-SMT-PD protocols.*
2. *For any values $r \ge 1$ and $0 \le \varepsilon \le 1$, it is impossible to construct $(r, 0)$-round (ε, δ)-SMT-PD protocols with $\delta < (1 - 1/|M|)/2$.*
3. *There is neither $(2, 1, 1)$-round nor $(2, 2, 0)$-round (ε, δ)-SMT-PD protocol with $\varepsilon + \delta < 1 - 1/|M|$.*
4. *There is no $(3, 1, 0)$-round (ε, δ)-SMT-PD protocol with $\varepsilon + \delta < 1 - 1/|M|$.*
5. *For any $0 \le \varepsilon \le 1$, there is no $(3, 0, 1)$-round (ε, δ)-SMT-PD protocol with $\delta < (1 - 1/|M|)/2$.*

In addition, we mention a technical lemma stated in [18], which we also use to derive our results.

Lemma 2. ([18]) *Let Π be an (ε, δ)-SMT-PD protocol and assume that $\mathcal{S}$ selects $M_S \leftarrow M$. Then no adversary $\mathcal{A}$ can correctly guess M_S with probability larger than $\varepsilon + 1/|M|$. That is,*

$$\Pr[M_A = M_S] \le \varepsilon + \frac{1}{|M|},$$

where M_A denotes the adversary's output and the probability is taken over the randomness of $\mathcal{S}$, $\mathcal{R}$ and $\mathcal{A}$.

4   Lower Bounds

As mentioned, we have a $(3,1,1)$-round SMT-PD protocol and the protocol is round-optimal due to Theorem 1. The protocol uses the bidirectional public channel. The implementation of the bidirectional public channel may be more expensive than that of the unidirectional public channel. So, SMT protocols with the unidirectional public channel may be advantageous even if the round complexity becomes larger. Thus, we consider the possibility of SMT protocol with the unidirectional public channel.

To discuss the possibility of r-round SMT protocol with the unidirectional public channel, we consider a sequence of randomized functions $(f_1, f_2, \ldots, f_r, g)$ which specifies the functionality of the protocol. The function f_i is used to generate the communication in the i-th round. The input of f_i consists of the received messages in the previous rounds and randomness of the caller of the function. For a party P, either $\mathcal{S}$ or $\mathcal{R}$, C_P denotes the randomness of P and M_P^i denotes the set of all the messages received by P during the first i rounds. That is, $\mathsf{M}_S^0 = \{M_S\}$ and $\mathsf{M}_R^0 = \varnothing$. If the initiator of the i-th round is P, we write $P_i X_i Y_i = f_i(\mathsf{M}_P^{i-1}, C_P)$ to denote the random variable corresponding to the communication in the i-th round, where P_i denotes the communication over the public channel, X_i the communication over the corrupted wires, Y_i the communication over the uncorrupted wires. If the initiator does not invoke the public channel, then P_i should be null. The function g denotes an output function. That is, $\mathcal{R}$ outputs $M_R = g(\mathsf{M}_R^r, C_R)$ at the end of the protocol.

Now, we are ready to mention our results.

Theorem 2. *Suppose that $n \leq 2t$. For any $r \geq r' \geq 1$, there is no $(r, r', 0)$-round (ε, δ)-SMT-PD protocol with $\varepsilon + \delta < 1 - 1/|\mathbf{M}|$.*

Proof. We prove the statement by contradiction. Suppose that there exists an $(r, r', 0)$-round (ε, δ)-SMT-PD protocol Π with $\varepsilon + \delta < 1 - 1/|\mathbf{M}|$. We will construct an adversary $\mathcal{A}$ who can violate either the privacy or the reliability.

First of all, we give a proof sketch. We will show that for each execution of Π where $\mathcal{S}$ sends a message m to $\mathcal{R}$, there exists another execution (called *swapped execution*) where $\mathcal{S}$ sends the message m but $\mathcal{A}$ impersonates $\mathcal{R}$ such that $\mathcal{S}$ receives the identical communication in the two executions and so the two executions cannot be distinguished. The views of $\mathcal{R}$ and $\mathcal{A}$ are however swapped in the two executions, and so if $\mathcal{R}$ outputs $M_R = M_S$ in one of the two executions then $\mathcal{A}$ outputs $M_A = M_S$ in the swapped execution and so $\Pr[M_A = M_S] \geq \Pr[M_R = M_S]$. Using Lemma 2, we derive $\varepsilon + \delta \geq 1 - 1/\mathbf{M}$, which is a contradiction.

Without loss of generality, we assume that wires are labeled by $1, 2, \ldots, 2n$. We also assume that $n = 2t$. (Note that if there exists an (ε, δ)-SMT-PD protocol for $n' < 2t$, the same protocol can be run for $n = 2t$ by neglecting the last $n - n'$ wires. Thus an impossibility result for $n = 2t$ still holds for $n' < 2t$.)

Before going into the details of the proof, we need a bit more preparation. We partition $\mathcal{A}$'s randomness C_A into four parts $(C_{M_A}, C_{A0}, C_{A1}, C_{A2})$. C_{A0} is a single bit to indicate which subset of t wires to corrupt. We assume that $C_{A0} = 0$ (resp., $C_{A0} = 1$) specifies that the first (resp., the last) t wires are corrupted. C_{A1} is used to generate the forged communication to replace the communication sent by $\mathcal{S}$ with, and C_{A2} is used to generate the forged communication to replace the communication sent by $\mathcal{R}$ with. C_{M_A} is used to select uniformly a message from M to impersonate $\mathcal{S}$.

By using these notations, we can describe $\mathcal{A}$'s behavior as follows. Without loss of generality, we assume that $C_{A0} = 1$.

- In the j-th round with $1 \leq j \leq r$,
  - When $\mathcal{S}$ sends $X_j Y_j$ or $P_j X_j Y_j$, $\mathcal{A}$ blocks Y_j. Then $\mathcal{R}$ receives X_j or $P_j X_j$, respectively.
  - When $\mathcal{R}$ sends $X_j Y_j$, $\mathcal{A}$ computes $X'_j Y'_j = f_j(\mathsf{M}_A^{j-1}, C_{A2})$, then replaces Y_j with Y'_j. (Here M_A^{j-1} denotes the messages eavesdropped by $\mathcal{A}$ during the first $j-1$ rounds.) Then, $\mathcal{S}$ receives $X_j Y'_j$.
- Finally, $\mathcal{A}$ outputs $M_A = g(\mathsf{M}_A^r, C_{A2})$.

Due to the above strategy, we may assume that $C_{M_A} = \perp$ and $C_{A1} = \perp$.

Let E be the set of executions of Π, where each execution is determined by the message M_S and the randomness C_S, C_R and C_A used for $\mathcal{S}$, $\mathcal{R}$ and $\mathcal{A}$, respectively. We define a binary relation $W_1 \subseteq E \times E$ to specify a pair $(E, \hat{E})$ of executions as follows: $(E, \hat{E}) \in W_1$ if (1) (M_S, C_S) are the same for both executions, (2) $C_{\hat{A}0} \oplus C_{A0} = 1$, and (3) $C_{A2} = C_{\hat{R}}$ and $C_R = C_{\hat{A}2}$.

Lemma 3. *Let $(E, \hat{E}) \in W_1$. Then the following properties hold.*

(a) *The view of $\mathcal{S}$ in E is identical to her view in $\hat{E}$, that is, $\mathsf{M}_S^r = \mathsf{M}_{\hat{S}}^r$.*

(b) *The view of $\mathcal{A}$ in $\hat{E}$ is identical to the view of $\mathcal{R}$ in E, that is, $\mathsf{M}_R^r = \mathsf{M}_{\hat{A}}^r$. Thus the output of $\mathcal{R}$ in E is the same as the output of $\mathcal{A}$ in $\hat{E}$. That is, $M_R = M_{\hat{A}}$ holds.*

Proof. We show the statements by induction with respect to the rounds. At the beginning of the protocol, i.e., in the 0-th round, the views of $\mathcal{S}$, $\mathcal{R}$, and $\mathcal{A}$ are $\mathsf{M}_S^0 = \mathsf{M}_{\hat{S}}^0 = \{M_S\}$, $\mathsf{M}_R^0 = \mathsf{M}_{\hat{R}}^0 = \varnothing$, and $\mathsf{M}_A^0 = \mathsf{M}_{\hat{A}}^0 = \varnothing$, respectively. These imply that the statements (a) and (b) hold in the base case $j = 0$.

Next, we show the inductive step. That is, we assume that the statements (a) and (b) hold during the first $(j - 1)$ rounds and prove that they hold during the first j rounds.

First, we observe that the protocol is a combination of the following three transmissions:

(α) transmissions from $\mathcal{S}$ to $\mathcal{R}$ over wires;
(β) transmissions from $\mathcal{R}$ to $\mathcal{S}$ over wires;
(γ) transmissions from $\mathcal{S}$ to $\mathcal{R}$ over wires and the public channel.

Assume that $\mathcal{S}$ initiates the j-th round. From the inductive hypothesis, we have $M_S^{j-1} = M_{\hat{S}}^{j-1}$. Since C_S is the same in the two executions E and $\hat{E}$, $\mathcal{S}$ in E invokes the public channel if and only if $\mathcal{S}$ in $\hat{E}$ invokes the public channel. Thus, we can consider the inductive step in the cases (α), (β), and (γ) individually.

Case (α): From the inductive hypothesis, we have $\mathsf{M}_S^{j-1} = \mathsf{M}_{\hat{S}}^{j-1}$ and $\mathsf{M}_R^{j-1} = \mathsf{M}_{\hat{A}}^{j-1}$. In this case, $\mathcal{S}$ does not get any new information. This implies that $\mathsf{M}_S^{j-1} = \mathsf{M}_S^j$ and

$\mathsf{M}_{\hat{S}}^{j-1} = \mathsf{M}_{\hat{S}}^{j}$. Thus, we have $\mathsf{M}_{S}^{j} = \mathsf{M}_{\hat{S}}^{j}$. That is, the statement (a) holds. On the other hand, $\mathcal{R}$ and $\mathcal{A}$ get something new in the j-th round. In the execution E, $\mathcal{R}$ gets X_j and $\mathcal{A}$ gets Y_j. Thus, we have $\mathsf{M}_{R}^{j} = \mathsf{M}_{R}^{j-1} \cup \{X_j\}$ and $\mathsf{M}_{A}^{j} = \mathsf{M}_{A}^{j-1} \cup \{Y_j\}$. In the execution $\hat{E}$, the corrupted wires and the randomness are swapped, we have $\mathsf{M}_{\hat{R}}^{j} = \mathsf{M}_{\hat{R}}^{j-1} \cup \{Y_j\}$ and $\mathsf{M}_{\hat{A}}^{j} = \mathsf{M}_{\hat{A}}^{j-1} \cup \{X_j\}$. Thus, we have $\mathsf{M}_{R}^{j} = \mathsf{M}_{\hat{A}}^{j}$. That is, the statement (b) holds.

Case (β): From the inductive hypothesis, we have $\mathsf{M}_{S}^{j-1} = \mathsf{M}_{\hat{S}}^{j-1}$ and $\mathsf{M}_{R}^{j-1} = \mathsf{M}_{\hat{A}}^{j-1}$. In this case, $\mathcal{S}$ gets new information in the j-th round. In the execution E, $\mathcal{S}$ gets $X_j Y_j'$, where X_j is sent by $\mathcal{R}$ and Y_j' is transmission altered by $\mathcal{A}$. More specifically, we can write that $X_j Y_j = f_j(\mathsf{M}_{R}^{j-1}, C_R)$ and $X_j' Y_j' = f_j(\mathsf{M}_{A}^{j-1}, C_{A2})$. In the execution $\hat{E}$, $\mathcal{S}$ gets the same information $X_j Y_j'$. But $\mathcal{S}$ gets X_j from $\mathcal{A}$ and Y_j' from $\mathcal{R}$. Note that we can write $f_j(\mathsf{M}_{\hat{A}}^{j-1}, C_{\hat{A}2}) = f_j(\mathsf{M}_{R}^{j-1}, C_R) = X_j Y_j$ and $f_j(\mathsf{M}_{\hat{R}}^{j-1}, C_{\hat{R}}) = f_j(\mathsf{M}_{A}^{j-1}, C_A) = X_j' Y_j'$. Thus, we have $\mathsf{M}_{S}^{j} = \mathsf{M}_{\hat{S}}^{j}$, that is, the statement (a) holds. On the other hand, $\mathcal{R}$ and $\mathcal{A}$ do not get anything new. Thus, we have $\mathsf{M}_{R}^{j} = \mathsf{M}_{\hat{A}}^{j}$. That is, the statement (b) holds.

Case (γ): This case is similar to Case (α). From the inductive hypothesis, we have $\mathsf{M}_{S}^{j-1} = \mathsf{M}_{\hat{S}}^{j-1}$ and $\mathsf{M}_{R}^{j-1} = \mathsf{M}_{\hat{A}}^{j-1}$. We can show that the statement (a) holds as well as in Case (α). In the execution E, $\mathcal{R}$ gets $P_j X_j$ and $\mathcal{A}$ gets $P_j Y_j$. Thus, we have $\mathsf{M}_{R}^{j} = \mathsf{M}_{R}^{j-1} \cup \{P_j X_j\}$ and $\mathsf{M}_{A}^{j} = \mathsf{M}_{A}^{j-1} \cup \{P_j Y_j\}$. In the execution $\hat{E}$, the corrupted wires and the randomness are swapped, we have $\mathsf{M}_{\hat{R}}^{j} = \mathsf{M}_{\hat{R}}^{j-1} \cup \{P_j Y_j\}$ and $\mathsf{M}_{\hat{A}}^{j} = \mathsf{M}_{\hat{A}}^{j-1} \cup \{P_j X_j\}$. Thus, we have $\mathsf{M}_{R}^{j} = \mathsf{M}_{\hat{A}}^{j}$. That is, the statement (b) holds.

In any cases, we have shown that the statements (a) and (b) during the first j rounds hold. Thus, at the end of protocol, that is, after the r-th round, the statements (a) and (b) hold. Furthermore, we have $M_R = g(\mathsf{M}_{R}^{r}, C_R) = g(\mathsf{M}_{\hat{A}}^{r}, C_{\hat{A}2}) = M_{\hat{A}}$. This completes the proof of Lemma 3.   □

Let us proceed the proof of Theorem 2. Let $\boldsymbol{S}_1 \in \boldsymbol{E}$ be the set of all successful executions in which $\mathcal{R}$ outputs $M_R = M_S$, and p_E denotes the probability of execution E determined by the randomness of all the parties. Define $p_{\hat{E}}$ similarly. Then $\Pr[M_R = M_S] = \sum_{E \in \boldsymbol{S}_1} p_E$. By Lemma 3, if $E \in \boldsymbol{S}_1$, $\mathcal{A}$ will output M_S in the swapped execution of $\hat{E}$; therefore $\Pr[M_A = M_S] \geq \sum_{E \in \boldsymbol{S}_1} p_{\hat{E}}$.

Additionally, by the definition of $\boldsymbol{W}_1$ and the observation that $C_{MA} = C_{A1} = \bot$, we have

$$p_E = \frac{1}{|\boldsymbol{M}|} 2^{-r_S - r_R - r_{A2} - 1} = p_{\hat{E}},$$

where r_S, r_R, r_{A2} denote the length of the randomness of C_S C_R, C_{A2} used by $\mathcal{S}$, $\mathcal{R}$ and $\mathcal{A}$ respectively.

From the above equation and Lemma 2, it follows that

$$1 - \delta \leq \Pr[M_R = M_S] \leq \Pr[M_A = M_S] \leq \frac{1}{|\boldsymbol{M}|} + \varepsilon.$$

Therefore, we have $1 - 1/|\boldsymbol{M}| \leq \varepsilon + \delta$, contradicting the assumption on Π.   □

Theorem 3. *Suppose that* $n \leq 2t$. *For any* $r \geq r' \geq 1$ *and* $0 \leq \varepsilon \leq 1$, *there is no* $(r, 0, r')$-*round* (ε, δ)-*SMT-PD protocol with* $\delta < (1 - 1/|\boldsymbol{M}|)/2$.

Proof. We prove the statement by contradiction. Suppose that there exists an $(r, 0, r')$-round (ε, δ)-SMT-PD protocol Π with $\delta < (1 - 1/|M|)/2$. We will construct an adversary $\mathcal{A}$ who can violate the reliability of Π.

We will show that the *reliability* of Π will be violated. This is by showing that for every successful execution there exists an unsuccessful one and so probability of success is at most 1/2.

We describe $\mathcal{A}$'s behavior as follows. Without loss of generality, we assume that $C_{A0} = 1$.

- In the j-th round with $1 \leq j \leq r$,
  - When $\mathcal{R}$ sends $X_j Y_j$ or $P_j X_j Y_j$, $\mathcal{A}$ blocks Y_j. Then $\mathcal{S}$ receives X_j or $P_j X_j$, respectively.
  - When $\mathcal{S}$ sends $X_j Y_j$, $\mathcal{A}$ computes $X'_j Y'_j = f_j(\mathsf{M}_A^{j-1}, C_{A1})$, then replaces Y_j with Y'_j. Then, $\mathcal{R}$ receives $X_j Y'_j$.
- Finally, $\mathcal{A}$ outputs $M_A = g(\mathsf{M}_A^r, C_{A1})$.

Due to the above strategy, we may assume that $C_{A2} = \bot$. For simplicity, we abuse the notation M_A here to denote the uniformly selected message of $\mathcal{A}$ using the randomness C_{M_A}.

Let E and p_E be as defined as in the proof of Theorem 2 and consider a binary relation $\mathbf{W}_2 \subseteq E \times E$ where $(E, \hat{E}) \in \mathbf{W}_2$ if (1) C_R is the same in the two executions; (2) $C_{\hat{A}0} \oplus C_{A0} = 1$; and (3) $C_{A1} = C_{\hat{S}}$, $C_S = C_{\hat{A}1}$; (4) $M_S = M_{\hat{A}}$ and $M_A = M_{\hat{S}}$. We denote by $\mathbf{S}_2$ the set of *successful* executions in which $\mathcal{R}$ outputs $M_R = M_S$ under the condition that $M_A \neq M_S$.

Lemma 4. *Let $(E, \hat{E}) \in \mathbf{W}_2$. Then the following properties hold.*

(a) *The views of $\mathcal{R}$ in E and $\hat{E}$ are identical, that is, $\mathsf{M}_R^r = \mathsf{M}_{\hat{R}}^r$.*

(b) *The view of $\mathcal{A}$ in $\hat{E}$ is identical to the view of $\mathcal{S}$ in E, that is, $\mathsf{M}_S^r = \mathsf{M}_{\hat{A}}^r$. Therefore, if $E \in \mathbf{S}_2$ is a successful execution, then $\hat{E} \notin \mathbf{S}_2$ is a failed execution.*

Proof. We show the statements by induction with respect to the rounds. At the beginning of the protocol, i.e., in the 0-th round, the views of $\mathcal{R}$ are the same in E and $\hat{E}$, since $\mathsf{M}_R^0 = \mathsf{M}_{\hat{R}}^0 = \varnothing$. The views of $\mathcal{S}$ and $\mathcal{A}$ are $\mathsf{M}_S^0 = \mathsf{M}_{\hat{A}}^0 = \{M_S\}$ and $\mathsf{M}_A^0 = \mathsf{M}_{\hat{S}}^0 = \{M_A\}$. These imply that the statements (a) and (b) hold in the base case $j = 0$.

Next, we show the inductive step. That is, we assume that the statements (a) and (b) hold during the first $(j - 1)$ rounds and prove that they hold during the first j rounds.

First, we observe that the protocol is a combination of the following three transmissions:

(α) transmissions from $\mathcal{S}$ to $\mathcal{R}$ over wires;
(β) transmissions from $\mathcal{R}$ to $\mathcal{S}$ over wires;
(γ) transmissions from $\mathcal{R}$ to $\mathcal{S}$ over wires and the public channel.

Assume that $\mathcal{R}$ initiates the j-th round. From the inductive hypothesis, we have $M_R^{j-1} = M_{\hat{R}}^{j-1}$. Since C_R is the same in the two executions E and $\hat{E}$, $\mathcal{R}$ in E invokes the public channel if and only if $\mathcal{R}$ in $\hat{E}$ invokes the public channel. Thus, we can consider the inductive step in the cases (α), (β), and (γ) individually.

Case (α): From the inductive hypothesis, we have $\mathsf{M}_R^{j-1} = \mathsf{M}_{\hat{R}}^{j-1}$ and $\mathsf{M}_S^{j-1} = \mathsf{M}_{\hat{A}}^{j-1}$. In this case, $\mathcal{R}$ gets something new. In the execution E, $\mathcal{R}$ gets $X_j Y_j'$, where X_j is sent by $\mathcal{S}$ and Y_j' is a forged transmission by $\mathcal{A}$. More specifically, we can write that $X_j Y_j = f_j(\mathsf{M}_S^{j-1}, C_S)$ and $X_j' Y_j' = f_j(\mathsf{M}_A^{j-1}, C_{A1})$. In the execution $\hat{E}$, $\mathcal{S}$ gets the same information $X_j Y_j'$. But $\mathcal{R}$ gets X_j from $\mathcal{A}$ and Y_j' from $\mathcal{S}$. Note that we can write $f_j(\mathsf{M}_{\hat{A}}^{j-1}, C_{\hat{A}1}) = f_j(\mathsf{M}_S^{j-1}, C_R) = X_j Y_j$ and $f_j(\mathsf{M}_{\hat{S}}^{j-1}, C_{\hat{S}}) = f_j(\mathsf{M}_A^{j-1}, C_A) = X_j' Y_j'$. Thus, we have $\mathsf{M}_R^j = \mathsf{M}_R^{j-1} \cup \{X_j Y_j'\} = \mathsf{M}_{\hat{R}}^{j-1} \cup \{X_j Y_j'\} = \mathsf{M}_{\hat{R}}^j$, that is, the statement (a) holds. On the other hand, since $\mathcal{S}$ and $\mathcal{A}$ do not get anything new, we have $\mathsf{M}_S^j = \mathsf{M}_S^{j-1} = \mathsf{M}_{\hat{A}}^{j-1} = \mathsf{M}_{\hat{A}}^j$. That is, the statement (b) holds.

Case (β): From the inductive hypothesis, we have $\mathsf{M}_R^{j-1} = \mathsf{M}_{\hat{R}}^{j-1}$ and $\mathsf{M}_S^{j-1} = \mathsf{M}_{\hat{A}}^{j-1}$. In this case, $\mathcal{R}$ does not get anything new in the j-th round. This implies that $\mathsf{M}_R^{j-1} = \mathsf{M}_R^j$. Thus, we have $\mathsf{M}_R^j = \mathsf{M}_{\hat{R}}^j$. That is, the statement (a) holds. On the other hand, $\mathcal{S}$ and $\mathcal{A}$ get something new in the j-th round. In the execution E, $\mathcal{S}$ gets X_j and $\mathcal{A}$ gets Y_j. Thus, we have $\mathsf{M}_S^j = \mathsf{M}_S^{j-1} \cup \{X_j\}$ and $\mathsf{M}_A^j = \mathsf{M}_A^{j-1} \cup \{Y_j\}$. In the execution $\hat{E}$, the corrupted wires and the randomness are swapped, we have $\mathsf{M}_{\hat{S}}^j = \mathsf{M}_{\hat{S}}^{j-1} \cup \{Y_j\}$ and $\mathsf{M}_{\hat{A}}^j = \mathsf{M}_{\hat{A}}^{j-1} \cup \{X_j\}$. Thus, we have $\mathsf{M}_S^j = \mathsf{M}_{\hat{A}}^j$. That is, the statement (b) holds.

Case (γ): This case is similar to Case (β). From the inductive hypothesis, we have $\mathsf{M}_R^{j-1} = \mathsf{M}_{\hat{R}}^{j-1}$ and $\mathsf{M}_S^{j-1} = \mathsf{M}_{\hat{A}}^{j-1}$. We can show that the statement (a) holds as well as in Case (β). In the execution E, $\mathcal{S}$ gets $P_j X_j$ and $\mathcal{A}$ gets $P_j Y_j$. Thus, we have $\mathsf{M}_S^j = \mathsf{M}_S^{j-1} \cup \{P_j X_j\}$ and $\mathsf{M}_A^j = \mathsf{M}_A^{j-1} \cup \{P_j Y_j\}$. In the execution $\hat{E}$, the corrupted wires and the randomness are swapped, we have $\mathsf{M}_{\hat{S}}^j = \mathsf{M}_{\hat{S}}^{j-1} \cup \{P_j Y_j\}$ and $\mathsf{M}_{\hat{A}}^j = \mathsf{M}_{\hat{A}}^{j-1} \cup \{P_j X_j\}$. Thus, we have $\mathsf{M}_S^j = \mathsf{M}_{\hat{A}}^j$. That is, the statement (b) holds.

In any cases, we have shown that the statements (a) and (b) during the first j rounds hold. Thus, at the end of protocol, that is, after the r-th round, the statements (a) and (b) hold.

In the executions E and $\hat{E}$, M_S and M_A are swapped. If, in E, $\mathcal{R}$ outputs $M_R = g(\mathsf{M}_R^r, C_R) = M_S$, then in $\hat{E}$, $\mathcal{R}$ outputs $M_{\hat{R}} = g(\mathsf{M}_{\hat{R}}^r, C_{\hat{R}}) = g(\mathsf{M}_R^r, C_R) = M_S = M_{\hat{A}}$. Thus, if $M_A \neq M_S$ and $E \in \mathbf{S}_2$ then $E \notin \mathbf{S}_2$. This completes the proof of Lemma 4.   □

To complete Theorem 3, we need one more lemma.

Lemma 5

(c) *The occurrence probability of any two swapped executions $(E, \hat{E}) \in \mathbf{W}_2$ is the same, that is, $p_E = p_{\hat{E}}$.*

(d) *When $M_S \neq M_A$, the failure probability of $\mathcal{R}$ in recovering the secret message is not less than the success probability of $\mathcal{R}$, that is,*

$$\Pr[M_R = M_S \mid M_S \neq M_A] \leq \Pr[M_R \neq M_S \mid M_S \neq M_A],$$

where the probability is taken over the randomness and messages selected by $\mathcal{S}$, $\mathcal{R}$ and $\mathcal{A}$.

Proof. (c) Note that an execution $E \in \boldsymbol{E}$ is completely determined by the randomness and messages selected by all the parties. Then for each $E \in \boldsymbol{E}$, we have $p_E = 2^{-r_S - r_R - r_A} / |\boldsymbol{M}|$, where r_S, r_R and r_A denote the length of the randomness of C_S, C_R and C_A respectively. Similarly, we have $p_E = 2^{-r_S - r_R - r_A} / |\boldsymbol{M}|$.

Since $C_{A2} = \bot$, we have $r_A = r_{MA} + r_{A0} + r_{A1}$, where r_{MA}, r_{A0}, r_{A1} denote the length of C_{MA}, C_{A0}, C_{A1} respectively. Similarly, we have $r_{\hat{A}} = r_{M\hat{A}} + r_{\hat{A}0} + r_{\hat{A}1}$.

Note that $r_{A0} = r_{\hat{A}0} = 1$ and $r_{MA} = r_{M\hat{A}} = \lceil \log |\boldsymbol{M}| \rceil$. From the definition of W_2, we have that $r_R = r_{\hat{R}}$, $r_S = r_{\hat{A}1}$ and $r_{A1} = r_{\hat{S}}$. Hence we have $r_S + r_R + r_A = r_{\hat{S}} + r_{\hat{R}} + r_{\hat{A}}$, and $p_E = p_{\hat{E}}$ holds.

(d) Let $\bar{\boldsymbol{S}}_2 = \boldsymbol{E} \setminus \boldsymbol{S}_2$ denote the set of *failed* executions. Since $\hat{E} \in \bar{\boldsymbol{S}}_2$ holds for any $E \in \boldsymbol{S}_2$, and the one-to-one correspondence of E and $\hat{E}$, we get that $|\boldsymbol{S}_2| \leq |\bar{\boldsymbol{S}}_2|$. The probability that Π fails when $M_A \neq M_S$ can be computed as

$$\Pr[M_R \neq M_S \mid M_S \neq M_A]$$
$$= \Pr[E \in \bar{\boldsymbol{S}}_2] \geq \sum_{E \in \boldsymbol{S}_2} p_{\hat{E}} = \sum_{E \in \boldsymbol{S}_2} p_E = \Pr[M_R = M_S \mid M_S \neq M_A]. \qquad \square$$

Let us proceed the proof of Theorem 3. From Lemma 5, we must have $\Pr[M_R \neq M_S \mid M_A \neq M_S] \geq \frac{1}{2}$. Hence

$$\Pr[M_R \neq M_S] \geq \Pr[M_R \neq M_S \mid M_S \neq M_A] \Pr[M_S \neq M_A] \geq \frac{1}{2}\left(1 - \frac{1}{|\boldsymbol{M}|}\right).$$

On the other hand, since Π is a δ-reliable protocol, we have $\Pr[M_R \neq M_S] \leq \delta$. It follows that $\delta \geq (1 - 1/|\boldsymbol{M}|)/2$, which contradicts the assumption on Π. $\qquad \square$

Theorem 2 says that any number of invocations of unidirectional public channel is not helpful. On the other hand, Theorem 3 does not mention the privacy of SMT-PD protocols. Thus, there seems to be room to be improved. In the next section, we will see that Theorem 3 achieves almost optimal in a weak sense.

5   Upper Bound

In this section, we give an upper-bounding SMT-PD protocol where only $\mathcal{R}$ invokes the public channel. Since the protocol uses universal hash functions, we give the definition and see some property.

Definition 2. Suppose that $m > \ell$. A function family $\boldsymbol{H} = \{h : \{0,1\}^m \to \{0,1\}^\ell\}$ is called strongly universal$_2$ hash function family if for any distinct $a_1, a_2 \in \{0,1\}^m$ and any $b_1, b_2 \in \{0,1\}^\ell$, it holds that $\Pr_{h \in \boldsymbol{H}}[h(a_1) = b_1 \wedge h(a_2) = b_2] \leq 2^{-2\ell}$.

Proposition 1. *Let* $\boldsymbol{H} = \{h : \{0,1\}^m \to \{0,1\}^\ell\}$ *be a strongly universal$_2$ hash function family. Then, for any* $(a_1, c_1) \neq (a_2, c_2) \in \{0,1\}^m \times \{0,1\}^\ell$, *it holds that* $\Pr_{h \in \boldsymbol{H}}[c_1 \oplus h(a_1) = c_2 \oplus h(a_2)] \leq 2^{-\ell}$.

We can use "almost" strongly universal$_2$ hash functions instead of strongly universal$_2$ hash functions to reduce the communication complexity. But, in this paper, we do not optimize the communication complexity.

Now, we give an upper-bounding SMT-PD protocol Π_1 as follows. The following protocol is an adaptation of (3,2)-round SMT-PD protocol proposed in [18].

---

1. $(\mathcal{S} \to \mathcal{R})$: For $i = 1, \ldots, n$, $\mathcal{S}$ randomly selects $r_i \in \{0,1\}^\ell$ and $R_i \in \{0,1\}^m$ and sends the pair (r_i, R_i) to $\mathcal{R}$ along wire i.
2. $(\mathcal{S} \leftarrow \mathcal{R})$: For $i = 1, \ldots, n$, if $\mathcal{R}$ correctly receives a pair (r_i', R_i') along wire i (i.e., $r_i' \in \{0,1\}^\ell$ and $R_i' \in \{0,1\}^m$), $\mathcal{R}$ selects $h_i \leftarrow \boldsymbol{H}$ and computes $T_i' = r_i' \oplus h_i(R_i')$; otherwise, wire i is assumed *corrupted*. $\mathcal{R}$ then constructs an indicator bit string $B = b_1 b_2 \cdots b_n$ where $b_i = 1$ if the wire i is corrupted and $b_i = 0$ otherwise. Finally, $\mathcal{R}$ sends $(B, (H_1, \ldots, H_n))$ over the public channel, where $H_i = (h_i, T_i')$ if $b_i = 0$; and H_i is empty, otherwise.
3. $(\mathcal{S} \to \mathcal{R})$: $\mathcal{S}$ ignores the wires with $b_i = 1$. For $i = 1, \ldots, n$, if $b_i = 0$, $\mathcal{S}$ computes $T_i = r_i \oplus h_i(R_i)$ and checks $T_i' = T_i$; if $T_i' = T_i$, wire i is assumed *consistent*; otherwise, wire i is corrupted. $\mathcal{S}$ constructs an indicator bit string $V = v_1 v_2 \cdots v_n$, where $v_i = 1$ if wire i is considered consistent; otherwise $v_i = 0$. $\mathcal{S}$ sends the pair $(V, C = M_S \oplus \{\bigoplus_{v_i=1} R_i\})$ along every consistent wire j. $\mathcal{R}$ receives (V_j, C_j) along wire j and set $\boldsymbol{V} = \{(V_j, C_j)\}$. $\mathcal{R}$ chooses some pair $(V', C') \in \boldsymbol{V}$ as follows.
   - All the elements in $\boldsymbol{V}$ can be enumerated as $(U_1, D_1), \ldots, (U_d, D_d)$ according to some order, where $d = |\boldsymbol{V}|$.
   - Let d_i be the number of indices j such that $(V_j, C_j) = (U_i, D_i)$.
   - If $d_i < n - t$ then reset $d_i = 0$.
   - Choose $(V', C') = (U_k, D_k)$ with probability $d_k / (\sum_i d_i)$.
   Finally, $\mathcal{R}$ recovers $M_R = C' \oplus \{\bigoplus_{v_i'=1} R_i'\}$ and outputs it, where $V' = v_1' \cdots v_n'$.

Theorem 4. *The protocol Π_1 is a $(3, 0, 1)$-round $(0, \delta)$-SMT-PD protocol, where*

$$\delta \leq \frac{t}{n} + \frac{n-t}{n} \cdot t \cdot 2^{-\ell}.$$

Proof. Let $\boldsymbol{B} = \{i : \text{wire } i \text{ is corrupted}\}$ and $\boldsymbol{G} = \{i : \text{wire } i \text{ is consistent}\}$.

First, we upper-bound the reliability parameter. If $\mathcal{S}$ can detect all corrupted wires with $(r_i', R_i') \neq (r_i, R_i)$ and $(V, C) = (V', C')$, then the protocol is perfectly reliable.

The probability of the event when $(V, C) = (V', C')$ is at least $(n - t)/n$. We consider that the reliability is always violated when $(V, C) \neq (V', C')$. Hence in the following we assume that $(V, C) = (V', C')$.

In the second round the wires with $b_i = 1$ are detected as corrupted and are ignored in the third round. Hence in the following we only consider wires with $b_i = 0$. For wire i, the wire is called *bad* if $(r_i', R_i') \neq (r_i, R_i)$ but $r_i \oplus h_i(R_i) = r_i' \oplus h(R_i')$. Bad wires are always included in $\boldsymbol{G}$. Using Proposition 1 and noting that r_i, R_i, r_i', R_i' are fixed before the second round and then h_i is selected with uniform distribution, we have

$$\Pr[\text{wire } i \text{ is bad}] = \Pr[r_i \oplus h_i(R_i) = r_i' \oplus h(R_i') \wedge (r_i, R_i) \neq (r_i', R_i')]$$
$$\leq \Pr[r_i \oplus h_i(R_i) = r_i' \oplus h(R_i') \mid (r_i, R_i) \neq (r_i', R_i')] \leq 2^{-\ell},$$

where the probability is over the randomness of all the parties. Then, the probability of unreliable message transmission is

$$
\Pr[M_R \neq M_S] = \Pr\left[\bigoplus_{j \in G} R_j \neq \bigoplus_{j \in G} R'_j\right] \leq \Pr[\exists j \in G \text{ s.t. } R_j \neq R'_j]
$$

$$
\leq \Pr[\exists \text{ at least one bad wire}]
$$

$$
\leq \sum_{j \in B} \Pr[\text{wire } j \text{ is bad}] \leq t \cdot 2^{-\ell},
$$

where the probability is over the random coins of all the parties.

Thus, putting altogether, we can say that the reliability δ satisfies

$$
\delta \leq \frac{t}{n} + \frac{n-t}{n} \cdot t \cdot 2^{-\ell}.
$$

Next, we estimate the privacy parameter. The adversary can obtain transmissions related to M_S only from the corrupted wires in the third round. Thus, the situation is completely the same as (3,2)-round SMT-PD protocol by Shi et al.[18]. Thus, the proof of the perfect privacy for their protocol also works in our case.   □

Let us consider the case where $n = 2t$. Then, if ℓ is large enough, then the reliability parameter in Theorem 4 comes close to $1/2$. In this sense, the gap between the lower bound in Theorem 3 and the upper bound in Theorem 4 is slight.

6 Concluding Remarks

We have considered the secure message transmission with unidirectional public channel. We have shown that any $(r, r', 0)$-round protocol must satisfy that $\varepsilon + \delta \geq 1 - 1/|M|$. It says that there is no useful $(r, r', 0)$-round SMT-PD protocol. We have also shown that any $(r, 0, r')$-round protocol must satisfy that $\delta \geq (1 - 1/|M|)/2$. It says that there may exist an $(r, 0, r')$-round $(0, 1/2)$-SMT-PD protocol. Actually, if $n = 2t$ then the protocol in Theorem 4 satisfies that $\delta \approx 1/2$. However, there is still a gap in general. In other words, either the lower bound in Theorem 3 or the upper bound in Theorem 4 may be further improved.

Anyway, we may say that SMT-PD protocols require the bidirectional public channel.

References

1. Araki, T.: Almost secure 1-round message transmission scheme with polynomial-time message decryption. In: Safavi-Naini, R. (ed.) ICITS 2008. LNCS, vol. 5155, pp. 2–13. Springer, Heidelberg (2008)
2. Agarwal, S., Cramer, R., de Haan, R.: Asymptotically optimal two-round perfectly secure message transmission. In: Dwork, C. (ed.) CRYPTO 2006. LNCS, vol. 4117, pp. 394–408. Springer, Heidelberg (2006)

3. Ben-Or, M., Goldwasser, S., Wigderson, A.: Completeness theorems for non-cryptographic fault-tolerant distributed computation. In: Proc. 20th Annual ACM Symposium on Theory of Computing, pp. 1–10 (1988)
4. Berman, P., Garay, J.A.: Fast consensus in networks of bounded degree. Distributed Computing 2(7), 62–73 (1993)
5. Chandran, N., Garay, J.A., Ostrovsky, R.: Improved fault tolerance and secure computation on sparse networks. In: Abramsky, S., Gavoille, C., Kirchner, C., Meyer auf der Heide, F., Spirakis, P.G. (eds.) ICALP 2010 (Part 2). LNCS, vol. 6199, pp. 249–260. Springer, Heidelberg (2010)
6. Chaum, D., Crépeau, C., Damgård, I.: Multiparty unconditionally secure protocols. In: Proc. 20th Annual ACM Symposium on Theory of Computing, pp. 11–19 (1988)
7. Dolev, D., Dwork, C., Waarts, O., Yung, M.: Perfectly secure message transmission. J. ACM 40(1), 17–47 (1993)
8. Dwork, C., Peleg, D., Pippenger, N., Upfal, E.: Fault tolerance in networks of bounded degree. SIAM J. Comput. 17(5), 975–988 (1988)
9. Desmedt, Y., Wang, Y.: Perfectly secure message transmission revisited. IEEE Trans. Information Theory 54(6), 2582–2595 (2008)
10. Fitzi, M., Franklin, M.K., Garay, J.A., Simhadri, H.V.: Towards optimal and efficient perfectly message transmission. In: Vadhan, S.P. (ed.) TCC 2007. LNCS, vol. 4392, pp. 311–322. Springer, Heidelberg (2007)
11. Franklin, M.K., Wright, R.N.: Secure communication in minimal connectivity models. J. Cryptology 13(1), 9–30 (2000)
12. Garay, J.A., Ostrovsky, R.: Almost-everywhere secure computation. In: Smart, N.P. (ed.) EUROCRYPT 2008. LNCS, vol. 4965, pp. 307–323. Springer, Heidelberg (2008)
13. Garay, J.A., Givens, C., Ostrovsky, R.: Secure message transmission with small public discussion. In: Gilbert, H. (ed.) EUROCRYPT 2010. LNCS, vol. 6110, pp. 177–196. Springer, Heidelberg (2010)
14. Kurosawa, K., Suzuki, K.: Truly efficient 2-round perfectly secure message transmission scheme. IEEE Transactions on Information Theory 55(11), 5223–5232 (2009)
15. Kurosawa, K., Suzuki, K.: Almost secure (1-round, n-channel) message transmission scheme. IEICE Trans. Fundamentals of Electronics Communications and Computer Sciences E92-A(1), 105–112 (2009)
16. Rabin, T., Ben-Or, M.: Verifiable secrete sharing and multiparty protocols with honest majority. In: Proc. 21st Annual ACM Symposium on Theory of Computing, pp. 73–85 (1989)
17. Sayeed, H., Abu-Amara, H.: Efficient perfectly secure message transmission in synchronous networks. Information and Computation 126(1), 53–61 (1996)
18. Shi, H., Jiang, S., Safavi-Naini, R., Tuhin, M.A.: Optimal secure message transmission by public discussion. In: Proc. IEEE International Symposium on Information Theory 2009, pp. 1313–1317 (2009)
19. Srinathan, K., Narayanan, A., Rangan, C.P.: Optimal perfectly secure message transmission. In: Franklin, M. (ed.) CRYPTO 2004. LNCS, vol. 3152, pp. 545–561. Springer, Heidelberg (2004)
20. Srinathan, K., Patra, A., Choudhary, A., Rangan, C.P.: Probabilistic perfectly reliable and secure message transmission — possibility, feasibility and optimality. In: Srinathan, K., Rangan, C.P., Yung, M. (eds.) INDOCRYPT 2007. LNCS, vol. 4859, pp. 101–122. Springer, Heidelberg (2007)
21. Upfal, E.: Tolerating a linear number of faults in networks of bounded degree. Information and Computation 115(2), 312–320 (1994)
22. Wegman, M., Carter, J.: New hash functions and their use in authentication and set equality. J. Computer and System Sciences 22(2), 265–279 (1981)

Scalar Product-Based Distributed Oblivious Transfer

Christian L.F. Corniaux and Hossein Ghodosi

James Cook University, Townsville QLD 4811, Australia
{chris.corniaux,hossein.ghodosi}@jcu.edu.au

Abstract. In a distributed oblivious transfer (DOT) the sender is replaced with m servers, and the receiver must contact k ($k \leq m$) of these
servers to learn the secret of her choice. Naor and Pinkas introduced
the first unconditionally secure DOT for a sender holding two secrets.
Blundo, D'Arco, Santis, and Stinson generalized Naor and Pinkas's protocol, in the case that the sender holds n secrets, in the first so-called
(k, m)-DOT-$\binom{n}{1}$ protocol. Such a protocol should be secure against a
coalition of less than k parties. However, Blundo et al. have shown that
this level of security is impossible to achieve in one-round polynomial-
based constructions.

In this paper, we show that if communication is allowed amongst the
servers, we are able to construct an unconditionally secure, polynomial-
based (k, m)-DOT-$\binom{n}{1}$ protocol with the highest level of security. More
precisely, in our construction, a receiver who contacts k servers and corrupt up to $k - 1$ servers (not necessarily from the set of the contacted
servers) cannot learn more than one secret.

Keywords: Oblivious Transfer, Unconditional Security, Secret Sharing
Scheme.

1 Introduction

Oblivious Transfer (OT) is a cryptographic protocol which allows two parties to
exchange, in total privacy, one or more secret messages. The first OT, introduced
by Rabin [10], enables a sender to transmit a message to a receiver in such a
way that the receiver gets the message with probability $\frac{1}{2}$ while the sender does
not know whether the message was received. Even, Goldreich and Lempel [7]
introduced a variant of the original OT for a contract signature application.
This OT, identified as OT-$\binom{2}{1}$, is an exchange protocol between a receiver and a
sender who has two secret messages; the receiver chooses one of the two messages
and the sender transmits the chosen message to the receiver. At the end of the
protocol, the sender does not know which message was selected and the receiver
knows nothing of the other message.

A major drawback with OT-$\binom{2}{1}$ and with the more general OT-$\binom{n}{1}$ described
by Brassard, Crépeau and Roberts [5] is the restriction in the availability of the
secret messages, because if the unique sender is unavailable, the receiver cannot

K.-H. Rhee and D. Nyang (Eds.): ICISC 2010, LNCS 6829, pp. 338–354, 2011.

obtain any of the messages. To increase the availability of messages, the sender distributes them to m servers. However, to keep the messages confidential, a server is only provided with parts – called *shares* – of the original messages. Then, the receiver needs to communicate with at least k servers to gain enough shares to reconstruct a chosen message. In this distributed model, the sender does not intervene in the rest of the protocol once he has transmitted the partial secret messages to the servers.

In 2000, Naor and Pinkas [8] introduced the first distributed model for an OT in an unconditionally secure setting. In this Distributed Oblivious Transfer (DOT) protocol, the parties encompass a *sender* who has two secrets, m *servers* owning shares of the secrets, and a *receiver* whose purpose is to obtain one of the secrets. The protocol itself is composed of two phases: (i) the *initialization phase* and (ii) the *transfer phase*. During the initialization phase, the sender generates a bivariate polynomial Q, determines shares from this polynomial and sends to each of the m servers a different set of shares. In the transfer phase, the receiver chooses the index of a secret and selects the k servers she intends to contact. Then, she generates a univariate polynomial S and sends to each of the k selected servers a value determined by the polynomial S and the identifier of the server. Each contacted server generates a response based on its program and the received value from the receiver. The response is sent back to the receiver. After receiving k responses, the receiver is able to determine the chosen secret.

In [3,4], Blundo, D'Arco, De Santis and Stinson generalize Naor and Pinkas's protocol to n secrets, and define a security model composed of four fundamental conditions that every DOT must satisfy:

$C1$. Correctness – The receiver is able to determine the chosen secret once she receives information from the k contacted servers.

$C2$. Receiver's privacy – A coalition of up to $k-1$ servers cannot obtain any information on the choice of the receiver.

$C3$. Sender's privacy with respect to $k-1$ servers and the receiver – A coalition of up to $k-1$ servers with the receiver does not obtain any information about the secrets.

$C4$. Sender's privacy with respect to a "greedy" receiver – Given the transcript of the interaction with k servers, a coalition of up to $k-1$ dishonest servers and the receiver does not obtain any information about secrets which were not chosen by the receiver.

As it has been pointed out by Blundo et al. in [3,4], the protocol introduced by Naor and Pinkas only satisfies conditions $C1$ and $C2$. Their own protocol satisfies conditions $C1$, $C2$ and $C3$ only. Actually, they have proven that condition $C4$ cannot be guaranteed with a one-round DOT protocol – a round being defined as a set of consistent requests/responses exchanged between the receiver and k servers.

In this paper, we show that allowing communication amongst the servers enables us to devise an unconditionally secure DOT protocol that satisfies all the above conditions. We have chosen to assess the security of our protocol against

Blundo et al.'s security model because this model covers other unconditionally secure DOT security models [8,9].

Note that allowing communication amongst the servers is a silent condition in existing DOT protocols. Indeed, the impossibility result of [3,4], in achieving condition $C4$, indicates that, given the transcript of the interaction with k servers, a coalition of the receiver and *only* one dishonest server is sufficient for the receiver to learn all secrets. Therefore, in their protocol there must be a mechanism for ensuring that the receiver does not contact more than k servers (see [8]). This kind of mechanism could be implemented with communication amongst the servers.

The organization of the paper is as follows. In Sect. 2 we review the works related to ours. After providing definitions and notations in Sect. 3, we briefly describe our model in Sect. 4, and we list the components involved in our system in Sect. 5. Section 6 is devoted to the detailed description of the model. In Sect. 7, we evaluate our proposed scheme and demonstrate that it guarantees conditions $C1, \ldots, C4$.

2   Related Works

There have been few publications related to unconditionally secure polynomial-based DOT protocols and the basic principles underlying these protocols are conceptually similar. In the original DOT protocol [8] introduced by Naor and Pinkas, as well as in its generalization [3,4] presented by Blundo et al., a sender distributes some information amongst m servers so that, by contacting k servers, a receiver is able to learn only one of the secrets held by the sender. A simplified overview of the (k, m)-DOT-$\binom{n}{1}$ presented in [4] may be described as follows (operations are executed in a finite field $\mathbb{K} = \mathbb{F}_p$, where p is a prime number):

1. The sender, who has n secrets $\omega_0, \ldots, \omega_{n-1}$ generates a sparse n-variate polynomial function Q defined by

$$Q(x, y_1, \ldots, y_{n-1}) = \sum_{i=1}^{k-1} a_i x^i + \omega_0 + \sum_{i=1}^{n-1} (\omega_i - \omega_0) \times y_i,$$

   where the coefficients a_i $(1 \leq i \leq k-1)$ are numbers randomly selected in $\mathbb{K}$. We note that $\omega_0 = Q(0, \ldots, 0)$ and, for $\ell \in \{1, \ldots, n-1\}$, $\omega_\ell = Q(0, \ldots, 0, 1, 0, \ldots, 0)$, where the number 1 is in position $\ell + 1$.

2. Then, to each server S_j $(1 \leq j \leq m)$, the sender transmits the $(n-1)$-variate polynomial function F_j defined by

$$F_j(y_1, \ldots, y_{n-1}) = Q(j, y_1, \ldots, y_{n-1}) .$$

3. In the oblivious transfer phase, the receiver chooses the identifier ℓ of one secret and generates univariate polynomial functions Z_i $(1 \leq i \leq n-1)$ of degree at most $k-1$ such that $(Z_1(0), \ldots, Z_{n-1}(0))$ is an $(n-1)$-tuple of zeros if the receiver is interested in ω_0 (i.e., $\ell = 0$), or an $(n-1)$-tuple of zeros and a single one in position ℓ if the receiver is interested in ω_ℓ (where $\ell \in \{1, \ldots, n-1\}$).

4. Then, the receiver selects a subset $\mathcal{I}_k \subset \{1,\ldots,m\}$ of k indices and sends to each server S_i $(i \in \mathcal{I}_k)$ a request $(i, Z_1(i),\ldots,Z_{n-1}(i))$. When a server S_i receives such a request, it replies with the share $F_i(Z_1(i),\ldots,Z_{n-1}(i))$.
5. After receiving k responses, the receiver interpolates a univariate polynomial R from the k points $(i, F_i(Z_1(i),\ldots,Z_{n-1}(i)))$ and calculates the chosen secret: $\omega_\ell = R(0)$.

In this scheme, each server S_j knows the value $\omega_i - \omega_0$ $(1 \le i \le n-1)$ as the coefficient of y_i in the polynomial function F_j received from the sender. After execution of the protocol, the receiver who has learned a secret ω_ℓ of her choice can request the values $\omega_i - \omega_0$ $(1 \le i \le n-1)$ from a corrupt server and compute all other secrets. Therefore, a coalition of the receiver and only one dishonest server enables the receiver to learn all secrets. Indeed, in [3,4], each secret ω_i is masked by multiplying it with a random value r_i, but at the end of protocol the receiver learns all masks as well.

Our observation is that, in [8,3,4], the amount of information given to each server is too high. A fundamental requirement for achieving condition $C4$ is that not only each server, but also any coalition of up to $k-1$ servers, should not gain a linear combination of any of two secrets. To meet this requirement, a solution consists, for the sender, in storing the secret values in the servers using a (k,m)-threshold scheme (see Sect. 5.1).

3   Preliminaries

Definition 1. *A (k,m)-DOT-$\binom{n}{1}$ protocol is $(k-1)$-private if, after completion of the protocol, any subset of up to $k-1$ servers cannot learn which secret has been chosen by the receiver.*

Definition 2. *A (k,m)-DOT-$\binom{n}{1}$ protocol is $(k-1)$-secure if, after completion of the protocol, any subset composed of up to $k-1$ servers and the receiver cannot learn information about the secrets in addition to the information already learned by the receiver.*

Throughout this paper, all operations are executed in a finite field $(\mathbb{K} = \mathbb{F}_p, +, \times)$, where $p \in \mathbb{N}$ is a prime number, $+$ is the additive law of composition, and $\times$ is the multiplicative law of composition of the field. We assume that $p > \max(n, \omega_1,\ldots,\omega_n, m)$, where n is the number of secrets, m is the number of servers, and $\omega_1,\ldots,\omega_n$ are the n secrets in the system.

Definition 3. *If $(\mathbb{K}[X], +, \times)$ is the ring of polynomials over $\mathbb{K}$ and $(\mathbb{K}_t[X], +)$ the group of polynomials of degree at most t over $\mathbb{K}$, we say that a polynomial $F = \sum_{i=0}^{t} f_i X^i$ of $\mathbb{K}_t[X]$ is quasi-random, if the coefficients f_i $(1 \le i \le t)$ are randomly selected in $\mathbb{K}$ and the constant term f_0 has a predefined value.*

Definition 4. *By an abuse of language, if $F = \sum_{i=0}^{r} f_i X^i$ $(r \in \mathbb{N}, f_i \in \mathbb{K})$ is a polynomial of $\mathbb{K}[X]$, we define the polynomial function $F \colon \mathbb{K} \to \mathbb{K}$, $x \mapsto \sum_{i=0}^{r} f_i x^i$. Note that the additive and multiplicative laws of composition in $\mathbb{K}[X]$ are defined such that, for $P, Q, R \in \mathbb{K}[X]$ and for $x \in \mathbb{K}$*

- If $R = P + Q$, then the corresponding polynomial functions satisfy the relationship $R(x) = P(x) + Q(x)$,
- Similarly, if $R = P \times Q$, then the corresponding polynomial functions satisfy the relationship $R(x) = P(x) \times Q(x)$.

Definition 5. Let $\boldsymbol{\Phi} = (F_1, \ldots, F_n)$ be a vector of polynomials of $\mathbb{K}_t[X]$. We define $\boldsymbol{\Phi}(x) = (F_1(x), \ldots, F_n(x))$ as the vector of corresponding polynomial functions. For any element $\alpha \in \mathbb{K}$, we denote $\boldsymbol{\Phi}_\alpha = \boldsymbol{\Phi}(\alpha) = (F_1(\alpha), \ldots, F_n(\alpha))$ the vector of polynomial functions $F_1(x), \ldots, F_n(x)$ evaluated for the argument $x = \alpha$. By homogeneity, we denote F_α the value of a polynomial function F applied to the argument $x = \alpha$.

Definition 6. Let $\boldsymbol{\Phi} = (F_1, \ldots, F_n)$ and $\boldsymbol{\Omega} = (Z_1, \ldots, Z_n)$ be two vectors of polynomials of $\mathbb{K}_{k-1}[X]$. We define the scalar product between $\boldsymbol{\Phi}$ and $\boldsymbol{\Omega}$ as

$$\boldsymbol{\Phi} \cdot \boldsymbol{\Omega} = \sum_{i=1}^{n} F_i \times Z_i \ .$$

Note that we assume $p \geq 2k - 1$, and thus $F_i \times Z_i$ is a polynomial of $\mathbb{K}_{2k-2}[X]$.

Lemma 1. Let F be a quasi-random polynomial of $\mathbb{K}_k[X]$ and G a (quasi-random) polynomial of $\mathbb{K}_k[X]$. Then, $P = F + G$ is a quasi-random polynomial of $\mathbb{K}_k[X]$.

Proof. The degree of the sum of two polynomials is lower or equal to the maximum degree of the two original polynomials. Thus $\deg P \leq \max(\deg F, \deg G) \leq k$. It follows $P \in \mathbb{K}_k[X]$.

The quasi-random polynomials F, G, and their sum, P, may be written under the form:

$$F = \sum_{i=0}^{k} f_i X^i, \qquad G = \sum_{i=0}^{k} g_i X^i, \quad \text{and} \quad P = \sum_{i=0}^{k} (f_i + g_i) X^i,$$

where $f_i \in_R \mathbb{K}$ and $g_i \in_R \mathbb{K}$ $(1 \leq i \leq k)$. Therefore, for $1 \leq i \leq k$, the coefficient $f_i + g_i$ is random (the sum of a random element of $\mathbb{K}$ with another element of $\mathbb{K}$ is random) and the constant term $f_0 + g_0$ has a predetermined value. Thus, P is a quasi-random polynomial. $\square$

Theorem 1 (Addition of shares). Let F and G be two quasi-random polynomials of $\mathbb{K}_{k-1}[X]$ generated by Shamir's (k, m)-threshold schemes (see Sect. 5.1) to share respectively the secret α and the secret β amongst m users. We also assume that $\mathcal{I}_k \subset \{1, \ldots, m\}$ is a set of k indices. For each index $i \in \mathcal{I}_k$, we define the shares F_i and G_i. If P is the sum of F and G, then:

(i) P is a quasi-random polynomial of $\mathbb{K}_{k-1}[X]$,
(ii) $P(0) = \alpha + \beta$ and for $i \in \mathcal{I}_k$, $P(i) = F_i + G_i$.

In other words, P may be considered as a polynomial generated by Shamir's (k, m)-threshold scheme to share the secret $\alpha + \beta$.

Proof

(i) Immediate with the application of Lemma 1.
(ii) From the definition 4 of the additive law of composition in $\mathbb{K}[X]$, it holds
$P(x) = F(x)+G(x)$, for $x \in \mathbb{K}$. Consequently, for $x = 0$, $P(0) = F(0)+G(0)$.
As F and G are the sharing polynomials for the secrets α and β, we have
$F(0) = \alpha$ and $G(0) = \beta$. It follows $P(0) = \alpha + \beta$.
Still from the relationship $P(x) = F(x)+G(x)$, we obtain for $i \in \mathcal{I}_k$, $P(i) = F(i) + G(i)$. From the definition 5 of shares, it follows $P(i) = F_i + G_i$.   □

Theorem 2 (Product of shares). *Let F and G be two quasi-random polynomials of $\mathbb{K}_{k-1}[X]$ generated by Shamir's (k, m)-threshold schemes to share respectively the secret α and the secret β. We assume that $m \geq 2k - 1$. For each index $i \in \mathcal{I}_m$, we define the shares F_i and G_i. If Q is the product of F and G, then:*

(i) $Q \in \mathbb{K}_{2k-2}[X]$,
(ii) $Q(0) = \alpha \times \beta$ and for $i \in \mathcal{I}_m$, $Q(i) = F_i \times G_i$.

Proof.

(i) If $F = \sum_{i=0}^{k-1} f_i X^i$ ($f_0 = \alpha$ and $f_i \in_R \mathbb{K}$, $1 \leq i \leq k-1$) and $G = \sum_{i=0}^{k-1} g_i X^i$ ($g_0 = \beta$ and $g_i \in_R \mathbb{K}$, $1 \leq i \leq k-1$), then from the explicit form of the multiplicative law of composition in $\mathbb{K}[X]$, it follows

$$Q = \sum_{\ell=0}^{2k-2} \left(\sum_{i+j=\ell} f_i g_j \right) X^\ell$$

where $f_i = 0$ if $i \geq k$ and $g_j = 0$ if $j \geq k$. The degree of Q is at most $2k - 2$, so $Q \in \mathbb{K}_{2k-2}[X]$.
(ii) Using the definition 4 of the multiplicative law of composition in $\mathbb{K}[X]$, we can write $Q(x) = F(x)\times G(x)$ for $x \in \mathbb{K}$. In particular, $Q(0) = F(0)\times G(0) = \alpha \times \beta$. Of course, the relationship is also true for $x \in \mathcal{I}_m$.

The polynomial Q cannot be considered as quasi-random because it is not irreducible. However, if we add Q to a quasi-random polynomial of $\mathbb{K}_{2k-2}[X]$ whose constant term is zero, the resulting polynomial is quasi-random (see Lemma 1) and may be considered as a polynomial generated by Shamir's $(2k - 1, m)$-threshold scheme to share the secret $\alpha \times \beta$.   □

Corollary 1. *Let $\Omega = (F_1,\ldots,F_n)$ and $\Phi = (G_1,\ldots,G_n)$ be two vectors of quasi-random polynomials of $\mathbb{K}_{k-1}[X]$, each polynomial being generated by a Shamir's (k, m)-threshold scheme to share secrets $(\omega_1,\ldots,\omega_n)$ and $(\varphi_1,\ldots,\varphi_n)$. We assume that $m \geq 2k - 1$. For each index $i \in \mathcal{I}_m$, we define the shares $(F_{1i},\ldots,F_{ni})$ and $(G_{1i},\ldots,G_{ni})$. If Q is the scalar product of Ω and Φ, then:*

(i) $Q \in \mathbb{K}_{2k-2}[X]$,
(ii) $Q(0) = \sum_{i=1}^{n} \omega_i \times \varphi_i$ and for $\ell \in \mathcal{I}_m$, $Q(\ell) = \sum_{i=1}^{n} F_{i\ell} \times G_{i\ell}$.

By simplification, we consider that the polynomial Q is quasi-random. However, to obtain a quasi-random polynomial, we would have to add Q to a quasi-random polynomial of $\mathbb{K}_{2k-2}[X]$ whose constant term is zero. The resulting polynomial would be quasi-random (see Lemma 1) and could be considered as a polynomial generated by Shamir's $(2k-1,\ m)$-threshold scheme to share the secret $\sum_{i=1}^{n} \omega_i \times \varphi_i$.

Proof

(i) From Theorem 2(i), $F_i \times G_i \in \mathbb{K}_{2k-2}[X]$ $(1 \le i \le n)$. Then, by generalization of Theorem 1(i), $\sum_{i=1}^{n} F_i \times G_i \in \mathbb{K}_{2k-2}[X]$. Consequently, $Q \in \mathbb{K}_{2k-2}[X]$.

(ii) From Theorems 2(ii) and 1(ii), it holds $Q(x) = \sum_{i=1}^{n} F_i(x) \times G_i(x)$ for $x \in \mathbb{K}$. In particular, $Q(0) = \sum_{i=1}^{n} F_i(0) \times G_i(0) = \sum_{i=1}^{n} \omega_i \times \varphi_i$. Of course, the relationship is also true for $\ell \in \mathcal{I}_m$, i.e., $Q(\ell) = \sum_{i=1}^{n} F_i(\ell) \times G_i(\ell)$. As $F_i(\ell) = F_{i\ell}$ and $G_i(\ell) = G_{i\ell}$ for $i \in \mathcal{I}_m$ and $1 \le \ell \le n$, it follows $Q(\ell) = \sum_{i=1}^{n} F_{i\ell} \times G_{i\ell}$. □

Notations

The Kronecker's symbol, δ_{ij}, is equal to 0 if $i \neq j$ and equal to 1 if $i = j$.

By $\alpha \in_R \mathbb{K}$, we mean that α is chosen randomly from all possible elements of $\mathbb{K}$.

Addition and multiplication of vectors (of the same variety) are defined naturally. For example, if $U = (u_1, \ldots, u_n)$ and $V = (v_1, \ldots, v_n)$, then $U + V = (u_1 + v_1, \ldots, u_n + v_n)$ and $U \times V = (u_1 \times v_1, \ldots, u_n \times v_n)$. Similarly, the product $\alpha \times U$ between an element α of $\mathbb{K}$ and a vector $U = (u_1, \ldots, u_n)$ of n elements of $\mathbb{K}$ is defined as the vector $(\alpha \times u_1, \ldots, \alpha \times u_n)$.

Each server participating in the protocol is identified by an index selected in $\mathcal{I}_m = \{1, \ldots, m\}$. Thus, the server associated with index $i \in \mathcal{I}_m$ is identified as S_i.

4   Our Model

The setting of our model is similar to the setting of DOT protocols in [8,3,4], i.e., it encompasses a sender $\mathcal{S}$ who owns n secrets $\omega_1, \ldots, \omega_n$ $(n > 1)$, a receiver $\mathcal{R}$ who wishes to learn a secret ω_σ $(\sigma \in \{1, \ldots n\})$, and m servers. In addition to these parties, we also need to take into account an adversary whose characteristics are defined below.

Like other DOT protocols, our protocol is composed of two phases: a setup phase and an oblivious transfer phase. During the setup step, the sender generates shares of the n secrets he holds and distributes them to the m servers. The sender does not intervene in the rest of the protocol. During the oblivious transfer phase, the receiver has to contact k servers $(1 < k \le m)$ to collect enough shares to construct ω_σ.

In our scheme, however, the inequality $m \ge 2k - 1$ must be satisfied. This is because, condition $C4$ allows the receiver to corrupt up to $k-1$ servers, in addition to the k servers chosen by the receiver for gaining shares of the chosen secret.

4.1   The Adversary

In our DOT, the receiver contacts k servers during the execution of the protocol. In addition, she is allowed to corrupt up to $k-1$ servers, possibly amongst the contacted servers. That is, the receiver plays the role of a passive adversary who wishes to breach the sender's security by learning more than one secret from the protocol. Like in other DOT schemes (see [8]), we assume the existence of a mechanism preventing the receiver from contacting more than k servers in one round.

On the other hand, up to $k-1$ servers may collaborate to learn the choice σ of the receiver. In this scenario, the coalition of servers may be considered as an adversary who wishes to breach the privacy of the receiver.

4.2   Overview of the Protocol

The key idea of our model is that if a sender holds a vector $\boldsymbol{u} = (\omega_1, \ldots, \omega_n)$ of n secrets ($n > 1$) and if a receiver wishes to learn the secret ω_σ ($1 \leq \sigma \leq n$), then the receiver contributes with a vector $\boldsymbol{v} = (\delta_{\sigma 1}, \ldots, \delta_{\sigma n})$, and the servers respond with the scalar product of these two vectors, i.e., $\boldsymbol{u} \bullet \boldsymbol{v} = \sum_{i=1}^{n} \omega_i \times \delta_{\sigma i}$, which is the requested secret.

To guarantee the security of the sender and the privacy of the receiver, the vectors involved in the scalar product are shared thanks to Shamir's [11] threshold schemes. That is, the sender transmits to each server S_i ($1 \leq i \leq m$) a vector $\boldsymbol{u_i} = (F_1(i), \ldots, F_n(i))$ of n shares, where F_i is the sharing polynomial related to ω_i. In the same way, to obtain a secret ω_σ, the receiver selects a subset $\mathcal{I}_k \subset \mathcal{I}_m$ of k indices, sends to each server S_j ($j \in \mathcal{I}_k$) a vector $\boldsymbol{v_j} = (Z_1(j), \ldots, Z_n(j))$ of n shares (Z_i is the sharing polynomial related to $\delta_{\sigma i}$) and receives k shares of the chosen secret ω_σ. The shares are associated with a polynomial μ of degree $k-1$ and so, by interpolation, the receiver is able to determine μ and to calculate the chosen secret $\omega_\sigma = \mu(0)$.

5   Components of the System

Our protocol is mainly based on two components.

The first one is a secret sharing scheme, allowing on one hand the sender to generate and distribute shares of the secrets he holds, and on the other hand the receiver to generate and distribute shares of the identifier of the secret she wishes to learn.

The second one is a mechanism which enables a set of users to redistribute a secret to another set of users. This component requires the availability of private communication channels between any two users involved in the protocol. We assume that these communication channels are secure, i.e., any party is unable to eavesdrop on them and they guarantee that communications cannot be tampered with.

5.1 Secret Sharing Scheme

The underlying secret sharing scheme of our DOT protocol is Shamir's threshold scheme. In [11], Shamir shows how to share a secret $w \in \mathbb{K}$ amongst m users $u_1, \ldots, u_m$ such that w can be reconstructed from any set of k ($k \leq m$) shares. Shamir suggests the following algorithm for constructing such secret sharing schemes, called (k, m)-threshold schemes.

1. A dealer, $\mathcal{D}$, who has a secret w, chooses m distinct and non-zero elements of $\mathbb{K}$, denoted $x_1, \ldots, x_m$ and sends x_i to u_i via a public channel. For convenience, we assume that $x_i = i$.
2. $\mathcal{D}$ secretly chooses $k - 1$ random elements of $\mathbb{K}$, denoted $a_1, \ldots, a_{k-1}$ and forms the polynomial function $f(x) = \sum_{\ell=1}^{k-1} a_\ell x^\ell + w$.
3. $\mathcal{D}$ gives (in private) the share $f(i)$ to the user u_i.

In the secret reconstruction phase, a set of at least k participants uses the Lagrange interpolation formula to recover the secret. Without loss of generality, let $u_1, \ldots, u_k$ be the set of collaborating users in the secret reconstruction phase. Then the secret can be recovered, using

$$w = \sum_{i=1}^{k} f(i) \prod_{\substack{j=1 \\ j \neq i}}^{k} \frac{j}{j - i} .$$

It is well-known that Shamir's threshold scheme is perfect, i.e., the knowledge of $k - 1$ or less shares leaves w completely undetermined (in the sense that all its possible values are equally likely).

5.2 Transformation from One to Another Threshold Scheme

There are many applications that require redistribution of a secret from one set of users to another set of users, with possibly a different threshold. Desmedt and Jajodia [6] have considered this problem and proposed some protocols. The specific case of the reduction of the threshold is commonly discussed in multiparty computation studies. Indeed, if two secrets are shared (Shamir's scheme) thanks to two polynomials P_1 and P_2 of $\mathbb{K}_{k-1}[X]$, then from the polynomial $Q = P_1 \times P_2$ can be generated shares of the product of the two secrets (See Theorem 2). But the degree of Q (up to $2k - 2$) corresponds to a threshold $2k - 1$, and so a degree reduction is necessary to obtain a sharing polynomial corresponding to a threshold k. Such a reduction was introduced, for example, by Ben-Or, Goldwasser and Wigderson [2].

Below we present a combined method, allowing a set of k_1 or more users holding shares generated by a Shamir's (k_1, m_1)-threshold scheme to distribute them under the form of shares generated by a Shamir's (k_2, m_2)-threshold scheme, where $k_1 \geq k_2$, to a set of m_2 users.

Let a secret w be shared amongst the group of users $\mathcal{U} = \{u_1, \ldots, u_{m_1}\}$ thanks to a polynomial f generated in a (k_1, m_1)-threshold scheme ($k_1 \leq m_1$). Each user u_i holds the share $f(i)$. Also, let assume that the users $u_1, \ldots, u_{k_1}$ wish to redistribute the secret ω over the set of users $\mathcal{V} = \{v_1, \ldots, v_{m_2}\}$ using a (k_2, m_2)-threshold scheme $(k_2 \leq m_2)$. Note that $\mathcal{U}$ and $\mathcal{V}$ are two arbitrary groups of users. Each user $u_i \in (u_1, \ldots, u_{k_1})$ generates a polynomial H_i such that

$$H_i = \sum_{r=1}^{k_2-1} h_{i,r} X^r + \overline{L}_i \times f(i),$$

where $h_{i,r} \in_R \mathbb{K}$ $(1 \leq i \leq k_1, 1 \leq r \leq k_2 - 1)$ and $\overline{L}_i$ is the truncated Lagrange basis polynomial L_i mod X^{k_2} where

$$L_i = \prod_{\substack{j=1 \\ j \neq i}}^{k_1} \frac{X - j}{i - j} \ .$$

Truncating the polynomial L_i $(1 \leq i \leq k_1)$ allows each user $u_i \in (u_1, \ldots, u_{k_1})$ to generate a polynomial H_i of degree $k_2 - 1$ and not $k_1 - 1$ (when $k_2 < k_1$), by application of the technique described by Beaver [1]. Then, for all $1 \leq j \leq m_2$, the user u_i privately sends the value $H_i(j)$ to the user $v_j \in \mathcal{V}$. Once a user $v_j \in \mathcal{V}$ obtains k_1 partial shares, sent by the users $u_i \in \mathcal{U}$, he computes a new share $\rho_j = \sum_{i=1}^{k_1} H_i(j)$ as his share of the secret ω.

With this method, the original sharing polynomial f is replaced with a sharing polynomial g such that

$$g = \sum_{i=1}^{k_1} \left(\sum_{r=1}^{k_2-1} h_{i,r} X^r + f(i) \times \overline{L}_i \right)$$

$$= \sum_{i=1}^{k_1} \sum_{r=1}^{k_2-1} h_{i,r} X^r + \sum_{i=1}^{k_1} f(i) \times \overline{L}_i$$

$$= \sum_{i=1}^{k_1} \sum_{r=1}^{k_2-1} h_{i,r} X^r + \overline{f}$$

where $\overline{f}$ is the truncated polynomial f mod X^{k_2}.

The degree of the polynomial g is at most $k_2 - 1$ and $g(0) = \overline{f}(0) = f(0) = \omega$. Therefore, with k_2 values $\rho_i = g(i)$, any set of k_2 users in $\mathcal{V}$ is able to interpolate g and to determine the secret ω.

Theorem 3. *If $k_2 \leq k_1$, the proposed protocol for transforming a Shamir (k_1, m_1)-threshold scheme to a Shamir (k_2, m_2)-threshold scheme is secure.*

Proof. To demonstrate the security of ω, we show that after the new shares have been obtained, no coalition of $k_2 - 1$ users or less can infer any information about ω.

After the redistribution of shares, a coalition of $k_2 - 1$ users holds at most $k_2 - 1$ shares $f(i)$ and $k_2 - 1$ shares $g(i)$. We analyze the worst case, i.e., when all the members of the coalition belong to $\mathcal{U}$. First, the coalition cannot interpolate f and calculate $f(0)$ since k_1 shares $f(i)$ would be necessary, and the coalition only holds $k_2 - 1 \le k_1 - 1 < k_1$ shares. Second, with only $k_2 - 1$ values $g(i)$, and g being a polynomial of degree $k_2 - 1$, the coalition cannot either interpolate g and calculate $g(0)$. Consequently, a coalition of $k_2 - 1$ users $u_i \in \mathcal{V}$ is unable to gain any information on the secret ω.   □

6   A Secure (k, m)-DOT-$\binom{n}{1}$ Protocol

In this section we present our (k, m)-DOT-$\binom{n}{1}$ protocol, which is $(k-1)$-private and $(k-1)$-secure. Our protocol, depicted in Fig. 1, is composed of five steps, described in the following sections.

6.1   Setup Phase

This step is straightforward. The sender, $\mathcal{S}$, distributes the shares of the secrets $\omega_1, \ldots, \omega_n$ amongst the servers $S_1, \ldots, S_m$, using Shamir's (k, m)-threshold scheme. That is, for each secret ω_i $(1 \le i \le n)$, the sender generates a quasi-random polynomial $F_i \in \mathbb{K}_{k-1}[X]$ such that

$$F_i = \sum_{j=1}^{k-1} f_{i,j} X^j + \omega_i, \qquad \text{where } f_{i,j} \in_R \mathbb{K}, 1 \le i \le n, 1 \le j \le k - 1,$$

and computes the shares of ω_i for all servers S_ℓ $(\ell \in \mathcal{I}_m)$. Then, $\mathcal{S}$ transmits to each server S_ℓ the vector $\Omega_\ell = (F_1(\ell), \ldots, F_n(\ell))$. The sender does not intervene in the rest of the protocol.

6.2   Elaboration of the Receiver's Requests

The receiver, $\mathcal{R}$, chooses the index σ of the secret she wishes to obtain, as well as a subset $\mathcal{I}_k \subset \mathcal{I}_m$ of k indices of the servers she intends to contact. Then, the receiver generates n quasi-random polynomials $Z_i \in \mathbb{K}_{k-1}[X]$ such that

$$Z_i = \sum_{j=1}^{k-1} z_{i,j} X^j + \delta_{\sigma i}, \qquad \text{where } z_{i,j} \in_R \mathbb{K}, 1 \le i \le n, 1 \le j \le k - 1,$$

and transmits to each server S_ℓ $(\ell \in \mathcal{I}_k)$ the vector $\Phi_\ell = (Z_1(\ell), \ldots, Z_n(\ell))$.

6.3   Redistribution of the Receiver's Input

In this step the k contacted servers, which have received from $\mathcal{R}$ shares generated by a (k, k)-threshold scheme, redistribute them as shares considered as generated by a $(k, 2k-1)$-threshold scheme. To perform this redistribution, the servers S_ℓ $(\ell \in \mathcal{I}_k)$ select a subset $\mathcal{I}_{2k-1} \subset \mathcal{I}_m$ of servers. Then for every

Let $S_1, \ldots, S_m$ be m servers.

Input   The sender $\mathcal{S}$, contributes with n secrets $\omega_1, \ldots, \omega_n \in \mathbb{K}$

The receiver $\mathcal{R}$, chooses an index $\sigma \in \{1, \ldots, n\}$, and contributes with n private values $\delta_{\sigma 1}, \ldots, \delta_{\sigma n} \in \{0, 1\}$

Output $\mathcal{R}$ receives ω_σ, while $\mathcal{S}$ receives nothing.

Step 1 – Setup phase

1. $\mathcal{S}$ generates a vector $\boldsymbol{\Omega} = (F_1, \ldots, F_n)$ of n quasi-random polynomials $F_i \in \mathbb{K}_{k-1}[X]$, where the constant term of F_i is ω_i $(1 \le i \le n)$.
2. $\mathcal{S}$ transmits to the server S_ℓ $(\ell \in \mathcal{I}_m)$ the vector $\boldsymbol{\Omega}_\ell = (F_1(\ell), \ldots, F_n(\ell))$.

Step 2 – Elaboration of the receiver's request

1. $\mathcal{R}$ chooses $\sigma \in \{1, \ldots, n\}$ and $\mathcal{I}_k \subset \mathcal{I}_m$.
2. $\mathcal{R}$ generates a vector $\boldsymbol{\Phi} = (Z_1, \ldots, Z_n)$ of n quasi-random polynomials $Z_i \in \mathbb{K}_{k-1}[X]$, where the constant term of Z_i is $\delta_{\sigma i}$.
3. $\mathcal{R}$ transmits to server S_ℓ $(\ell \in \mathcal{I}_k)$ the vector $\boldsymbol{\Phi}_\ell = (Z_1(\ell), \ldots, Z_n(\ell))$.

Step 3 – Redistribution of the receiver's input

1. The contacted servers S_ℓ $(\ell \in \mathcal{I}_k)$ select a subset $\mathcal{I}_{2k-1} \subset \mathcal{I}_m$ of $2k - 1$ servers such that $\mathcal{I}_k \subset \mathcal{I}_{2k-1}$.
2. Each server S_ℓ $(\ell \in \mathcal{I}_k)$ generates two polynomial vectors:
   (a) $\boldsymbol{\Theta}^{\ell 1} = (G_{\ell,1}, \ldots, G_{\ell,n})$ such that $G_{\ell,i}$ is a quasi-random polynomial of $\mathbb{K}_{k-1}[X]$ with a null constant term,
   (b) $\boldsymbol{\Theta}^{\ell 2} = \prod_{\substack{u \in \mathcal{I}_k \\ u \ne \ell}} \frac{X-u}{\ell-u} \times \boldsymbol{\Phi}_\ell$,
   and builds a vector $\boldsymbol{\Psi}^\ell = \boldsymbol{\Theta}^{\ell 1} + \boldsymbol{\Theta}^{\ell 2}$ of n polynomials.
3. The value $\boldsymbol{\Psi}^\ell_j$ is sent to the server S_j $(j \in \mathcal{I}_{2k-1})$.
4. Each server S_j, once it holds k values $\boldsymbol{\Psi}^\ell_j$ $(\ell \in \mathcal{I}_k)$, calculates $\boldsymbol{\Phi}'_j = \sum_{\ell \in \mathcal{I}_k} \boldsymbol{\Psi}^\ell_j$, as its shares from the receiver's input related to $(k, 2k-1)$-threshold schemes.

Step 4 – Computation of the requested secret

1. Each server S_ℓ $(\ell \in \mathcal{I}_{2k-1})$ calculates the scalar product $\lambda_\ell = \boldsymbol{\Omega}_\ell \bullet \boldsymbol{\Phi}'_\ell$, which is its share of the requested secret, $\boldsymbol{\Omega}_0 \bullet \boldsymbol{\Phi}_0$, related to a polynomial of degree $2k - 2$. So, the collaborating set of at least $2k - 1$ servers involved in the computation of $\boldsymbol{\Omega}_0 \bullet \boldsymbol{\Phi}_0$, redistributes its shares using a (k, k)-threshold scheme.
2. S_ℓ $(\ell \in \mathcal{I}_{2k-1})$ generates two polynomials:
   (a) $H^{\ell 1}$, a quasi-random polynomial of $\mathbb{K}_{k-1}[X]$ with a null constant term,
   (b) $H^{\ell 2} = \overline{L_\ell} \times \lambda_\ell$, where $\overline{L_\ell}$ is the truncated Lagrange basis polynomial L_ℓ mod X^k, where $L_\ell = \prod_{\substack{j \in \mathcal{I}_{2k-1} \\ j \ne i}} \frac{X-j}{\ell-j}$,
   and builds a polynomial $H^\ell = H^{\ell 1} + H^{\ell 2}$. The value H^ℓ_j is sent to the server S_j $(j \in \mathcal{I}_k)$.
3. A server S_j, once it holds $2k - 1$ values H^ℓ_j $(\ell \in \mathcal{I}_{2k-1})$, calculates the new share $\mu_j = \sum_{\ell \in \mathcal{I}_{2k-1}} H^\ell_j$.

Step 5 – Oblivious transfer of the requested secret

1. S_j $(j \in \mathcal{I}_k)$ sends μ_j to $\mathcal{R}$.
2. $\mathcal{R}$ interpolates a polynomial μ of degree at most $k - 1$, and calculates $\omega_\sigma = \mu(0)$.

Fig. 1. A secure (k, m)-DOT-$\binom{n}{1}$ protocol

secret value $\delta_{\sigma i}$ shared thanks to a polynomial Z_i $(1 \leq i \leq n)$, they execute the redistribution protocol described in Sect. 5.2, and reshare that secret value using a $(k, 2k-1)$-threshold scheme. This process is shown in Fig. 1 for vectors of secret values. The new shares associated with a secret value $\delta_{\sigma i}$ $(1 \leq i \leq n)$ are considered as generated from a new polynomial that we denote Z_i' (we also denote $\boldsymbol{\Phi'} = (Z_1', \ldots, Z_n')$ the vector of sharing polynomials related to the vector $\boldsymbol{\Phi} = (\delta_{\sigma 1}, \ldots, \delta_{\sigma n})$ of private values). Therefore, each server S_ℓ $(\ell \in \mathcal{I}_{2k-1})$ obtains a vector $\boldsymbol{\Phi'_\ell} = (Z_1'(\ell), \ldots, Z_n'(\ell))$ of shares of $(\delta_{\sigma 1}, \ldots, \delta_{\sigma n})$.

6.4   Computation of the Requested Secret

Now, each server S_ℓ $(\ell \in \mathcal{I}_{2k-1})$ holds a vector $\boldsymbol{\Omega_\ell} = (F_1(\ell), \ldots, F_n(\ell))$ as its share of the secrets $(\omega_1, \ldots, \omega_n)$, and a vector $\boldsymbol{\Phi'_\ell} = (Z_1'(\ell), \ldots, Z_n'(\ell))$ as its share of $(\delta_{\sigma 1}, \ldots, \delta_{\sigma n})$.

The elements of the two vectors $\boldsymbol{\Omega}$ and $\boldsymbol{\Phi'}$ are polynomials belonging to $\mathbb{K}_{k-1}[X]$. It follows (Corollary 1) that $\boldsymbol{\Omega} \bullet \boldsymbol{\Phi'}$ is a polynomial of degree at most $2k-2$, that $(\boldsymbol{\Omega} \bullet \boldsymbol{\Phi'})(0) = \omega_\sigma$ and that for $i \in \mathcal{I}_{2k-1}$, $\lambda_\ell = \boldsymbol{\Omega_\ell} \bullet \boldsymbol{\Phi'_\ell}$ is a share of ω_σ, generated by the sharing polynomial $\boldsymbol{\Omega} \bullet \boldsymbol{\Phi'}$.

So, each server S_ℓ $(\ell \in \mathcal{I}_{2k-1})$ calculates the new share $\lambda_\ell = \boldsymbol{\Omega_\ell} \bullet \boldsymbol{\Phi'_\ell}$. Since at least $2k-1$ servers participate in the computation, they redistribute the resulting shares λ_ℓ, $(\ell \in \mathcal{I}_{2k-1})$, thanks to a (k, k)-threshold scheme (using the method described in Sect. 5.2), to the servers $S_j (j \in \mathcal{I}_k)$.

6.5   Oblivious Transfer of the Requested Secret

After redistribution of the secret shares in the previous step, the set of contacted servers S_j $(j \in \mathcal{I}_k)$ collectively owns the value of $\boldsymbol{\Omega_0} \bullet \boldsymbol{\Phi_0}$, under the form of shares generated by a (k, k)-threshold scheme. Each server S_j $(j \in \mathcal{I}_k)$ responds to the receiver's request with the value μ_j, which is its share of $\boldsymbol{\Omega_0} \bullet \boldsymbol{\Phi_0}$ generated from a sharing polynomial μ of degree at most $k-1$. The receiver $\mathcal{R}$ interpolates a $(k-1)$-degree polynomial corresponding to the k responses, and obtains ω_σ.

7   Evaluation of the Protocol

In this section we demonstrate that the proposed protocol satisfies all desirable conditions listed in [3,4] (i.e., conditions $C1$, $C2$, $C3$, and $C4$).

7.1   Correctness

We demonstrate that $\mu(0) = \omega_\sigma$.

The degree of the polynomial μ is at most $k-1$. So, the k shares μ_j $(j \in \mathcal{I}_k)$ are sufficient to interpolate μ.

The redistribution procedure, in Sect. 6.4, does not modify the shared secret. Thus, the sharing polynomials $\boldsymbol{\Omega} \bullet \boldsymbol{\Phi'}$ (before the redistribution) and μ (after the redistribution) are such that $(\boldsymbol{\Omega} \bullet \boldsymbol{\Phi'})(0) = \mu(0)$. Because $(\boldsymbol{\Omega} \bullet \boldsymbol{\Phi'})(0) = \omega_\sigma$, it follows $\mu(0) = \omega_\sigma$.

Therefore, condition $C1$ is guaranteed.

7.2   Receiver's Privacy

Theorem 4. *The proposed (k, m)-DOT-$\binom{n}{1}$ protocol, which is depicted in Fig. 1, is $(k-1)$-private.*

Proof. The index σ chosen by the receiver is represented under the form of a vector $(\delta_{\sigma 1}, \ldots, \delta_{\sigma n})$ of private values, where $\delta_{\sigma i} = 0$ if $\sigma \neq i$ and $\delta_{\sigma i} = 1$ if $\sigma = i$. The receiver's input to the protocol consists of shares of these values. That is, each element $\delta_{\sigma i}$, which is either zero or one, is distributed amongst the set of k servers S_ℓ ($\ell \in \mathcal{I}_k$), using Shamir's (k, k)-threshold scheme. This is achieved by generating a vector $\boldsymbol{\Phi} = (Z_1, \ldots, Z_n)$ of n quasi-random polynomials of $\mathbb{K}_{k-1}[X]$, such that $Z_i(0) = \delta_{\sigma i}$.

In order to breach the privacy of the receiver, a set of $k-1$ colluding servers (along the execution of the protocol, or after completion of the protocol) should be able to determine at least one of the values $\delta_{\sigma i}$. The set of $k-1$ collaborating servers, however, owns $k-1$ shares corresponding to each values $\delta_{\sigma i}$ associated with a Shamir's (k, k)-threshold scheme. Due to the perfectness of Shamir's threshold scheme, every set of $k-1$ shares provides the coalition with absolutely no information about the relevant secret. That is, the inputs of the receiver guarantees her privacy, during Step 2 of the protocol. Although in the next step of the protocol, the set of k servers redistribute the private values $\delta_{\sigma i}$, to $2k-1$ servers, this transformation does not leak any information to a set of $k-1$ servers (see Theorem 3). In the same way, the redistribution presented in Sect. 6.4 preserves the privacy of the receiver.

One may argue that, after completion of the protocol, the knowledge of the output ω_σ may help the set of $k-1$ servers to determine the choice of the secret. However, this is not the case, since the shares held by the $k-1$ servers after the redistribution may be considered as generated from a (k, m)-threshold scheme, and thus $k-1$ shares of any secret provides no information about the secret.   □

Consequently, the condition $C2$ is guaranteed.

7.3   Sender's Security with Respect to $k-1$ Servers and the Receiver

At the level of the servers, the role of the vector $\boldsymbol{\Omega}$ is the same as the role of the vector $\boldsymbol{\Phi}'$. That is, a set of $k-1$ colluding servers cannot learn any of the secrets $\omega_1, \ldots, \omega_n$ (see previous section). Note that, in this scenario, there is no advantage for the coalition of $k-1$ servers to collude with the receiver to breach the sender's security, since the receiver has no input to contribute in this attack. Indeed, even if the receiver initiates the protocol by contacting only $k-1$ servers (no matter what her input to the kth selected server is), she cannot learn any of the secrets. This is because, the $k-1$ shares she receives from the servers are considered as generated by a (k, m)-threshold scheme, and thus provide no information about the secrets.

Remark 1. Actually, this level of security, which is expressed as condition $C3$, is achievable easily. Our claim, however, is that the security of the sender can be guaranteed against a coalition of $k-1$ servers and the receiver, before and after the protocol is executed. In this scenario, the receiver can examine the transcripts of the protocol, the secret she has obtained, and all information held by $k-1$ corrupt servers (see the next section).

Remark 2. One may note that, contrary to the protocol presented in [3,4], our protocol does not require the use of masks. Indeed, in our scheme, a corrupt server cannot provide the receiver with a linear combination of the secrets $\omega_1, \ldots, \omega_n$ because it never owns such a combination.

7.4   Sender's Security with Respect to a "Greedy" Receiver

Let $\mathcal{I}_C$ be the set of indices of $k-1$ corrupt servers. A corrupt server S_ℓ ($\ell \in \mathcal{I}_C$) may be one of the servers selected by the receiver in the second step of the protocol. In this case, $\ell \in \mathcal{I}_k$. We assume the worst scenario from the sender's point of view (the best scenario from the "greedy" receiver's point of view), i.e., that all the corrupt servers are selected by the receiver in Step 2 of the protocol ($\mathcal{I}_C \subset \mathcal{I}_k$). Thus, we assume that the receiver has access to all information of these $k-1$ corrupt servers along the execution of the protocol.

In the initialization step, each server S_c ($c \in \mathcal{I}_C$) receives from the sender a vector $\boldsymbol{\Omega}_c = (F_1(c), \ldots, F_n(c))$ of n elements, where the element $F_i(c)$ is the share allocated to S_c for the secret ω_i. Clearly, the knowledge of $k-1$ shares of any secret ω_i, provides the receiver with no useful information about the secret ω_i.

In Step 2, the receiver transmits her inputs to servers S_ℓ ($\ell \in \mathcal{I}_k$). Obviously, she cannot learn any information about $\omega_1, \ldots, \omega_n$ from her inputs.

In Step 3, the contacted servers redistribute the receiver's private values $\delta_{\sigma i}$ ($1 \le i \le n$) amongst the servers S_ℓ ($\ell \in \mathcal{I}_{2k-1}$), using the method described in Sect. 5.2. The receiver has access to all partial shares received by servers S_c ($c \in \mathcal{I}_C$), and has also access to all partial shares sent by servers S_c ($c \in \mathcal{I}_C$) to servers S_ℓ ($\ell \in \mathcal{I}_{2k-1}$). At the end of this step, the receiver's private values, $\delta_{\sigma i}$ ($1 \le i \le n$), are shared amongst the servers S_ℓ ($\ell \in \mathcal{I}_{2k-1}$). The shares are considered as generated from a (k, $2k-1$)-threshold scheme. Although the receiver, for each private value $\delta_{\sigma i}$, has only access to $k-1$ shares owned by servers S_c ($c \in \mathcal{I}_C$), she can figure out all the $2k-1$ shares elaborated by servers S_ℓ ($\ell \in \mathcal{I}_{2k-1}$). Indeed, for $i = 1, \ldots, n$, the receiver knows $\delta_{\sigma i}$ and $k-1$ shares of $\delta_{\sigma i}$; so she can reconstruct the associated sharing polynomial generated in the (k, $2k-1$)-threshold scheme, and from the polynomial, the corresponding shares held by servers S_ℓ ($\ell \in \mathcal{I}_{2k-1}$). That is, the receiver is able to find out all the elements of the vector $\boldsymbol{\Phi}'_\ell$ ($\ell \in \mathcal{I}_{2k-1}$).

In Step 4, each server S_ℓ ($\ell \in \mathcal{I}_{2k-1}$) computes $\lambda_\ell = \boldsymbol{\Omega}_\ell \bullet \boldsymbol{\Phi}'_\ell$. The receiver is able to multiply the corresponding shares of two vectors $\boldsymbol{\Omega}_c$ and $\boldsymbol{\Phi}'_c$ ($c \in \mathcal{I}_C$), and thus learns $k-1$ shares of $\omega_i \times \delta_{\sigma i}$ ($1 \le i \le n$). She also knows the value of $\omega_i \times \delta_{\sigma i}$ ($1 \le i \le n$), because if $\delta_{\sigma i} = 0$, then the $\omega_i \times \delta_{\sigma i} = 0$, and if $\delta_{\sigma i} = 1$, then $\omega_i \times \delta_{\sigma i} = \omega_\sigma$, which is the secret that the receiver learns at the end of the protocol. However, contrary to the previous step, knowing a value $\omega_i \times \delta_{\sigma i}$

$(1 \leq i \leq n)$ and $k - 1$ of its shares is not sufficient to learn all other shares. This is because in Step 4, the value and the shares are associated with a $(2k - 2)$-degree polynomial (see Sect. 6.4), generated in a $(2k - 1, 2k - 1)$-threshold scheme. Consequently, the receiver cannot learn any useful information about ω_i $(1 \leq i \leq n, i \neq \sigma)$. Note that the redistribution of secrets from a $(2k-1, 2k-1)$-threshold scheme to a (k, k)-threshold scheme is completely secure (see Sect. 5.2) and therefore provides the receiver with no information about $\omega_1, \ldots, \omega_n$.

Finally, in Step 5, the data held by servers S_c $(c \in \mathcal{I}_C)$ are not modified; The servers S_j $(j \in \mathcal{I}_k)$ transmit to the receiver k shares of ω_σ, generated by a (k, k)-threshold scheme and the receiver learns the secret ω_σ. That is, there is no way for the receiver to learn more than the requested secret, and thus the following is proved.

Theorem 5. *The proposed (k, m)-DOT-$\binom{n}{1}$ protocol, which is depicted in Fig. 1, is $(k - 1)$-secure.*

7.5   Efficiency Considerations

In this paper, we have presented a new polynomial-based unconditionally secure (k, m)-DOT-$\binom{n}{1}$ protocol. Our result significantly improves the security of DOT protocols. If we compare the efficiency of our protocol to the efficiency of the (k, m)-DOT-$\binom{n}{1}$ protocol presented in [3,4], we observe that our protocol requires more computation on the servers' sides, but less computation on the sender and receiver's sides, as shown in Table 1.

Table 1. Efficiency of DOT protocols

Party	Blundo et al.'s protocol	Our protocol
Sender $\mathcal{S}$	1 distribution of n secrets $c_0\omega_0, \ldots, c_{n-1}\omega_{n-1}$, n masks $c_0, \ldots, c_{n-1}$ and $2 \times (n - 1)$ masks $r_1^1, \ldots, r_{n-1}^1, r_1^2, \ldots, r_{n-1}^2$	1 distribution of n secrets $\omega_1, \ldots, \omega_n$
Receiver $\mathcal{R}$	1 distribution of n private values (0 or 1) and 4 polynomial interpolations	1 distribution of n private values $\delta_{\sigma i}$ $(1 \leq i \leq n)$ and one polynomial interpolation
Server S_ℓ (worst case)	1 generation of 2 shares from degree 1 polynomials	n transformations from a (k, k) to a $(k, 2k - 1)$-threshold scheme, 1 transformation from a $(2k - 1, 2k - 1)$- to a (k, k)-threshold scheme, 1 generation of the scalar product of 2 vectors of length n

Acknowledgments. We would like to thank the anonymous referees for their helpful comments.

References

1. Beaver, D.: Multiparty protocols tolerating half faulty processors. In: Brassard, G. (ed.) CRYPTO 1989. LNCS, vol. 435, pp. 560–572. Springer, Heidelberg (1990)

2. Ben-Or, M., Goldwasser, S., Wigderson, A.: Completeness theorems for non-cryptographic fault-tolerant distributed computation. In: Proceedings of the Twentieth Annual ACM Symposium on Theory of Computing - STOC 1988, pp. 1–10. ACM, New York (1988)

3. Blundo, C., D'Arco, P., Santis, A.D., Stinson, D.R.: New results on unconditionally secure distributed oblivious transfer. In: Nyberg, K., Heys, H.M. (eds.) SAC 2002. LNCS, vol. 2595, pp. 291–309. Springer, Heidelberg (2003)

4. Blundo, C., D'Arco, P., Santis, A.D., Stinson, D.R.: On unconditionally secure distributed oblivious transfer. J. Cryptology 20(3), 323–373 (2007)

5. Brassard, G., Crépeau, C., Robert, J.M.: All-or-nothing disclosure of secrets. In: Odlyzko, A.M. (ed.) CRYPTO 1986. LNCS, vol. 263, pp. 234–238. Springer, Heidelberg (1987)

6. Desmedt, Y.G., Jajodia, S.: Redistributing secret shares to new access structures and its applications. Technical report, George Mason University (1997)

7. Even, S., Goldreich, O., Lempel, A.: A randomized protocol for signing contracts. Communications of the ACM 28, 637–647 (1985)

8. Naor, M., Pinkas, B.: Distributed oblivious transfer. In: Okamoto, T. (ed.) ASIACRYPT 2000. LNCS, vol. 1976, pp. 205–219. Springer, Heidelberg (2000)

9. Nikov, V., Nikova, S., Preneel, B., Vandewalle, J.: On unconditionally secure distributed oblivious transfer. In: Menezes, A., Sarkar, P. (eds.) INDOCRYPT 2002. LNCS, vol. 2551, pp. 395–408. Springer, Heidelberg (2002)

10. Rabin, M.O.: How to exchange secrets with oblivious transfer. Technical report, Aiken Computation Lab, Harvard University (1981)

11. Shamir, A.: How to share a secret. Communications of the ACM 22(11), 612–613 (1979)

Unconditionally Secure Rational Secret Sharing in Standard Communication Networks[*]

Zhifang Zhang[1,2] and Mulan Liu[1]

[1] Key Laboratory of Mathematics Mechanization, Academy of Mathematics and Systems Science, CAS, Beijing, China
[2] State Key Laboratory of Information Security, Beijing, China
zfz@amss.ac.cn

Abstract. Rational secret sharing protocols in both the two-party and multi-party settings are proposed. These protocols are built in standard communication networks and with unconditional security. Namely, the protocols run over standard point-to-point networks without requiring physical assumptions or simultaneous channels, and even a computationally unbounded player cannot gain more than ϵ by deviating from the protocol. More precisely, for the 2-out-of-2 protocol the ϵ is a negligible function in the size of the secret, which is caused by the information-theoretic MACs used for authentication. The t-out-of-n protocol is $(t-1)$-resilient and the ϵ is exponentially small in the number of participants. Although secret recovery cannot be guaranteed in this setting, a participant can at least reduce the Shannon entropy of the secret to less than 1 after the protocol. When the secret-domain is large, every rational player has great incentive to participate in the protocol.

Keywords: rational secret sharing, ϵ-Nash equilibrium, unconditional security.

1 Introduction

Secret sharing [2,18] is an important tool in cryptography. The widely used t-out-of-n scheme is that a dealer holding a secret distributes shares among n players such that any group of t or more players can recover the secret from their shares while any group of fewer than t players can not. In 2004 Halpern and Teague [8] studied the problem in a game theoretic sense and proposed *rational secret sharing* which is to fulfill the task among rational players who only act in their own self-interest. As Halpern and Teague pointed out that no rational player would broadcast his share in a deterministic recovering process, since keeping silence can guarantee him a utility that is equal to and sometimes even higher than the utilities of other players (because he might be the only one who gets the secret). Therefore most previous secret sharing schemes fail in the rational setting which requires to design a protocol such that all rational

---

[*] Research supported in part by the National Natural Science Foundation of China (No.60821002/F02, No.11001254) and the Foundation of President of AMSS, CAS.

K.-H. Rhee and D. Nyang (Eds.): ICISC 2010, LNCS 6829, pp. 355–369, 2011.

players have the incentive for participation. Furthermore, it is more desirable to design a protocol where no player has an incentive to deviate as long as the other players follow the protocol. This requirement is captured by the notion of *equilibrium* in game theory. Although many rational secret sharing schemes [1,14,12,13,15,5,7,8,9,10,11,20] have been developed achieving kinds of equilibria, they are less satisfactory in some of the following aspects:

Notions of Equilibria. Halpern and Teague [8] first proposed achieving a Nash equilibrium *surviving iterated deletion of weakly dominated strategies*. But Kol and Naor [10] later pointed out that some "intuitively bad" strategies cannot be deleted anyway, then they proposed the notion of *strict Nash equilibrium* requiring that each player's strategy is his unique best response to other players' strategies. Although strict Nash equilibrium is a more appealing notion, it is too restrictive to be achieved in many cases. Kol and Naor only achieved strict Nash equilibrium in the two-party case assuming the existence of simultaneous broadcast channels [1]. In non-simultaneous channels, only an approximate equilibrium (i.e. ϵ-*Nash equilibrium*) was achieved. Recently, Fuchsbauer et al. [5] proposed *computational strict Nash equilibrium* and *computational Nash equilibrium that is stable with respect to trembles*. Efficient schemes achieving these equilibria were built in standard communication networks, but only computational security was guaranteed during the protocols. Moreover, equilibria concerning about sequential rationality, such as *everlasting Nash equilibrium* [10] and *sequential equilibrium* [20], were also achieved in the simultaneous channel.

Communication Models. Halpern and Teague [8] first assumed private channels, the simultaneous broadcast channel as well as an on-line dealer. Gordon and Katz [7] removed the on-line dealer by using a secure multi-party computation protocol among players, but the simultaneous broadcast channel was still necessary. Actually, many rational secret sharing protocols [1,14,15,20] rely on the assumption of simultaneous channels. Besides, some protocols [9,12,13] use even stronger assumptions such as secure envelopes and ballot boxes.

Coalition-Resilience. The main drawback of Kol and Naor's construction [10] is that it cannot resist the collusion attack of even two players. But coalition-resilience is an important requirement for t-out-of-n secret sharing. Previous protocols achieved good resilience in either simultaneous broadcast channels [1] or in the computational setting [5,11].

Unconditional/Computational Security. In the computational setting, equilibria with good properties (e.g. coalition-resilience [11]) could be achieved and more efficient protocols could run in standard communication networks [5], but it works in the condition that all players are computationally bounded. When higher security is required or players' computing power is unclear, a rational secret

---

[1] When using simultaneous broadcast channels, players must decide on what value (if any) to broadcast in a given round before observing the values broadcast by other players.

sharing protocol with unconditional security (i.e., in the information theoretic setting, such as [10]) is more reliable.

It can see that there is a tradeoff between the above aspects. In this work we focus on rational secret sharing that is coalition-resilient in the information theoretic setting and standard communication networks, at the cost of achieving ϵ-Nash equilibria only. But we will see that the "ϵ" is quite small and mostly acceptable.

1.1   Our Results and Main Ideas

We first design a 2-out-of-2 rational secret sharing protocol with unconditional security in standard communication networks. The main idea is distributing to player P_1 (resp. P_2) a list of length l_1 (resp. l_2) where $l_2 \leq l_1 \leq l_2 + 1$. Each cell of the lists contains a value, and all the values jointly determine the secret. The recovering phase consists of at most $l_1 + 1$ iterations. In each iteration, say, the j-th iteration, P_1 first broadcasts the value in his j-th cell, then P_2 does similarly. Since the two cases $l_1 = l_2 + 1$ and $l_1 = l_2$ both are possible, P_1 and P_2 cannot know which case really happens before the protocol ends. Therefore each player still has an incentive to broadcast the value even if it comes to his last cell. This protocol achieves an ϵ-Nash equilibrium, where ϵ is a negligible function in the size of the secret and is caused by the information-theoretic MACs used inside.

Then we build a t-out-of-n rational secret sharing protocol that is $(t-1)$-resilient. Since in the information theoretic setting with non-simultaneous channels, a coalition of $t-1$ players can easily get the secret earlier than other players and leave the protocol early, we try to insure that the innocent players (i.e. players who follow the protocol) get as much information as possible. The main idea is to divide each cell into two parts where two values are stored respectively, and the two values are both possible to be the secret if the secret appears in this cell. In each iteration, players first broadcast the first part of the current cell in some order, then the second part. The index indicating whether the current value is the secret or not is to be revealed only after the next value has been recovered. More precisely, suppose the secret appears in the j-th cell which contains s_j^0 and s_j^1 respectively in the two parts. Even the players in a $(t-1)$-coalition at most know that $\mathsf{Prob}[s = s_j^0] = q$ and $\mathsf{Prob}[s = s_j^1] = 1 - q$ for some constant q before seeing the index I_j^1 (i.e. $I_j^1 = 0$ if $s = s_j^1$, and $I_j^1 = 1$ if $s = s_j^0$). But I_j^1 is to be revealed only after recovering s_j^1 (by that time s_j^0 has already been recovered). Therefore after the coalition determines the secret s and leaves the protocol, the rest players at least know $s = s_j^0$ or s_j^1, which is also a pleasant result when the secret-domain is large. On the other hand, the extra gain of the deviating coalition is at most ϵ, where ϵ is exponentially small in the number of participants in the recovering process.

1.2   Related Work

Table 1 displays comparisons in some aspects between our protocols in this paper and those in some previous work.

Table 1.

	equilibrium	channel	coalition resilience	security
KN-[10]	strict Nash	simultaneous	1-resilient	unconditional
	ϵ-Nash	non-simultaneous	1-resilient	unconditional
ADGH-[1]	ϵ-Nash	simultaneous	k-resilient	computational/ unconditional
FKN-[5]	strict Nash	non-simultaneous	$(t-1)$-resilient	computational
This paper	ϵ-Nash	non-simultaneous	$(t-1)$-resilient	unconditional

Kol and Naor [10] provided constructions in both simultaneous and non-simultaneous channels in the information theoretic setting. Our constructions are similar to theirs in that shares are both in the form of lists with different length and the recovering is accomplished by revealing the lists cell by cell. But our 2-out-of-2 protocol is more efficient because shorter lists are involved and simpler cells are contained. Details will be found in the remarks after Theorem 1. General k-resilience was discussed in [1] where it achieved unconditional security for $k < \frac{n}{3}$ and computational security for $k < n$. But the protocols in [1] relied on simultaneous channels. Efficient protocols with optimal coalition resilience in standard communication networks were designed in [5]. Most importantly, it achieved equilibria with appealing properties, such as strict Nash, and stability with respect to trembles. But only computational security was guaranteed from the beginning of the recovering process.

2 Preliminaries

In this section it introduces notions about rational secret sharing and information-theoretic MACs, as well as concepts of the equilibrium to be achieved in this work.

2.1 Secret Sharing and Players' Utilities

In a t-out-of-n secret sharing scheme, a dealer (denoted as Dealer hereafter) holding a secret distributes shares among n players such that the following two conditions are satisfied:

1. **Recoverability.** Any group of t or more players puting their shares together can uniquely determine the secret.
2. **Secrecy.** Any group of fewer than t players cannot recover the secret.

It usually assumes that Dealer is the trusted third party and each player is either honest or malicious. In a game theoretic view, it is more realistic to view each player as a rational party who acts only in his interest. To model rationality, we define for each player P_i a real-valued utility function u_i such that everyone's interest is to maximize his utility. The commonly used assumptions for defining utilities in rational secret sharing are as follows [8]:

- Each player always prefers to learn the secret than to not learn it;
- Secondarily, each player prefers that the fewer of the other players who get it, the better.

In particular, we define four utility values for each player P_i :

(1) $u_i = a$ if P_i gets the secret while P_j does not for any $j \neq i$;
(2) $u_i = b$ if P_i gets the secret and so does P_j for some $j \neq i$;
(3) $u_i = c$ if P_i does not get the secret and neither does P_j for any $j \neq i$;
(4) $u_i = d$ if P_i does not get the secret while P_j does for some $j \neq i$.

From the common assumptions on utilities, it obviously holds that $a > b > c > d$. Let S denote the secret-domain and $|S|$ be the cardinality of S. Then by guessing the secret uniformly from S, a player at most gets the utility

$$U_{random} = \frac{1}{|S|}a + (1 - \frac{1}{|S|})c .$$

To make every player has the incentive to participate in a protocol for secret recovering, it requires $b > U_{random}$.

Concerning about coalitions, for simplicity we additionally assume that

- Once a player joins a coalition, he will never leave the coalition before the protocol ends;
- Players in the same coalition always share all information they jointly have.

Given an execution of a protocol, let $\mathcal{C}(i)$ denote the coalition that P_i joined in. Thus all players in $\mathcal{C}(i)$ have the same utility as P_i. As an extension, we similarly define the four utility values a, b, c, d for each player P_i as in (1)-(4) just replacing "$j \neq i$" with "$j \notin \mathcal{C}(i)$".

When no coalition is formed, namely, $\mathcal{C}(i) = \{i\}$ for any $i \in \{1, ..., n\}$, the problem is much easier [10]. In this work we deal with the most general coalitions in t-out-of n secret sharing, i.e. $1 \leq |\mathcal{C}(i)| \leq t - 1$.

2.2   Notions of Equilibria

In the recovering process of a secret sharing scheme, view the interaction between players as a game among the n players. Let $\sigma = (\sigma_1, ..., \sigma_n)$ denote a strategy profile of players, where σ_i is P_i's strategy for $1 \leq i \leq n$. Usually, we let σ_{-i} denote the strategy profile of all players except P_i and $\sigma_{\mathcal{C}}$ denote the strategy profile constricted to the coalition $\mathcal{C} \subseteq \{1, ..., n\}$. Given a strategy profile σ, it induces the utility $u_i(\sigma)$ for each player P_i. Referring to the definitions in [1,5,10,11], we give some notions of equilibria as follows:

Definition 1. *A strategy σ induces an ϵ-Nash equilibrium if for any player P_i and any strategy σ'_i of P_i, it holds that*

$$u_i(\sigma'_i, \sigma_{-i}) \leq u_i(\sigma_i, \sigma_{-i}) + \epsilon .$$

When $\epsilon = 0$ it is the well-known Nash equilibrium [16]. In some cases, a Nash equilibrium in the strict sense is hard to compute [3], while computing the ϵ-approximate Nash equilibrium is much easier [4]. Therefore, the ϵ-Nash equilibrium is also an appealing notion for a small ϵ.

Definition 2. *A strategy σ induces an k-resilient ϵ-Nash equilibrium if for any coalition $\mathcal{C}$ of at most k players (i.e. $|\mathcal{C}| \leq k$) and for any strategy profile $\sigma'_{\mathcal{C}}$ of the coalition $\mathcal{C}$, it holds that*

$$u_i(\sigma'_{\mathcal{C}}, \sigma_{\overline{\mathcal{C}}}) \leq u_i(\sigma_{\mathcal{C}}, \sigma_{\overline{\mathcal{C}}}) + \epsilon \quad \text{for any } i \in \mathcal{C} ,$$

where $\overline{\mathcal{C}}$ denotes the complement of $\mathcal{C}$.

When $k = 1$ it is the ϵ-Nash equilibrium just defined. In this work, we realize the resilience for $k = t - 1$ in a t-out-of-n secret sharing scheme. Obviously, this is the optimal coalition resilience in the t-out-of-n case.

2.3   Information-Theoretic MACs

We refer to [6] for the description of information theoretically secure message authentication codes (MACs). A message authentication code consists of three polynomial-time algorithms (Gen,Mac,Vrfy). The key-generation algorithm Gen takes as input the security parameter 1^m and outputs a key k. The message authentication algorithm Mac takes as input a key k and a message $M \in \{0,1\}^{\leq m}$, and outputs a tag t; we write this as $t = \mathsf{Mac}_k(M)$. The verification algorithm Vrfy takes as input a key k, a message M and a tag t, and outputs a bit b; i.e., $b = \mathsf{Vrfy}_k(M, t)$. We regard $b = 1$ as acceptance and $b = 0$ as rejection, and require that for all m, all k output by $\mathsf{Gen}(1^m)$, all $M \in \{0,1\}^{\leq m}$, it holds that $\mathsf{Vrfy}_k(M, \mathsf{Mac}_k(M)) = 1$.

Definition 3. *(Gen,Mac,Vrfy) is an information-theoretic MAC if for any $M \in \{0,1\}^{\leq m}$, $k = \mathsf{Gen}(1^m)$, $t = \mathsf{Mac}_k(M)$, and for any (computationally unbounded) adversary $\mathcal{A}$, the following probability is negligible in m:*

$$\mu(m) = \mathsf{Prob}\left[(M', t') \leftarrow \mathcal{A}(M, t) : \mathsf{Vrfy}_k(M', t') = 1 \bigwedge M' \neq M\right] .$$

For example, an information-theoretic MAC can be built as follows [17,19]: Let $\mathbb{F}$ be a finite field, the key is $(\alpha, \beta) \in \mathbb{F}^2$. For a message $M \in \mathbb{F}$, the tag is generated as $t = \beta - \alpha M \in \mathbb{F}$.

3   Rational Secret Sharing: The 2-Out-of-2 Case

In this section we give a 2-out-of-2 rational secret sharing protocol in standard communication networks (i.e. point-to-point and non-simultaneous channel) and with unconditional security. Denote the protocol by Π, we describe Π in terms of Dealer's protocol and players' protocol separately. Actually, Dealer's protocol corresponds to the distributing phase, and players' protocol corresponds to the recovering phase where only players are active.

Let $S = \{0,1\}^m$ be the secret-domain and $s \in S$ be the secret. For player P_1 and P_2, let a, b, c, d be the utility values as defined in Section 2.1. Suppose (Gen,Mac,Vrfy) is an information-theoretic MAC.

Dealer's Protocol

1. Choose an integer $l \in \mathbb{N}$ according to a geometric distribution with parameter p [2], where p is a constant to be determined later (in Theorem 1).

2. Determine the two integers l_1 and l_2 such that $l_1 + l_2 = l + 1$ and $l_2 \leq l_1 \leq l_2 + 1$.

3. Randomly select $a_1, ..., a_{l_1} \in S$ and $b_1, ..., b_{l_2} \in S$ such that

$$(\oplus_{i=1}^{l_1} a_i) \oplus (\oplus_{i=1}^{l_2} b_i) = s .$$

4. Generate secret keys $\alpha_1, ..., \alpha_{l_2+1}$ and $\beta_1, ..., \beta_{l_1}$ for the MAC by Gen(1^m). Construct two lists L_1 and L_2 of length l_1 and l_2 respectively, where for $1 \leq i \leq l_1$ (resp. $1 \leq i \leq l_2$) the i-th cell of L_1 (resp. L_2) contains a_i, $\mathsf{Mac}_{\alpha_i}(a_i)$ and β_{i-1} (resp. contains b_i, $\mathsf{Mac}_{\beta_i}(b_i)$ and α_i).

5. Send the list L_1 and the secret key β_{l_1} (resp. the list L_2 and the secret key α_{l_2+1}) to P_1 (resp. P_2).

Players' Protocol

It consists of l_1 or $l_1 + 1$ iterations. For $1 \leq j \leq l_1 + 1$, the j-th iteration goes along the following two rounds:

1. Denote by $(b'_{j-1}, t^{(b)}_{j-1})$ the message that P_1 received from P_2 in last round. Player P_1 first checks if it holds $\mathsf{Mac}_{\beta_{j-1}}(b'_{j-1}) = t^{(b)}_{j-1}$ (Note for $j = 1$ this check is not needed). If it holds, then P_1 sends $(a_j, \mathsf{Mac}_{\alpha_j}(a_j))$ to P_2; otherwise, P_1 quits and outputs $(\oplus_{i=1}^{j-1} a_i) \oplus (\oplus_{i=1}^{j-2} b'_i)$ as the secret.

2. Denote by $(a'_j, t^{(a)}_j)$ the message that P_2 received from P_1 in last round. Player P_2 checks if it holds $\mathsf{Mac}_{\alpha_j}(a'_j) = t^{(a)}_j$. If it holds, P_2 sends $(b_j, \mathsf{Mac}_{\beta_j}(b_j))$ to P_1; otherwise, P_2 quits and outputs $(\oplus_{i=1}^{j-1} a'_i) \oplus (\oplus_{i=1}^{j-1} b_i)$ as the secret.

If a player's list comes to the end, i.e., the j-th cell of his list is empty, then after verifying the message just received from the opposite, he sends the message "end" in the j-th iteration. After that both players stop running and set the secret to be the XOR of all the values revealed so far.

In brief, the recovering process is accomplished by letting the two players alternately reveal their lists cell by cell, while P_1 goes first. Figure 1 describes the recovering process when $l_1 = l_2$.

Then we give some intuition as to why the recovering process of Π (i.e. players' protocol) is an ϵ-Nash equilibrium for an appropriate choice of p, where $\epsilon = \epsilon(m)$ is a negligible function in length of the secret.

---

[2] Suppose in each coin toss, the Head appears with probability p. Then l is the number of independent tosses needed until the first Head turns up.

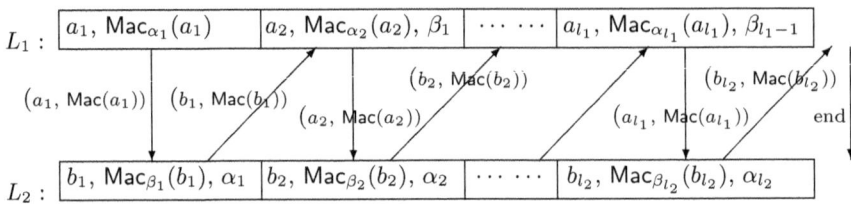

Fig. 1. The recovering process when $l_1 = l_2$

(a) P_1 has no incentive to deviate in the first iteration.
Since $l_1 + l_2 = l + 1 > 1$, it must have $l_2 \geq 1$. Namely, P_2 at least holds a value that contributes to determining s. P_1 cannot get this value if his message broadcast in the first iteration does not pass verification of the MAC. So by deviating, P_1 can get utility at most $\mu(m)a + (1 - \mu(m))U_{random}$, where $\mu(m)$ is the probability of successfully forging an MAC as defined in Definition 3 and $U_{random} = \frac{1}{|S|}a + (1 - \frac{1}{|S|})c$ is an upperbound of the utility that a player can get by guessing the secret uniformly from S. By requiring

$$\mu(m)a + (1 - \mu(m))U_{random} < b \qquad (1)$$

P_1 has no incentive to deviate in this iteration.

(b) For $2 \leq j \leq l_1$, P_1 has no incentive to deviate in the j-th iteration.
Similarly to the analysis in (a), P_1 has no incentive to deviate through iteration 2 to $l_1 - 1$. Achieving the l_1-th iteration, with probability p it holds that $l_2 = l_1 - 1$, i.e. P_2's list has run out. In this situation, P_1 can get utility at most a by deviation. But if $l_2 = l_1$ which happens with probability $1 - p$, P_1 get at most $\mu(m)a + (1 - \mu(m))U_{random}$. Therefore P_1 will not deviate by requiring

$$pa + (1 - p)(\mu(m)a + (1 - \mu(m))U_{random}) < b . \qquad (2)$$

Note that inequality (2) implies inequality (1).

(c) For $1 \leq j \leq l_2$, P_2 has no incentive to deviate in the j-th iteration.
The analysis is similar to that of (b).

(d) P_1 (resp. P_2) cannot increase his utility more than ϵ by deviating in the $(l_1 + 1)$-th (resp. the $(l_2 + 1)$-th) iteration.
In the $(l_1 + 1)$-th iteration and after verifying the MAC, P_1 already knows that $l_2 = l_1$ and he can determine $s = (\oplus_{i=1}^{l_1} a_i) \oplus (\oplus_{i=1}^{l_2} b_i')$. But P_2 still does not know whether P_1's list is longer than his or not. P_1 can deceive P_2 by continuing to send a fake value in the $(l_1 + 1)$-th iteration which passes verification of the MAC under the secret key $\alpha_{l_1+1} = \alpha_{l_2+1}$, and the success probability is at most $\mu(m)$ due to security of the MAC. Thus P_1 can get utility at most $\mu(m)a + (1 - \mu(m))b$. Therefore,

$$\epsilon(m) = \mu(m)a + (1 - \mu(m))b - b = \mu(m)(a - b) .$$

The analysis of P_2's $(l_2 + 1)$-th iteration is similar.

From the analysis (a)-(d), it immediately has the following theorem.

Theorem 1. *If the parameter p satisfies the inequality (2), then the protocol Π for 2-out-of-2 rational secret sharing induces an ϵ-Nash equilibrium with $\epsilon = \mu(m)(a - b)$, where $\mu(m)$ is the negligible probability of successfully forging an information-theoretic MAC.*

Remark 1. The 2-out-of-2 protocol in [10] used lists of length $l' - 1$ and $l' + d' - 1$ respectively, where l' and d' both were chosen according to a geometric distribution with parameter β. Our protocol Π uses lists of length l_1 and l_2 respectively where $l_1 + l_2 - 1$ is chosen according to a geometric distribution with parameter p. Since both β and p are determined by the utility values under the similar inequalities, we can simply regard $\beta = p$. Then the expected length of lists in [10] are $\frac{1}{p} - 1$ and $\frac{2}{p} - 1$, while our lists are both of length about $\frac{1}{2p}$. That is, we only need the list that is almost half as long as the shorter list in [10], which means the expected size of shares in our protocol is smaller.

Remark 2. Since in [10] the shorter list was just a prefix of the longer one and every value alone could possibly be the secret, a player can certainly determine the secret if he finds all his remain cells contain the same value. To fix this problem, it masked each value by a random number for each cell. Thus the cells in [10] contained both the masked value and share of the mask. But in our protocol, the secret is jointly determined by all values contained in the two lists, a player cannot determine the secret even if he sees all values in his list. Therefor no mask is needed in our protocol and our lists consist of simpler cells.

4 Rational Secret Sharing: The t-Out-of-n Case

We now construct a t-out-of-n rational secret sharing protocol in the information theoretic setting. Since it is in non-simultaneous channels and $(t - 1)$-resilience is required, the protocol is not a simple extension of the protocol Π constructed in Section 3. Denote the t-out-of-n protocol by Π'. We still describe Π' in terms of Dealer's protocol and players' protocol separately.

Dealer's Protocol

1. Choose integers l^* and d according to a geometric distribution with parameter p', where p' is a constant to be determined later (in Theorem 2).

2. Randomly select $\sigma \in \{0, 1\}$ such that $\mathsf{Prob}[\sigma = 0] = q$, where q is a constant to be determined later (in Theorem 2).

3. Construct a list of length $l = l^* + d$. For $1 \le j \le l$, the j-th cell contains:

- Main: $(s_j^0, s_j^1) \in S^2$, where S is the secret-domain. In particular, it requires $s_{l^*}^\sigma = s$ and the other values are randomly chosen.
- Index: $(I_j^0, I_j^1) \in \{0, 1\}^2$ where

$$I_j^0 = \begin{cases} 1, & \text{if } j - 1 = l^* \text{ and } \sigma = 1 \\ 0, & \text{otherwise} \end{cases}, I_j^1 = \begin{cases} 1, & \text{if } j = l^* \text{ and } \sigma = 0 \\ 0, & \text{otherwise} \end{cases}.$$

For consistence, fix $I_1^0 = 0$.

- Permutation: $\pi_j \in \Pi_n$ where Π_n denotes the set of all permutations on $\{1, ..., n\}$ [3].

4. Randomly select a permutation $\pi_0 \in \Pi_n$, and send π_0 to all players.

5. Suppose $i_0 \in \{1, ..., n\}$ appears first in the permutation π_{l^*-1}. Construct n lists, denoted by $L_1, ..., L_n$, where L_{i_0} is of length l^* and the other $n-1$ lists are of length l. For $1 \le i \le n$ and $1 \le j \le l$, the j-th cell of L_i contains: (Note the list L_{i_0} ends after the l^*-th cell)

- Share of main: s_{ji}^0 and s_{ji}^1, where s_{ji}^0 (resp. s_{ji}^1) is a (t, n)-share [4] of s_j^0 (resp. s_j^1).
- Share of index: I_{ji}^0 and I_{ji}^1, where I_{ji}^0 (resp. I_{ji}^1) is a (t, n)-share of I_j^0 (resp. I_j^1).
- Share of permutation: π_{ji} which is a (t, n)-share of π_j.
- Authentication information: The tags

$$\left\{ \mathsf{Mac}_{\alpha_{j,i,h}}(s_{ji}^0), \mathsf{Mac}_{\alpha'_{j,i,h}}(s_{ji}^1), \mathsf{Mac}_{\beta_{j,i,h}}(I_{ji}^0), \mathsf{Mac}_{\beta'_{j,i,h}}(I_{ji}^1), \mathsf{Mac}_{\gamma_{j,i,h}}(\pi_{ji}) \ \middle| \ \begin{array}{c} 1 \le h \le n, \\ h \ne i \end{array} \right\}$$

and the keys $\{\alpha_{j,h,i}, \alpha'_{j,h,i}, \beta_{j,h,i}, \beta'_{j,h,i}, \gamma_{j,h,i} \mid 1 \le h \le n, h \ne i\}$. We note that the key $\alpha_{j,h,i}$ is used to verify a tag of s_{jh}^0 and is stored in the j-th cell of L_i.

6. For $1 \le i \le n$, send the list L_i to player P_i.

Players' Protocol

Suppose k $(k \ge t)$ players are to jointly recover the secret. The recovering process consists of at most l iterations. In the j-th iteration for $1 \le j \le l$, if the protocol does not end, the players do the following:

1. Recover s_j^0. In the order determined by the permutation π_{j-1}, each player (say, P_i) sends to the other players $(s_{ji}^0, \mathsf{Mac}(s_{ji}^0))$. Hereafter we usually omit the key in the MAC because it is clearly determined by the message and the receiver. Players verify the MACs after receiving messages. If all messages pass the verification, then each player recovers s_j^0.

2. Recover I_j^0. Still in the order of π_{j-1} players send their shares along with MACs, and then recover I_j^0.

3. Recover s_j^1. Same as above.

4. Recover I_j^1. Same as above.

5. Recover π_j. Same as above.

In any of the above five steps, a player quits from the protocol at encountering any one of the following situations.

- His list has run out. Then he quits and sets the secret to be the last value he recovered. For example, if his list is of length l' and the protocol does not end after the first l' iterations, then he quits in the $(l'+1)$-th iteration and sets $s = s_{l'}^1$.

---

[3] Precisely, the permutation π_j denotes an order in which players send messages in the $(j+1)$-th iteration.

[4] The share can be generated by Shamir's (t, n)-threshold secret sharing scheme.

- Find some index $I_j^\delta = 1$. Then he quits and sets $s = s_{j-1+\delta}^{1-\delta}$.
- Find someone cheats in recovering s_j^0. Then he quits and sets $s = s_{j-1}^1$.
- Find someone cheats in recovering I_j^0. Then he quits and sets $s = s_{j-1}^1$ with probability $1 - q$ and $s = s_j^0$ with probability q.
- Find someone cheats in recovering s_j^1. Then he quits and sets $s = s_j^0$.
- Find someone cheats in recovering I_j^1. Then he quits and sets $s = s_j^0$ with probability q and $s = s_j^1$ with probability $1 - q$.
- Find someone cheats in recovering π_j. Then he quits and sets $s = s_j^1$.

Now we give some analysis to explain why the recovering process of Π' induces an ϵ-Nash equilibrium with $(t-1)$-resilience. For simplicity, we neglect the negligible part of ϵ caused by successfully forging the MAC. As a warm-up, we first show that any single player has no incentive to deviate from the protocol. For a single player P_i, there are two cases:

(a) P_i holds a list of length l.

It is important to note that P_i cannot know he is holding the long list until the protocol ends or it comes to his last cell (i.e. the l-th cell). Therefore, for $1 \le j < l$, P_i guesses $l^* = j$ and deviates in the j-th iteration, then he can get utility at most $p'a + (1 - p')U_{random}$. P_i has no incentive to deviate if it holds

$$p'a + (1 - p')U_{random} < b . \tag{3}$$

When it comes to the last cell (i.e. the l-th cell) and P_i is not the first one to send messages according to π_{l-1}, then P_i knows that $l^* = l - 1$ and $s = s_{l-1}^1$. Actually, every other player can also conclude $s = s_{l-1}^1$ no matter what P_i does in the l-th iteration. Thus P_i has no incentive to deviate.

(b) P_i holds a list of length l^*.

Similarly, it can see that P_i has no incentive to deviate in the j-th iteration for $1 \le j \le l^* - 1$, if the inequality (3) holds. When it comes to the l^*-th iteration P_i knows he is holding the short list because he is the first to send messages in that iteration. Since P_i is the first one to talk in the l^*-th iteration, when P_i determines for sure what the secret is, so do the other players. Thus P_i has no incentive to deviate.

Then we give some intuition as to why the recovering process of Π' is $(t - 1)$-resilient. For any coalition $\mathcal{C}$ with $1 < |\mathcal{C}| \le t - 1$, there are two cases:

(c) The short list holder is contained in $\mathcal{C}$.

Since the lists are of different length, players in $\mathcal{C}$ can easily determine l^* in advance. Thus ignoring the negligible probability of forging the MAC successfully, the best option for players in $\mathcal{C}$ is to get as much information about $\{s_{l^*}^0, s_{l^*}^1, I_{l^*}^1\}$ as possible and secondarily, to make players outside $\mathcal{C}$ know as little as possible. It is easy to see that if the inequality (3) holds $\mathcal{C}$ has no incentive to deviate before the l^*-th iteration. In the l^*-th iteration,

- If $\mathcal{C}$ deviates in recovering $s_{l^*}^0$, the best result for $\mathcal{C}$ is that they get $s_{l^*}^0$ while no one else does. Thus $\mathcal{C}$ guesses $s = s_{l^*}^0$ and the other players set

$s = s^1_{l_*-1}$. Since Dealer set $s = s^0_{l_*}$ with probability q, $\mathcal{C}$ guesses wrong with probability $1 - q$. Therefore by deviating players in $\mathcal{C}$ get utility at most $qa + (1 - q)c$. Requiring

$$qa + (1 - q)c < b ,\tag{4}$$

then $\mathcal{C}$ has no incentive to deviate.

- When recovering $I^0_{l_*}$, since $I^0_{l_*}$ only indicates whether $s^1_{l_*-1}$ is the secret or not which $\mathcal{C}$ has already known. Besides, at this time players outside $\mathcal{C}$ already get $s^0_{l_*}$ which means they also have opportunity to get the right secret even if $\mathcal{C}$ deviates. Based on the inequality (4), $\mathcal{C}$ has no incentive to deviate.
- If $\mathcal{C}$ deviates in recovering $s^1_{l_*}$, then players in $\mathcal{C}$ set $s = s^0_{l_*}$ with probability q and set $s = s^1_{l_*}$ with probability $1 - q$. By the protocol Π', after detecting someone cheats in recovering $s^1_{l_*}$, each of the players outside $\mathcal{C}$ sets $s = s^0_{l_*}$ and quits. If Dealer set $\sigma = 0$ (which happens with probability q), then with probability q all players get the right secret and with probability $1 - q$ players in $\mathcal{C}$ guess wrong while others guess right. If Dealer set $\sigma = 1$ (which happens with probability $1 - q$), then players outside $\mathcal{C}$ get the wrong secret, while $\mathcal{C}$ guesses right with probability $1 - q$.

  Thus deviation in recovering $s^1_{l_*}$ makes players in $\mathcal{C}$ get utility at most

  $$q(qb + (1 - q)d) + (1 - q)(qc + (1 - q)a) = (1 - q)^2 a + q^2 b + q(1 - q)(c + d) .$$

  By requiring

  $$(1 - q)^2 a + q^2 b + q(1 - q)(c + d) < b ,\tag{5}$$

  $\mathcal{C}$ has no incentive to deviate.
- If $\mathcal{C}$ deviates in recovering $I^1_{l_*}$, we will show that players in $\mathcal{C}$ can increase the utility by at most $\epsilon = O(\lambda^k)$ where k is the number of participants in the recovering process and $\lambda < 1$ is a constant determined by q. After deviation players in $\mathcal{C}$ can determine the secret, while each player outside $\mathcal{C}$ sets $s = s^0_{l_*}$ with probability q and $s = s^1_{l_*}$ with probability $1 - q$. Suppose $|\mathcal{C}| = c$, then there are $k - c$ players outside $\mathcal{C}$. If Dealer set $\sigma = 0$, then the probability that none of the $k - c$ players outputs the right secret is $(1 - q)^{k-c}$, while if $\sigma = 1$, this probability is q^{k-c}. Thus by deviation players in $\mathcal{C}$ get utility at most

  $$U_D = q((1 - q)^{k-c}a + (1 - (1 - q)^{k-c})b) + (1 - q)(q^{k-c}a + (1 - q^{k-c})b)$$
  $$= (q(1 - q)^{k-c} + (1 - q)q^{k-c})a + (1 - q(1 - q)^{k-c} - (1 - q)q^{k-c})b .$$

  Therefore $\epsilon = U_D - b = (q(1 - q)^{k-c} + (1 - q)q^{k-c})(a - b)$. Denote $\lambda = max\{q, 1 - q\}$, then $\epsilon \leq \lambda^{k-c}(a - b) = O(\lambda^k)$.
- Neglecting the negligible probability of successfully forging a MAC, $\mathcal{C}$ has no incentive to deviate after recovering $I^1_{l_*}$, because $\mathcal{C}$ has already known the secret and players outside $\mathcal{C}$ can also output the right secret.

(d) The short list holder is not contained in $\mathcal{C}$.
    Then the coalition $\mathcal{C}$ can only know $l^* \leq l - 1$ in advance. By the analysis similar to that of (a), $\mathcal{C}$ has no incentive to deviate in the j-th iteration for $1 \leq j < l - 1$. In the $(l-1)$-th iteration, similar to the analysis of the fourth situation in (c), $\mathcal{C}$ can only increase the utility by at most $\lambda^{k-c}(a-b)$ if they deviates from the protocol.

From the analysis (a)-(d) above, we can get the following theorem.

Theorem 2. *Let the parameters p', q and the utility values satisfy the inequalities (3)-(5), then the protocol Π' for t-out-of-n rational secret sharing induces a $(t-1)$-resilient ϵ-Nash equilibrium with $\epsilon < \lambda^{k-t+1}(a-b)$, where $\lambda = max\{q, 1-q\}$ and k is the number of participants in the recovering process.*

Remark 3. Note that the inequality (4) and (5) may not simultaneously hold for some values of a, b, c, d. This can be solved by making some additional assumptions on the utility values. For example, assume that $a - b < b - c$, then the inequality (4) and (5) are satisfied for $\frac{a-b}{a-c} < q < \frac{b-c}{a-c}$. Actually, the assumption $a - b < b - c$ is implied from the natural requirement of $U_{random} < b$ for $|S| = 2$, i.e. each player still has an incentive to participate in the protocol for recovering even if the secret is just one bit.

Remark 4. It can see that the ϵ is exponentially small in the number of participants. When a large number of players participate in the recovering process or the utility values a and b are very close, a coalition of $(t-1)$ players cannot gain much by deviation form Π'. Actually, as pointed out in [10] a gain by a $(t-1)$-coalition is inevitable in the information theoretic setting. We leave it as an open problem to determine the lower bound of ϵ at achieving $(t-1)$-resilience in standard communication networks.

On the other side, although some players quit from the protocol after they get the secret, leaving the other players (who honestly follow the protocol so far, thus we call them "innocent players") cannot determine what the secret is, the innocent player can at least be sure that the secret must be one of the two values he has already recovered. Thus in innocent players' view the Shannon entropy of the secret reduces to less than 1. When $|S|$ is very large, every rational player has great incentive to participate in the protocol Π' even if he might encounter a coalition of $t-1$ players.

5   Conclusions

In the information theoretic setting of rational secret sharing, only approximate Nash equilibrium can be achieved in standard communication networks. We realize ϵ-Nash both for the 2-out-of-2 case and the t-out-of-n case. The 2-out-of-2 protocol is more efficient than previous ones and the ϵ is a negligible function in the size of the secret. This negligible function is due to the information-theoretic

MAC used inside. The t-out-of-n protocol is $(t-1)$-resilient and the ϵ is exponentially small in the number of participants. We leave it as an open problem to determine the lower bound of ϵ in both cases.

References

1. Abraham, I., Dolev, D., Gonen, R., Halpern, J.: Distributed computing meets game theory: robust mechanisms for rational secret sharing and multiparty computation. In: 25th ACM Symposium Annual on Principles of Distributed Computing, pp. 53–62. ACM Press, New York (2006)
2. Blakley, G.R.: Safeguarding cryptographic keys. In: Proceedings of the National Computer Conference, American Federation of Information Processing Societies Proceedings, vol. 48, pp. 313–317 (1979)
3. Daskalakis, C., Goldberg, P., Papadimitriou, C.: The complexity of computing a Nash equilibrium. In: Proc. STOC 2006, pp. 71–78. ACM Press, New York (2006)
4. Daskalakis, C., Mehta, A., Papadimitriou, C.: A note on approximate Nash equilibria. In: International Workshop on Internet and Network Economics, pp. 297–306 (2006)
5. Fuchsbauer, G., Katz, J., Naccache, D.: Efficient Rational Secret Sharing in Standard Communication Networks. In: Micciancio, D. (ed.) TCC 2010. LNCS, vol. 5978, pp. 419–436. Springer, Heidelberg (2010)
6. Gordon, S.D., Hazay, C., Katz, J., Lindell, Y.: Complete fairness in secure two-party computation. In: Proc. STOC 2008, pp. 413–422. ACM Press, New York (2008)
7. Gordon, S.D., Katz, J.: Rational secret sharing, revisited. In: De Prisco, R., Yung, M. (eds.) SCN 2006. LNCS, vol. 4116, pp. 229–241. Springer, Heidelberg (2006)
8. Halpern, J., Teague, V.: Rational secret sharing and multiparty computation. In: Proc. of 36th STOC, pp. 623–632. ACM Press, New York (2004)
9. Izmalkov, S., Micali, S., Lepinski, M.: Rational secure computation and ideal mechanism design. In: 46th Annual Symposium on Foundations of Computer Science (FOCS), pp. 585–595. IEEE, Los Alamitos (2005)
10. Kol, G., Naor, M.: Games for exchanging information. In: STOC 2008, pp. 423–432. ACM, New York (2008)
11. Kol, G., Naor, M.: Cryptography and game theory: Designing protocols for exchanging information. In: Canetti, R. (ed.) TCC 2008. LNCS, vol. 4948, pp. 320–339. Springer, Heidelberg (2008)
12. Lepinski, M., Micali, S., Peikert, C., Shelat, A.: Completely fair SFE and coalitionsafe cheap talk. In: 23rd ACM Symposium Annual on Principles of Distributed Computing, pp. 1–10. ACM Press, New York (2004)
13. Lepinski, M., Micali, S., Shelat, A.: Collusion-free protocols. In: 37th Annual ACM Symposium on Theory of Computing (STOC), pp. 543–552. ACM Press, New York (2005)
14. Lysyanskaya, A., Triandopoulos, N.: Rationality and adversarial behavior in multiparty computation. In: Dwork, C. (ed.) CRYPTO 2006. LNCS, vol. 4117, pp. 180–197. Springer, Heidelberg (2006)
15. Maleka, S., Shareef, A., Rangan, C.P.: The deterministic protocol for rational secret sharing. In: IEEE International Symposium on Parallel and Distributed Processing, IPDPS 2008, pp. 1–7 (2008)

16. Osborne, M., Rubinstein, A.: A Course in Game Theory. MIT Press, Cambridge (2004)
17. Rabin, T., Ben-Or, M.: Verifiable Secret Sharing and Multiparty Protocols with Honest Majority. In: Proceedings of the 21th Annual ACM Symposium on Theory of Computing (STOC), pp. 73–85 (1989)
18. Shamir, A.: How to share a secret. Communications of the ACM 22(11), 612–613 (1979)
19. Wegman, M., Carter, L.: New hash functions and their use in authentication and set equality. Journal of Computer and System Sciences 22, 265–279 (1981)
20. Zhang, Z.: Rational secret sharing as extensive games,
    http://eprint.iacr.org/2010/184

Oblivious Transfer with Complex Attribute-Based Access Control*

Lingling Xu and Fangguo Zhang

School of Information Science and Technology
Sun Yat-sen University, Guangzhou 510275, China
xulingling810710@163.com, isszhfg@mail.sysu.edu.cn

Abstract. In this paper, we present oblivious transfer with complex attribute-based access control policies. The protocol allows a database server to directly enforce "*and*" and "*or*" access control policies ($c_{1_1} \wedge c_{1_2} \wedge \ldots c_{1_{n_1}}) \vee (c_{2_1} \wedge c_{2_2} \wedge \ldots c_{2_{n_2}}) \vee \ldots \vee (c_{t_1} \wedge c_{t_2} \wedge \ldots c_{t_{n_t}})$ on each message in a database without duplication of the message as in Camenisch *et al.*'s AC-OT. To realize this protocol, we present the blind attribute-based encryption (ABE) scheme as a building block. Combining the blind ABE with a credential signature scheme, a generic construction for the oblivious transfer with complicated access control is presented. We also give a concrete scheme for the construction in which the policy is provided by an access tree which is represented by a formula involving "*and*($\wedge$)" and "*or*($\vee$)" boolean operators.

Keywords: Oblivious Transfer, Access Control, Attribute-Based Encryption.

1 Introduction

With the growth of the Internet, the need to provide privacy to users accessing sensitive information is increasing. The traditional approach for protecting user privacy is to implement an anonymous protocol [13]. Another approach is to implement oblivious transfer (OT) protocol presented by Rabin [29] in which user privacy is protected in such a way that a user makes requests to a server, and at the end obtains the messages of his choices without the server learning anything about the choices. That is, OT addresses the problem of hiding the data choices of users rather than user anonymity. So far, plenty of OT protocols have been proposed to provide user privacy, such as [26,16,20] et.al. As we know, in traditional OT, a user can arbitrarily retrieve data of his choices from a server without any restrictions, while this rules out many practical applications, such as medical or financial data access, pay-per-view TV. In these cases, on one hand, the database server wants to enforce access control policies on the database such that each data can be only available, on request, to the users who meet

---

* This work was supported by the National Natural Science Foundation of China (Nos. 60773202, 61070168) and 973 Program (No. 2006CB303104).

K.-H. Rhee and D. Nyang (Eds.): ICISC 2010, LNCS 6829, pp. 370–395, 2011.

the corresponding policies. On the other hand, the users do not want to reveal what part of the data they are retrieving or more personal information than is absolutely necessary.

Specially, consider the following motivating example: a paper database server wants only students in College A or teachers in College B to download master's theses, and wants only teachers in College C to download the PhD theses, namely, the access control condition for master's theses is $(Student \wedge Col.A) \vee (Tea \wedge Col.B)$, and the condition for PhD theses is $(Teacher \wedge Col.C)$. Meanwhile, the users want to protect their identities and choices during the access to the database, that is, all electronic transactions performed between the server and the users will not reveal more personal information than is absolutely necessary. To achieve this functionality, some papers [17,7,1,22,12,10] have given their solutions.

In 1999, Crescenzo *et al.* [17] proposed conditional oblivious transfer. Later, in 2004, Blake *et al.* [7] presented strong conditional oblivious transfer. In 2001, priced OT [1] was proposed in which each user can buy goods if and only if the price of the goods is less than the user's balance. However, neither of the above schemes provided the user anonymity, and they can only achieve simple access control such as " $=$ " or " $\geq$ ".

In [22], Herranz proposed a primitive called restricted adaptive oblivious transfer in which the policies defines which subsets of entries of the database can be available, on request, to the different users. However, the second scheme is not efficient due to the amount of computational and communication effort which is required in each execution of the protocol.

To the best of our knowledge, so far there have been two papers that consider oblivious transfer with access control. One is Coull *et al.*'s protocol [12] using stateful anonymous credential which permits a database server to restrict which messages each user may access, without learning anything about users' identities or message choices. However, since the user credential must be re-issued when each user requests a message each time, the protocol is not very efficient. Another one is Camenisch *et al.*'s OT protocol with access control (AC-OT) [10] for anonymous access to a database where the different records have different control permissions. Camenisch *et al.* show that the AC-OT can be implemented using Coull *et al.*'s protocol and it is more efficient. However, in AC-OT, for each message which is associated with a category set, if and only if one user has all these categories in the set, they can obtain the message by queries. Namely, for each message, AC-OT just directly achieves "*and*" access control policy. The "*or*" policy can be realized by duplicating the messages in the database with a second set of categories. For example, for a database in which the complicated policies for each message such as $(c_{1_1} \wedge c_{1_2} \wedge \ldots c_{1_{n_1}}) \vee (c_{2_1} \wedge c_{2_2} \wedge \ldots c_{2_{n_2}}) \vee \ldots \vee (c_{t_1} \wedge c_{t_2} \wedge \ldots c_{t_{n_t}})$ is requested, then using the method to realize "*or*" policy in AC-OT, the message must be duplicated for t times, each with a policy $(c_{i_1} \wedge c_{i_2} \wedge \ldots c_{i_{n_i}})$. Moreover, when the server initializes the database, he must encrypt the message for t times under different "*and*" policies, and the initialized database will also increase greatly.

Our Contributions. In this paper, we present an oblivious transfer with complex attribute-based access control permissions. The protocol can directly achieve "$and(\wedge)$" and "$or(\vee)$" policies on each message without the duplication of the message. More concretely, for each message, we can directly enforce such type of access control permissions as $(c_{1_1} \wedge c_{1_2} \wedge \ldots c_{1_{n_1}}) \vee (c_{2_1} \wedge c_{2_2} \wedge \ldots c_{2_{n_2}}) \vee \ldots \vee (c_{t_1} \wedge c_{t_2} \wedge \ldots c_{t_{n_t}})$ on it, where each c_{i_j} is an attribute. That is, the message can only be available, on request, to the users who possess at least one attribute set $(c_{i_1}, c_{i_2}, \ldots, c_{i_{n_i}})$, $i \in [1, t]$.

To realize the functionality of the protocol, we present a new primitive called blind (ciphertext policy) attribute-based encryption(CP-ABE) as a building block. Combining the blind ABE with an anonymous credential scheme, a generic construction for the protocol is proposed. In this construction, a server, n users and a credential issuer are included. Assume that Ω is an attribute universe where $|\Omega| = l$. Each user first authenticates to obtain the credentials for his entitled attributes from the issuer. The server initializes the database $(M_1, M_2, \ldots, M_N)$ by encrypting the messages under the associated access control policies so that the users must obtain the private keys for the attributes to decrypt out the messages. In the transfer phase, to obtain the allowed messages according to the policies, each user first makes requests to the server for the private keys of his entitled attributes, and simultaneously executes a proof of knowledge to convince the server that he possesses a valid credential generated by the issuer for the requested attributes. From the requests, even if the server colludes with the credential issuer, they cannot learn anything about the attributes or identity of the user. Then after the user obtains the private keys, he can arbitrarily decrypt out the messages whose associated policies are satisfied by the user's attributes.

If we let k be the number of messages a user can access in a database according to the access control policies, then in our construction, the user needs to interact with the server for just one time to obtain the private keys to decrypt all the k allowed messages. Therefore, the communication cost of our construction is $O(N + l)$.

To present a concrete oblivious transfer protocol with complex access control policies, we first construct a new blind ABE scheme which achieves "and" and "or" policies. The access control structure is provided by an access tree in which leaves are attributes and inner nodes are "$\wedge$" and "$\vee$" boolean operators.

Organization. The rest of this paper is organized as follows. In section 2, we introduce some preliminaries. In section 3, we present the functionality and security definition for the oblivious transfer with attribute-based access control. In section 4, a generic construction for oblivious transfer with attribute-based access control is proposed based on blind attribute-based encryption and credential signature scheme. In section 5, a concrete scheme for the construction is given. We give some analysis and extensions in section 6. Finally, we conclude in section 7.

2   Preliminaries

Notation. We denote by κ the security parameter and by PPT the property of an algorithm of running in probabilistic polynomial-time. A function $\mathsf{negl}(\cdot)$ is negligible in κ if for every polynomial $p(\cdot)$ there exists a value N such that for all $n > N$ it holds that $\mathsf{negl}(\cdot) < 1/p(n)$.

2.1   Bilinear Maps

Let G, G_T be two (multiplicative) cyclic groups such that $|G| = |G_T| = p$, where p is a large prime. Let g be a generator of G, and e be an admissible bilinear map: $G \times G \to G_T$, satisfying that (1) for all $a, b \in Z_p$ it holds that $e(g^a, g^b) = e(g, g)^{ab}$; (2) $e(g, g) \neq 1$; and (3) it is efficiently computable.

2.2   Complexity Assumptions

Definition 1. *(Computational Diffie-Hellman(CDH) Assumption.) Suppose a challenger chooses $a, b, c, z \in Z_p$ at random. The DBDH assumption is that no polynomial-time adversary is to be able to distinguish the tuple $(g^a, g^b, g^c, e(g, g)^{abc})$ from $(g^a, g^b, g^c, e(g, g)^z))$ with more than a negligible advantage.*

Definition 2. *(Decision Bilinear Diffie-Hellman(DBDH) Assumption[6].) Suppose a challenger chooses $a, b, c, z \in Z_p$ at random. The DBDH assumption is that no polynomial-time adversary is to be able to distinguish the tuple $(g^a, g^b, g^c, e(g, g)^{abc})$ from $(g^a, g^b, g^c, e(g, g)^z))$ with more than a negligible advantage.*

Definition 3. *(q-Strong Diffie-Hellman(q-SDH) Assumption[4]). Suppose a challenger chooses $x \in Z_p$ at random. The q-SDH assumption is that no polynomial-time adversary is to be able to output a pair $(c, g^{1/(x+c)})$ where $c \in Z_p$ from the tuple $(g, g^x, \dots, g^{x^q})$ with more than a negligible advantage.*

2.3   Zero-Knowledge Proof

Throughout the paper, we use known techniques for proving statements about discrete logarithms such as (1) proof of knowledge of a discrete logarithm modulo a prime [30,16], (2) proof of knowledge of a pairing preimage [16], (3) proof of the disjunction or conjunction of any ones of the previous [11,16]. According to Cramer *et al.* [11], this class of zero-knowledge (ZK) proof systems (1)(2) can be combined into ZK proof systems (3) for any disjunctive and conjunctive formula. Furthermore, using the Fiat-Sahmir heuristic, this class of ZK proof systems can be converted to non-interactive ZKs at no extra cost. We will use the notation, for example $PoK\{(x, r) : y = g^x h^r \wedge y' = g'^x\}$ [18] to denote a ZK proof of knowledge of integers x and r such that both $y = g^x h^r$ and $y' = g'^x$ hold. All values not enclosed in ()'s are assumed to be known to the verifier.

2.4   Credential Signature Scheme

A credential signature scheme [14,13] includes three participants: an issuer, a user and a verifier. It consists of the following four algorithms:

- ISetup(1^κ) The credential issuer sets up the key pair (pk_I, sk_I) for issuing credentials for users, and publishes pk_I.
- IssueCred(sk_I, m) For a message m, the issuer generates the corresponding credential $cred_m$ using his secret key sk_I.
- VerifyCred($pk_I, m, cred_m$) The user verifies the credential $(m, cred_m)$ using pk_I.
- ProveCred($m, cred_m$) The user executes a proof of knowledge with the verifier

$$PoK\{(m, cred_m) : cred_m \in \mathsf{IssueCred}(sk_I, m)\}.$$

$PoK\{(m, cred_m) : cred_m \in \mathsf{IssueCred}(sk_I, m)\}$ denotes a zero-knowledge proof of knowledge of a valid credential $cred_m$ for m.

The credential signature scheme we will use later is the scheme [2] as follows. The scheme is unforgeable under adaptively chosen message attack based on q-SDH assumption, where q is the number of signature queries, and that the associated proof of knowledge is complete, sound and perfect honest-verifier zero-knowledge.

- ISetup(1^k) The credential issuer sets up the parameters as follows: G, G_T are two cyclic groups of order p, and $e : G \times G \to G_T$. He chooses at random $x_I \in Z_p$ and a number of random bases $g_0, y_1, ..., y_l, y_{l+1} \in G$, and computes $y_I \leftarrow g_0^{x_I}$. He lets $sk_I \leftarrow x_I$, and $pk_I \leftarrow (g_0, y_1, ..., y_l, y_{l+1}, y_I)$.
- IssueCred($sk_I, (m_1, \ldots, m_l)$) For $m_1, ..., m_l \in Z_p$, the issuer chooses random $r, s \in Z_p$, and computes $\sigma \leftarrow (g_0 y_1^{m_1} ... y_l^{m_l} y_{l+1}^r)^{1/(x_I+s)}$. The tuple (σ, r, s) is the credential for the tuple $m_1, ..., m_l$.
- VerifyCred($pk_I, (m_1, \ldots, m_l), (\sigma, r, s)$) A user verifies the correctness of a credential by checking whether the equality $e(\sigma, g_0^s y_I) = e(g_0 y_1^{m_1} ... y_l^{m_l} y_{l+1}^r, g_0)$ holds.
- ProveCred($(m_1, \ldots, m_l), (\sigma, r, s)$) A user proves that he possesses such a credential (σ, r, s) for $m_1, \ldots, m_l$ by choosing random $A, B \in G$, $v, w \in Z_p$, computing $\tilde{A} \leftarrow \sigma A^v, \tilde{B} \leftarrow B^v A^w$, and executing a proof of knowledge as follows:
  $PoK\{(\alpha, \beta, v, w, m_1, ..., m_l, s, r) : \tilde{B} = B^v A^w \bigwedge 1 = \tilde{B}^{-s} B^\alpha A^\beta \bigwedge$
  $\frac{e(\tilde{A}, y_I)}{e(g_0, g_0)} = e(\tilde{A}, g_0)^{-s} e(A, y_I)^v e(A, g_0)^\alpha e(y_{l+1}, g_0)^r \prod_{i=1}^l e(y_i, g_0)^{m_i}\}$,

where $\alpha = sv$ and $\beta = sw$.

3   OT with Attribute-Based Access Control

3.1   Definition

For the oblivious transfer with complex attribute-based access control(CAC-OT), we want to implement such functionality: assume that an attribute universe

Ω of size l is $\Omega = \{a_1, a_2, \ldots, a_l\}$, where each element of Ω is a descriptive attribute. Each user $\mathcal{U}_i$ is entitled to a subset of attributes $\omega_i \subseteq \Omega$. A server maintains a database $DB = \{m_1, \ldots, m_N\}$, and associates each message m_i with an attribute-based access control structure $\tau_i \subseteq \Omega$. Each structure τ_i specifies which combination of attributes can obtain the corresponding message m_i. The server requests that only the users whose attributes satisfy τ_i are able to have access to the message m_i. That is, for every $j \in [1, N]$ and every user $\mathcal{U}_i$, only if $\omega_i \models \tau_j$, the message m_j can be available, on request, to $\mathcal{U}_i$. Meanwhile, the server should not get to know any information about each user's identity, attributes or message choices.

In the following, we will give a solution to fulfill the functionality of CAC-OT. It includes these participants: n users $\mathcal{U}_1, \mathcal{U}_2, \ldots, \mathcal{U}_n$, a server $\mathcal{S}$ and an issuer $\mathcal{I}$. The protocol works as follows.

- IssueSetup(1^κ)
  The issuer $\mathcal{I}$ generates his key pair (pk_I, sk_I) for generating credentials for users, and publishes pk_I as the system-wide parameter.
- DB-Initialization($\Omega, pk_I, m_1, \ldots, m_N, \tau_1, \ldots, \tau_N$)
  For a database containing messages $m_1, \ldots, m_N$, the algorithm outputs a key pair (pk_{DB}, sk_{DB}) and encrypted messages $C_1, \ldots, C_N$ under the access control policies $\tau_1, \ldots, \tau_N$. The server keeps the key sk_{DB} secret, and publishes $(C_1, \ldots, C_N, pk_{DB})$ to make it available to all users.
- ObtainCred($\Omega, sk_I; \omega_i$)
  Each user interacts with the issuer to obtain the credentials for his attributes that he is entitled to access. Assuming each user $\mathcal{U}_i$'s attribute set is $\omega_i \subseteq \Omega$, at the end of the algorithm, $\mathcal{U}_i$ will obtain the credentials $cred_{\omega_i}$ for all attributes of ω_i from the issuer.
- Transfer($sk_{DB}; \omega_i, cred_{\omega_i}$)
  It is an interactive algorithm between the server $\mathcal{S}$ and users. The input for $\mathcal{S}$ is his private key sk_{DB}. The inputs for each user $\mathcal{U}_i$ includes his entitled attributes ω_i and the corresponding credentials. At the end of the protocol, $\mathcal{U}_i$ can decrypt out a message subset $\phi_i \subseteq \{m_1, \ldots, m_N\}$, where the policies for the messages in ϕ_i are all satisfied by ω_i.

During the protocol, when users request credentials from the issuer, the issuer will know users' identities. However, when users interact with the server, the server will know neither their identities nor their attributes. That is, the communication links between the users and the issuer are authenticated and the links between the users and the server are anonymous.

3.2   Security

We first discuss the security properties the CAC-OT should satisfy:

- User privacy: After each user executes the protocol with the server, the server does not learn the user's identity or attributes, nor does it learn which messages the user obtains. Even if the server colludes with the issuer, they cannot tell the identity or attributes of the user.

- Server security: After the protocol, any collection of possibly cheating users cannot obtain the messages which none of them would have been able to obtain individually according to the access control policies.

In the following, we will formally define the security of the oblivious transfer with access control policies. We first define an ideal functionality $\mathcal{F}_{CAC-OT}$ for the protocol.

Functionality $\mathcal{F}_{CAC-OT}$

Parameterized with (N, n, l), and running with a server $\mathcal{S}$, n users $\{\mathcal{U}_1, \ldots, \mathcal{U}_n\}$ and an issuer $\mathcal{I}$, $\mathcal{F}_{CAC-OT}$ works as follows:
$\mathcal{F}_{CAC-OT}$ maintains an initially empty set Att_i for each user $\mathcal{U}_i$.

- On input a message $(init, \Omega, m_1, \tau_1, \ldots, m_N, \tau_N)$ from $\mathcal{S}$, it stores $(\Omega, m_1, \tau_1, \ldots, m_N, \tau_N)$ and sends $(init, \tau_1, \ldots, \tau_N)$ to all users.
- On input a message $(issue, id_i, a)$ from $\mathcal{U}_i$, it sends $(issue, id_i, a)$ to $\mathcal{I}$, and receives a bit b from $\mathcal{I}$. If $b = 1$ then $\mathcal{I}$ adds a to Att_i and sends b to $\mathcal{U}_i$, otherwise it simply sends b to $\mathcal{U}_i$.
- On input a message $(transfer, id_i, \omega_i)$ from $\mathcal{U}_i$, where $\omega_i \subseteq \Omega$, it sends $transfer$ to $\mathcal{S}$ and receives a bit b'. For all $j \in \{1, \ldots, N\}$, if $(b' = 1) \wedge (\omega_i \subseteq Att_i) \wedge (\omega_i \models \tau_j)$, then it sends m_j to $\mathcal{U}_i$.

In the real experiment, a server, n users and an issuer works as defined in section 3.1, and in the ideal experiment, $\mathcal{F}_{CAC-OT}$ works as defined above, with a server, n users and an issuer. Assume that the outputs of the real and ideal experiment are $Real_{\mathcal{A}}(\kappa)$ and $Ideal_{\mathcal{A}'}(\kappa)$ respectively, where κ is a security parameter. In terms of the experiments, the formal security is defined as follows:

Server-Security: We say that CAC-OT is server-secure if for every PPT real-world adversary $\mathcal{A}$ who corrupts a collection of users $\{\hat{\mathcal{U}}_1, \ldots, \hat{\mathcal{U}}_t\}$, there exists a PPT ideal-world adversary $\mathcal{A}'$ who corrupts the the same participants, such that for κ (which is a security parameter), and every PPT distinguisher $\mathcal{D}$:

$$| Pr[Real_{\mathcal{A}}(\kappa) = 1] - Pr[Ideal_{\mathcal{A}'}(\kappa) = 1] |$$

is negligible in κ.

User-Security: We say that CAC-OT is user-secure if for every PPT real-world adversary $\mathcal{A}$ who corrupts $\hat{\mathcal{S}}, \hat{\mathcal{I}}$ and a collection of users $\{\hat{\mathcal{U}}_1, \ldots, \hat{\mathcal{U}}_t\}$, there exists a PPT ideal-world adversary $\mathcal{A}'$ who corrupts the the same participants, such that for κ (which is a security parameter), and every PPT distinguisher $\mathcal{D}$:

$$| Pr[Real_{\mathcal{A}}(\kappa) = 1] - Pr[Ideal_{\mathcal{A}'}(\kappa) = 1] |$$

is negligible in κ.

4   A Generic Construction for CAC-OT

In this part, we will present a generic construction for our CAC-OT built from the blind (ciphertext policy) attribute-based encryption which is introduced below. The construction can be proved secure in the security model described in section 3. In the following, we first present the definition and security notions of the blind attribute-based encryption, and next using the encryption as a building block, give a generic construction for CAC-OT.

4.1   Blind Attribute-Based Encryption

To make the blind attribute-based encryption more expressive, we first briefly introduce the traditional ABE scheme and its security definition. A CP-ABE scheme [8] consists of four algorithms Setup, KeyGen, Encrypt and Decrypt. There is a private key generation center ($\mathcal{KGC}$) who is responsible for the generation of private keys for users' attributes.

- Setup(1^κ)   The algorithm takes no input other than the implicit security parameter κ. It outputs the public parameters pk and a mater key sk.
- KeyGen($\mathcal{KGC}(pk, sk), \mathcal{U}(pk, \omega)) \rightarrow (\omega, sk_\omega)$ An honest user $\mathcal{U}$ with an attributes set ω makes requests to $\mathcal{KGC}$ and obtains the corresponding secret key sk_ω from $\mathcal{KGC}$.
- Encrypt(pk, τ, m)   The algorithm returns a ciphertext c_τ to a message m corresponding to the access control structure τ, such that only users who have the secret key generated from the attributes that satisfy τ will be able to decrypt the message m.
- Decrypt(c_τ, sk_ω)   The algorithm outputs a message m on input a ciphertext c_τ, a secret key sk_ω associated with ω.

ABE can be seen as a generalized identity-based encryption (IBE). In this work, the blind ABE we present is analogously a generalized blind IBE proposed in [20]. In the blind ABE, after each user extracts the secret key corresponding to his attribute set from the $\mathcal{KGC}$, $\mathcal{KGC}$ will not obtain anything about the user's attribute set.

The blind ABE includes four algorithms Setup, BlindKeyGen, Encrypt and Decrypt. The Setup, Encrypt and Decrypt algorithms are the same as those in traditional CP-ABE, and the BlindKeyGen algorithm is described as follows.

- BlindKeyGen($\mathcal{KGC}(pk, sk), \mathcal{U}$ $(pk, \omega)) \rightarrow (nothing, sk_\omega)$ An honest user $\mathcal{U}$ with an attribute set ω makes request to $\mathcal{KGC}$ and obtains the corresponding secret key sk_ω from $\mathcal{KGC}$. It includes three sub-algorithms (Blind, BKeyGen and Unblind): the user first runs Blind(pk, ω) algorithm to blind his attribute set ω to ω' and sends ω' to $\mathcal{KGC}$; then $\mathcal{KGC}$ performs BKeyGen(sk, ω') to generate the private key $sk_{\omega'}$ for ω'; and finally the user obtains the private key sk_ω for the attribute set ω by executing Unblind($sk_{\omega'}$) algorithm. At the end, the BlindKeyGen algorithm outputs the private key sk_ω for ω for the user and nothing for the $\mathcal{KGC}$.

In [20], the IND-sID-CPA security of blind IBE was defined. Similarly, we give the security definitions for blind ABE as follows.

Definition 4. *(Secure Blind ABE) A blind ABE* Π *=(Setup, BlindKeyGen, Encrypt, Decrypt) is called IND-sAtt-CPA-secure (resp. IND-Att-CPA) if and only if: (1) the CP-ABE* Π' *=(Setup, KeyGen, Encrypt, Decrypt) is IND-sAtt-CPA secure (resp. IND-Att-CPA), and (2) BlindKeyGen is leak free and selective-failure blind.*

Next, we will give the definition of IND-sAtt-CPA security for CP-ABE, and leak freeness and selective-failure blindness for BlindKeyGen protocol.

Definition 5. *(Selective-Attribute Secure CP-ABE (IND-sAtt-CPA))[15] An CP-ABE* Π' *=(Setup, KeyGen, Encrypt, Decrypt) is called IND-sAtt-CPA-secure if every PPT adversary* $\mathcal{A}$ *has only an advantage negligible in* κ *(which is a security parameter) for the following game carried out between the adversary* $\mathcal{A}$ *and a challenger* $\mathcal{C}$*:*

- *Initialization* $\mathcal{A}$ *chooses a target access tree* τ^* *and gives it to* $\mathcal{C}$*.*
- *Setup* $\mathcal{C}$ *runs Setup(1^κ) algorithm to obtain* (pk, sk)*, and give pk to* $\mathcal{A}$*.*
- *Phase 1* $\mathcal{A}$ *may query private keys for attribute sets* $\omega_1, \ldots, \omega_{q_l}$*, where each attribute set* ω_i *does not satisfy* τ^**.*
- *Challenge* $\mathcal{A}$ *outputs two messages* m_0, m_1*, where the length of them is the same.* $\mathcal{C}$ *Selects a random bit b and encrypts* m_b *to* τ^**. The resulting ciphertext* c^* *is given to* $\mathcal{A}$*.*
- *Phase 2* $\mathcal{A}$ *may continue to query private keys for attribute sets* $\omega_{q_{l+1}}, \ldots, \omega_q$ *as in* *Phase 1.*
- *Guess* $\mathcal{A}$ *outputs* $b' \in \{0, 1\}$*.*

We define $\mathcal{A}$*'s advantage in the above game as* $| Pr[b' = b] - \frac{1}{2} |$*.*

A secure BlindKeyGen protocol should satisfy two properties:

- **Leak-freeness** A possibly cheating user cannot learn anything by executing the BlindKeyGen protocol with an honest $\mathcal{KGC}$ except for the necessarily known knowledge.
- **Selective-failure blindness** A possibly cheating $\mathcal{KGC}$ cannot learn anything about the user's attributes during the BlindKeyGen protocol. Moreover, the $\mathcal{KGC}$ cannot cause the BlindKeyGen protocol to fail selectively depending on the user's attributes.

The formal definitions for leak-freeness and selective-failure blindness of Blind-KeyGen protocol associated with an ABE scheme are described in Appendix B.

4.2   The Generic Construction for CAC-OT

In this part, a generic construction for CAC-OT from a blind ABE and a credential signature scheme will be presented and it is proved to be secure in the standard model.

Construction Overview. We import a credential issuer external to the protocol who is responsible to issuing credentials for each user's attributes into the protocol. A server maintains a database $DB = \{m_1, \ldots, m_N\}$, and associates an attribute-based access control structure τ_i with each message m_i. Each user $\mathcal{U}_i$ is entitled to an attribute subset $\omega_i \subseteq \Omega$ and an identity id_i.

The server first generates the parameters for a blind ABE. Then he initializes the database by encrypting each message m_i under τ_i such that the message can only be decrypted by the users who can obtain the private keys corresponding to an attribute subset satisfying τ_i. The encrypted database is published to all users. In the transfer phase, each user makes queries for the private keys corresponding to his entitled attributes by running BlindKeyGen algorithm with the server. Simultaneously, the user must make a proof of knowledge that he possesses a valid credential signature for the requested attributes issued by the issuer. If the user is verified, he will obtain the private keys for the requested attributes from the server. Then he can use the keys to arbitrarily decrypt the messages available to him according to the policies. We assume that at the end of the protocol, each user will output a message subset that includes all the messages available to them.

The above gives the basic construction idea for CAC-OT. After each user obtains the private keys for the requested attributes, he can check the correctness of the keys. So the protocol is resistant against the possibly cheating server who may cause the selective-failure attacks. However, a problem during the simulation of a cheating user comes along with this property. The simulator of a collection of possibly cheating users works as a server in the real word. It must encrypt N random values in the DB-Initialization phase since it does not know the correct messages, and in the Transfer phase open some of these values to the corresponding correct messages received from the trusted party during simulation. To solve the problem, we can use a commitment scheme or programming a random oracle here. In addition, a zero-knowledge proof $PoK\{(sk_{DB}) : (pk_{DB}, sk_{DB}) \in \mathsf{Setup}(1^\kappa, pk_I)\}$ is also needed in the simulation of a cheating server.

The Construction. Next, we will describe the solution similar to [20], by using a secure commitment scheme such as Pedersen's scheme [28] $\mathcal{COM} = (\mathsf{CSetup}, \mathsf{Commit}, \mathsf{Decommit})$ and present a secure generic construction for CAC-OT as follows. In the commitment scheme, CSetup is the system parameters generation algorithm which generates public parameters ρ. Taking a message m as input, $\mathsf{Commit}(m, \rho)$ algorithm outputs $(\mathcal{C}, \mathcal{D})$. The Decommit algorithm outputs 1 if $\mathcal{D}$ decommits $\mathcal{C}$ to m, or 0 otherwise. Moreover, for the commitment scheme, we require that the knowledge of a decommitment $\mathcal{D}$ can be proved efficiently with respect to $(\rho, m, \mathcal{C})$. In the following, we assume that the credential signature scheme used is $\mathcal{CS} = (\mathsf{ISetup}, \mathsf{IssueCred}, \mathsf{VerifyCred}, \mathsf{ProveCred})$ and the blind ABE used is $\mathcal{BABE} = (\mathsf{Setup}, \mathsf{BlindKeyGen}, \mathsf{Encrypt}, \mathsf{Decrypt})$, where BlindKeyGen algorithm consists of three sub-algorithms Blind, BKeyGen and Unblind. The parameters for the commitment scheme and a collision-resistant hash function H can be generated by a trusted party.

- IssueSetup(1^κ)
    1. $\mathcal{I}$ does as follows:
        (a) generates $(pk_I, sk_I) \leftarrow$ ISetup(1^κ);
        (b) publishes pk_I as the system-wide parameters.

The credential issuer generates the parameters for the credential signature scheme, and makes his public key pk_I as the system-wide parameters.

- DB-Initialization:
    1. $\mathcal{S}$:
        (a) generates $(pk_{DB}, sk_{DB}) \leftarrow$ Setup($1^k, pk_I$);
        (b) computes $C_j \leftarrow$ Encrypt(pk_{DB}, m_j, τ_j), $j = 1, \ldots, N$;
        (c) computes $(\mathcal{C}, \mathcal{D}) \leftarrow$ Commit($H(C_1, \ldots, C_N)$);
        (d) publishes $(pk_{DB}, \mathcal{C})$ to all users, and simultaneously executes a proof of knowledge

$$PoK_1\{(sk_{DB}) : (pk_{DB}, sk_{DB}) \in Setup(1^k, pk_I)\};$$

The server $\mathcal{S}$ first generates the parameters for the blind ABE scheme. Then he encrypts each message in the database by running Encrypt(pk_{DB}, m_j, τ_j) to C_j and commits to the ciphertexts $(C_1, \ldots, C_N)$ by using the commitment scheme. Finally he publishes the commitment and public key pk_{DB} to all users, and simultaneously conducts a proof of knowledge of sk_{DB}. Moreover, this proof will enable to decrypt the messages of the database in the security proof.

- ObtainCred($\Omega, sk_I; \omega_i, id_i$)
    1. $\mathcal{U}_i$: authenticates his identity and attributes (ω_i, id) to $\mathcal{I}$;
    2. $\mathcal{I}$:
        (a) for each attribute $a_j \in \omega_i$, computes $cred_{i_{a_j}} \leftarrow$ IssueCred($sk_I, (id_i, a_j)$);
        (b) sends $cred_{\omega_i} = \{cred_{i_{a_j}}\}_{a_j \in \omega_i}$ to $\mathcal{U}_i$;
    3. $\mathcal{U}_i$: verifies each credential by running VerifyCred($pk_I, (id_i, a_j), cred_{i_{a_j}}$) algorithm.

For each attribute $a_j \in \omega_i$, the issuer runs IssueCred($sk_I, (id_i, a_j)$) to output $cred_{i_{a_j}}$ as the credential of a_j for $\mathcal{U}_i$. Alternatively, the issuer can issue the credentials for $\mathcal{U}_i$'s all attributes at once by running IssueCred($sk_I, (id_i, \{a_j\}_{a_j \in \omega_i})$) algorithm. By linking each attribute a_j with the user's identity id_i in the credential, the protocol can be resistant against multiple users' collusion attacks. Since in the Transfer phase below, when each user requests messages, he must make a proof that the credentials for the requested attributes are valid and linked with one identity. So if two or more users collude by pooling their credentials, due to the soundness of knowledge proof they cannot conduct such a knowledge proof which can convince the server that the credentials are linked with one identity. Note that id_i only appears in the user's credentials, and is not involved in the blind ABE scheme as a new attribute.

- Transfer$(sk_{DB}; \omega_i)$
  1. $\mathcal{U}_i$: verifies the PoK_1 and aborts if the verification fails;
  2. $\mathcal{U}_i$: for each $a_j \in \omega_i$, computes $a'_j \leftarrow$ Blind(pk_{DB}, a_j), sends a'_j to $\mathcal{S}$, and simultaneously conducts a proof of knowledge

$$PoK_2\{(\omega_i, id, cred_i) : a'_j \in \text{Blind}(pk_{DB}, a_j) \wedge cred_{ij} \in \text{IssueCred}(sk_I, (id, a_j)), \forall a_j \in \omega_i\};$$

  3. $\mathcal{S}$:
     (a) verifies the PoK_2, and aborts if the verification fails;
     (b) computes $sk'_{\omega_i} \leftarrow$ BKeyGen$(sk_{DB}, \{a'_j\}_{a_j \in \omega_i})$;
     (c) sends $(sk'_{\omega_i}, C_1, \ldots, C_N)$ to $\mathcal{U}_i$ and conducts a proof of knowledge

$$PoK_3\{(\mathcal{D}) : \text{Decommit}(H(C_1, \ldots, C_N), \mathcal{C}, \mathcal{D}) = 1\};$$

  4. $\mathcal{U}_i$:
     (a) verifies the PoK_3 and aborts if the verification fails;
     (b) computes $sk_{\omega_i} \leftarrow$ Unblind(sk'_{ω_i}) as the private keys for ω_i and verifies the correctness of the keys. If they do not verify, aborts;
     (c) if for all τ_j, ω_i does not satisfy τ_j, returns "$\perp$", otherwise for each message m_j that satisfies $\omega_i \models \tau_j$, runs Decrypt(C_j, sk_{ω_i}) to decrypt out m_j. At the end outputs a message subset $\phi_i \subseteq \{m_1, \ldots, m_N\}$, where the access control structure for each message in ϕ_i is satisfied by ω_i.

To retrieve the private keys to an attribute subset, each user and the server engage in the BlindKeyGen protocol for the requested attribute subset. Simultaneously, the user must prove to the server by PoK_2 that he possesses the valid credentials for the requested attributes and the credentials are linked with one identity. If the BlindKeyGen protocol and anonymous credential scheme are both secure, and the proof is zero-knowledge, then the server will not learn any information about the requested attributes or the user's identity. Finally, the user will obtain the private keys requested from the server, and decrypt arbitrarily the messages allowed for him by using the private keys.

Security Analysis. If the based blind ABE is IND-sAtt-CPA secure, then any colluding users cannot combine their private keys to decrypt out a message that none of them would have been able to obtain individually. Therefore, from the construction of the CAC-OT, we can say that the CAC-OT can also be resistant against this type of colluding users.

In the following, we will show that the generic construction for CAC-OT is server-secure and user-secure under the security model presented in section 3.

Theorem 1. *If the based blind ABE is IND-sAtt-CPA secure, the commitment scheme is secure, and the knowledge proof PoK_2 and PoK_3 is zero-knowledge, then the generic construction for CAC-OT is server-secure.*

We give a proof of theorem 1 in Appendix A.

Theorem 2. *If the based blind ABE is IND-sAtt-CPA secure and the knowledge proof PoK_1 is zero-knowledge, then the generic construction for CAC-OT is user-secure.*

We give a proof of theorem 2 in Appendix A.

Note that, we only describe a secure generic construction for CAC-OT secure in the standard model by using a commitment scheme. In fact, we can also apply a random oracle to achieve a secure construction for CAC-OT in the random oracle model. The technique is similar to that in [20].

5   A Concrete Scheme for CAC-OT

In section 5, we combine the blind ABE with the credential signature scheme to present a construction for CAC-OT. To give a concrete scheme, we have to construct a blind ABE and combine the blind ABE with a credential signature scheme to achieve this point: when each user extracts the private keys for his attributes, he must make a proof of knowledge to convince the server that he has the credentials for the requested attributes. However, it seems that presenting a blind ABE based on the existing ABE schemes such as [8,27,21,31,15] and finding a credential scheme to make such a knowledge proof as above are infeasible. This is in part due to the fact that either the indexes of attributes are hashed into an element or a secret value for each attribute is selected unknown to the user in the existing ABE schemes. The fact makes the BlindKeyGen protocol not to be realized and makes the proof of knowledge techniques unwieldy.

In the following, we first present a new ABE based on the IBE scheme [3], and then a blind ABE using the similar technique to that in [20]. Then we combine the blind ABE with the credential signature [2] to give a concrete CAC-OT protocol.

5.1   A Concrete Blind ABE

Blind ABE We first present the basic ABE scheme, and next give the Blind-KeyGen protocol. The technique in encryption phase is similar to that in [23].

We give a description of the access control structure τ_i used in our protocol. The access control structure is a n-ary tree, in which leaves are attributes and inner nodes are "$and(\wedge)$" and "$or(\vee)$" boolean operations.

- Setup(1^k)
    1. Select a generator g of G, where $|G| = p$, a random α, and set $g_1 = g^\alpha$. Then pick a random element $g_2 \in G$.
    2. Generate the attribute set $\Omega = \{a_1, a_2, \ldots, a_l\} \subseteq Z_p^*$, for some integer l, and the random elements $h_1, h_2, \ldots, h_l \in G$.

    The public key is $pk = (g, g_1, g_2, h_1, \ldots, h_l)$, and the master key is $sk = g_2^\alpha$.
- KeyGen(sk, ω)
    1. Select a random value $r \in Z_p^*$, and compute $d_0 = g^r$;

2. Compute $d_j = g_2^\alpha F_j(a_j)^r$, for each $a_j \in \omega$, where $F_j(a_j) = g_1^{a_j} h_j$;
3. Return the secret key $sk_\omega = (d_0, \forall a_j \in \omega : d_j)$.

- Encrypt(m, τ, pk) To encrypt a message $m \in G_T$, the algorithm proceeds as follows.
  1. First select a random element $s \in Z_p^*$, and compute $C_0 = e(g_1, g_2)^s \cdot m$.
  2. Set the value of the root node of τ to be s, mark all child nodes as un-assigned, and mark the root node assigned.
     (a) If the symbol is " $\wedge$ " and its child nodes are marked un-assigned, for each child node except the last one, we assign a random value s_i where $1 \le s_i \le p-1$, and to the last child node assign the value $s_t = (s - \Sigma s_i) \bmod p$. Mark this node assigned.
     (b) If the symbol is "$\vee$", set the values of each child node to be s. Mark this node assigned.
  3. For each leaf attribute $a_{j,i} \in \tau$, compute $C_{j,i} = (C_{j,i}^1, C_{j,i}^2)$, where $C_{j,i}^1 = g^{s_i}$, $C_{j,i}^2 = F_j(a_j)^{-s_i}$, where i denotes the index of the attribute in the access tree τ.
  4. Return the ciphertext $C_\tau = (\tau, C_0, (C_{j,i}^1, C_{j,i}^2) : a_{j,i} \in \tau)$.

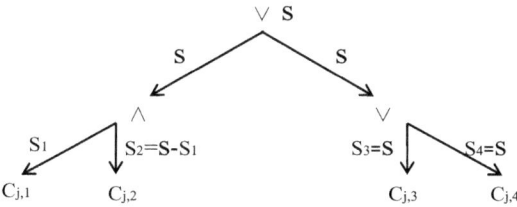

Fig. 1.

- Decrypt(C_τ, sk_ω) The algorithm chooses the smallest set $\omega' \subseteq \omega$ (we assume that this can be computed efficiently by the decryptor) that satisfies τ and performs as follows.
  1. For every attribute $a_j \in \omega'$, compute

$$\prod_{a_j \in \omega'} e(d_0, C_{j,i}^2) e(d_j, C_{j,i}^1)$$

$$= \prod_{a_j \in \omega'} e(g^r, F_j(a_j)^{-s_i}) e(g_2^\alpha F_j(a_j)^r, g^{s_i})$$

$$= \prod_{a_j \in \omega'} e(g^r, F_j(a_j)^{-s_i}) e(g_2^\alpha, g^{s_i}) e(F_j(a_j)^r, g^{s_i})$$

$$= \prod_{a_j \in \omega'} e(g_2^\alpha, g^{s_i})$$

$$= e(g_1, g_2)^s$$

  2. Compute $m' = \dfrac{C_0}{e(g_1, g_2)^s}$.

Next, we will present the BlindKeyGen protocol. In [20], a blind IBE was presented based on the IBE [3]. Since the ABE we give above is an extension of the IBE [3], the BlindKeyGen protocol can also be seemed as a simple extension of the BlindExtract protocol in [20]. Without loss of generality, we assume that the user is entitled to an attribute set $\omega = \{a_1, \ldots, a_t\}$. The BlindKeyGen proceeds as follows.

- Blind(ω, pk)
    1. $\mathcal{U}$ picks a random $r_1, \ldots, r_t \in Z_p$;
    2. computes $h'_j = g_1^{a_j} h_j g^{r_j}$, $j = 1, \ldots, t$;
    3. conducts $PoK\{(r_1, \ldots, r_t, a_1, \ldots, a_t) : \bigwedge_{j=1}^{t} h'_j = g_1^{a_j} h_j g^{r_j}\}$.
- BKeyGen$(sk, h'_j, j = 1, \ldots, t)$
    1. $\mathcal{KGC}$ verifies the proof. If the proof fails, abort;
    2. chooses a random $r \in Z_p$;
    3. computes $d'_0 \leftarrow g^r$, $d'_j \leftarrow h'^r_j g_2^{\alpha}$, $j = 1, \ldots, l$;
    4. sends $sk'_\omega = (d'_0, d'_1, \ldots, d'_t)$ to $\mathcal{U}$.
- Unblind(sk'_ω)
    1. $\mathcal{U}$ checks that $e(g_1, g_2)e(d'_0, h'_j) = e(d'_j, g)$, for $j = 1, \ldots, t$;
    2. If the check passes, chooses a random $z \in Z_p$, otherwise, outputs " $\perp$ " and aborts;
    3. computes $d_0 \leftarrow d'_0 g^z$, $d_j \leftarrow (d'_j / d'^{r_j}_0) F_j(a_j)^z$, $j = 1, \ldots, t$;
    4. outputs $sk_\omega = (d_0, d_1, \ldots, d_t)$.

Security. We will show that the blind ABE above is IND-sAtt-CPA secure by the following theorems.

Theorem 3. *The blind ABE above is both leak-free and selective-failure blind.*

We sketch a proof of theorem 3 in Appendix B.

Theorem 4. *The basic ABE scheme above is IND-sAtt-CPA secure based on DBDH assumption.*

We give a proof of theorem 4 in Appendix B.

5.2   A Concrete Construction for CAC-OT

We will combine the credential signature scheme in [2] with the above blind ABE to present a concrete CAC-OT scheme. The Pedersen commitment [28] is also used. The scheme operates on the same group as the blind attribute-based encryption that we present and the credential signature scheme [2], the knowledge of the value $\mathcal{D}$ can be proved using schnorr's technique [30], and hence Pedersen commitment scheme is well-suited for the concrete scheme for our CAC-OT. In the protocol below, the parameters (H, h) used for the commitment scheme can be generated by a trusted party.

Assume that the attributes set is $\Omega = \{a_1, a_2, \ldots, a_l\}$, where each $a_j \in Z_p$. Without loss of generality, we assume that the user $\mathcal{U}_i$ has identity id_i and attributes $\omega_i = \{a_1, \ldots, a_m\}$, where $m \leq l$. In the protocol, the number of each user's attributes may be different. Then during the Transfer phase, if the server colludes with the issuer, then they may guess out some users' identities from the numbers of these users' attributes. So the number of each user's attributes should be protected against the server and issuer. We give a solution to this problem as follows: in the Transfer phase, for the user $\mathcal{U}_i$, we model his attributes subset ω_i as a tuple of l attributes $\omega_i^* = \{a_1^*, a_2^*, \ldots, a_m^*, \underbrace{a_{m+1}^*, \ldots, a_l^*}\} \leftarrow$

$(a_1, a_2, \ldots, a_m, \underbrace{a_1, \ldots, a_1})$. Thus in the Transfer phase, $\mathcal{U}_i$ can make queries to the server and prove that he has the credentials for the requested attributes without revealing the number of the attributes. And meanwhile, the user can only obtain the private keys for his entitled attributes.

- IssueSetup(1^κ)
  1. $\mathcal{I}$:
     (a) generates the keys $(G, G_T, p, e, g_0, y_l, y_2, y_3, y_I; x_I) \leftarrow \mathsf{ISetup}(1^\kappa)$;
         $pk_I \leftarrow (G, G_T, p, e, g_0, y_1, y_2, y_3, y_I)$; $sk_I \leftarrow x_I$; $y_I \leftarrow g_0^{x_I}$;
     (b) publishes pk_I as the system-wide parameters.
- DB-Initialization($\Omega, m_1, \ldots, m_N, \tau_1, \ldots, \tau_N$)
  1. $\mathcal{S}$:
     (a) generates $(g, g_1, g_2, h_1, \ldots, h_l, \alpha) \leftarrow \mathsf{Setup}(1^\kappa, pk_I)$;
         $pk_{DB} \leftarrow (g, g_1, g_2, h_1, \ldots, h_l)$; $sk_{DB} \leftarrow \alpha$;
     (b) for each $m_j \in G_T$, computes $C_j \leftarrow \mathsf{Encrypt}(pk_{DB}, m_j, \tau_j)$, $j = 1, \ldots, N$, chooses a random value $z \in Z_p$ and computes $\mathcal{C} \leftarrow g^{H(C_1, \ldots, C_N)} h^z$;
     (c) publishes $(\mathcal{C}, pk_{DB})$ to all users, and does a proof of knowledge

$$PoK_1\{(\alpha) : g_1 = g^\alpha\};$$

- ObtainCred(Ω, x_I, ω)
  1. $\mathcal{U}_i$: verifies the PoK_1, and aborts if the verification fails;
  2. $\mathcal{U}_i$: authenticates his identity and attributes (id_i, ω_i) to $\mathcal{I}$;
  3. $\mathcal{I}$: generates the credentials for ω_i as follows:
     (a) for each attribute $a_j \in \omega_i$, chooses $r_{a_j}, s_{a_j} \in Z_p$ at random, and computes $\sigma_{a_j} \leftarrow (g_0 y_1^{a_j} y_2^{id_i} y_3^{r_{a_j}})^{1/(x_I + s_{a_j})}$;
     (b) sends $\{(\sigma_{a_j}, r_{a_j}, s_{a_j})\}_{a_j \in \omega_i}$ to $\mathcal{U}_i$ as the credential for ω_i;
  4. $\mathcal{U}_i$: checks whether $e(\sigma_{a_j}, g_0^{s_{a_j}} y_I) = e(g_0 y_1^{a_j} y_2^{id_i} y_3^{r_{a_j}}, g_0)$ holds.
- Transfer
1. $\mathcal{U}_i$:
(a) models his attribute subset as a tuple $\omega_i^* = \{a_1^*, a_2^*, \ldots, a_m^*, a_{m+1}^*, \ldots, a_l^*\}$; chooses values $r_1, \ldots, r_l \in Z_p$ at random, and for $j = 1, \ldots, l$, computes $h_j' \leftarrow g_1^{a_j^*} g^{r_j}$;

(b) for each a_j^*, chooses values $v_j, w_j \in Z_p$ and $A_j, B_j \in G$ at random, computes
$$\widetilde{A}_j \leftarrow \sigma_{a_j^*} A_j^{v_j}, \ \widetilde{B}_j \leftarrow B_j^{v_j} A_j^{w_j};$$

(c) executes the following proof of knowledge
$$PoK_2\{(r_j, \alpha_j, \beta_j, s_{a_j^*}, v_j, w_j, a_1^*, ..., a_l^*, id_i, r_{a_j^*})_{j=1,...,l} :$$
$$\bigwedge_{j=1}^l \left(h_j' = g_1^{a_j^*} g^{r_j} \bigwedge \widetilde{B}_j = B_j^{v_j} A_j^{w_j} \bigwedge 1 = \widetilde{B}_j^{-s_{a_j^*}} B_j^{\alpha_j} A_j^{\beta_j} \right)$$
$$\bigwedge_{j=1}^l \left(\frac{e(\widetilde{A}_j, y_I)}{e(g_0, g_0)} = e(\widetilde{A}_j, g_0)^{-s_{a_j^*}} e(A_j, y_I)^{v_j} e(A_j, g_0)^{\alpha_j} e(y_3, g_0)^{r_{a_j^*}} e(y_2, g_0)^{id_i} e(y_1, g_0)^{a_j^*} \right) \},$$

where $\alpha_j = s_{a_j^*} v_j$ and $\beta_j = s_{a_j^*} w_j$;

2. $\mathcal{S}$:

(a) verifies the PoK_2, and aborts if the verification fails;

(b) otherwise chooses a random $r \in Z_p$, and computes
$$d_0' \leftarrow g^r;$$
$$d_j' \leftarrow g_2^\alpha (h_j' h_j)^r, \text{ for } j = 1, \dots, l;$$
$$sk_{\omega_i^*}' \leftarrow (d_0', d_1', \dots, d_l');$$

(c) sends $(sk_{\omega_i^*}', C_1, \dots, C_N)$ to $\mathcal{U}_i$ and conducts a proof of knowledge

$$PoK_3\{(z) : \mathcal{C} = g^{H(C_1, \dots, C_N)} h^z\};$$

3. $\mathcal{U}_i$:

(a) verifies the PoK_3, and aborts if the verification fails;

(b) parses sk_{ω^*} as $(d_0', d_1', \dots, d_l')$, and for each j, checks whether $e(g_1, g_2) e(d_0', h_j' h_j) = e(d_j', g)$ holds, and if the check does not pass, outputs " $\perp$ " and aborts;

(c) otherwise chooses a random $z \in Z_p$, computes $d_0 \leftarrow d_0' g^z$, $d_j \leftarrow (d_j'/(d_0')^{r_j}) F_j(a_j)^z$, $j = 1, \dots, l$, and lets $sk_{\omega_i} \leftarrow (d_0, d_1, \dots, d_l)$;

(d) if for all τ_j, ω_i does not satisfy τ_j, returns "$\perp$", otherwise for each message m_j that satisfies $\omega_i \models \tau_j$, runs $\mathsf{Decrypt}(C_j, sk_{\omega_i})$ to decrypt out m_j. At the end outputs a message subset $\phi_i \subseteq \{m_1, \dots, m_N\}$, where the access tree for each message in ϕ_i is satisfied by ω_i.

Theorem 5. *The concrete CAC-OT scheme satisfies server security and user security.*

Proof. Since the blind ABE, the credential signature and Pedersen commitment scheme are all proved to be secure, and the associated proofs of knowledge are zero-knowledge, we can conclude that the CAC-OT scheme satisfies server security and user security.

6   Analysis and Extensions

AC-OT [10] can just directly achieve "*and*" access control. For "*or*" condition, the message in the database has to be duplicated with a second access control structure. For a database in which complicated "*or*" and "*and*" access control policies like $(c_{1_1} \wedge c_{1_2} \wedge \ldots c_{1_{n_1}}) \vee (c_{2_1} \wedge c_{2_2} \wedge \ldots c_{2_{n_2}}) \vee \ldots \vee (c_{t_1} \wedge c_{t_2} \wedge \ldots c_{t_{n_t}})$ are enforced to each message, AC-OT does not work efficiently. Since the message must be duplicated for t times, each with a policy $(c_{i_1} \wedge c_{i_2} \wedge \ldots c_{i_{n_i}})$. Moreover, when the server initializes the database, he must encrypt the message for t times under different "*and*" policies, and the initialized database will also increase greatly. However, our CAC-OT directly achieves flexible "*or*" and "*and*" access control policies such that it can be efficiently applied in the databases that request complex access control permissions.

Since in the transfer phase of our construction, to obtain the private keys for his certificates, the user needs to interact with the server for just one round. Therefore, if we let k be the number of messages that a user can access in a database according to the access control policies, then the communication cost of our construction is $O(N + l)$. However, in Camenisch *et al.*'s AC-OT [10], each time the user requests a message, the communication cost is $O(N+l)$. Then to obtain all the k allowed messages, the user has to interact with the server for k times and correspondingly the communication cost is $O(N + kl)$. So in the case that k is a large number, our protocol works efficiently in communications.

In the protocol, we combine the blind ABE with credential signature scheme to present the CAC-OT. Under our construction, the CAC-OT achieves the same access control policies as the blind ABE. However, in the concrete blind ABE scheme we presented, the ciphertext size increases linearly with the number of "*or*" policies, and the access policies are just "*and*" and "*or*". As we know, there have been some efficient ABE schemes which have achieved more complicated access control structures such as [32]. In the scheme [32], access control is expressed by a Linear Secret Sharing Scheme(LSSS) matrix over the attributes in the system. Presenting a more efficient blind ABE with more complicated access control policies, and applying it to our construction for CAC-OT is our future work.

7   Conclusion

In this paper we presented the oblivious transfer with complex access control policies which directly achieves "*and*" and "*or*" policies. To realize the protocol, we first presented a primitive called blind ABE using the similar technique to that in the blind IBE [20], and then gave a generic construction combining the blind ABE with a credential signature scheme. Moreover, a new blind ABE scheme was proposed in which the access control structure is provided by access trees, and based on it, a concrete CAC-OT protocol is presented.

References

1. Aiello, W., Ishai, Y., Reingold, O.: Priced oblivious transfer: How to sell digital goods. In: Pfitzmann, B. (ed.) EUROCRYPT 2001. LNCS, vol. 2045, pp. 119–135. Springer, Heidelberg (2001)
2. Au, M.H., Susilo, W., Mu, Y.: Constant-size dynamic k-TAA. In: De Prisco, R., Yung, M. (eds.) SCN 2006. LNCS, vol. 4116, pp. 111–125. Springer, Heidelberg (2006)

3. Boneh, D., Boyen, X.: Efficient selective-ID secure identity-based encryption without random oracles. In: Cachin, C., Camenisch, J.L. (eds.) EUROCRYPT 2004. LNCS, vol. 3027, pp. 223–238. Springer, Heidelberg (2004)
4. Boneh, D., Boyen, X.: Short signatures without random oracles. In: Cachin, C., Camenisch, J.L. (eds.) EUROCRYPT 2004. LNCS, vol. 3027, pp. 56–73. Springer, Heidelberg (2004)
5. Brassard, G., Crépeau, C., Robert, J.M.: All-or-nothing disclosure of secrets. In: Odlyzko, A.M. (ed.) CRYPTO 1986. LNCS, vol. 263, pp. 234–238. Springer, Heidelberg (1987)
6. Boneh, D., Franklin, M.: Identity based encryption from the Weil pairing. In: Kilian, J. (ed.) CRYPTO 2001. LNCS, vol. 2139, pp. 213–229. Springer, Heidelberg (2001)
7. Blake, I.F., Kolesnikov, V.: Strong conditional oblivious transfer and computing on intervals. In: Lee, P.J. (ed.) ASIACRYPT 2004. LNCS, vol. 3329, pp. 515–529. Springer, Heidelberg (2004)
8. Bethencourt, J., Sahai, A., Waters, B.: Ciphertext-policy attribute-based encryption. In: IEEE Symposium on Security and Privacy, pp. 321–334 (2007)
9. Canetti, R.: Security and composition of multi-party cryptographic protocols. Journal of Cryptology 13(1), 143–202 (2000)
10. Camenisch, J., Dubovitskaya, M., Neven, G.: Oblivious transfer with access control. In: ACM CCS 2009, pp. 131–140 (2009)
11. Cramer, R., Damgård, I., Schoenmakers, B.: Proofs of partial knowledge and simplified design of witness hiding protocols. In: Desmedt, Y.G. (ed.) CRYPTO 1994. LNCS, vol. 839, pp. 174–187. Springer, Heidelberg (1994)
12. Coull, S., Green, M., Hohenberger, S.: Controlling access to an oblivious database using stateful anonymous credentials. In: Jarecki, S., Tsudik, G. (eds.) PKC 2009. LNCS, vol. 5443, pp. 501–520. Springer, Heidelberg (2009)
13. Chaum, D.: Security without identification: Transaction systems to make big brother obsolete. Communications of ACM 28(10), 1030–1044 (1985)
14. Camenisch, J., Lysyanskaya, A.: A signature scheme with efficient protocols. In: Cimato, S., Galdi, C., Persiano, G. (eds.) SCN 2002. LNCS, vol. 2576, pp. 268–289. Springer, Heidelberg (2003)
15. Cheung, L., Newport, C.: Provably secure ciphertext policy ABE. In: ACM CCS 2007, pp. 456–465 (2007)
16. Camenisch, J., Neven, G., Shelat, A.: Simulatable adaptive oblivious transfer. In: Naor, M. (ed.) EUROCRYPT 2007. LNCS, vol. 4515, pp. 573–590. Springer, Heidelberg (2007)
17. Di Crescenzo, G., Ostrovsky, R., Rajagopalan, S.: Conditional oblivious transfer and timed-release encryption. In: Stern, J. (ed.) EUROCRYPT 1999. LNCS, vol. 1592, pp. 74–89. Springer, Heidelberg (1999)
18. Camenisch, J., Stadler, M.: Efficient group signature schemes for large groups. In: Kaliski Jr., B.S. (ed.) CRYPTO 1997. LNCS, vol. 1294, pp. 410–424. Springer, Heidelberg (1997)
19. Even, S., Goldreich, O., Lempel, A.: A randomized protocol for signing contracts. Communications of the Association for Computing Machinery 28(6), 637–647 (1985)
20. Green, M., Hohenberger, S.: Blind identity-based encryption and simulatable oblivious transfer. In: Kurosawa, K. (ed.) ASIACRYPT 2007. LNCS, vol. 4833, pp. 265–282. Springer, Heidelberg (2007)
21. Goyal, V., Pandey, O., Sahai, A., Waters, B.: Attribute based encryption for fine-grained access control of encrypted data. In: ACM CCS 2006, pp. 89–98 (2006)

22. Herranz, J.: Restricted adaptive oblivious transfer, Cryptology ePrint Archive, 2008/182
23. Ibraimi, L., Tang, Q., Hartel, P., Jonker, W.: Efficient and provable secure ciphertext-policy attribute-based encryption schemes. In: Bao, F., Li, H., Wang, G. (eds.) ISPEC 2009. LNCS, vol. 5451, pp. 1–12. Springer, Heidelberg (2009)
24. Kalai, Y.T.: Smooth projective hashing and two-message oblivious transfer. In: Cramer, R. (ed.) EUROCRYPT 2005. LNCS, vol. 3494, pp. 78–95. Springer, Heidelberg (2005)
25. Lindell, Y.: Efficient fully-simulatable oblivious transfer. In: Malkin, T. (ed.) CT-RSA 2008. LNCS, vol. 4964, pp. 52–70. Springer, Heidelberg (2008)
26. Ogata, W., Kurosawa, K.: Oblivious keyword search. Journal of Complexity 20(2-3), 356–371 (2004)
27. Ostrovsky, R., Sahai, A., Waters, B.: Attribute-based encryption with non-monotonic access structures. In: ACM CCS 2007, pp. 195–203 (2007)
28. Pedersen, T.P.: Non-interactive and information-theoretic secure verifiable secret sharing. In: Feigenbaum, J. (ed.) CRYPTO 1991. LNCS, vol. 576, pp. 129–140. Springer, Heidelberg (1992)
29. Rabin, M.O.: How to exchange secrets by oblivious transfer, Technical Report TR-81, Harvard Aiken Computation Laboratory (1981)
30. Schnorr, C.: Efficient signature generation for smart cards. Journal of Cryptology 4(3), 239–252 (1991)
31. Sahai, A., Waters, B.: Fuzzy identity-based encryption. In: Cramer, R. (ed.) EUROCRYPT 2005. LNCS, vol. 3494, pp. 457–473. Springer, Heidelberg (2005)
32. Waters, B.: Ciphertext-Policy Attribute-Based Encryption: An Expressive, Efficient, and Provably Secure Realization, Cryptology ePrint Archive, 2008/290 (2008)

A   Security Proofs for OT with Attribute-Based Access Control

A.1   Proof of Theorem 1

In the proof of server security, we do not consider the case that the issuer and a collection of users collude. Since if the issuer colludes with any user, then the user may obtain the private keys for all attributes in Ω to decrypt out all messages in the database. So we assume that the issuer will not collude with any user. In the following, we will only consider a collection of possibly cheating users.

For any real world adversary $\mathcal{A}$ who corrupts a collection of cheating users $\{\hat{\mathcal{U}}_1, \ldots, \hat{\mathcal{U}}_t\}$, we can construct an ideal world adversary $\mathcal{A}'$ who corrupts the same participants $\{\hat{\mathcal{U}}'_1, \ldots, \hat{\mathcal{U}}'_t\}$ such that for any PPT distinguisher $\mathcal{D}$, $|\, Pr[\mathsf{Real}_{\mathcal{A}}(\kappa) = 1] - Pr[\mathsf{Ideal}_{\mathcal{A}'}(\kappa) = 1]\,|$ is negligible in κ.

We construct the adversary $\mathcal{A}'$ as follows. $\mathcal{A}'$ plays simultaneously roles of $\{\hat{\mathcal{U}}_1, \ldots, \hat{\mathcal{U}}_t\}$, and performs an honest server $\mathcal{S}$ in the real world. $\mathcal{A}'$ first generates (pk_{DB}, sk_{DB}) by running $\mathsf{Setup}(1^\kappa, pk_I)$. It generates a random commitment in DB-Initialization phase. In the Transfer phase, if the PoK does not verify, then $\mathcal{A}'$ aborts, otherwise uses an extractor of knowledge proof to extract (id, ω_i) of $\hat{\mathcal{U}}_i$, and sends (id, ω_i) to $\mathcal{T}$. If $\mathcal{T}$ outputs " $\perp$ ", then $\mathcal{A}'$ causes Transfer to fail. Otherwise, $\mathcal{A}'$ obtains a message set ϕ_i from $\mathcal{T}$. For $j = 1, \ldots, N$, $\mathcal{A}'$ sets $C'_j \leftarrow \mathsf{Encrypt}(pk_{DB}, \tau_j, m_j)$ for $m_j \in \phi_i$, and remains other ciphertexts to be encryptions of random messages. At last $\mathcal{A}'$ sends $(C'_1, \ldots, C'_N)$ to $\hat{\mathcal{U}}_i$ along with a simulated proof of decommitment.

We consider a sequence of distributions Game-0,..., Game-3 to prove the indistinguishability between the real and ideal worlds. Let Game i be the output of the Game-i.

Game-0: The adversary $\mathcal{A}$ interacts with the honest server $\mathcal{S}$ exactly as in the real world. Clearly

$$Pr[\text{Game } 0 = 1] = Pr[\mathsf{Real}_{\mathcal{A}}(\kappa) = 1].$$

Game-1: The extractor for the PoK_2 is used to extract (id', w'_i). If the extractor fails or different identities are extracted, then output " $\perp$ ". (If different identities are extracted, there must be cheating users that collude with each other and try to combine their credentials to retrieve the messages they have not access to.) The difference between the two output distributions is given by the knowledge error of the PoK_2,

$$|\, Pr[\text{Game } 1 = 1] - Pr[\text{Game } 0 = 1] \,| \le O(2^{-\kappa}).$$

Game-2: If in the extracted w'_i, there is at least one attribute $a \in w'_i$ which is not an element of Ω (i.e. $a \notin \Omega$), then output " $\perp$ ". The difference between Game-1 and Game-2 is given by the probability of forging a valid credential signature,

$$|\, Pr[\text{Game } 2 = 1] - Pr[\text{Game } 1 = 1] \,| \le O(2^{-\kappa}).$$

Game-3: We replace the commitment $\mathcal{C}$ with a commitment to a random value, and replace the final proof of knowledge of decommitment with a simulated proof. For a secure commitment scheme and zero-knowledge proof, the difference between this game and Game-1 is given by $\mathcal{D}$'s negligible advantage in correctly distinguishing $\mathcal{C}$ from a valid commitment to $H(C_1, \ldots, C_N)$, and the simulated proof from a valid proof,

$$|\, Pr[\text{Game } 3 = 1] - Pr[\text{Game } 2 = 1] \,| \le O(2^{-\kappa}).$$

Game-4: We denote the set ϕ_i as the set of the messages whose access trees are all satisfied by w_i, and denotes I_{w_i} as the index of these messages. We alter the ciphertext vector $(C_1, \ldots, C_N)$ to produce a new vector $(C'_1, \ldots, C'_N)$ as follows: if $j \notin I_{w_i}$, set $C'_j \leftarrow \mathsf{Encrypt}(pk_{DB}, \tau_j, m')$, where m' is selected randomly in the message space, and otherwise $C'_j \leftarrow C_j$. Then we have

$$|\, Pr[\text{Game } 4 = 1] - Pr[\text{Game } 3 = 1] \,| \le O(2^{-\kappa}).$$

From the construction of the adversary $\mathcal{A}'$, we have that

$$Pr[\text{Game } 4 = 1] = Pr[\mathsf{Ideal}_{\mathcal{A}'}(\kappa) = 1].$$

Summing the differences between the above games, we can conclude that

$$| Pr[\mathsf{Real}_\mathcal{A}(\kappa) = 1] - Pr[\mathsf{Ideal}_{\mathcal{A}'}(\kappa) = 1] | \leq O(2^{-\kappa}).$$

Lemma 1. $| Pr[\mathsf{Game\ 2} = 1] - Pr[\mathsf{Game\ 1} = 1] | \leq O(2^{-\kappa})$ *holds based on* l*-SDH assumption.*

Proof. If Game 1 and Game 2 are distinguishable, then we can construct a forger $\mathcal{A}$ who can break the existential unforgeability under chosen-message attack of the credential signature scheme [2].

The challenger of the credential signature scheme first generates the system parameters and sends the public parameters to $\mathcal{A}$. Then $\mathcal{A}$ can make signature queries to the challenger. Here we let $\mathcal{A}$ make queries for the messages in Ω where $|\Omega| = l$. Simultaneously $\mathcal{A}$ runs a server and plays the Game 2 with the user. In the transfer phase, $\mathcal{A}$ extracts the $(id', \omega'_i, cred'_i)$ from the zero-knowledge proof conducted with the user. If Game 1 and Game 2 are distinguishable, then it means that $\mathcal{A}$ can extract at least one credential $(a, cred_a)$ where $a \in \omega'_i$ but $a \notin \Omega$. Finally $\mathcal{A}$ outputs $(a, cred_a)$ to the challenger as a forgery. Since the credential signature scheme has been proved to be existentially unforgeable under chosen-message attack under the l-SDH assumption, we have the conclusion in the lemma 1.

Lemma 2. *(Indistinguishability of Ciphertexts)* $| Pr[\mathsf{Game\ 4}{=}1] - Pr[\mathsf{Game\ 3}{=}1] | \leq O(2^{-\kappa})$ *if the based blind ABE is IND-sAtt-CPA secure and leak-free and the PoK$_1$ of sk_{DB} is zero-knowledge.*

Proof. If a PPT distinguisher $\mathcal{D}$ can distinguish Game 3 and Game 4 with a non-negligible probability, then we can construct an adversary $\mathcal{A}$ that wins the IND-sAtt-CPA game against the blind ABE with the same probability. We use a hybrid proof as follows.

We define a series of hybrids such that $\mathsf{Hybrid}_0{=}$Game 3 and $\mathsf{Hybrid}_N{=}$Game 4. Hybrid_{j-1} and Hybrid_j only differ in the distribution of j_{th} ciphertext vector, where $(1 \leq j \leq N)$. If Game 3 and Game 4 can be distinguished by $\mathcal{D}$, then there must exist a j such that $\mathcal{D}$ can distinguish Hybrid_{j-1} and Hybrid_j. We construct $\mathcal{A}$ as follows. $\mathcal{A}$ outputs $\tau^* = \tau_j$. It runs $\mathcal{D}$ and conducts the protocol with $\hat{\mathcal{R}}$ as in Game 3. Then $\mathcal{A}$ selects a random message m^* from the message space, outputs (m_j, m^*) to the challenger and obtains a challenge ciphertext C^*. Then it constructs a ciphertext vector $(C'_1, \ldots, C'_N)$ as in Game 3 expect that at the j_{th} position, sets $C'_j \leftarrow C^*$.

When the distinguisher $\mathcal{D}$ returns a bit b, the adversary $\mathcal{A}$ returns b to the challenger as its answer.

In addition, since the PoK of sk_{DB} is zero-knowledge and the blind AEB scheme is leak-free, the distinguisher clearly cannot distinguish the two games.

A.2   Proof of Theorem 2

For any real world adversary $\mathcal{A}$ who corrupts the issuer $\hat{\mathcal{I}}$, the server $\hat{\mathcal{S}}$ and a collection of cheating users $\{\hat{\mathcal{U}}_1, \ldots, \hat{\mathcal{U}}_t\}$, we can construct an ideal world adversary $\mathcal{A}'$ who corrupts the same participants such that for any PPT distinguisher $\mathcal{D} \mid Pr[\mathsf{Real}_{\mathcal{A}}(\kappa) = 1] - Pr[\mathsf{Ideal}_{\mathcal{A}'}(\kappa) = 1] \mid$ is negligible in κ.

We construct an adversary $\mathcal{A}'$ as follows. $\mathcal{A}'$ plays simultaneously roles of the issuer $\hat{\mathcal{I}}'$, $\hat{\mathcal{S}}'$ and a collection of cheating users $\{\hat{\mathcal{U}}'_1, \ldots, \hat{\mathcal{U}}'_t\}$, and performs a collection of honest users $\{\hat{\mathcal{U}}_{t+1}, \ldots, \hat{\mathcal{U}}_n\}$ in the real world. $\mathcal{A}'$ runs $\mathcal{A}$ to obtain $\hat{\mathcal{I}}$'s public key pk_I and $\mathsf{Commit}(\mathsf{H}(C_1, \ldots, C_N))$ from $\hat{\mathcal{S}}$. After receiving $(\mathcal{U}'_i, \omega_i)$ from $\mathcal{T}$ in the ideal world, it executes $\mathsf{ObtainCred}$ with $\mathcal{A}$ on $(\mathcal{U}', \omega)$ in the real world. If the resulting credentials are all valid, $\mathcal{A}'$ works as the issuer in the ideal world to return $b{=}1$ to $\mathcal{T}$, otherwise return $b{=}0$. $\mathcal{A}'$ extracts sk_{DB} from $\mathcal{A}$ in the proof of knowledge in the DB-Initialization phase, and in the Transfer phase, simulates honest users by requesting for ω' for which it has the credentials. Upon receiving the ciphertexts $(C_1, \ldots, C_N)$, $\mathcal{A}'$ decrypts the messages $(m_1, \ldots, m_N)$, and then sends $\{(m_i, \tau_i)\}_{i=1,\ldots,N}$ to $\mathcal{T}$. If the transfer succeeds, $\mathcal{A}'$ then sends $b'{=}1$ to $\mathcal{T}$, otherwise sends $b'{=}0$.

We consider a sequence of distributions Game-0,..., Game-3 to prove the indistinguishability between the real and ideal worlds. Let Game i be the output of the Game-i.

Game-0: In this game, $\mathcal{A}$ interacts with the honest users exactly in the real world. Clearly

$$Pr[\mathsf{Game}\ 0 = 1] = Pr[\mathsf{Real}_A(\kappa) = 1].$$

Game-1: The extractor for PoK_1 of sk_{DB} is used to extract sk_{DB}. If extractor fails or outputs invalid sk_{DB}, outputs " $\perp$ ". Since the knowledge proof PoK_1 is zero-knowledge, the extractor fails with probability negligible in κ, and there is

$$\mid Pr[\mathsf{Game}\ 1 = 1] - Pr[\mathsf{Game}\ 0 = 1] \mid \leq O(2^{-\kappa}).$$

Game-2: In the Transfer phase, let the user algorithm request an attribute set ω for which it has the credentials. If the requests succeed, set a bit $b'{=}1$, otherwise $b'{=}0$. Since we assume that the blind ABE is selective-failure blind and the knowledge proof PoK_2 is zero-knowledge, then we have

$$\mid Pr[\mathsf{Game}\ 2 = 1] - Pr[\mathsf{Game}\ 1 = 1] \mid \leq O(2^{-\kappa}).$$

From the construction of the adversary $\mathcal{A}'$, we have that

$$Pr[\mathsf{Game}\ 2 = 1] = Pr[\mathsf{Ideal}_{\mathcal{A}'}(\kappa) = 1].$$

Summing the differences between the above games, we can conclude that

$$\mid Pr[\mathsf{Real}_A(\kappa) = 1] - Pr[\mathsf{Ideal}_{\mathcal{A}'}(\kappa) = 1] \mid \leq O(2^{-\kappa}).$$

Lemma 3. $\mid Pr[\textit{Game 2} = 1] - Pr[\textit{Game 1} = 1] \mid \leq O(2^{-\kappa})$ *if the based blind ABE is selective-failure blind.*

Proof. If there exists a distinguisher $\mathcal{D}$ who can distinguish Game 1 and Game 2, then we can construct an adversary $\mathcal{A}$ who can win the selective-failure blindness game. We use a hybrid proof as follows.

Note that the Game 1 and Game 2 differ only in the distribution of extracted attribute set. We define a series of hybrids such that $\mathsf{Hybrid}_0=\mathsf{Game}$ 1 and $\mathsf{Hybrid}_l=\mathsf{Game}$ 2, where $l = |\omega|$ and $\omega \subseteq \Omega$ is an attribute set. Hybrid_{j-1} and Hybrid_j only differ in the j_{th} attribute. If $\mathcal{D}$ can distinguish Game 1 and Game 2, then there must exist a $j \in \{1,\ldots,l\}$, such that $\mathcal{D}$ can distinguish Hybrid_{j-1} and Hybrid_j. Then we can construct $\mathcal{A}$ as follows.

$\mathcal{A}$ runs $\mathcal{D}$ and conducts the protocol with the real world server $\hat{\mathcal{S}}$ as in Game 1 except that $\mathcal{A}$ chooses a "real" attribute a_j (or a random $a'_j \in \Omega$) for Hybrid_{j-1}. Then $\mathcal{A}$ outputs $(params, a_j, a'_j)$ and sends the outputs of the first oracle $\mathcal{U}_b$ to $\hat{\mathcal{S}}$. After the BlindKeyGen protocol ends, $\mathcal{A}$ returns $\hat{\mathcal{S}}$ in the selective-failure blindness game. When $\mathcal{D}$ finally outputs a bit b', $\mathcal{A}$ outputs b' as its guess. We assume that the probability that $\mathcal{D}$ outputs 1 when presented with Hybrid_{j-1} is a, and the probability when presented with Hybrid_j is b, then the probability that $\mathcal{A}$ wins the selective-failure blindness game is $\frac{|b-a|}{2}$.

B   Security Proofs for Blind ABE

B.1   Definition of Leak-Freeness and Selective-Failure Blindness

Similarly to [20], we present the definition of leak-freeness for BlindKeyGen algorithm associated with an ABE.

Definition 6. *(Leak-Freeness) A protocol BlindKeyGen associated with an ABE scheme (Setup, KeyGen, Encrypt, Decrypt) is leak-free if for any efficient adversary $\mathcal{A}$, there exists an efficient simulator $\mathcal{S}im$ such that for any efficient distinguisher $\mathcal{D}$, the probability to distinguish real game and ideal game is negligible:*

- *Real game: $\mathcal{A}$ chooses an attributes set ω and interacts with $\mathcal{KGC}$ by running BlindKeyGen on ω. As many times as $\mathcal{D}$ wants, $\mathcal{A}$ repeats the actions above. Then $\mathcal{A}$ outputs a list of attributes set and the corresponding private keys extracted.*
- *Ideal game: $\mathcal{S}im$ chooses an attributes set ω and sends it to a trusted party $\mathcal{T}$ to obtain the output of KeyGen on ω. As many times as $\mathcal{D}$ wants, $\mathcal{A}$ repeats the actions above. Then $\mathcal{S}im$ outputs a list of attributes set and the corresponding private keys extracted.*

Next, we present the definition of selective-failure blindness similarly to [16,20].

Definition 7. *(Selective-Failure Blindness) A protocol $P(\mathcal{A}(\cdot), \mathcal{U}(\cdot, \cdot))$ is said to be selective-failure blind if for every PPT adversary $\mathcal{A}$ has a negligible advantage in the following game: First, $\mathcal{A}$ outputs params and two attributes set $\omega_0, \omega_1 \in \Omega$. A random bit $b \in \{0, 1\}$ is chosen. $\mathcal{A}$ is given black-box access to two oracles $\mathcal{U}(params, \omega_b)$ and $\mathcal{U}(params, \omega_{1-b})$. $\mathcal{U}$ algorithms produce local output sk_b and sk_{1-b} respectively. If $sk_b \neq 1$ and $sk_{1-b} \neq 1$ then $\mathcal{A}$ receives (sk_0, sk_1). If $sk_b = \perp$ and $sk_{1-b} \neq \perp$ then $\mathcal{A}$ receives $(\perp, \varepsilon)$. If $sk_b \neq \perp$ and $sk_{1-b} = \perp$ then $\mathcal{A}$ receives $(\varepsilon, \perp)$. If $sk_b = \perp$ and $sk_{1-b} = \perp$ then $\mathcal{A}$ receives $(\perp, \perp)$. Finally, $\mathcal{A}$ outputs its guess b'. We define $\mathcal{A}$'s advantage in the above game as $| Pr[b' = b] - \frac{1}{2} |$.*

B.2 Proof of Theorem 3

We give a proof sketch of theorem 3 as follows.

Leak-Freeness. For any adversary $\mathcal{A}$ who is interacting with $\mathcal{KGC}$ running the BlindKeyGen protocol in the real game, we can construct a simulator $\mathcal{Sim}$ who interacts with a trusted party executing the ideal KeyGen protocol, such that no efficient distinguisher can distinguish the real game and idea game.

$\mathcal{Sim}$ plays the role of $\mathcal{KGC}$ in the real game, interacting with $\mathcal{A}$, and simultaneously interacts with $\mathcal{T}$ in the ideal game. Each time $\mathcal{A}$ engages $\mathcal{Sim}$ in a BlindKeyGen protocol, $\mathcal{Sim}$ behaves as follows. $\mathcal{A}$ sends $(h'_1, \ldots, h'_t)$ to $\mathcal{Sim}$, and conducts a proof of knowledge $PoK\{(r_1, \ldots, r_t, a_1, \ldots, a_t) : \bigwedge_{j=1}^{t} h'_j = g_1^{a_j} h_j g^{r_j}\}$. If the PoK does not verify, $\mathcal{Sim}$ aborts. Otherwise, $\mathcal{Sim}$ extracts $(r_1, \ldots, r_t, \omega)$ and submits ω to $\mathcal{T}$. Then $\mathcal{T}$ returns the private keys $sk_\omega = (d_0, d_1, \ldots, d_t)$ to $\mathcal{Sim}$, where $d_0 \leftarrow g^r$, $d_j \leftarrow F_j(a_j)^r g_2^\alpha$, $j = 1, \ldots, l$ for some random $r \in Z_q$. Finally, $\mathcal{Sim}$ computs $sk_{\omega'} = (d'_0, d'_1, \ldots, d'_t)$, where $d'_0 \leftarrow d_0 g^z$, $d'_j \leftarrow d_j F_j(a_j)^z d_0^{r_j}$, $j = 1, \ldots, l$ for some random $z \in Z_q$.

We can see that sk'_ω is correctly formed and has the same distribution as that of $\mathcal{KGC}$. Hence any efficient distinguisher cannot distinguish the real game and ideal game.

Selective-Failure Blindness. The adversary $\mathcal{A}$ plays the game defined in definition 4. $\mathcal{A}$ first outputs $params$ and two attributes set ω_0 and ω_1. A random bit b is chosen. $\mathcal{A}$ is given black-box access to two oracles $\mathcal{U}(params, \omega_b)$ and $\mathcal{U}(params, \omega_{1-b})$. Then $\mathcal{U}$ algorithms conduct BlindKeyGen protocol with $\mathcal{A}$ who is playing the role of $\mathcal{KGC}$. Finally $\mathcal{A}$ receives the outputs of $\mathcal{U}$ algorithms (The outputs may be one of the four forms defined in definition 4). $\mathcal{A}$ then returns a bit b' as its answer. Since in the BlindKeyGen protocol, the blinded attributes are $h'_1, \ldots, h'_t$, where $h'_j = g_1^{a_j} h_j g^{r_j}$, $j = 1, \ldots, t$. Clearly, h'_j is uniformly distributed in G. Moreover, the PoK of $(r_1, \ldots, r_t, a_1, \ldots, a_t)$ is zero-knowledge, so $\mathcal{A}$ cannot determine which attributes the user requests, and $\mathcal{A}$ cannot cause failure depending on the user's attributes. Furthermore, since there is a random value z in the unblinded private keys, we can finally conclude that $\mathcal{A}$ cannot distinguish between $\mathcal{U}(params, \omega_b)$ and $\mathcal{U}(params, \omega_{1-b})$ with non-negligible probability.

B.3 Proof of Theorem 4

Suppose $\mathcal{A}$ has advantage ϵ in attacking the ABE system. We show how to use the adversary $\mathcal{A}$ to build an algorithm $\mathcal{B}$ that solves the DBDH problem with advantage $\epsilon/2$. Algorithm $\mathcal{B}$ is given as input a random tuple (g, g^a, g^b, g^c, Z), where $Z = e(g, g)^{abc}$ or is sampled from G_T. Algorithm $\mathcal{B}$ works by interacting with $\mathcal{A}$ in a IND-sAtt-CPA game as follows:

- Initialization $\mathcal{A}$ chooses the challenge access tree τ^* and sends it to $\mathcal{B}$.
- Setup $\mathcal{B}$ selects random $\alpha_1, \ldots, \alpha_l, t \in Z_p$. If $a_j \in \tau^*$, $\mathcal{B}$ sets $t_j \leftarrow a_j$, otherwise $t_j \leftarrow (a_j - t)$, $\forall a_j \in \Omega$. Let $h_j \leftarrow g_1^{t_j} g^{\alpha_j}$, then

$$F_j(a_j) = g_1^{a_j} h_j = \begin{cases} g^{\alpha_j}, & a_j \in \tau^*, & (1a) \\ g_1^t g^{\alpha_j}, & a_j \notin \tau^*, . & (1b) \end{cases}$$

- **Phase 1** $\mathcal{A}$ makes secret key requests for any set of attributes $\omega = \{a_j \mid a_j \in \Omega\}$ with the restriction that $\omega \nVdash \tau^*$. On each request $\mathcal{B}$ chooses a random $r \in Z_p$, and computes

$$d_0 = g_2^{(-1/t)} g^r = g^{r-(b/t)},$$

$$d_j = g_2^{(-\alpha_j/t)} F_j(a_j)^r = g_2^{(-\alpha_j/t)} (g_1^t g^{\alpha_j})^r = g_2^a F_j(a_j)^{r-(b/t)}, \forall a_j \in \omega.$$

Let $\tilde{r} = r - (b/t)$, then $d_0 = g^{\tilde{r}}$, $d_j = g_2^\alpha F_j(a_j)^{\tilde{r}}$, $\forall a_j \in \omega$.

- **Challenge** $\mathcal{A}$ outputs two messages $m_0, m_1 \in G_T$. $\mathcal{B}$ picks a random bit $b \in \{0,1\}$, and returns the encryption of m_b. The encryption is generated as follows.

1. Let $C_0 = m_b \cdot Z$.
2. Set the value of the root node of τ^* to be g^c, mark all child nodes as un-assigned, and mark the root node assigned.
   - If the symbol is "$\wedge$" and its child nodes are marked un-assigned, for each child $a_{j,i}$ except the last one $\mathcal{B}$ chooses random values $s_i \in Z_p^*$, and assigns $C_{j,i}^1 = g^{s_i}$, $C_{j,i}^2 = (g^{-s_i})^{\alpha_j}$ to them, and to the last child it assigns $C_{j,i}^1 = g^{s_v} = g^c/(\prod_{i=1}^{v-1} g^{s_i})$, $C_{j,i}^2 = (g^{s_v})^{-\alpha_j} = F_j(a_j)^{-s_v}$. Since $F_j(a_j) = g^{\alpha_j}$ when $a_j \in \tau^*$, the equality above holds.
   - If the symbol is "$\vee$", set the values of each child node to be g^c. Mark this node assigned.

   Hence, if $Z = e(g,g)^{abc} = e(g_1, g_2)^c$, then $C = (C_0, \forall a_{j,i} \in \tau^* : C_{j,i}^1, C_{j,i}^2)$ is a valid encryption of m_b.

- **Phase 2** $\mathcal{A}$ continues secret key requests with the same restriction as in Phase 1.
- **Guess** Finally, $\mathcal{A}$ outputs a guess $b' \in \{0,1\}$. $\mathcal{B}$ outputs a bit $u \in \{0,1\}$ as the response to the DBDH problem.
  If $b = b'$, $\mathcal{B}$ lets $u = 1$ meaning $Z = e(g,g)^{abc}$, otherwise lets $u = 0$ meaning $Z \neq e(g,g)^{abc}$.
  Then the overall advantage of $\mathcal{B}$ to solve DBDH problem is:

$$\frac{1}{2} Pr[u = 1 \mid Z = e(g,g)^{abc}] + \frac{1}{2}[u = 0 \mid Z \neq e(g,g)^{abc}] - \frac{1}{2}$$

$$= \frac{1}{2} Pr[b = b' \mid Z = e(g,g)^{abc}] + \frac{1}{2} Pr[b \neq b' \mid Z \neq e(g,g)^{abc}] - \frac{1}{2}$$

$$= \frac{\epsilon}{2}.$$

Fault Attacks on the Montgomery Powering Ladder

Jörn-Marc Schmidt[1] and Marcel Medwed[1,2]

[1] Graz University of Technology
Institute for Applied Information Processing and Communications
Inffeldgasse 16a, A–8010 Graz, Austria
joern-marc.schmidt@iaik.tugraz.at
[2] Université catholique de Louvain, Crypto Group, Belgium
marcel.medwed@uclouvain.be

Abstract. Security-aware embedded devices which are likely to operate in hostile environments need protection against physical attacks. For the RSA public-key algorithm, protected versions of the Montgomery powering ladder have gained popularity as countermeasures for such attacks.

In this paper, we present a general fault attack against RSA implementations which use the Montgomery powering ladder. In a first step, we discuss under which realistic fault assumptions our observation can be used to attack basic implementations. In a second step, we extend our attack to a scenario, where the message is blinded at the beginning of the exponentiation algorithm. To the best of our knowledge this is the first fault attack on a blinded Montgomery powering ladder.

Keywords: Montgomery Powering Ladder, Fault Attack, Blinded Exponentiation, Quadratic Residue.

1 Introduction

In order to judge the security of the practical realization of a cryptographic algorithm, the way it is implemented is as important as the theoretical security of the algorithm itself. This is because an adversary can try to manipulate the computation and reveal secrets from the erroneous output of the device. These so-called fault attacks were first described against public-key schemes by Boneh, DeMillo and Lipton [1]. They showed for example how to recover an RSA secret exponent by disturbing one of the two exponentiations of a CRT-RSA computation. In the following years, further fault attacks on the RSA algorithm were published [2,3]. In particular, those two works target the square-and-multiply exponentiation algorithm. They iteratively disturb and skip respectively the square operation in order to recover the secret key bit-by-bit.

In practice, a straightforward implementation of the square-and-multiply algorithm is also weak against other side-channel attacks like timing attacks [4] and Simple Power Analysis [5]. Therefore, the square-and-multiply algorithm is often replaced by more regular algorithms like the Montgomery powering ladder. This replacement does not only thwart some side-channel attacks but also

K.-H. Rhee and D. Nyang (Eds.): ICISC 2010, LNCS 6829, pp. 396–406, 2011.

hardens an adversary's life for fault attacks. For instance, in [6] it is stated that higher-order faults (several faults per algorithm invocation) are needed to apply their attack to the Montgomery ladder.

An attack that combines side-channel analysis and fault attacks was presented by Park et at. [7]. In their work, they induce single faults either during the multiplication or during the squaring and check which future intermediate results are affected by looking at the device's power consumption.

Our Contribution: In this paper, we present a general fault attack against the Montgomery powering ladder. We show that various fault models can be used to put the attack into practice, such as tampering with the intermediate variables or with the program flow. Moreover, we show how our approach can be extended to blinded implementations. In particular, our attack can defeat the blinded implementation of the Montgomery ladder proposed by Fumaroli and Vigilant [8]. To the best of our knowledge this is the first fault attack on a blinded Montgomery powering ladder.

The remaining paper is organized as follows: Section 2 gives a brief introduction into RSA and the Montgomery powering ladder. The general attack method is detailed in Section 3. Afterwards, Sections 4 and 5 show how to launch the attack in a realistic scenario. Finally, we extend the attack to work on blinded implementations in Section 6. The complexity of the attacks is discussed in Section 7. Conclusion is drawn in Section 8.

2   Preliminaries

In this section, we first discuss the RSA public-key algorithm. Afterwards, we look at the Montgomery powering ladder and its protected version as proposed by Fumaroli and Vigilant [8]. Finally, we revisit the Jacobi symbol.

The security of RSA is based on the hardness of factoring the product of two large primes. Let p and q be such primes, $n = pq$ their product, and $\varphi(\cdot)$ denote Euler's totient function. All computations of the RSA take place in the ring $\mathbf{Z}_n$. The public exponent e is an element of $\mathbf{Z}^*_{\varphi(n)}$, its corresponding secret exponent is $d = e^{-1} \mod \varphi(n)$. Due to this construction, $m = (m^e)^d \mod n$ holds for any $m \in \mathbf{Z}_n$. The owner of the secret key can sign messages by computing $s = m^d \mod n$ or decrypt ciphertexts by calculating $m = c^d \mod n$. By giving away the public key (N, e), he enables everybody else to either verify signatures ($m = s^e \mod n$) or to encrypt messages ($c = m^e \mod n$). In order to omit side-channel attacks on the exponentiation, it is recommended to use regular exponentiation algorithms, which leak as little information as possible about the secret exponent. As mentioned before, one such regular algorithm to compute a modular exponentiation is the Montgomery powering ladder [9]. It is depicted in Algorithm 1.

In each iteration of the ladder, one intermediate is assigned the product of both, the other one is squared. If the current bit is one, R_0 is set to $R_0 \cdot R_1$ and R_1 is squared, and vice versa if the bit is zero. Let $d = (d_{t-1}, \ldots, d_0)_2 = [d_L, d_i, d_T]$. Here, d_i denotes the bit which is currently processed and d_L the

Algorithm 1. Montgomery ladder [9]	**Algorithm 2.** Protected ladder [8]
Require: n, $d = (d_{t-1}, \ldots, d_0)_2$, $m \in \mathbf{Z}_n$	**Require:** $d = (d_{t-1}, \ldots, d_0)_2$, $m \in \mathbf{Z}_n$
Ensure: $m^d \bmod n$	**Ensure:** $m^d \bmod n$
	$r = rand() \in \mathbf{Z}_n^*;\ R_2 = r^{-1} \bmod n$
$R_0 = 1$	$R_0 = r$
$R_1 = m$	$R_1 = r \cdot m \bmod n$
for $i = t - 1$ **downto** 0 **do**	**for** $i = t - 1$ **downto** 0 **do**
$\quad R_{\bar{d}_i} = R_0 \cdot R_1 \bmod n$	$\quad R_{\bar{d}_i} = R_0 \cdot R_1 \bmod n$
$\quad R_{d_i} = R_{d_i}^2 \bmod n$	$\quad R_{d_i} = R_{d_i}^2 \bmod n$
	$\quad R_2 = R_2^2 \bmod n$
end for	**end for**
return R_0	return $R_0 \cdot R_2 \bmod n$

already processed (Leading) bits. The remaining (Trailing) bits are denoted by d_T. The intermediates after processing the bit d_i are

$$
(R_0, R_1) = \begin{cases} (m^{2 \cdot d_L} \bmod n, & m^{2 \cdot d_L + 1} \bmod n) \ \text{for}\ d_i = 0 \\ (m^{2 \cdot d_L + 1} \bmod n, m^{2 \cdot d_L + 2} \bmod n) \ \text{for}\ d_i = 1. \end{cases}
$$

A basic property of the Montgomery ladder is that the quotient $\frac{R_1}{R_0}$ is constant. This property is important for our attack. In a correct execution of the RSA algorithm this quotient equals m.

In order to achieve further side-channel resistance, Fumaroli and Vigilant suggested a blinded version [8]. Their proposal is depicted in Algorithm 2. It includes a random mask r, which is a factor of both intermediates, R_0 and R_1. Hence, r is squared in each step in both variables. As a consequence, the result includes the factor r^{2^t}. By squaring the inverse of r in every iteration we get r^{-2^t} at the end of the algorithm. Thus the result can be unblinded by calculating $R_0 \cdot R_2 \bmod n$.

To counteract fault attacks, they suggested to include a checksum which is updated at the end of every iteration. If one or more iterations (key bits) are skipped due to a fault attack, the checksum is invalid at the end of the algorithm. This prevents an adversary from tampering with the loop counter. However, this checksum cannot detect our attack. Therefore, it is not included in Algorithm 2.

Finally, we use the Jacobi symbol for the attack on the blinded Montgomery ladder. The Jacobi symbol is a generalization of the Legendre symbol for composite moduli and is defined as

$$
\left(\frac{c}{k}\right) = \left(\frac{c}{p_1}\right)^{\alpha_1} \left(\frac{c}{p_2}\right)^{\alpha_2} \cdots \left(\frac{c}{p_k}\right)^{\alpha_k} \quad \text{where } k = p_1^{\alpha_1} p_2^{\alpha_2} \cdots p_k^{\alpha_k}
$$

with $\left(\frac{c}{p}\right)$ denoting the Legendre symbol for primes p

$$
\left(\frac{c}{p}\right) = \begin{cases} 0 \ \text{if}\ c = 0 \pmod{p} \\ +1 \ \text{if}\ c \neq 0 \pmod{p} \text{ and } \exists x \text{ s.t. } x^2 = c \pmod{p} \\ -1 \ \text{otherwise.} \end{cases}
$$

The Jacobi symbol can be efficiently computed using the law of quadratic reciprocity (see e.g. [10]), even if the factorization of the modulus is not known. Note that the Jacobi symbol only gives a guarantee for quadratic non-residues by evaluating to -1, but a result of 1 does not imply that c is a quadratic residue modulo k.

3   General Attack Method

As already mentioned, the aim of a fault attack is to deduce information on secrets involved in the internal computation of the device. In most cases, this is done by manipulating the device and analyzing its erroneous output[1]. Normally, an adversary is assumed to have access to the attacked device and can manipulate it. Hence, the inputs of the device are chosen by the adversary.

In order to successfully attack a device, it is important to know the algorithm that is computed internally. Furthermore, it is necessary to make assumptions about the faults that occur and to set up relations between the intermediates processed inside the device and the erroneous output.

In this paper, we discuss such relations for the Montgomery powering ladder. A fault attack on its implementation can exploit a general observation: For two arbitrary values in R_0 and R_1 the algorithm behaves as in Table 1. For $a = 1$

Table 1. The Montgomery powering ladder starting with R_0 and R_1 set to arbitrary values. $(\cdot)^d$ denotes the application of the ladder with the exponent d of length t.

Step	R_0	R_1
	a	b
$=$	a	$a \cdot \frac{b}{a}$
$(\cdot)^d$	$a^{2^t} \cdot (\frac{b}{a})^d$	$a^{2^t} \cdot (\frac{b}{a})^{d+1}$
$=$	$a^{2^t - d} \cdot b^d$	$a^{2^t - d} \cdot b^d \cdot \frac{b}{a}$

and $b = m$ this evaluates directly to m^d. In a fault attack an adversary would set one of the two intermediates to a random (or partially random) value during the exponentiation. As before, let $d = (d_{t-1}, \ldots, d_0)_2 = [d_L, d_i, d_T]$. After the computation of d_L, one intermediate is modified to contain a random value z. As a consequence, the intermediate values develop as depicted in Table 2. This basic property can be exploited to retrieve the secret exponent. Therefore, the fault must be injected in a way it is predictable. What is left is to reduce the number of unknown bits of the exponent in the equation by exploiting the property of RSA that $d = e^{-1} \pmod{\varphi(n)}$. In the following, we are considering different

---

[1] Note that for some fault attacks the behavior of the device itself, after a fault is injected, is sufficient [11].

Table 2. Fault injection during Montgomery ladder computation

| Step | Fault in R_0 | | Fault in R_1 | |
	R_0	R_1	R_0	R_1
	1	m	1	m
$(\cdot)^{d_L}$	m^{d_L}	m^{d_L+1}	m^{d_L}	m^{d_L+1}
Fault	z	m^{d_L+1}	m^{d_L}	z
Output	$m^{(2^i \cdot d_i + d_T)\cdot(d_L+1)} \cdot z^{2^i-(2^i \cdot d_i + d_T)}$		$m^{(2^i-(2^i \cdot d_i + d_T))\cdot d_L} \cdot z^{(2^i \cdot d_i + d_T)}$	

fault models that are suitable for an attack and show how to proceed in these cases. Furthermore, we discuss how to mount an attack on blinded versions of the algorithm.

4   Fault Model: A Guessable Fault

In the first model, we assume that an intermediate variable of an RSA implementation using the Montgomery powering ladder is modified by a fault at a known point in time. Furthermore, this fault influences the variable in a way that the result is within a limited range that can be searched exhaustively.

In order to inject such a fault, an adversary can use a focused laser beam to flip/set some bits in a register [12]. Another method to cause a fault is to interrupt the loading of a word into a register, e.g. by injecting glitches or spikes [13]. Hence, the size of the fault depends on the word size of the microcontroller. Both methods can control the point in time when the fault is injected very precisely.

In particular, if fault w is injected into R_0 after processing the exponent bits of d_L (a fault in R_1 leads to similar equations), the intermediate variables (R_0, R_1) contain $(m^{d_L} \oplus w, m^{d_L+1})$. As a consequence, the final (erroneous) signature is

$$\tilde{S} = m^{(2^i \cdot d_i + d_T)\cdot(d_L+1)} \cdot (m^{d_L} \oplus w)^{2^{i+1}-(2^i \cdot d_i + d_T)} \pmod{n}.$$

Since we assume that the fault is injected at the beginning of the computation, d_L is small and hence guessable, while d_T is not. However, we can substitute $d_T + 2^i \cdot d_i = d - 2^{i+1} \cdot d_L$:

$$\tilde{S} = m^{(d-2^{i+1}\cdot d_L)\cdot(d_L+1)} \cdot (m^{d_L} \oplus w)^{2^{i+1}-(d-2^{i+1}\cdot d_L)} \pmod{n}.$$

From this we can eliminate d because we know e and we further know that $ed = 1 \pmod{\varphi(n)}$. Raising $\tilde{S}$ to the power of e delivers an expression in which only w and d_L remain unknown:

$$\tilde{S}^e = m^{(1-2^{i+1}\cdot e\cdot d_L)\cdot(d_L+1)} \cdot (m^{d_L} \oplus w)^{2^{i+1}i\cdot e+2^{i+1}\cdot d_L\cdot e-1} \pmod{n}. \qquad (1)$$

Now we can test hypotheses for w and d_L. For correctly guessed values equation (1) must hold.

The observations above allow an iterative attack: Inject an appropriate fault while the first bits of d are processed. Using the public exponent, the result can be transferred into a value that depends on the message and a few unknown bits. Now the hypotheses for the fault and d_L can be tested against this value. A correct hypothesis increases the knowledge about d. The attack is repeated until the whole exponent is known.

Note that the same attack is possible injecting a fault into R_1 by substituting R_1 by $R_1 \oplus w$ in Table 2. If a register that stores either R_0 or R_1 is attacked, both possibilities have to be checked simultaneously.

5   Fault Model: Skipping an Instruction

Another possible fault model is based on a modification of the program flow. Instead of manipulating the data directly by flipping bits, an instruction is not executed. This reduces the overhead generated by guessing the flipped bits, since only the position of the skipped instruction is required, which depends on the point in time the fault is injected[2]. Using the public exponent e the same way as in the previous attack delivers a value which contains only a small part of the unknown secret exponent. In this way, the whole exponent can be determined iteratively.

Let m be a message to be signed using the exponent $d = [d_L, d_i, d_T]$ with d_i the bit that is processed as the squaring is skipped. The resulting equation depends on d_i, because if it is zero, a squaring of R_0 is skipped, while for a one the squaring of R_1 is left out.

First, assume $d_i = 0$. By skipping the squaring, R_0 stays unchanged and R_1 contains the value $m^{2 \cdot d_L + 1}$. This can be seen as skipping d_i and changing the quotient to $d_L + 1$. Together with the last line of Table 2 and $d = 2^{i+1} \cdot d_L + d_T$ this results in:

$$\tilde{S} = m^{(2^i - d_T) \cdot d_L + (2d_L + 1) \cdot d_T} \quad (\text{mod } n)$$

$$= m^{2^i \cdot d_L + d - 2^{i+1} \cdot d_L + d_L \cdot (d - 2^{i+1} \cdot d_L)} \quad (\text{mod } n)$$

$$\Rightarrow \tilde{S}^e = m^{1 + d_L - e \cdot 2^i \cdot d_L \cdot (1 + 2 \cdot d_L)} \quad (\text{mod } n).$$

Table 3 details the content of the intermediate variables and the quotient for a skipped squaring of $d_i = 0$. After the quotient between R_0 and R_1 is changed by the fault, it stays constant for the rest of the computation. For $d_i = 1$, R_1 stays constant and R_0 changes to $2 \cdot d_L + 1$. Together with $d = 2^{i+1} \cdot d_L + 2^i + d_T$, we get:

$$\tilde{S} = m^{d_T \cdot (d_L + 1) + (2^i - d_T) \cdot (2d_L + 1)} \quad (\text{mod } n)$$

$$= m^{2^i \cdot (1 + d_L \cdot (3 + 2 \cdot d_L)) - d_L \cdot d} \quad (\text{mod } n)$$

$$\Rightarrow \tilde{S}^e = m^{e \cdot 2^i \cdot (1 + d_L \cdot (3 + 2 \cdot d_L)) - d_L} \quad (\text{mod } n).$$

---

[2] In this model, we allow the fault injection to be imprecise, since it is possible to check whether the fault is exploitable for our attack.

Table 3. Intermediates of a Montgomery Powering ladder with a skipped squaring instruction while $d_i = 0$ was processed

Step		R_0	R_1	Quotient
after d_L		m^{d_L}	m^{d_L+1}	m
after d_i		m^{d_L}	$m^{2 \cdot d_L+1}$	m^{d_L+1}
next bit	$d_{i+1} = 0$	$(m^{d_L})^2$	$m^{3 \cdot d_L+1}$	m^{d_L+1}
	$d_{i+1} = 1$	$m^{3 \cdot d_L+1}$	$m^{4 \cdot d_L+2}$	m^{d_L+1}
Erroneous Output		$m^{(2^i - d_T) \cdot d_L + (2d_L+1) \cdot d_T}$		

The same equations can be set up for a skipped multiplication:

$$\tilde{S}^e = \begin{cases} m^{1-d_l \cdot (1-2^{i+1} \cdot d_l \cdot e)} \quad (\bmod\ n) & \text{for } d_i = 0 \\ m^{(2+d_L) \cdot (1-2^{i+1} \cdot e \cdot (1+d_L))} \quad (\bmod\ n) & \text{for } d_i = 1. \end{cases}$$

Hence, a Montgomery powering ladder can be attacked by iteratively skipping either squarings or multiplications and calculating the expected values. If they do not match, the fault was not injected in the intended way.

Note that the attack works analogously starting from the least-significant bits of the exponent. Furthermore, the attack can be applied to algorithms that are based on ECC. Moreover, for an ECDSA implementation, guessing a block of several bits for each ephemeral key and building a lattice is also possible, like it is done in [6] for a double-and-add algorithm.

6   Attack on a Blinded Implementation

In order to provide a side-channel secure implementation, Fumaroli and Vigilant proposed a blinded version of the Montgomery ladder [8]. Their suggestion also includes a signature for preventing an adversary from tampering with the loop counter. In this section, we assume the same fault model as in the previous one. The only difference is that a precise fault injection is required. On the one hand, each fault that is injected into only one operation i.e. the squaring or the multiplication is not detected by the checksum. On the other hand, such a fault produces an unpredictable output, since a part of the random mask is still included in the return value. This makes a direct guessing of bit chunks as in the previous attacks impossible. But there is still one bit of exploitable information left in the output, namely if the result is a quadratic residue. More precisely, we can compute the Jacobi symbol of the result, which indicates a quadratic non-residue if it is negative[3]. A schematic view of the attack on a blinded implementation is given in Algorithm 3.

---

[3] Note that a positive result does not imply a quadratic residue.

Algorithm 3. Schematic of the Attack

Require: A device that can be manipulated and uses the blinded Montgomery ladder
to produce faulty signatures $\tilde{S}$.
Ensure: The exponent $d = (d_{t-1}, \ldots, d_0)_2$ that is used by the device.
   Set $d_{t-1} = 1$ (leading zeros are neglected)
   for $i = t - 2$ **downto** 0 **do**
      Choose $m \in \mathbf{Z}_n$ with $\left(\dfrac{m}{n}\right) = -1$
      Calculate $\tilde{S}$ with the ith squaring operation skipped
      if $\left(\dfrac{\tilde{S}}{n}\right) = -1$ **then**
         $d_i = d_{i+1}$
      else
         $d_i = 1 \oplus d_{i+1}$
      end if
   end for
   return d

Taking a closer look at the result shows that if a squaring is skipped during
the processing of Algorithm 2, the result $\tilde{S}$ is

$$\tilde{S} = R_2^{(2^t)} \cdot r^{2^{t-1} + 2^{i-1} \cdot u} \cdot m^{\tilde{d}} \text{ with}$$

$$u = \begin{cases} d_T & \text{for } d_i = 0 \\ 2^i - d_T & \text{for } d_i = 1 \text{ and} \end{cases}$$

$$\tilde{d} \cdot e = \begin{cases} 1 + d_L - e \cdot 2^i \cdot d_L \cdot (1 + 2 \cdot d_L) \pmod{\varphi(n)} & \text{for } d_i = 0 \\ e \cdot 2^i \cdot (1 + d_L \cdot (3 + 2 \cdot d_L)) - d_L \pmod{\varphi(n)} & \text{for } d_i = 1. \end{cases}$$

Hence, the result can be split up into an unknown part, which includes the
random mask and another one that depends on the input message, on the exponent, and of the position of the fault. Raising the resulting $\tilde{S}$ to the power e
cancels the unknown bits of d_T out. If the fault is chosen in a way that only d_i is
unknown and d_L is known, the whole message-dependent part of the signature
depends on the one bit d_i. Furthermore, it follows that it directly depends on
this bit, whether the result is a quadratic residue assuming that m is a quadratic
non-residue. This is because the remaining part of the random mask is always a
quadratic residue due to its exponent, which is a multiple of two. In detail, if m
is chosen with a Jacobi symbol $\left(\dfrac{m}{n}\right) = -1$, $\tilde{S}$ is a quadratic non-residue with
$\left(\dfrac{\tilde{S}}{n}\right) = -1$, iff $\tilde{d}$ is odd. Moreover, whether $\tilde{d}$ is even or not depends on the last
bit of d_L and d_i. Since d_L is known, the knowledge of the Jacobi symbol of $\tilde{S}$
determines d_i. This is because $\varphi(n)$ is always even and d is always odd in the
case of RSA. Thus, computing the Jacobi symbol leads to an attack similar to
the one presented by Boreale on square-and-multiply [2]. In contrast, our result

is not probabilistic, if m can be chosen with a negative Jacobi symbol. Since the loop itself is not manipulated, a checksum cannot prevent this attack.

6.1   Practical Considerations

The possibility of influencing the program flow by means of spike attacks was demonstrated in [3]. In their work, the authors used the resulting fault model for an attack on a square-and-multiply implementation. It turned out that in a practical attack the probability of a successfully injected fault is smaller one. Thus, a method to check whether the output of the device is the intended result or not is favorable if not mandatory. For the unblinded version of the Montgomery ladder, this can be easily done by checking if the output corresponds to one of the desired (precalculated) results. For the blinded version, this is not possible because a multiple of the random mask is still a part of the result. Thus, an adversary cannot tell from the result whether computing the Jacobi symbol delivers information about one bit of the exponent or not. Fortunately, there is another kind of information that makes it possible to recognize a successful fault injection. By measuring the time a computation takes, an adversary can judge whether a whole multiplication was skipped. Thus, the adversary can sweep through the algorithm and identify the positions where multiplications are invoked. Additionally, there is a way of telling the three different multiplications of one loop iteration of the ladder apart: A skipped multiplication and the skipped consecutive squaring respectively yield the same Jacobi symbol. For a skipped squaring of the mask, the result has always a negative Jacobi symbol, since d is odd. This leaves the adversary with the necessary tools for a successful attack.

7   Complexity of the Attacks

Table 4 overviews the efforts the different attacks require. The first two attacks allow to determine the exponent in chunks of several bits. This reduces the number of fault injections but increases the computational effort of the attack. This is because an exponentiation is required for each possible value of such a chunk to determine the corresponding exponent bits. Hence, it is possible to trade fault

Table 4. Required effort for the attacks with t denoting the bit-length of the secret exponent and c the bit-size of the chunks the exponent is recovered in

Attack Method (presented in Section)	Fault Injections	Exponentiations
Fault of Bit-Size v (4)	t/c	$t/c \cdot 2^{(c+v)}$
Skipping an Instruction (5)	t/c	$t/c \cdot 2^c$
Attack on Blinded Implementation (6)	t	-

injections for computational effort. In particular, if we assume a chunk of bit-size c and an exponent of bit-size t, t/c fault injections are required. For each faulty computation 2^c values have to be guessed and tested with the corresponding formulas, which requires an exponentiation each. In the first attack, the maximum possible bit-size v of the injected fault also needs to be considered. Each unknown bit doubles the number of required tests.

The attack on a blinded implementation of the algorithm recovers the exponent bit-by-bit. Thus, the number of injected faults equals the bit length of the exponent. Since the test involves only the computation of the Jacobi symbol, it does not require extra exponentiations.

8 Conclusion

In this paper, we presented new fault attacks on the Montgomery powering ladder. We demonstrated that our attacks are feasible for two realistic fault models: (1) for random register faults that can be guessed and (2) for a manipulation of the program flow. For both models, we discussed how to determine the secret exponent of an unprotected implementation in bit-chunks. In addition, it is possible to recognize a successful fault injection by the output of the device.

For the latter fault model, we also showed how to mount an attack on a blinded implementation. In this attack, the exponent was recovered bit-wise by measuring the execution time and checking the Jacobi symbol of the output. To the best of our knowledge, this is the first fault attack on a blinded Montgomery ladder.

The presented results show that fault attacks on the blinded Montgomery ladder are possible. All attacks require the adversary to know the plaintext. For the attack on the blinded version, knowing the Jacobi symbol of the plaintext is sufficient.

We conclude that blinding in combination with a loop-checksum does not prevent all fault attacks on the Montgomery powering ladder. Therefore, additional protection by exponent blinding or by a check whether the quotient between the two intermediate variables is correct should be implemented.

Acknowledgements. The work described in this paper has been supported in part through the Austrian Science Fund (FWF) under grant number P22241-N23. The information in this document reflects only the authors views, is provided as is and no guarantee or warranty is given that the information is fit for any particular purpose. The user thereof uses the information at its sole risk and liability.

References

1. Boneh, D., DeMillo, R.A., Lipton, R.J.: On the Importance of Checking Cryptographic Protocols for Faults (Extended Abstract). In: Fumy, W. (ed.) EUROCRYPT 1997. LNCS, vol. 1233, pp. 37–51. Springer, Heidelberg (1997)

2. Boreale, M.: Attacking right-to-left modular exponentiation with timely random faults. In: Breveglieri, L., Koren, I., Naccache, D., Seifert, J.-P. (eds.) FDTC 2006. LNCS, vol. 4236, pp. 24–35. Springer, Heidelberg (2006)

3. Schmidt, J.M., Herbst, C.: A Practical Fault Attack on Square and Multiply. In: Breveglieri, L., Gueron, S., Koren, I., Naccache, D., Seifert, J.P. (eds.) Proceedings of Fifth International Workshop on Fault Diagnosis and Tolerance in Cryptography, FDTC 2008, August 10, pp. 53–58. IEEE Computer Society, Washington DC, USA (2008)

4. Kocher, P.C.: Timing attacks on implementations of diffie-hellman, RSA, DSS, and other systems. In: Koblitz, N. (ed.) CRYPTO 1996. LNCS, vol. 1109, pp. 104–113. Springer, Heidelberg (1996)

5. Kocher, P.C., Jaffe, J., Jun, B.: Differential power analysis. In: Wiener, M. (ed.) CRYPTO 1999. LNCS, vol. 1666, pp. 388–397. Springer, Heidelberg (1999)

6. Schmidt, J.M., Medwed, M.: A Fault Attack on ECDSA. In: Naccache, D., Oswald, E. (eds.) Procceedings of Sixth International Workshop on Fault Diagnosis and Tolerance in Cryptography, FDTC 2009, Lausanne, Switzerland, September 6, pp. 93–99. IEEE-CS Press, Los Alamitos (2009)

7. Park, J., Bae, K., Moon, S., Choi, D., Kang, Y., Ha, J.: A New Fault Cryptanalysis on Montgomery Ladder Exponentiation Algorithm. In: Proceedings of the 2nd International Conference on Interaction Sciences: Information Technology, Culture and Human. ACM International Conference Proceeding Series, vol. 403, pp. 896–899. ACM Press, New York (2009)

8. Fumaroli, G., Vigilant, D.: Blinded fault resistant exponentiation. In: Breveglieri, L., Koren, I., Naccache, D., Seifert, J.P. (eds.) FDTC 2006. LNCS, vol. 4236, pp. 62–70. Springer, Heidelberg (2006)

9. Joye, M., Yen, S.M.: The Montgomery Powering Ladder. In: Kaliski Jr., B.S., Koç, Ç.K., Paar, C. (eds.) CHES 2002. LNCS, vol. 2523, pp. 291–302. Springer, Heidelberg (2003)

10. Menezes, A.J., van Oorschot, P.C., Vanstone, S.A.: Handbook of Applied Cryptography. Series on Discrete Mathematics and its Applications. CRC Press, Boca Raton (1997), http://www.cacr.math.uwaterloo.ca/hac/, ISBN 0-8493-8523-7

11. Yen, S.M., Joye, M.: Checking Before Output May Not Be Enough Against Fault-Based Cryptanalysis. In: IEEE Transactions on Computers, vol. 49, pp. 967–970. IEEE Computer Society, Los Alamitos (2000)

12. Skorobogatov, S.P., Anderson, R.J.: Optical Fault Induction Attacks. In: Kaliski Jr., B.S., Koç, Ç.K., Paar, C. (eds.) CHES 2002. LNCS, vol. 2523, pp. 2–12. Springer, Heidelberg (2003)

13. Bar-El, H., Choukri, H., Naccache, D., Tunstall, M., Whelan, C.: The Sorcerer's Apprentice Guide to Fault Attacks. Cryptology ePrint Archive Report 2004/100 (2004), http://eprint.iacr.org/

First Principal Components Analysis: A New Side Channel Distinguisher

Youssef Souissi[1], Maxime Nassar[1,2], Sylvain Guilley[1],
Jean-Luc Danger[1], and Florent Flament[1]

[1] TELECOM ParisTech, CNRS LTCI (UMR 5141),
46 rue Barrault
75 634 Paris Cedex, France
[2] BULL TrustWay
Rue Jean Jaurès, B.P. 68
78 340 Les Clayes-sous-Bois, France

Abstract. Side Channel Analysis (SCA) are of great concern since they
have shown their efficiency in retrieving sensitive information from secure
devices. In this paper we introduce First Principal Components Analysis (FPCA) which consists in evaluating the relevance of a partitioning
using the projection on the first principal directions as a distinguisher.
Indeed, FPCA is a novel application of the Principal Component Analysis (PCA). In SCA like Template attacks, PCA has been previously
used as a pre-processing tool. The originality of FPCA is to use PCA
no more as a preprocessing tool but as a distinguisher. We conducted all
our experiments in real life context, using a recently introduced practice-
oriented SCA evaluation framework. We show that FPCA is more performant than first-order SCA (DoM, DPA, CPA) when performed on
unprotected DES architecture. Moreover, we outline that FPCA is still
efficient on masked DES implementation, and show how it outperforms
Variance Power Analysis (VPA) which is a known successful attack on
such countermeasures.

Keywords: Principal Component Analysis (PCA), Data Encryption
Standard (DES), Side Channel Attacks (DoM, DPA, CPA, VPA), Masking countermeasures.

1 Introduction

Different forms of technologies, which require an adequate level of security, are
extensively manipulated around the world. Any violation of such systems could
lead to the loss of sensitive and personal information. In this context, Side Channel Analysis (SCA) pose a real threat to these technologies since they are non
intrusive, low cost and easily mounted in practice [16]. Actually, SCA exploit
the information leaked from cryptographic devices during the encryption or decryption process to extract the secret information referred to as *secret key*. This
information is retrieved by analysing the power consumption or the electromagnetic (EM) radiations of the device under attack. SCA are based on statistical

K.-H. Rhee and D. Nyang (Eds.): ICISC 2010, LNCS 6829, pp. 407–419, 2011.

computations to exhibit the secret. Indeed, the leaked information can be statistically modeled by a continuous random variable following an unknown or uncertain probability law P_{law}.

The main challenge of SCA is to make a sound estimation of P_{law} relevant features without loss of information. The more accurate this estimation is, the greater the efficiency of SCA is. Basically, random variables are measured and analyzed in term of their statistical and probabilistic features [7]. In the case of SCA, calculations based on the first and second order statistics seem to be good ways to quantify the secret information. For instance, Differential Power Analysis (DPA) is mainly based on computations related to the first-order statistic, the "mean". Moreover, Variance Power Analysis (VPA) [30, 18] which is based on the variance, has shown its efficiency on masked implementations.

Recently, a new powerful variant of SCA so-called MIA [8] has been presented to the cryptographic community. This attack is based on mutual information theory which requires a reliable estimation of the probability density function of P_{law}. Basically, an accurate probabilistic measure, such as the entropy, describes better one random variable than other statistics [24]. However, the optimal accuracy is hardly achieved specially when the probability law is unknown. As a matter of fact, the probability density of an unknown law is quite difficult to properly estimate when the available data to be studied is limited [7]. Statisticians are used to calculate quantities easier to estimate. These quantities are the moments of a probability distribution like the mean, the variance or the kurtosis. By analogy to the cryptographic domain, statisticians are identified to attackers and the available data to power or EM consumption signals. Indeed, the attacker is often required to conduct its attack under certain constraints. Actually, according to the security levels as defined by Abraham et al. [2], secure devices could be classified into seven levels of security. According to each level, the attacker behaves in different manners. In the real life, the attacker has to perform the attack by considering the external environment of the device under attack which depends on the factory and the type of the circuit (FPGA, ASIC, ...). For instance, some security measures could be employed to limit the acquisition of power consumption signals (*traces*). Thus, the attacker would not be free to acquire as much traces as he wants. In addition to that, we believe that any cryptographic design could be attacked by exploiting its sensitivity against one chosen statistic, denoted by CS, that could be the mean, the variance or any other statistic describing one P_{law}. The higher the sensitivity is, the greater is the vulnerability of the implementation against attacks based on the considered CS. This is true since an ideal cryptographic implementation could not really exist, in accordance with the fact that real life application could not fit exactly the theory.

In this paper, we outline the way how Principal Component Analysis (PCA [12]) could be used to extract the value of the secret key. PCA is a multivariate data analytic technique [24, 26] that has found application in fields such as computer vision [28, 15], robotics [34], sociology and economics [27]. It is a way of identifying patterns in multidimensional data set, and visualising these data into a lower dimensional space, in order to highlight their similarities and differences. In the SCA techniques portfolio, PCA has already revealed its efficiency on Template attacks [23]. Basically, Template attacks are considered very powerful since they can break cryptographic implementations which security is dependent on the assumption that an attacker cannot obtain more than one or a limited number of side channel traces. Moreover, these attacks require that an attacker has access to a clone device on which he can perform trials to get trained. As described in [23], PCA improves the class of Template attacks by pre-processing the leakage traces before performing the attack on the real cryptographic device. Indeed, in the pre-processing phase the attacker builds templates in order to profile the clone device. Then, those templates are used to mount an attack on the real device. Our attack uses PCA no more as pre-processing tool but as a distinguisher. Moreover, it follows the usual steps of differential power analysis (DoM [20], DPA [5] or CPA [6]) that consists of only one phase and does not require a clone device for profiling, which makes the task of the attacker easier.

The rest of the paper is organized as follows. First, Section 2 attempts to give some elementary background that is required to understand the process of PCA. Second, this background knowledge is taken advantage of in section 3 to outline the way how PCA could be exploited to mount an efficient attack. This section goes through the different steps needed to perform the FPCA. Section 4 is devoted to experiments on unprotected and protected DES implementations. This section highlights the efficiency of FPCA by making a comparative analysis with existing attacks (DoM, DPA, CPA, VPA). The conclusions and perspectives are in section 5.

2   Principal Component Analysis: Background Knowledge

Let a data set of M quantitative variables describing N samples, arranged respectively in rows and columns. The goal of PCA is to ensure a better representation of the N samples by describing the data set with a smaller number M' of new variables. Technically speaking, PCA proposes to seek a new representation of the N samples in a subspace of the initial space by defining M' new variables which are linear combinations of the M original variables, and that are called principal components. Generally speaking, reducing the number of variables used to describe data will lead to some loss of information. PCA operates in a way that makes this loss minimal. For PCA to work properly, the data set should be centred. PCA starts by computing the covariance matrix of the data set in order to find the eigenvectors and eigenvalues which permit the capture of the existing dispersion in variables. In other words, it makes a change of orthogonal reference frame, the new variables being replaced by the Principal Components which are totally characterized by the associations of the eigenvectors and eigenvalues. But, more importantly, these associations reveal the hidden dynamics of the data set. Determining this fact allows the attacker to discern which dynamics are important and which are just redundant. The first component can be expected to account for a fairly large amount of the total variance. Each succeeding component will account for progressively smaller amounts of variance. In practice, the attacker sorts eigenvectors by their eigenvalues, from the highest to the lowest. This gives the components in order of significance. Most of the time, only few M' components account for meaningful amount of variance. Thus, only these first M' components will be retained. The decision on the number of the M' best components could be achieved by performing some deciding tests such as the Kaiser criterion, the scree test or the cumulative variance criteria [13].

3   FPCA: The Attack Process

In the SCA field, PCA has often been used as pre-processing tool to minimize the coding complexity by reducing the dimensionality of recorded traces [3, 29]. By contrast, our approach is different in the sense that PCA is used as an attack tool to retrieve the secret information. Indeed, FPCA uses the projection on the first principal components to tell good *secret key* candidates from incorrect ones. FPCA shares some key points with first-order SCA, differential and correlation power analysis (DPA and CPA). As stated before, FPCA does not require a detailed knowledge about the cryptographic device to be performed. It exploits data dependency of the power consumption of the device under attack. The main difference with first-order SCAs resides in the way to distinguish the behaviour of the good key hypothesis. In fact, we remind that each attack has its own statistical test, referred to as distinguisher [30, 9], which allows the attacker to detect the value of the secret key. In this context, FPCA comes with a new distinguisher for side channel analysis. In the rest of this section, we detail the different steps needed to perform a FPCA, while introducing our notations at the same time. One schematic description of FPCA attack is depicted in Fig. 1.

3.1   Preliminary Preparation Phase

This phase is common with differential and correlation power attacks. Suppose that T power consumption traces are recorded while a cryptographic device is performing an encryption or a decryption operation. Collected traces are L-dimensional time vectors. The attacker chooses an intermediate result of the cryptographic algorithm that is processed by the cryptographic implementation. The intermediate value denoted by $f(d,k)$ is a function that takes two parameters. The first parameter denoted by d is a known data value that can be either the plain text or the cipher text. The number of data values is equal to T, the number of recorded traces. These known data values are represented by a vector $D_{vect} = (d_1, d_2, \ldots, d_D)$ of size D. The second parameter, denoted by k, is secret, hence unknown. Indeed, k is a small part of the cryptographic key and can take K possible values referred to as key hypotheses that we write as a vector $K_{vect} = (k_1, k_2, \ldots, k_K)$.

Thus, the *trace* can be written as a matrix of size $D \times L$. Given vectors D_{vect} and K_{vect}, the attacker is able to compute, without difficulties, the hypothetical intermediate value $f(d,k)$ for all K key hypotheses and for all T executed cryptographic operation. Then the attacker builds a matrix V of size $K \times T$:

$V_{i,j} = f(d_i, k_j)$ with $1 \leq i \leq T$ and $1 \leq j \leq K$. For each value $V_{i,j}$, the attacker computes a hypothetical power consumption value $h_{i,j}$ based on a power
consumption model. The most commonly used power models are the Hamming
distance (HD) and the Hamming weight (HW) [6]. R being the number of possible values that the power consumption model could take, the traces are arranged
in X ($X \leq R$) different partitions for each key hypothesis k_j. We denote these
partitions as a vector $P_{k_j} = (P_{k_j,1}, P_{k_j,2}, \ldots, P_{k_j,X})$ with $1 \leq j \leq K$. For instance, suppose that our power consumption model is the HD and that it can
take integral values from 0 to 4: $HD = \{0,1,2,3,4\} = \{HD_i\}_{i=1}^5$. The trivial
partitioning is to associate each HD_i value to one partition. Thus $X = R = 5$.
One other possibility is to build only $X=3$ partitions in this way : First partition
for $HD > 2$, second for $HD = 2$ and third for $HD < 2$. Intuitively, the more
accurate the used power model is, the better our description of the secret information will be. Many papers are dealing with the investigation of new power
models and techniques for traces classification [1, 21]. The optimal choice of the
power consumption model, including the partitioning process, is out of the scope
of this paper. In what follows, our study will focus on the Hamming distance
model as it is one of the most commonly used, and often one of the most efficient.

3.2   References Computation

Once traces are arranged in X partitions for each key hypothesis k_j, we propose
to compute for each partition a statistical trace based on one CS and referred
to as *reference*. For instance, if CS is the "mean" then the *reference* would be
the average of all traces that belong to the considered partition. Actually, the X
references of one key hypothesis k_j will be used by PCA as criterions to highlight
differences between the X partitions. For references computation, we notice that
the same CS (the mean, the variance ...) is used for all partitions and for all
key hypotheses k_j. One reference is an L-dimensional time vector. Thus we have
one dataset of X references, for each k_j. We denote this set by $V_{ref_{k_j}}$. In what
follows, our study will focus on analysing each dataset $V_{ref_{k_j}}$ corresponding to
each key hypothesis k_j. This analysis will allow the attacker to discriminate the
behavior of the secret key with regards to all other key hypotheses. Moreover, it
will reduce the computational complexity of the PCA step.

3.3   FPCA Distinguisher

For one key hypothesis k_j, the dependencies between references are made more
eligible by PCA, when the references are projected to the new axes system
composed by the principal components. The PCA is used to analyze these dependencies by measuring the dispersion of the references in the new coordinate
space. Indeed, the larger the eigenvalue, denoted by λ, corresponding to one
eigenvector is, the greater is the dispersion of the references on this eigenvector.
As stated by equation (1), the total variance of one $V_{ref_{k_j}}$ is equal to the sum
of all eigenvalues corresponding to all principal components:

$$V_{tot} = \sum_{j=1}^{L} \lambda_j \, . \tag{1}$$

Given a valid power consumption model and one CS, there are two cases to be discussed regarding the fluctuation of the total variance when increasing the number of recorded traces. The first case is the one for which the cryptographic implementation is not sensitive to the considered CS. In this case, PCA could not discriminate references of the secret key as well for the other key hypotheses.

The second case happens when the implementation is sensitive to the chosen CS. In this case, V_{tot} related to $V_{ref_{k_{secret\ key}}}$ is getting high by increasing the number of recorded traces. This can be explained by the fact that the secret key partitioning is the one for which the references are the most different. Intuitively, for an infinity of traces, V_{tot} converges towards the leakage value. By contrast, V_{tot} corresponding to one false key approaches the zero value when increasing the number of traces. This is due to the fact that PCA is not able to discriminate the references.

In order to highlight the dispersion of the references related to the secret key with regards to false keys, we carried out an experiment on DES [19] power consumption traces that are made freely available on line, in the context of the first version of DPA CONTEST competition [33]. The DES algorithm that has been selected for the competition is unprotected and easily breakable by first-order SCA. More details about this implementation could be found in [11]. For this purpose, we fixed the "mean" as CS and the Hamming distance as power consumption model. Fig. 2 shows the dispersion of references related to the secret key and one false key, when projected to the first and the second principal components. These principal components are the most significant given that they cover a high rate of the total variance so-called explained variance (EV). For the m-th principal component PC_m, this rate is defined by the following ratio:

$$EV(PC_m) = \lambda_m / V_{tot} \, ,$$

where λ_m is the eigenvalue corresponding to PC_m. For m' principal components, we introduce the cumulative explained variance (CEV) that is defined by:

$$CEV(PC_1, \ldots, PC_{m'}) = (\sum_{i=1}^{m'} \lambda_i) / V_{tot} \, .$$

In practice, last principal components are usually considered to be related to the noise contribution and only few m' components are retained for analysis.

The main idea behind using PCA is to reduce the dimensionality of power consumption traces in order to take account of the secret information for different time samples and thus to properly exploit the leakage. For this purpose we used the cumulative variance criteria to extract the significant components. For instance, we keep only the m' first components which explain more than 95% of the total variance, for each key hypothesis k_j.

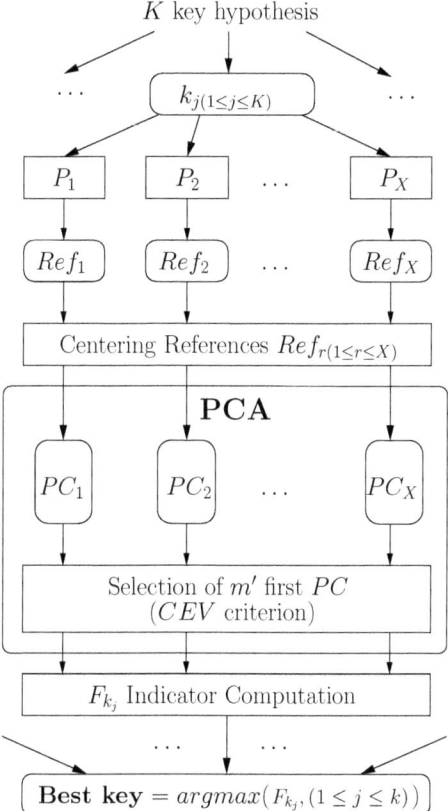

K key hypothesis

$k_{j(1 \leq j \leq K)}$

P_1 P_2 ... P_X

Ref_1 Ref_2 ... Ref_X

Centering References $Ref_{r(1 \leq r \leq X)}$

PCA

PC_1 PC_2 ... PC_X

Selection of m' first PC
(CEV criterion)

F_{k_j} Indicator Computation

...   ...

Best key $= argmax\big(F_{k_j}, (1 \leq j \leq k)\big)$

Fig. 1. FPCA description

Then, we propose to compute an indicator F_{k_j} that is defined as follows:

$$F_{k_j}^{CS} = \sum_{m=1}^{m'} \left(\lambda_m \cdot \Big| h(W, C^m)\Big|\right) = \sum_{m=1}^{m'} \left(\lambda_m \cdot \Big|\sum_{i=1}^{X} (w_i \cdot c_i^m)\Big|\right), \qquad (2)$$

where m' is the number of retained principal components, λ_m is the eigenvalue corresponding to PC_m, h is a linear combination function with $C^m = \{c_i^m\}_{i=1}^{X}$ is the centred coordinate vector of references when projected to PC_m and $W = \{w_i\}_{i=1}^{X}$ is the associated weight vector. Actually, this indicator takes two factors into consideration: the dispersion and the position of references in the new system coordinate which is composed by the principal components. The dispersion is quantified by the value of the eigenvalues λ_m and the position by the vector of weights W. The best key guess corresponds to the highest value of F_{k_j} regarding all key hypotheses $(argmax(F_{k_j}))$.

One other alternative is to consider only the factor of dispersion. This is useful in the case that the position factor is unknown. In fact, the dispersion factor

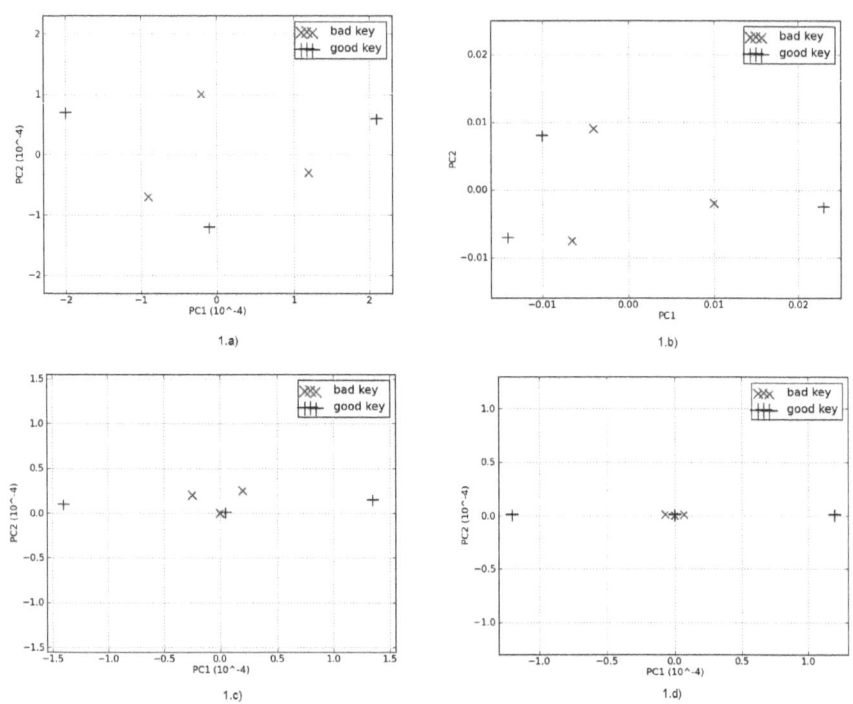

Fig. 2. References dispersion for different number of traces: (1.a) 100 traces,(1.b) 1000 traces, (1.c) 10000 traces, (1.d) 81000 traces

represents a global description of the leakage without the need of more detailed knowledge about the encryption process. The idea is that, if the key guess is correct, PCA applied to the different partitions should be able to explain a big proportion of the variance with only a few components. On the opposite, if the key guess is wrong, the power traces are sorted randomly, and PCA will need more components to explain the same proportion of variance. Thus, a reduced form of the indicator F_{k_j} is deduced from equation (2) and defined as follows:

$$Red\_F_{k_j}^{CS} = \sum_{m=1}^{m'} (\lambda_m).$$

This indicator can be used for Dual rail Precharge Logic (DPL) architectures, like WDDL [10], which aim at making the activity of the cryptographic process constant independently from the manipulated data. Statistically, the idea behind DPL is to force all references to have the same statistic and probabilistic features, for all made partitions. However, in the real life application, an ideal DPL implementation could not exist. In that sense, the reduced indicator $Red\_F_{k_j}^{CS}$ can exploit the leaked information without the knowledge of the position factor.

4   FPCA on DES Implementations

All our experiments were conducted on real power consumption traces recorded from three different hardware implementations of DES coprocessor. The first architecture is the unprotected DES of DPA CONTEST. The second and the third ones deal with two masking styles: USM and Masked-ROM [18] DES implementation which are configured in an Altera Stratix II FPGA on the SASEBO-B evaluation board provided by the RCIS [25]. Moreover, we note that the length of acquired side channel traces covers only the first two rounds for all investigated DES implementations.

Following the recent advances concerning the comparison of univariate side-channel distinguishers [30], Standaert et al. proposed two evaluation metrics [31] to assess the performance of different attacks. On one hand, the first-order success rate expresses the probability that, given a pool of traces, the attack's best guess is the correct key. On the other hand, the guessing entropy measures the position of the correct key in a list of key hypotheses ranked by a distinguisher.

In this paper, we deal with DoM, DPA, CPA, and VPA attacks. These attacks have shown their efficiency to break cryptographic implementations. Moreover, they are the basis of new derived distinguishers like the Spearman's rank correlation [4]; the correlation concept of Kendall is also of potential interest. Recently, Gierlichs et al. have presented an article dealing with the comparison of many existing distinguishers related to the aforementioned attacks [9]. For the reason that we aim at making a reliable evaluation for our attack (FPCA), the rest of the paper deals with experiments on unprotected and masked implementations which have been the target of the mentioned attacks.

4.1   FPCA on Unprotected DES

In order to mount a successful FPCA on unprotected DES we fixed the "mean" as CS, as it is shown that such implementation is very vulnerable against differential attacks which are generally based on the "mean" in their calculations. In fact, the leakage related to the mean is linearly correlated to the power consumption model $HD = \{0, 1, 2, 3, 4\}$. For this purpose, the weight vector W can be defined as follows: $W = \{-2, -1, 0, +1, +2\}$. One other alternative is to consider the probability that one trace belongs to one partition according to one power consumption model. Hence $W = \{-0.25, -1, 0, +1, +0.25\}$. Results regarding attacks on unprotected DES implementation are depicted in Fig. 3 and Fig. 4. Indeed, the first-order success rate shows a superior performance of FPCA attack. This can be explained by the fact that DoM, CPA, and DPA are implicitly taking into account only the position factor relatively to our proposed attack. According to Fig. 4, FPCA needs around 160 traces to perform a successful attack. Unsurprisingly, the guessing entropy metric depicted in Fig. 3 is in accordance with the first success rate results. One note is that FPCA is able to distinguish the *secret key* at an early stage. In fact, only 30 traces are required to get the *secret key* in the top ten of the key hypotheses rank list.

4.2   FPCA on Masked DES

Basically, masking technique is considered to be a powerful countermeasure against SCA. Indeed, it aims at masking the intermediate values that occur during encryption and decryption process. Many masking schemes have been proposed to the cryptographic community for symmetric encryption algorithms (DES, AES, ...) [32, 14, 22]. Basically, they differ in term of hardware design complexity. But, they all aim at fulfilling the same goal by ensuring the resistance against first-order SCA like DPA and CPA. Statistically, an ideal masking implementation is one for which all references, for all made partitions, are the same when using the mean as CS. However, it has been proved that masking technique is still susceptible to first-order SCA as long as glitches problem remains not completely resolved [32]. For instance, authors in [18], have shown that one masked structure so-called "Universal Substitution boxes with Masking" (USM) is vulnerable against DPA. Moreover, masked implementations are not resistant against new variants of SCA like VPA which is mainly based on the variance analysis. It is also shown that a full-fledged masked DES implementation using a ROM (Masked-ROM) is breakable by VPA attack, in spite of its high resistance against first-order attacks. In what follows, we use the same power consumption model as described in [18] to perform the FPCA on USM and Masked-ROM DES implementations.

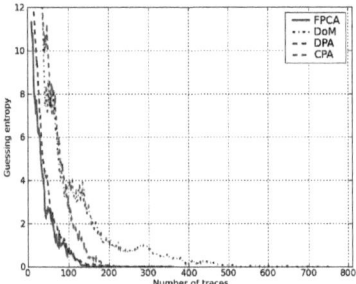

Fig. 3. Unprotected DES guessing entropy metric

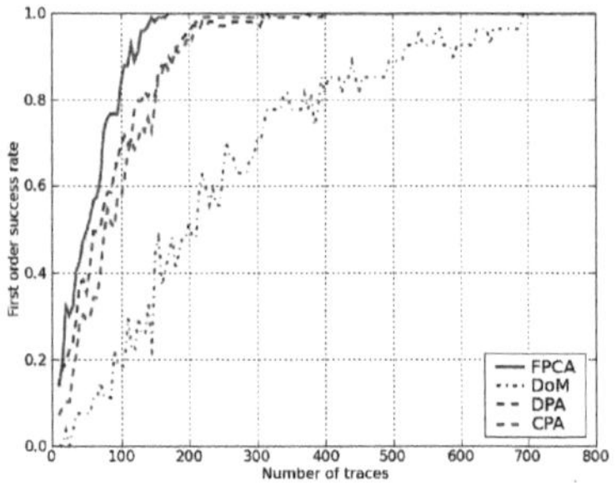

Fig. 4. Unprotected DES 1st-order success rate metric

First, in order to make a fair evaluation for our attack on USM DES structure we kept the "mean" as CS and we classified traces into five partitions for each key hypothesis k_j. For reasons of clarity, comparison is made between FPCA and DPA for which we realised the best performance with regards to DoM and CPA. Results are deduced from Fig. 5 and Fig. 6. Obviously, according to the first-order success rate metric shown in Fig. 6, FPCA is more efficient than DPA. Indeed, 15000 traces are needed for DPA to achieve a rate of 0.8. Whereas, for the same rate, FPCA attack requires only 10000 traces. The guessing entropy metric, is quite equivalent for both attacks. Second, we targeted a Masked-ROM DES implementation. For this purpose, we chose the variance as CS, as it has

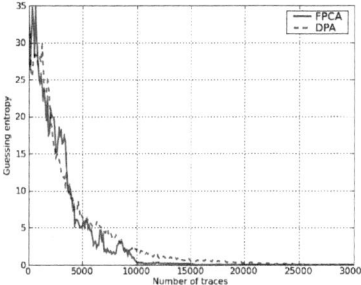

Fig. 5. USM DES guessing entropy metric

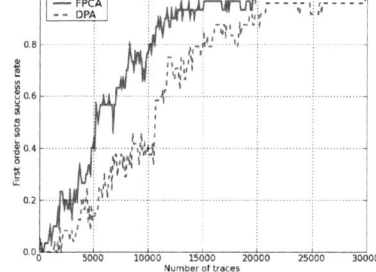

Fig. 6. USM DES 1st-order success rate metric

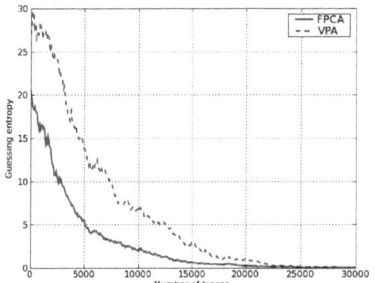

Fig. 7. Masked-ROM guessing entropy metric

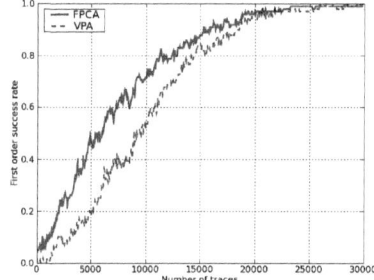

Fig. 8. Masked-ROM 1st-order success rate metric

shown that such implementation is sensitive to VPA, which is based on variance analysis [18,17]. Fig. 7 shows that the guessing entropy curve, related to FPCA, approaches the best rank (the zero rank) more rapidly than VPA. Moreover, the success rate metric depicted in Fig. 8 reveals noticeable differences between both attacks.

5   Conclusion and Outlooks

In this work we proposed a new variant of SCA called FPCA, which is mainly based on Principal Components Analysis (PCA), the powerful multivariate data analytic tool. We have shown the efficiency of FPCA on unprotected and protected cryptographic implementations. Moreover, we have empirically shown its superior performance with regards to existing attacks (DoM, DPA, CPA, VPA). Our future work consists in investigating new ways to improve FPCA. Actually, we are looking for new applications based on other multivariate data analytic tools such as the Linear Discriminant Analysis (LDA), PCA based Spearman correlation, Kernel PCA or Independent Component Analysis (ICA), which have

been proposed as new alternatives to the basic PCA. One other possible key point
that could be investigated is the improvement of our distinguisher by combining
different CS (the mean, the variance, the entropy . . .), in order to make more
eligible the description of the leakage related to the secret information.

References

1. Aabid, M.A.E., Guilley, S., Hoogvorst, P.: Template Attacks with a Power Model.
   Cryptology ePrint Archive, Report 2007/443 (December 2007),
   http://eprint.iacr.org/2007/443/
2. Abaraham, D.G., Dolan, G.M., Double, G.P., Stevens, J.V.: Transaction security
   system. IBM Systems Journal 30(2), 206–229 (1991)
3. Archambeau, C., Peeters, E., Standaert, F.-X., Quisquater, J.-J.: Template attacks in principal subspaces. In: Goubin, L., Matsui, M. (eds.) CHES 2006. LNCS,
   vol. 4249, pp. 1–14. Springer, Heidelberg (2006)
4. Batina, L., Gierlichs, B., Lemke-Rust, K.: Comparative Evaluation of Rank Correlation Based DPA on an AES Prototype Chip. In: Wu, T.-C., Lei, C.-L., Rijmen,
   V., Lee, D.-T. (eds.) ISC 2008. LNCS, vol. 5222, pp. 341–354. Springer, Heidelberg
   (2008)
5. Bevan, R., Knudsen, E.: Ways to Enhance Differential Power Analysis. In: Lee, P.J.,
   Lim, C.H. (eds.) ICISC 2002. LNCS, vol. 2587, pp. 327–342. Springer, Heidelberg
   (2003)
6. Brier, É., Clavier, C., Olivier, F.: Correlation Power Analysis with a Leakage Model.
   In: Joye, M., Quisquater, J.-J. (eds.) CHES 2004. LNCS, vol. 3156, pp. 16–29.
   Springer, Heidelberg (2004)
7. David, H., Nagaraja, H.N.: Order Statistics. Wiley, Chichester
8. Gierlichs, B., Batina, L., Tuyls, P., Preneel, B.: Mutual Information Analysis – A
   Generic Side-Channel Distinguisher. In: Oswald, E., Rohatgi, P. (eds.) CHES 2008.
   LNCS, vol. 5154, pp. 426–442. Springer, Heidelberg (2008)
9. Gierlichs, B., De Mulder, E., Preneel, B., Verbauwhede, I.: Empirical comparison of
   side channel analysis distinguishers on DES in hardware. In: IEEE (ed.) ECCTD.
   European Conference on Circuit Theory and Design, Antalya, Turkey, August 23-
   27, pp. 391–394 (2009)
10. Guilley, S., Chaudhuri, S., Sauvage, L., Hoogvorst, P., Pacalet, R., Bertoni, G.M.:
    Security Evaluation of WDDL and SecLib Countermeasures against Power Attacks.
    IEEE Transactions on Computers 57(11), 1482–1497 (2008)
11. Guilley, S., Hoogvorst, P., Pacalet, R.: A Fast Pipelined Multi-Mode DES Ar-
    chitecture Operating in IP Representation. Integration, The VLSI Journal 40(4),
    479–489 (2007), doi:10.1016/j.vlsi.2006.06.004
12. Jolliffe, I.T.: Principal Component Analysis. Springer Series in Statistics (2002)
    ISBN: 0387954422
13. Khattree, R., Naik, D.N.: Multivariate data reduction and descrimination (2000)
14. Koichi, I., Masahiko, T., Naoya, T.: Encryption secured against DPA, Fujitsu US
    Patent 7386130 (June 10, 2008),
    http://www.patentstorm.us/patents/7386130/fulltext.html
15. U. Kyungnam Kim Department of Computer Science University of Maryland. Face
    recognition using principal component analysis (February 26, 2002)
16. Le, T.-H., Canovas, C., Clédière, J.: An overview of side channel analysis attacks.
    In: ASIACCS, ASIAN ACM Symposium on Information, Computer and Commu-
    nications Security, Tokyo, Japan, pp. 33–43 (2008), doi:10.1145/1368310.1368319

17. Li, Y., Sakiyama, K., Batina, L., Nakatsu, D., Ohta, K.: Power Variance Analysis Breaks a Masked ASIC Implementation of AES. In: DATE 2010, Dresden, Germany, March 8-12, IEEE Computer Society, Los Alamitos (2010)
18. Maghrebi, H., Danger, J.-L., Flament, F., Guilley, S.: Evaluation of Countermeasures Implementation Based on Boolean Masking to Thwart First and Second Order Side-Channel Attacks. In: SCS, Jerba, Tunisia, November 6–8, IEEE, Los Alamitos (2009) Complete version available,
    http://hal.archives-ouvertes.fr/hal-00425523/en/
19. NIST/ITL/CSD. Data Encryption Standard. FIPS PUB 46-3 (October 1999),
    http://csrc.nist.gov/publications/fips/fips46-3/fips46-3.pdf
20. Kocher, P., Jaffe, J., Jun, B.: Differential power analysis. In: Wiener, M. (ed.) CRYPTO 1999. LNCS, vol. 1666, pp. 388–397. Springer, Heidelberg (1999)
21. Peeters, É., Standaert, F.-X., Quisquater, J.-J.: Power and electromagnetic analysis: Improved model, consequences and comparisons. Integration, The VLSI Journal, special issue on "Embedded Cryptographic Hardware" 40, 52–60 (2007),
    http://dx.doi.org/10.1016/j.vlsi.2005.12.013,
    doi:10.1016/j.vlsi.2005.12.013
22. Popp, T., Mangard, S.: Masked Dual-Rail Pre-charge Logic: DPA-Resistance Without Routing Constraints. In: Rao, J.R., Sunar, B. (eds.) CHES 2005. LNCS, vol. 3659, pp. 172–186. Springer, Heidelberg (2005)
23. Rechberger, C., Oswald, E.: Practical template attacks. In: Lim, C.H., Yung, M. (eds.) WISA 2004. LNCS, vol. 3325, pp. 440–456. Springer, Heidelberg (2005)
24. Saporta, G.: Probabilités analyse des données et statistiques (2008)
25. SASEBO board from the Japanese RCIS-AIST,
    http://www.rcis.aist.go.jp/special/SASEBO/index-en.html
26. Shlens, J.: A tutorial in Principal Component Analysis (December 10, 2005)
27. Kolenikov, S., Angeles, G.: The use of discrete data in PCA for socio-economic status evaluation (February 2, 2005)
28. Smith, L.I.: A tutorial in Principal Component Analysis (February 26, 2002)
29. Standaert, F.-X., Archambeau, C.: Using subspace-based template attacks to compare and combine power and electromagnetic information leakages. In: Oswald, E., Rohatgi, P. (eds.) CHES 2008. LNCS, vol. 5154, pp. 411–425. Springer, Heidelberg (2008)
30. Standaert, F.-X., Gierlichs, B., Verbauwhede, I.: Partition vs. comparison side-channel distinguishers: An empirical evaluation of statistical tests for univariate side-channel attacks against two unprotected cmos devices. In: Lee, P.J., Cheon, J.H. (eds.) ICISC 2008. LNCS, vol. 5461, pp. 253–267. Springer, Heidelberg (2009)
31. Standaert, F.-X., Malkin, T., Yung, M.: A unified framework for the analysis of side-channel key recovery attacks. In: Joux, A. (ed.) EUROCRYPT 2009. LNCS, vol. 5479, pp. 443–461. Springer, Heidelberg (2009)
32. Mangard, S., Schramm, K.: Pinpointing the Side-Channel Leakage of Masked AES Hardware Implementations. In: Goubin, L., Matsui, M. (eds.) CHES 2006. LNCS, vol. 4249, pp. 76–90. Springer, Heidelberg (2006)
33. TELECOM ParisTech SEN research group. DPA Contest 1st (edn.) (2008–2009)
    http://www.DPAcontest.org/
34. Hou, Z.G.: Principal component analysis (PCA) for data fusion and navigation of mobile robots. In: Kantor, P., Muresan, G., Roberts, F., Zeng, D.D., Wang, F.-Y., Chen, H., Merkle, R.C. (eds.) ISI 2005. LNCS, vol. 3495, pp. 610–611. Springer, Heidelberg (2005)

Fault Analysis on Stream Cipher MUGI

Junko Takahashi[1,2], Toshinori Fukunaga[1], and Kazuo Sakiyama[2]

[1] NTT Information Sharing Platform Laboratories, NTT Corporation,
3-9-11, Midori-cho Musashino-shi, Tokyo 180-8585, Japan
{takahashi.junko,fukunaga.toshinori}@lab.ntt.co.jp
[2] Department of Informatics, The University of Electro-Communications,
1-5-1, Chofugaoka Chofu, Tokyo 182-8585, Japan
{junko,saki}@ice.uec.ac.jp

Abstract. This paper proposes differential fault analysis, which is a well-known type of fault analysis, on a stream cipher MUGI, which uses two kinds of update functions of an intermediate state. MUGI was proposed by Hitachi, Ltd. in 2002 and it is specified as ISO/IEC 18033-4 for keystream generation. Fault analysis is a side-channel attack that uses the faulty output obtained by inducing faults into secure devices. To the best knowledge of the authors, this is the first paper that proposes applying fault analysis to MUGI. The proposed attack uses the relation between two kinds of the update functions that are mutually dependent. As a result, our attack can recover a 128-bit secret key using 12.54 pairs of correct and faulty outputs on average within 1 sec.

Keywords: Fault analysis, Differential fault analysis (DFA), Stream cipher, Side-channel analysis.

1 Introduction

Nowadays, side-channel attacks are considered to be serious attacks because the secret keys embedded in a secure computing device such as smart cards and RFID tags can be recovered within a feasible computational time. Fault analysis is one type of side-channel attacks which deduces the secret key by deliberately inducing faults into the secure device during its cryptographic computation. Differential fault analysis (DFA) proposed by Biham *et al.* [1] is the most well-known fault analysis. In their attack, the secret key of DES can be recovered by comparing the correct and faulty output results after injecting faults into the secure device. Previously, DFA on some symmetric ciphers was proposed with some success in recovering secret keys [1,2,3,4,5,6,7,8]. Recently, fault analyses against stream ciphers have been proposed in [9,10,11,12,13,14,15].

At FSE 2002, Hitachi, Ltd. proposed the pseudo-random number generator (PRNG) MUGI [16]. MUGI uses a 128-bit secret key and a 128-bit initial vector. It generates 64-bit random output and transforms the internal state [17]. MUGI is specified as ISO/IEC 18033-4 for keystream generation. Its structure is based on the PANAMA PRNG [18] the design for which targets suitability for both software and hardware implementations, and the design principle is based on a

K.-H. Rhee and D. Nyang (Eds.): ICISC 2010, LNCS 6829, pp. 420–434, 2011.

block cipher. Therefore, the evaluation techniques for a block cipher are considered to be applicable to MUGI. Some cryptographic evaluations of MUGI have been reported [19,20,21,22,23,24]; however, the success of side-channel analysis such as fault analysis on MUGI has not yet been proposed. Hoch [22] considers differential fault analysis on MUGI to be difficult because it is hard to obtain sufficient information to achieve a successful attack using only the evaluation techniques of the block ciphers.

In this paper, we propose DFA on MUGI. To the best knowledge of the authors, this is the first paper that proposes DFA on a stream cipher MUGI and the proposed attack uses the characteristics that two kinds of update functions are mutually dependent, i.e., each update function operates using another intermediate state as a parameter. In the proposed attack, we employ a random fault model in which the intermediate states are randomly corrupted and the attacker does not need to know the values of the faults. We note that the random fault model is more practical compared to the 1-bit flip fault injection frequently used in the DFA on other stream ciphers [12,14,15]. The proposed attack requires only 12.54 pairs of correct and faulty outputs on average to recover the complete internal states and the 128-bit secret key.

The remainder of this paper is organized as follows. Notations are defined in Sec. 2. We review the description of MUGI in Sec. 3. We describe the concept behind the proposed attack in Sec. 4. We describe the proposed attack in Sec. 5 and the evaluation of the proposed attack in Sec. 6. Finally, we conclude the paper in Sec. 7. Some additional calculations and evaluations are given in the appendix.

2   Notations

In this section, we give some notations used in this paper.

$a_i^{(t)}$: An 8-byte state in round t where $i = 0, \cdots, 2$

$b_i^{(t)}$: An 8-byte buffer in round t where $i = 0, \cdots, 15$

$X||Y$: Concatenation

$X \oplus Y$: Bitwise exclusive-OR operation

$\ggg_n$: Circular rotations of n bits to the right (in the 64-bit register)

$\lll_n$: Circular rotations of n bits to the left (in the 64-bit register)

3   Description of MUGI

In this section, we review the description of MUGI [17]. MUGI has two inputs as parameters, 128-bit secret key K and 128-bit initial vector I, which is a public parameter. It generates a 64-bit length random bit string for each round. The structure of MUGI is shown in Fig.1(a). The data size of MUGI is 64 bits, which is referred to as a unit. As shown in Fig.1(a), the internal state is divided into two parts: state a and buffer b. State a consists of 3 units, $a = a_0||a_1||a_2$, where each element a_i is 64 bits. Buffer b consists of 16 units, $b_0, \ldots, b_{15}$ where each

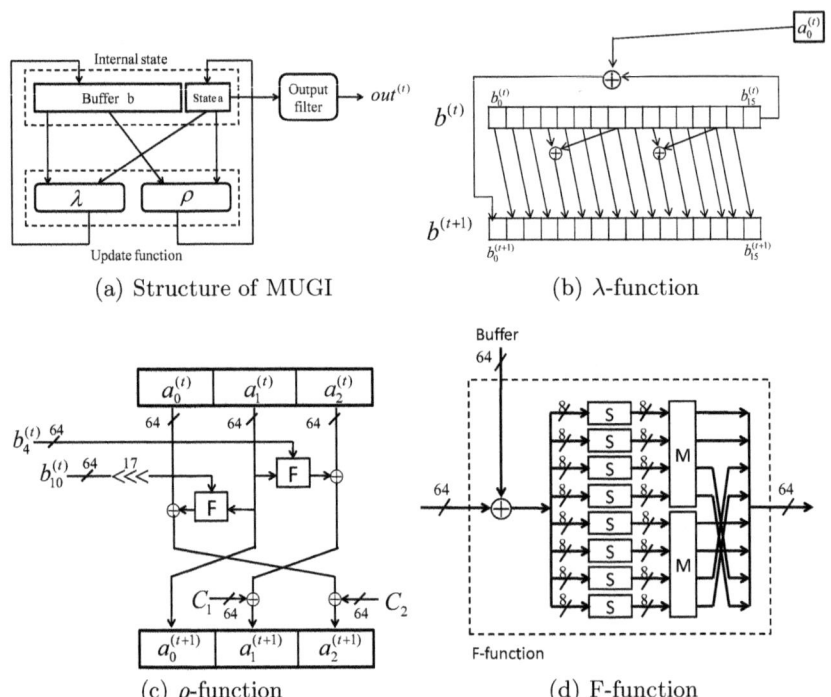

(a) Structure of MUGI   (b) λ-function

(c) ρ-function   (d) F-function

Fig. 1. Structure of MUGI , λ-function, ρ-function, and F-function

element b_i is 64 bits. The update function is described as a combination of ρ and λ functions and the update functions of state a and buffer b each of which uses another internal state as a parameter. The update function $Update$, of the entire internal state is expressed as

$$(a^{(t+1)}, b^{(t+1)}) = Update(a^{(t)}, b^{(t)}) = (\rho(a^{(t)}, b^{(t)}), \lambda(a^{(t)}, b^{(t)})).$$

Here, we call a step in which the update function is applied a round. In the following, we review the structure of ρ and λ, initialization, and the output filter.

λ Function. λ is the update function of buffer b and uses a part of state a as a parameter. λ is a linear transformation of b as shown in Fig.1(b).

ρ Function. ρ is the update function of state a. It is a kind of generalized Feistel structure with two F-functions and uses buffer b as a parameter. Figure 1(c) shows the structure of the ρ function where C_1 and C_2 are public constants. The F-function of MUGI uses the same S-box table and linear function, M, as those used in AES. Figure 1(d) shows the structure of the F-function in the ρ-function where S and M denote the S-box table and a 4×4 matrix, respectively.

Initialization and Output Filter. The initialization of MUGI comprises three steps. In the first step, buffer b is initialized with secret key K. We set secret key K to state a as

$$a_0 = K_0,$$
$$a_1 = K_1,$$
$$a_2 = (K_0 \lll 7) \oplus (K_1 \ggg 7) \oplus C_0,$$

where C_0 is a constant. Then, we iterate running only ρ and insert a part of each $a^{(t)}$ into buffer b

$$b_{15-i} = (\rho^{(i+1)}(a,0))_0, \tag{1}$$

where ρ^i denotes the i-th iteration and $\rho(a,0)$ denotes that the input from b is 0.

In the second step, mixed state a is set as

$$a(K,I)_0 = a(K)_0 \oplus I_0,$$
$$a(K,I)_1 = a(K)_1,$$
$$a(K,I)_2 = a(K)_2(K_0 \lll 7) \oplus (K_1 \ggg 7) \oplus C_0.$$

Then, state a is mixed again by 16-round iteration of ρ. So mixed state a is represented as $\rho^{16}(a(K,I),0)$.

The last step is a 16-round iteration of the entire update function, *Update*, i.e.

$$a^{(1)} = Update^{16}(\rho^{16}(a(K,I),0), b(K)),$$

where $b(K)$ represents buffer b initialized by secret key K.

After initialization, MUGI generates a 64-bit random number for each round and transforms the internal state in all rounds. The output from round t is given as

$$out^{(t)} = a_2^{(t)}.$$

In other words, MUGI outputs the lower 64 bits of state a at the beginning of the round process.

4   Concept Behind Proposed Attack

In this section, we describe the concept behind the proposed attack. In order to recover the 128-bit secret key, we need to recover three 8-byte states, a_0, a_1, and a_2, and sixteen 8-byte buffers, $b_0, \cdots, b_{15}$, in any round. Once we recover them, we can obtain the secret key by calculating the inverse of the ρ- and λ-functions and the inverse of the initialization.

In [22], DFA techniques for a block-cipher were simply applied to the ρ-function of MUGI. It was reported that only $a_1 \oplus b_{10} \lll_{17}$ were obtained by fault injections. This information is not sufficient to recover the secret key.

In order to obtain sufficient information to recover the secret key, we use the relation between the ρ- and λ-functions, which mutually interact. We find that we can recover a part of the state and buffer one-after-another by alternately using the algorithms for the ρ- and λ-functions. In the proposed attack, we recover a part of the buffer used in the λ- function using the faults injected into the state in the ρ-function. Then, we consider its propagation in the λ-function and we try to obtain another part of the buffer in the ρ-function using the propagated result. By iterating this process, we can recover all states and buffers to obtain the secret key.

5   Proposed Attack

In this section, we describe the details of the proposed attack.

5.1   Attack Assumptions

We describe the attack assumptions.

- We consider a transient fault, i.e., the attacker can reset the cryptographic device to its original state and then inject a fault into the same device during each new execution.
- An intermediate state is randomly corrupted by the fault injection, i.e., the attacker does not need to know the faulty value. The faulty value is uniformly distributed.
- The attacker knows the initial vector and he obtains pairs of correct and faulty keystreams calculated from the same key and the initial vector.
- The attacker can randomly modify any 8-byte value, $b_8^{(t)}$, $b_9^{(t)}$, ..., $b_{13}^{(t)}$, $a_0^{(t)}$ or $a_2^{(t)}$ during the keystream generation in any round, t. He has no control of the timing of the fault injection.

5.2   Attack Procedure

We propose an attack procedure to recover the 128-bit secret key using 12.54 pairs of correct and faulty outputs on average. We note that we know an 8-byte state, a_2, in all rounds because MUGI outputs an 8-byte state, a_2, at the beginning of each round process.

Step 0: Obtain a Correct Keystream. The attacker randomly selects an initial vector and obtains one correct keystream, (the correct keystream, a_2, in each round).

Step 1: Inject Fault into the ρ-Function and Obtain a Part of the Intermediate States of the ρ- and λ-Functions. The attacker obtains faulty keystreams by injecting faults during the generation of the keystreams. As an example, we consider the case that 8 bytes of $a_2^{(n)}$ ($n = t, ..., t + 5$) are randomly corrupted by the fault injection as shown in Fig. 2 (a). Figure 2 (a) shows the fault propagation when 8 bytes of $a_2^{(t)}$ are corrupted and the dotted lines indicate the fault propagation in this case. We note that the buffer

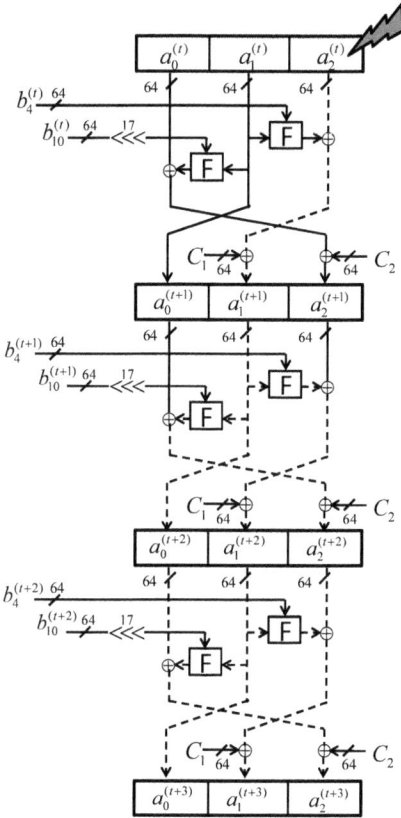

Fig. 2. Fault propagation when 8 bytes of $a_2^{(t)}$ are randomly corrupted. The fault propagation is shown as dotted lines.

is affected by a fault propagated to $a_0^{(t+2)}$ in round $(t+2)$ shown in Fig.2 because the λ-function uses a_0 as a parameter. However, this fact is not a problem for the attack.

The attacker can obtain the values of faults by calculating the difference between correct output, $a_2^{(n)}$, and faulty output, $\tilde{a}_2^{(n)}$ $(n = t, \ldots, t+5)$. Here, we define the difference between correct and faulty outputs in round n as $\Delta^{(n)}(= a_2^{(n)} \oplus \tilde{a}_2^{(n)})$ $(n = t, \ldots, t+5)$.

When the fault is injected into $a_2^{(n)}$, the equation for the difference of the S-box table in round $(n+1)$ is expressed as

$$S[\{a_1^{(n+1)} \oplus (b_{10}^{(n+1)} \lll 17)\}_l] \oplus S[\{a_1^{(n+1)} \oplus (b_{10}^{(n+1)} \lll 17)\}_l \oplus \Delta_l^{(n)}] = y_l \quad (2)$$

$$y = (y_0||y_1||y_2||y_3||y_4||y_5||y_6||y_7)$$
$$= M^{-1}(\Delta_4^{(n+1)}||\Delta_5^{(n+1)}||\Delta_2^{(n+1)}||\Delta_3^{(n+1)}) \, || \, M^{-1}(\Delta_0^{(n+1)}||\Delta_1^{(n+1)}||\Delta_6^{(n+1)}||\Delta_7^{(n+1)})$$

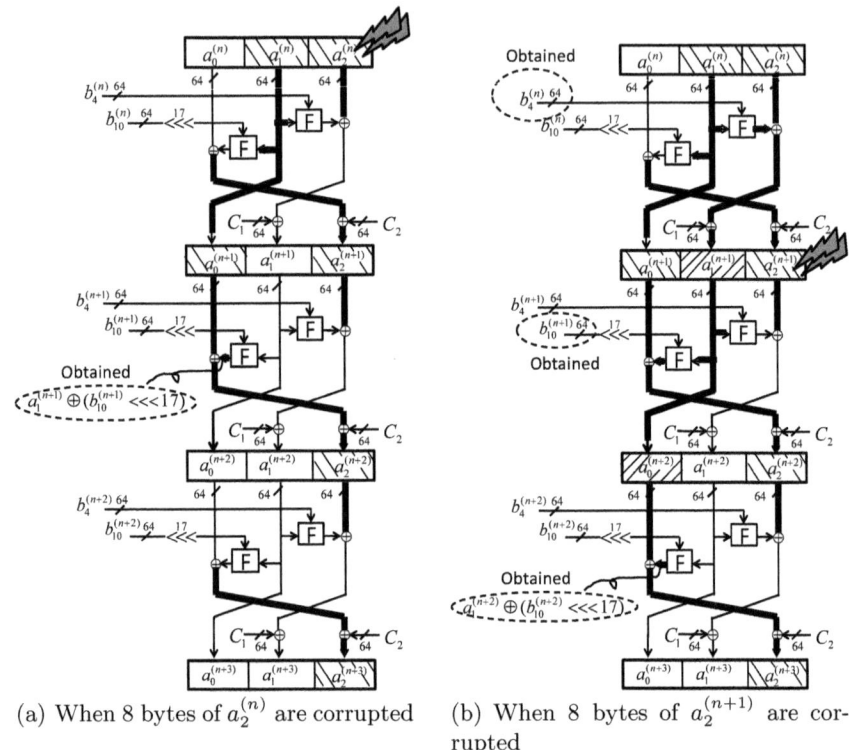

(a) When 8 bytes of $a_2^{(n)}$ are corrupted   (b) When 8 bytes of $a_2^{(n+1)}$ are corrupted

Fig. 3. Known values are shown in the heavy line and the known states are indicated as diagonal lines

In the above equation, y_l $(l = 0, \ldots, 7)$ is the l-th byte of y and $\Delta_l^{(n)}$ $(l = 0, \ldots, 7)$ is the l-th byte of $\Delta^{(n)}$ $(n = t, \ldots, t + 5)$. S is the S-box table and M^{-1} is the inverse of the matrix M.

Since $\Delta^{(n)}$ and $\Delta^{(n+1)}$ are known values, the attacker can solve the above equation and obtain the candidates for $a_1^{(n+1)} \oplus (b_{10}^{(n+1)} \lll 17)$ $(n = t, \ldots, t+5)$. The number of the solutions to (2) is 2 at 99.2% probability and 4 at 0.8% probability as shown in Appendix A.

In order to determine uniquely the solutions for (2), the attacker injects another 8-byte fault into the same location, $a_2^{(n)}$. Similarly, he obtains the candidates for $\{a_1^{(n+1)} \oplus (b_{10}^{(n+1)} \lll 17)\}_l$ $(l = 0, \ldots, 7, n = t, \ldots, t + 5)$ by solving the equation for the S-box. At this point, the attacker finds that one solution is the same as the solutions for (2) with 98.8% probability as shown in Appendix B, then, he finds $\{a_1^{(n+1)} \oplus (b_{10}^{(n+1)} \lll 17)\}_l$ $(l = 0, \ldots, 7, n = t, \ldots, t + 5)$. From Sec.6.4, the attacker can obtain 8 bytes of $a_1^{(n+1)} \oplus (b_{10}^{(n+1)} \lll 17)$ $(n = t, \ldots, t+5)$ using 2.09 pairs of correct and faulty outputs on average.

Since the attacker knows $a_1^{(n+1)} \oplus (b_{10}^{(n+1)} \lll 17)$ $(n = t, \ldots, t + 5)$, he also knows $a_1^{(n)} (= a_0^{(n+1)})$ $(n = t, \cdots, t + 5)$ by the characteristics of the ρ-function shown in Fig.3 (a).

Round $a_0\,a_1$ b_4 b_{10}

Round	a_0	a_1																
t		1					1											
$t+1$	1	1					1	2				1						
$t+2$	1	1					1	2	2			1	2					
$t+3$	1	1					1	2	2	2		1	2	2				
$t+4$	1	1					1	2	2	2	2	1	2	2	2			
$t+5$	1	1					4	2	2	2	2	2	1	2	2	2	2	
$t+6$	1	4	5	5	5	5	5	4	2	2	2	2	3	2	2	2	2	2
$t+7$	4	5		5	5	5	5	5	4	2	2	2	2	3	3	2	2	2	2
$t+8$	5	5			5	5	5	5	5	4	2	2	3	3	3	2	2	2
$t+9$	5	5				5	5	5	5	4	2	3	3	3	3	2	2
$t+10$	5	5				5	5	5	5	5	4	3	3	3	3	3	2

Fig. 4. The obtained bytes of sixteen 8-byte buffers b at any round by executing each attack step. Number i ($i = 1, \cdots, 5$) represents the bytes obtained at Step i.

Since the attacker knows $a_1^{(n)}$ and $a_1^{(n)} \oplus (b_{10}^{(n)} \lll 17)$ ($n = t+1, \ldots, t+5$), he can calculate $b_{10}^{(n)}$ ($n = t+1, \ldots, t+5$). By the characteristics of ρ-function, he also obtains $b_4^{(n-1)}$ ($n = t+1, \ldots, t+5$) shown in Fig.3 (b).

As a result, the attacker can obtain $b_4^{(n-1)}$ and $b_{10}^{(n)}$ ($n = t+1, \ldots, t+5$) with $12.54(= 2.09 \times 6)$ pairs of correct and faulty outputs. Each byte of the buffer obtained at each step is shown in Fig.4. Number i in Fig.4 means that the attacker obtains its location in each round at step i and the locations of the 8 bytes of b_4 and b_{10} are shown as heavy lines because the attacker only knows these bytes from the ρ-function.

Step 2: Obtain Intermediate State from the Shift Operation in the λ-Function. By the structure of the λ-function, the attacker obtains another parts of the buffer shown in Fig.4 using buffers obtained in Step 1.

Step 3: Obtain Intermediate State from the XORed Operation in the λ-Function. The attacker also obtains another byte of the buffer through the feed-back of the XORed operation of the λ-function. As an example, when the attacker knows $b_4^{(t)}$ and $b_{10}^{(t+2)}$, he also knows $b_{10}^{(t+6)}$ shown in Fig.5. Using this characteristic, he can obtain b_{10} also used in the ρ-function as a parameter by only calculating the λ-function.

Step 4: Obtain Intermediate State from the ρ-Function Since the attacker already knows $b_{10}^{(t+6)}$ by calculating the λ-function in Step 3, he immediately calculates $a_1^{(t+6)}$ from $a_1^{(t+6)} \oplus (b_{10}^{(t+6)} \lll 17)$ obtained in Step 1. Then, he also obtains $a_1^{(t+6)}(= a_0^{(t+7)})$ and $b_4^{(t+5)}$.

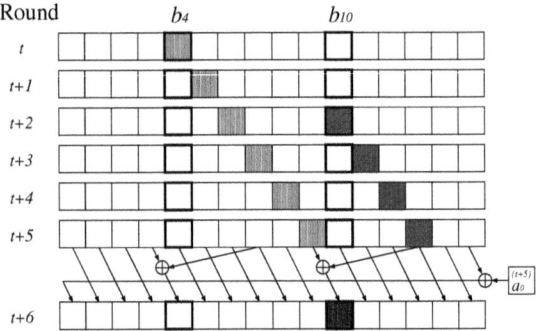

Fig. 5. The propagation in the λ-function

Step 5: Obtain Three 8 Bytes of State and Sixteen 8 Bytes of Buffer.

Since the attacker already knows $b_{10}^{(t+7)}$ in Step 3 and $a_0^{(t+7)}$ in Step 4, he also knows $a_1^{(t+7)} (= a_0^{(t+8)})$. Then, he obtains $b_4^{(t+6)}$. Similarly, he can obtain $b_4^{(t+7)}$ from $b_{10}^{(t+8)}$ in Step 3, $a_1^{(t+8)}$ and $a_0^{(t+8)}$.

Similarly, he can obtains $b_4^{(n)}$ ($n = t + 8, \ldots, t + 10$) because he can also calculate $b_{10}^{(n+6)} = b_4^{(n)} \oplus b_{10}^{(n+2)}$ ($n = t + 2, \ldots, t + 4$) by calculating the λ-function. Then, he also obtains $b_i^{(t+6)}$ ($i = 0, \ldots, 3$) by calculating the inverse of the λ-function.

The attacker can obtain three 8-byte states and sixteen 8-byte buffers in round $(t + 6)$. Once recovered, he can obtain the secret key by calculating the inverse of the ρ- and λ-functions until the initial state of the initialization. Therefore, The attacker obtains the secret key using a total of $12.54 (= 2.09 \times 6)$ pairs of correct and faulty outputs.

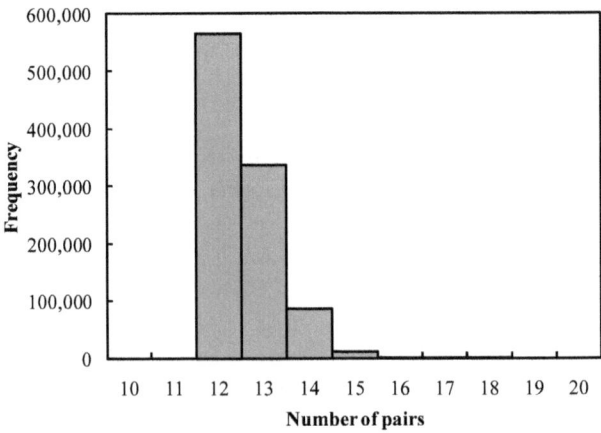

Fig. 6. Histogram of the numbers of pairs of correct and faulty outputs for a successful attack of 1,000,000 samples

5.3   Simulation Results

In order to verify the proposed attack and evaluate the number of correct and faulty outputs needed to recover the secret key, we implement the attack in C code and execute it on a PC with an Intel Core2duo 3.0 GHz CPU. In the simulation, we use a randomly chosen secret key and initial vector, and we assume that a random fault is injected into 8 bytes of $a_2^{(i)}$ $(i = 0, \cdots, 5)$. We execute the attack simulations 1,000,000 times. Figure 6 shows the histogram of the numbers of correct and faulty outputs for 1,000,000 samples. The simulation result shows that we need 12 pairs of correct and faulty outputs in the best cases and 18 pairs of them in the worst cases for the proposed attack. The average number of correct and faulty outputs is 12.55 pairs. This result agrees with the theoretical result that 12.54 pairs of them are required. In the simulation, we can recover all states and buffers and the 128-bit secret key within 1 sec.

6   Evaluation of Proposed Attack

In this section, we evaluate the proposed attack regarding aspects of the fault injection area, the number of the faulty bytes needed for a successful attack, the attacker's ability to control the fault injection, and the number of pairs of correct and faulty outputs needed for a successful attack.

6.1   Fault Injection Area

The proposed attack requires that 8 bytes of state a_2 are corrupted by some fault injections. Therefore, any fault injection that corrupts a_2 and does not affect intermediate values used in the attack can be used. We find that the fault injection that corrupts either 8 bytes of state a_0 or buffer $b_i (i = 8, \cdots, 13)$ can also be used for the attack.

For example, if 8 bytes of state $a_0^{(t)}$ are corrupted, the fault is only propagated to 8 bytes of state $a_2^{(t+1)}$ in the next round. This is absolutely the same situation as a fault injection that directly corrupts $a_2^{(t+1)}$. Another example is if 8 bytes of buffer $b_{13}^{(t)}$ are corrupted. The fault is propagated in the λ-function and corrupts $b_{10}^{(t+1)}$ and $b_{14}^{(t+1)}$. The corrupted $b_{10}^{(t+1)}$ corrupts $a_2^{(t+2)}$ and this situation is the same as the attack assumption. In this case, another fault is propagated via $b_{14}^{(t+1)}$. The next round that this fault affects the ρ-function is round $(t + 7)$ as $b_4^{(t+7)}$. Because of this timing delay, this fault propagation does not affect the attack procedure. Therefore, this fault injection can also be used for the attack.

6.2   Number of Faulty Bytes

By analyzing the differences between correct and faulty outputs, we can know the number of bytes affected by the fault injections. In the proposed attack, a total of 8 bytes of state or buffer is needed to be corrupted because there is no solution for the byte of the S-box equation (2) into which a fault is not injected. Then, in the case that a fault with fewer than 8 bytes is injected into each state, we need to inject faults until all bytes are corrupted. However, in the case that a fault with fewer than 8 bytes is injected into the buffer, we only need to inject faults into at least 1 byte in the lower 4 bytes of the buffer and in the upper 4 bytes of the buffer because 1-byte fault diffuses into 4 bytes based on the matrix M in the F-function.

6.3 Attacker's Ability to Control Fault Injection

We find that the attacker can verify a fault is injected into the desired location and timing from faulty outputs in the same way as other stream ciphers. Therefore, the attacker does not need to control the fault injection area and timing precisely. This fact makes the attack more practical.

Table 1 shows the number of rounds until the faulty output first appears due to fault injection, and patterns of the differences between the correct and faulty outputs when a fault is injected into each state or buffer. In the table, A and B denote kinds of the faulty patterns that are represented by the differences between the correct and faulty outputs in each round. T denotes a output without fault and $\boldsymbol{F}$ denotes that with fault. As an example, in the case that a fault is injected into $b_0^{(t)}$, the faulty output first appears after iterating the λ-function four times and iterating the ρ-function twice because $b_0^{(t)} = b_4^{(t+4)}$ and $b_4^{(t+4)}$ are used in the ρ-function. Therefore, the number of rounds until the faulty output first appears after the fault injection is $6(= 4 + 2)$. As another example, in the case that a fault is injected into $b_{11}^{(t)}$, the faulty output first appears after iterating the λ-function three times and iterating the ρ-function once because $b_{10}^{(t+4)}$ is affected by $b_{13}^{(t+2)}(= b_{11}^{(t)})$ and $b_{10}^{(t+4)}$ used in the ρ-function. Therefore, the number of rounds until the faulty output first appears after the fault injection is $4(= 3+1)$. For a successful attack, the attacker needs to inject a fault as pattern A. Then, the attacker needs to inject a fault into a_0, a_2, b_8, $\cdots$, or b_{13} as shown in Table 1. Therefore, the attacker can verify whether or not a fault is injected into the desired location and timing from Table 1 and he can select the output needed for a successful attack even if the attacker does not control the location and the timing of fault injection.

Table 1. Number of Rounds Until Faulty Output First Appears From the Fault Injection and Output Patterns. T Denotes a Output Without a Fault and $\boldsymbol{F}$ Denotes a Output With a Fault.

Fault location	a_0	a_1	a_2	b_0	b_1	b_2	b_3	b_4	b_5	b_6	b_7	b_8	b_9	b_{10}	b_{11}	b_{12}	b_{13}	b_{14}	b_{15}
Number of rounds	1	1	0	6	5	4	3	2	4	3	2	3	2	1	4	3	2	8	7
Output patterns	A	B	A	B	B	B	B	B	B	B	B	A	A	A	A	A	A	B	B

Output pattern A: $T, \cdots, T, \boldsymbol{F}, T, \boldsymbol{F}, \boldsymbol{F}, \boldsymbol{F}, \cdots$
Output pattern B: $T, \cdots, T, \boldsymbol{F}, \boldsymbol{F}, \boldsymbol{F}, \boldsymbol{F}, \boldsymbol{F}, \cdots$

We note that these characteristics are only shown in the stream cipher. In most cases of block cipher, it is hard for the attacker to know the fault location by analyzing the outputs when a fault is injected into the middle round because a fault diffuses and the block cipher generates an output after many iterations of rounds. Therefore, the attacker needs to execute the attack method and examine whether or not he can obtain the secret key to verify the fault injected into the desired area.

6.4   Number of Faulty Outputs

We investigate the number of pairs of correct and faulty outputs needed to determine uniquely 8 bytes of the solution of the S-box equation (2) in Sec.5.2 and the total number of pairs needed for a successful attack.

The number of pairs of correct and faulty outputs to determine uniquely 8 bytes of the solution of the S-box equation (2) on average is calculated as

$$N = 2 \times P^8 + 3 \times (1 - P^8) \times \{1 - (1 - P)^2\}^8 + \cdots \tag{3}$$

where P is the probability that the solution of the S-box equation (2) can be uniquely determined using 2 pairs of correct and faulty outputs and the coefficient starts from 2 because we need at least 2 pairs of correct and faulty outputs. As shown in Appendix B, P is calculated as 0.988. Ultimately, we calculate $N \approx 2.09$. In the attack, we need to obtain 6 rounds of 8 bytes of the XORed between the state and buffer, then, $12.54(= 2.09 \times 6)$ pairs of correct and faulty outputs are required for a successful attack.

7   Conclusions

We proposed a DFA on a stream cipher MUGI. To the best knowledge of the authors, this is the first paper that proposes a DFA on MUGI. We use the relation between two kinds of update functions that are mutually dependent. As a result, we can recover the secret key using 12.54 pairs of correct and faulty outputs on average within 1 sec.

We consider the technique of the proposed attack can be used to other stream ciphers that use two kinds of update functions that are mutually dependent such as linear and non-linear shift registers.

References

1. Biham, E., Shamir, A.: Differential Fault Analysis of Secret Key Cryptosystems, Technion - Computer Science Department - Technical Report CS0901.revised (1997)
2. Blömer, J., Seifert, J.-P.: Fault Based Cryptanalysis of the Advanced Encryption Standard (AES). In: Wright, R.N. (ed.) FC 2003. LNCS, vol. 2742, pp. 162–181. Springer, Heidelberg (2003)

3. Dusart, P., Letourneux, G., Vivolo, O.: Differential Fault Analysis on A.E.S., Cryptology eprint Archive Report 2003/010 (2003), http://www.iacr.org/
4. Piret, G., Quisquater, J.J.: A Differential Fault Attack Technique Against SPN Structures, with Application to the AES and KHAZAD. In: Walter, C.D., Koç, Ç.K., Paar, C. (eds.) CHES 2003. LNCS, vol. 2779, pp. 77–88. Springer, Heidelberg (2003)
5. Chen, C.-N., Yen, S.-M.: Differential Fault Analysis on AES Key Schedule and Some Countermeasures. In: Safavi-Naini, R., Seberry, J. (eds.) ACISP 2003. LNCS, vol. 2727, pp. 118–129. Springer, Heidelberg (2003)
6. Chen, H., Wu, W., Feng, D.: Differential Fault Analysis on CLEFIA. In: Qing, S., Imai, H., Wang, G. (eds.) ICICS 2007. LNCS, vol. 4861, pp. 284–295. Springer, Heidelberg (2007)
7. Takahashi, J., Fukunaga, T.: Improved Differential Fault Analysis on CLEFIA. In: FDTC 2008, pp. 25–39. IEEE-CS, Los Alamitos (2008)
8. Takahashi, J., Fukunaga, T.: Differential Fault Analysis on CLEFIA with 128, 192, and 256-Bit Keys. IEICE Transactions on Fundamentals of Electronics, Communications and Computer Sciences E93-A(1), 136–143 (2010)
9. Biham, E., Granboulan, L., Nguyen, P.Q.: Impossible Fault Analysis of RC4 and Differential Fault Analysis of RC4. In: Gilbert, H., Handschuh, H. (eds.) FSE 2005. LNCS, vol. 3557, pp. 359–367. Springer, Heidelberg (2005)
10. Hoch, J.J., Shamir, A.: Fault Analysis of Stream Ciphers. In: Joye, M., Quisquater, J.-J. (eds.) CHES 2004. LNCS, vol. 3156, pp. 240–253. Springer, Heidelberg (2004)
11. Debraize, B., Corbella, I.M.: Fault Analysis of the Stream Cipher Snow 3G. In: FDTC 2009, pp. 103–110. IEEE-CS, Los Alamitos (2009)
12. Hojsik, M., Rudolf, B.: Differential Fault Analysis of Trivium. In: Nyberg, K. (ed.) FSE 2008. LNCS, vol. 5086, pp. 158–172. Springer, Heidelberg (2008)
13. Berzati, A., Canovas, C., Castagnos, G., Debraize, B., Goubin, L., Gouget, A., Paillier, P., Salgado, S.: Fault Analysis of GRAIN-128. In: Proc. of the 2009 IEEE International Workshop on Hardware-Oriented Security and Trust, pp. 7–14. IEEE-CS, Los Alamitos (2009)
14. Kircanski, A., Youssef, A.M.: Differential Fault Analysis of Rabbit. In: Jacobson Jr., M.J., Rijmen, V., Safavi-Naini, R. (eds.) SAC 2009. LNCS, vol. 5867, pp. 197–214. Springer, Heidelberg (2009)
15. Berzati, A., Canovas-Dumas, C., Goubin, L.: Fault Analysis of Rabbit: Toward a Secret Key Leakage. In: Roy, B., Sendrier, N. (eds.) INDOCRYPT 2009. LNCS, vol. 5922, pp. 72–87. Springer, Heidelberg (2009)
16. Watanabe, D., Furuya, S., Yoshida, H., Takaragi, K., Preneel, B.: A New Keystream Generator MUGI. In: Daemen, J., Rijmen, V. (eds.) FSE 2002. LNCS, vol. 2365, pp. 179–194. Springer, Heidelberg (2002)
17. MUGI Pseudorandom Number Generator Specification Ver. 1.2, Hitachi, Ltd. (2001), This document is available at, http://www.sdl.hitachi.co.jp/crypto/mugi/index-e.html
18. Daemen, J., Clapp, C.: Fast Hashing and Stream Encryption with PANAMA. In: Vaudenay, S. (ed.) FSE 1998. LNCS, vol. 1372, pp. 60–74. Springer, Heidelberg (1998)
19. MUGI Pseudorandom Number Generator Self-Evaluation Report Ver. 1.1, Hitachi, Ltd. (2001), This document is available at http://www.sdl.hitachi.co.jp/crypto/mugi/index-e.html
20. Dawson, E., Carter, G., Gustafson, H., Henricksen, M., Millan, W., Simpson, L.: Evaluation of the MUGI Psuedo-Random Number Generator, Technical report, CRYPTREC, Information Technology Promotion Agency (IPA), Tokyo Japan (2002)

21. Henricksen, M., Dawson, E.: Rekeying Issues in the MUGI Stream Cipher. In: Preneel, B., Tavares, S. (eds.) SAC 2005. LNCS, vol. 3897, pp. 175–188. Springer, Heidelberg (2006)
22. Hoch, Y.: Fault Analysis of Stream Ciphers M.Sc. Thesis, Weizmann Institute of Science, Israel
23. Golić, J.D.: A Weakness of the Linear Part of Stream Cipher MUGI. In: Roy, B., Meier, W. (eds.) FSE 2004. LNCS, vol. 3017, pp. 178–192. Springer, Heidelberg (2004)
24. Biryukov, A., Shamir, A.: Analysis of the Non-linear Part of MUGI. In: Gilbert, H., Handschuh, H. (eds.) FSE 2005. LNCS, vol. 3557, pp. 320–329. Springer, Heidelberg (2005)

Appendix A: Probability of the Number of Solutions for the S-Box Equation in Sec.5.2

In order to obtain the probability that we obtain 2 or 4 solutions for the S-box equation (2) in Sec. 5.2, we describe the characteristics of the equation for the S-box used in MUGI. Let us consider the simple 1-byte S-box model shown in Fig.7. When we know a pair of inputs, $y_{i(8)}$ and $y_{j(8)}$, and know the output difference, $\delta_{ij(8)}$, we can obtain a set of unknown candidates of $x_{(8)}$ by solving the following equation.

$$S[y_{i(8)} \oplus x_{(8)}] \oplus S[y_{j(8)} \oplus x_{(8)}] = \delta_{ij(8)} \tag{4}$$

The number of key candidates depends on $y_{i(8)}$, $y_{j(8)}$, and $\delta_{ij(8)}$, and the structure of the S-box. By solving (4), we examine the size of the candidates of the

Table 2. S-Box Statistics

$|\langle x_{(8)}\rangle|$	NC	P	$P_{|\langle x_{(8)}\rangle|\neq 0}$	$E(|\langle x_{(8)}\rangle|)$
0	8,486,400	0.506	-	-
2	8,225,280	0.490	0.992	1.984
4	65,280	0.004	0.008	0.031
256	256	0.000	0.000	0.010
Total	16,777,216	1	1	2.024
	$(= 2^{24})$			$(= 2^{1.02})$

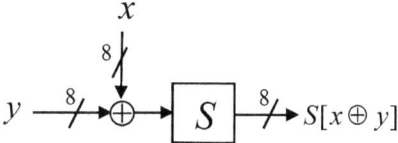

Fig. 7. One-byte S-box model

S-box for all combinations of $y_{i(8)}$, $y_{j(8)}$, and $\delta_{ij(8)}$. The total number of combinations of $y_{i(8)}$, $y_{j(8)}$, and $\delta_{ij(8)}$ is $2^{24} (= 16,777,216)$.

The results are shown in Table 2. In the table, $|\langle x_{(8)} \rangle|$ is the size of the candidates, and NC is number of case that the size of the candidates is $|\langle x_{(8)} \rangle|$ in all combinations of $(y_{i(8)}, y_{j(8)}, \delta_{ij(8)})$. P is the probability that the number of the candidates is $|\langle x_{(8)} \rangle|$ when 2 inputs of the S-box, $y_{i(8)}$, and $y_{j(8)}$, and the output difference, $\delta_{ij(8)}$, are randomly set, which is defined as $NC/2^{24}$. $P_{|\langle x_{(8)} \rangle| \neq 0}$ is probability except for $|\langle x_{(8)} \rangle| \neq 0$, which is defined as $NC_{|\langle x_{(8)} \rangle| \neq 0}/(2^{24} - NC_{|\langle x_{(8)} \rangle| = 0})$. $E(|\langle x_{(8)} \rangle|)$ is the expected value of $|\langle x_{(8)} \rangle|$ except for $|\langle x_{(8)} \rangle| \neq 0$, which is defined as $|\langle x_{(8)} \rangle| \cdot P_{|\langle x_{(8)} \rangle| \neq 0}$. Table 2 suggests that the number of solutions for (2) is 2 with 99.2% probability and 4 with 0.8 % probability when there is a candidate.

Appendix B: Probability That a Solution Can be Uniquely Determined Using Two Pairs of Correct and Faulty Outputs

In order to obtain the probability of a solution can be uniquely determined using 2 pairs of correct and faulty outputs, we examine the number of simultaneous equations in (5) for all combinations of (y_i, y_j, y_k, y_l) when answer x is fixed. The total number of all cases is 2^{32} because $\delta_{ij(8)}, \delta_{kl(8)}$ are calculated from all combinations of (y_i, y_j, y_k, y_l), and the fixed x. The results have no relation to the value of x.

$$S[x_{(8)} \oplus y_{i(8)}] \oplus S[x_{(8)} \oplus y_{j(8)}] = \delta_{ij(8)}$$
$$S[x_{(8)} \oplus y_{k(8)}] \oplus S[x_{(8)} \oplus y_{l(8)}] = \delta_{kl(8)} \tag{5}$$

Therefore, the probability that the number of solutions to the S-box equation (2) in Sec.5.2 is calculated as 98.8%.

Table 3. Number of Solutions and Probability for All Input Cases

Number of solutions	Number of cases	Probability
1	4,243,730,400	0.9881
over 2	51,236,896	0.0119
Total	4,294,967,296 $(= 2^{32})$	1

Author Index